Trees, Shrubs and Flowers to Know
in Washington & British Columbia

Trees, Shrubs & Flowers to Know
in Washington & British Columbia

LONE
PINE

C.P. LYONS
BILL MERILEES

The Publisher: Lone Pine Publishing

1808 B Street NW, Suite 140	202A, 110 Seymour Street	10145 – 81 Avenue
Auburn, WA 98001	Vancouver, BC V6B 3N3	Edmonton, AB T6E 1W9
USA	Canada	Canada

Website: http://www.lonepinepublishing.com

Canadian Cataloguing in Publication Data

Lyons, C.P. (Chester Peter), 1915–98
   Trees, shrubs and flowers to know in Washington and British Columbia
   Includes bibliographical references and index.
   Published also as: Trees, shrubs and flowers to know in British Columbia and Washington
   ISBN 13: 978-1-55105-062-1
   ISBN 10: 1-55105-062-5

   1. Botany—British Columbia—Handbooks, manuals, etc. 2.
Botany—Washington (State)—Handbooks, manuals, etc. 3.
Plants—Identification—Handbooks, manuals, etc. I. Title. II.
Title: Trees, shrubs and flowers to know in British Columbia
and Washington
QK203.B7L96 1995   581.9711   C95-910527-1

Editor-in-chief: Glenn Rollans
Editor: Volker Bodegom
Cover Design: Beata Kurpinski
Design and Production: Volker Bodegom
Colour separations & prepress: Piéce de Résistance Ltée.
Photography: Chess Lyons, Bill Merilees
Additional photography (Photo Key): Steve Cannings: *star-flowered Solomon's seal, mountain ladyslipper,
   sagebrush buttercup, brown-eyed Susan, pink fairies, bitterroot, wild bergamot, scarlet paintbrush;*
   Jack Gregson: *yellow ladyslipper, moss campion, pasqueflower;* Robert Le Blanc: *many-spined prickly pear
   cactus, blue-eyed grass, clustered broomrape.*

**Although this guide mentions medicinal and culinary uses of plants, it is not a 'how-to' reference for
using traditional plant-derived medicines or foods, nor does it advocate the use of such medicines or
foods. We do not recommend experimentation by readers, and we caution that many plants in our
region, including some with traditional medical or food uses—or look-alikes—can be poisonous or
harmful.**

The publisher gratefully acknowledges the support of Alberta Community Development, the Department
of Canadian Heritage and the Canada/Alberta Agreement on the cultural industries.

PC: P1

Canadä

• CONTENTS •

ABOUT THIS ILLUSTRATED GUIDE

CONCEPT

The idea for the original edition of this book on B.C. flora came about in the early 1950s when no popular guidebook existed. J.K. Henry had written a botanical treatise published in 1915, *Flora of Southern British Columbia*. Despite my background as a forester, it was unreadable (no reflection on the quality of the book). A thin pamphlet, *Trees and Shrubs of British Columbia*, from the University of British Columbia, was equally puzzling. Surely there was some way to understand, at least in part, the most colourful flora about us. Was there not some way to describe a buttercup without saying 'radical leaves, short-petioled, obovate-flabelliform, crenately toothed, cauline sessile....?' Such jargon is sheer joy to the skilled botanist, but most confusing for an amateur.

When I started to put a book together, someone asked who I was writing it for. Off the top of my head, I replied, 'Scouts and grandmothers!' Times have changed, but the concept is still the same—user-friendliness for all ages. Despite time limitations—I wrote the book as I travelled in connection with the blossoming B.C. provincial park system and I was forced to draw my own illustrations too—in 1952 a book did come off the press. It was, with one revision in 1965, in continuous publication for 40 years. A similar book on Washington flora was published in 1956.

Over the years, more and more people have become interested in the outdoors, our roads have expanded, parks have been made more accessible and the words 'ecology,' 'environment' and 'ecosystem' came into common usage. Meanwhile, botanists were finding new plants and revising species, so names had to be added to the book or changed. It became clear that it was more than time to press for another revision, a process that would require 6 years of rejections from the Ontario publisher, culminating in my contract release and a subsequent 2 years of production with Lone Pine Publishing.

What should be done? Revise the naming and add 100 new plants? That idea fell apart when I was shown the brand-new 4-volume *Vascular Plants of B.C.* produced by the B.C. Ministry of Forests and edited by George Douglas, Gerald Straley and Del Meindinger. Here was 'the Bible' with the latest information on B.C. flora, albeit in botanical language. In good part it depended on the background established for many years by the authoritative 5 volume set *The Vascular Plants of the Pacific Northwest*, by Hitchcock, Cronquist, Ownbey and Thomson.

Now it seemed that *Trees, Shrubs and Flowers* could be revised to include much more general information beyond merely describing one flower after another. How many paintbrushes are there in our area? How many cacti? How many waterlilies? A never-ending expansion appeared possible—but a decision was made to stress the commonly seen plants and thereby keep the book in its handy 'pick-up' size. And a colour section seemed necessary.

Many of the more common roadside plants are undesirable aliens: knapweeds and thistles, for example. Despite the stigma, they do catch the eye. Some of these plants are crowding out native species and so should be known. And what about including plants that are far from being wildflowers but are eye-catching nevertheless—duckweed, eelgrass and bulrush?

Plants do not recognize political boundaries, so there is a smooth transition between the flora of B.C. and that of Washington. All of the 11 principal ecosystems in Washington are also found in B.C. So, for practical as well as economic considerations, it seemed advantageous to treat this particularly beautiful region of the Pacific Northwest as one unit. One might point out that most of the plants in this book also relate closely to those in the northern third of Oregon, but how far can you go with a book title?

The author can claim no special skills in being a writer, illustrator, or botanist. There are bound to be omissions and discrepancies that the reader will discover. These discoveries will be welcomed as valuable material for a future new edition.

ACKNOWLEDGEMENTS

Conventionally, this is the last part of a preface, but for myself, depending on the kindness and knowledge of others, it is by far the most important. You are 'deep in the swamp' without your helpers. Although many decades have passed since preparation of the first book, I cannot forget George Hardy, the Provincial Botanist at the time. His help and encouragement made it all possible. Bill Merilees, naturalist with the Provincial Parks Branch, had long stirred the fires for a revision and a colour section. His interest and encouragement over a number of years provided a strong push for getting the project under way. Bill's own knowledge, his years of nature observation, loans of books and advice have been invaluable—and he is the main contributor to the colour section. Bill has been a constant advisor and helper on this project.

In stumbling around wondering how to get started, I met Alison Nicholson, a research ecologist in the Forest Research Department. She quickly assessed the situation, handed me the set of *Vascular Plants of B.C.* plus several related books and, in so doing, implied, 'It's all there, now get going.' One of the authors, Del Meindinger, also a research ecologist, could not escape and also provided helpful material in person.

Over many years, requests to identify plants that puzzled me (and that was easily done) have been presented to Dr. 'Bob' Ogilvie, Curator of Botany, B.C. Provincial Museum. His great knowledge of B.C. plants has saved me many of hours of puzzlement and frustration. But what does one do about Washington plants? Fortunately I met Joe ('No Problem') Antos, a plant ecologist and an adjunct assistant professor in the Department of Biology at the University of Victoria. Joe had obtained his doctorate at Corvallis, Oregon, based on his studies of forest plants after the eruption of Mt. St. Helens. So my 'unknowns' in Washington became text and drawings through his help. The university herbarium too was a valuable source of reference materials.

In one of these quirks of happy-chance, 'Trish' Ellison came during the summer as an unemployed university student to help me plant a lawn. She continued on to decipher and convert my many hundreds of pages of scribblings into computer files, for which she needed at least a remarkable gift of clairvoyance. Eventually, with remarkable patience, she helped bring order out of the chaos.

Ruth Kirk was a cheerful source of advice and master-minded the ecosystem section. My daughter, Susan, with years of experience in editing and printing, aided greatly in the preliminary problems of 'getting the show on the road.' Andy MacKinnon of the Forest Research Branch, a co-author of several recent botanical books, has given valuable help in reviewing the text. Dr. Gerald Straley, a research scientist at the University of British Columbia, also reviewed part of the text.

Fred King of Galiano Island and Malcolm Martin of Vernon gave their help in locating specimens in their areas. Good friends Steve Cannings, Jack Gregson and Robert LeBlanc helped with photos. Volker Bodegom of Lone Pine Publishing, through his careful editing and artwork, has given my writing, maps and charts a gloss of quality that they would otherwise not have had. To all those other friends who gave advice and encouragement, I truly hope you feel that you have been a valued part of this effort.

C.P. Lyons
Victoria, B.C.

ABOUT THE AUTHORS

C.P. LYONS

C.P. (Chess) Lyons spent all his boyhood in the southern Okanagan Valley around Penticton, only 66 km (40 miles) from the Washington border. His love of the outdoors was so strong that it became a matter of fine judgment to balance 'playing hookey' against the grade-passing requirements of high school. However, his parents shipped him off (via the Kettle Valley Railroad) to the University of B.C. where, in 1939, he graduated in Forest Engineering.

In 1940 he became the first technical employee of the embryonic B.C. Parks Branch, with the title of parks engineer. The first examinations and reports on the large and virtually unknown Wells Gray, Tweedsmuir and Garibaldi parks were made by Chess Lyons and side-kick Micky Trew. During this time, a campaign for greater recognition of the infant parks system was kept rolling through slides, films and talks.

This era was one when much of B.C. was considered wilderness and large parks could easily be established: the southern leg of Tweedsmuir and the large parks along the Alaska Highway, for example. And with a thought to the future, many small reserves were designated as picnic sites and campgrounds.

In charge of historic sites for the Parks and Recreation Division, he was responsible for the quick launching of the Barkerville restoration project in 1958, which he went on to manage for its first four years. The design and start of the now-famous 'Stop of Interest' signs were also his responsibility.

Upon leaving government service, he toured North America with his own films on wildlife and travel, under the auspices of the National Audubon Society and the *World Around Us* travel series. He was the main contributor to the popular and long-running Canadian Broadcasting Corporation program *Klahanie—The Great Outdoors*.

His films brought to public attention new activities and areas: the previously unknown West Coast Trail, the possibility of canoeing on the Bowron Lakes, as well as canoeing among the Broken Islands, now part of Pacific Rim National Park. In between there were many trips as tour guide to faraway lands.

He still canoes and hikes—he has never lost that 'itchy feet' compulsion to see around the next corner.

BILL MERILEES

Bill Merilees, the 'junior' author, is a native of Vancouver. The once-wild lands along the North Arm of the Fraser River were his natural history laboratory. As a youth, along with Scouting and a membership in the Junior Section of the Vancouver Natural History Society, Bill developed a keen interest in the flora and fauna of British Columbia, one that has never waned. As a student of zoology and botany at the University of British Columbia and later as a graduate student at Colorado State, he was able to integrate academic studies with his love for field observation.

He is a past president of the Vancouver Natural History Society, a past vice-president of the Federation of British Columbia Naturalists and has held a number of positions with environmental organizations in the Kootenays and on Vancouver Island. In addition to the book *Attracting Backyard Wildlife*, Bill has written more than 100 articles on natural history subjects.

These various experiences, paired with the knowledge gained from nature travel and tour-leading in our area, plus a keen photographic eye, has been a major factor in this collaboration. He is currently a visitor services coordinator with B.C. Parks on Vancouver Island.

HOW TO USE THIS BOOK

I. IS IT A TREE, SHRUB OR FLOWER?

TREES have strong trunks covered with bark. They are usually over 6 m (20') high, with trunks more than 5 cm (2") in diameter. There may be a single trunk or a number of them.

SHRUBS have tough, woody stems with a bark cover. Generally they are under 6 m (20') in height and less than 5 cm (2") in diameter and have several stems.

FLOWERS typically have soft stems that die each year, though some do remain green throughout the winter. Some species have a root or bulb that lives on, producing new growth each spring; these are perennials. Some flowers, annuals, always grow anew from seed, as they live no longer than 1 year. A biennial lives for 2 years, usually producing flowers and seed in the second year.

The descriptions of the various ecosystems (see pp. 15–27) include comprehensive lists of the most common trees, shrubs and flowers. They are both a quick way of separating trees from shrubs and an excellent overview of the flora for each region.

II. QUICK FLOWER IDENTIFICATION

No attempt has been made to group flowers by their botanical families, an arrangement understood only by fairly competent botanists. Emphasis is instead put on a simple and positive means of identification; flowers are first divided by colour and then by easily observed characteristics such as number of petals or shape of flower—whether tubular, pea-shaped, etc. Within each category they are arranged roughly in order of increasing size, with similar-looking species close to each other. Remember that flower colours can vary considerably, so you might have to check several colour possibilities. Additional key facts, such as height, type of leaves and stem, habitat, blooming season and range, should furnish all the clues necessary. Closely related and similar-looking plants are largely cross-referenced.

It is not practical to include drawings and full descriptions for the minor species. As a compromise, they have generally been grouped with more important plants to which they are related, even though their flowers might be of different colours.

A supplemental colour section of 450 photos, which includes many of our most-noticed plants, is keyed to the individual species descriptions (shrubs, flowers, irregular plants and ferns). You may want to look here first to find your specimen and then read about it. Conversely, in the write-ups, each species with a colour photo has a number beside its common name that matches its photo.

III. MAKE SURE YOU HAVE A TYPICAL SAMPLE

Young shoots and stems of some trees and shrubs have unnaturally large or misshapen leaves. So look around for a typical specimen. If a flower has gone to seed, look nearby in a cooler, shadier place for one that is still in bloom. Remember that a flower's colour may change with age, so compare several. Blues and purples are confusing colours to pin down. Check both if in doubt.

IV. STUDY THE APPROPRIATE KEY AND CHARTS

Keys to the trees are on pp. 58–59. A simplified grouping of the shrubs is on p. 105, and there is a key to the flowers on p. 156. A drawing on p. 10 shows some of the most common plants on coastal bluffs, sea shores, sand dunes, swamps and wet meadows. There is a comprehensive list of the flora in each ecosystem in the descriptions on pp. 15–27. Typical species for each elevation range in different parts of B.C. and Washington are listed on pp. 28–31.

V. COLOUR THE DRAWINGS!

Use coloured pencils or water colours to tone each drawing the exact colour you think it should be. A touch of colour works wonders!

NAMES

Scientific (or Latin, or botanical) names are known throughout the world while common names are often a matter of local usage. The Latin name is usually in two parts, like a person's. The following chart shows how a comparison might be made:

Anglo-saxon	Ericaceae
(a clan of people)	(a clan of plants)
Jones, Smith, Cark	***Vaccinium, Rhododendron, Gaultheria***
(families in the clan)	(families in the clan)
Jones, William	***Vaccinium ovalifolium***
(a certain individual)	(a certain blueberry)

In the above example, *Vaccinium* is the genus (plural is 'genera'), while *ovalifolium* identifies the particular species.

The following abbreviations are also used:

spp. **species**: Used with a genus name (e.g. *Vaccinium* spp.) to mean several species or all of them in that genus.

ssp. **subspecies**: Used after a species name (e.g., *Populus balsamifera* ssp. *trichocarpa*) to indicate a finer division of the species, often based on its geographical range.

var. **variety**: Used the same way as 'ssp.,' (e.g., *Pinus contorta* var. *latifolia*) but typically used to distinguish plants with a common range but slightly different characteristics (e.g., in leaf or flower).

To lessen confusion, botanical and common names have been largely correlated to *Vascular Plants of British Columbia*, which in turn is similar to *Vascular Plants of the Pacific Northwest*. Various references on Washington flora have been consulted to try and integrate the most common names.

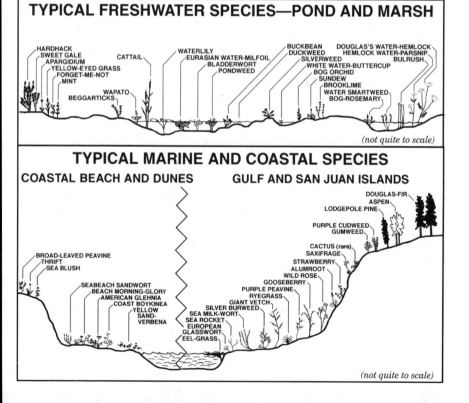

TYPICAL FRESHWATER SPECIES—POND AND MARSH

HARDHACK
SWEET GALE
APARGIDIUM
YELLOW-EYED GRASS
FORGET-ME-NOT
MINT
WAPATO
BEGGARTICKS
CATTAIL
WATERLILY
EURASIAN WATER-MILFOIL
BLADDERWORT
PONDWEED
BUCKBEAN
DUCKWEED
SILVERWEED
WHITE WATER-BUTTERCUP
BOG ORCHID
SUNDEW
BROOKLIME
WATER SMARTWEED
BOG-ROSEMARY
DOUGLAS'S WATER-HEMLOCK
HEMLOCK WATER-PARSNIP
BULRUSH

(not quite to scale)

TYPICAL MARINE AND COASTAL SPECIES

COASTAL BEACH AND DUNES

GULF AND SAN JUAN ISLANDS

BROAD-LEAVED PEAVINE
THRIFT
SEA BLUSH
SEABEACH SANDWORT
BEACH MORNING-GLORY
AMERICAN GLEHNIA
COAST BOYKINEA
YELLOW SAND-VERBENA
SEA MILK-WORT
SEA ROCKET
EUROPEAN GLASSWORT
EEL-GRASS
SILVER BURWEED
GIANT VETCH
PURPLE PEAVINE
RYEGRASS
GOOSEBERRY
WILD ROSE
ALUMROOT
STRAWBERRY
SAXIFRAGE
CACTUS (rare)
GUMWEED
PURPLE CUDWEED
LODGEPOLE PINE
DOUGLAS-FIR
ASPEN

(not quite to scale)

THE FLORA OF B.C. AND WASHINGTON

There is always the tendency to think that something more attractive and interesting lies just beyond our reach. For example, how colourfully pictured are the eastern hardwoods, the exotic, waving palms of California or the bristling cacti of New Mexico. But perhaps if we made a quick inventory of our own surroundings, we would appreciate what an outstanding variety of flora there is right at hand. Using a well-known forest classification scheme, we find that no fewer than 7 out of the 10 forest regions recognized in Canada and the United States occur in B.C. and Washington. By comparison, few Canadian provinces can claim more than 2 regions and few US states more than 3.

No other place in Canada has the parching desert climate of the southern Okanagan and the Similkameen Valley, a habitat more than matched by the sunbaked slopes and desert flatlands of central and eastern Washington. In these areas of limited rainfall grow the twisted antelope bush, the drab sagebrush and mats of spiny cactus. What a striking contrast to the sombre coastal rainforest with its forest giants, lush, green ferns and thick ground cover of bracken and salal.

The floral changes experienced between points separated by just a few hours' walk are amazing. Climb a mountain as it rises from the shadowy forests of giants at sea level, continue through the twisted, stunted trees and shrubs and brilliant flower meadows of the subalpine and go beyond, into the bleak, bold and beautiful alpine terrain, where it is finally capped by ice or snow. Here in this high land of solitude and silence, at elevations of 2250–3000 m (7500–10 000'), conditions are similar to those of the Arctic wastelands. Tiny flowers and mosses pinch themselves into sheltered niches, to smile briefly but brightly during the fleeting summer they know.

B.C. and Washington can lay claim to over 3000 different plants. And perhaps a full appreciation of our trees, shrubs and flowers is dulled because of our fortunate close association with them. Perhaps, like old and valued friends, they are taken too much for granted—or possibly they are strangers whose true worth is not realized.

Our famous scenery, beautiful lakes and streams, watersheds and commercial timber stands will remain of value only as long as the trees and plants are properly evaluated, so that our activities in them can be properly managed. Learn first to recognize the flora as individuals. Then notice the special conditions that each one requires. See for instance how hardhack masses in wet road ditches or around ponds and how skunk cabbage thrives in black, muddy soil. Every time you find hardhack or skunk cabbage, certain other plants will be discovered nearby. As more and more plants are identified, there comes the realization that every one is part of an intricate living community, a community in very delicate balance. It is a friend if understood and respected, but mishandling can upset this balance, causing floods, erosion or scars on the landscape.

You will find these forest friends almost as close as your next-door neighbour. Make their acquaintance and you can not help meeting more and more relations, each with some especially attractive feature to add pleasure to your every outing.

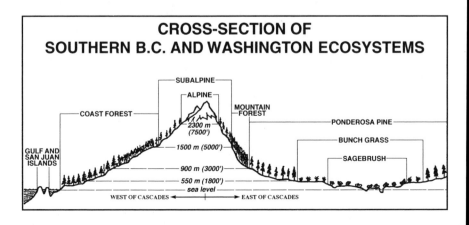

CROSS-SECTION OF SOUTHERN B.C. AND WASHINGTON ECOSYSTEMS

FLORAL AREAS OF B.C.

Note: Park brochures widely available.

1 **Strathcona Park**: Trail access to scenic areas and wildflower meadows.

2 **Pacific Rim National Park**: Road access to beaches and west-coast flora.

3 **Victoria Area**: Gulf Islands flora. Many parks. Spring wildflowers.

4 **Garibaldi Park**: Steep foot trails to exceptional scenery and floral areas.

5 **Manning Park**: Traversed by Hwy. 3. Side road to high country. Trails. Wide floral variety.

6 **Cathedral Lakes Park**: Side road from Hwy. 3, then trail or private jeep operated by lodge. High-mountain scenery and flowers.

7 **South Okanagan**: Highway and back roads. Sagebrush and ponderosa pine ecosystems.

8 **Kokanee Glacier Park**: Trails to high country. Scenic, with floral variety.

9 **Idaho Mountain**: From Sandon, near Hwy. 31A. Steep, rough road for 18 km (11 mi.). Subalpine, scenic.

10 **Mt. Revelstoke National Park**: Paved road from Hwy. 1 to outstanding floral area.

11 **Trophy Mountains, Wells Gray Park.**: Road off Hwy. 5, then easy trail to flower meadows.

12 **Mt. Robson Park**: Trail access from Hwy. 16. Glacier, falls, high mountain meadows.

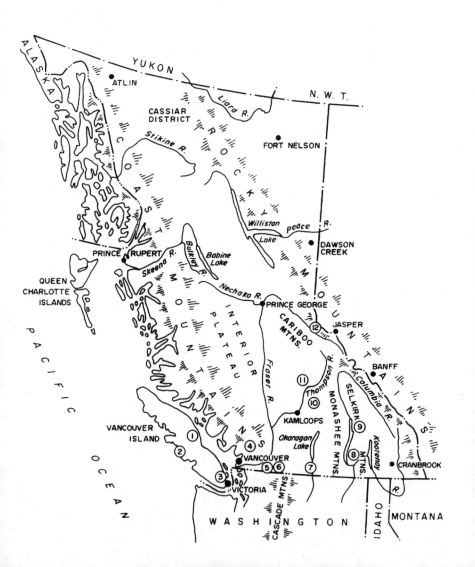

FLORAL AREAS OF WASHINGTON

1 **Hoh Rain Forest**: 29 km (18 mi.) pvd. rd. from U.S. 101. Museums, trails.

2 **Hurricane Ridge**: 29 km (18 mi.) pvd. rd. from Pt. Angeles. Scenery, flw. meadows.

3 **Nisqually Wildlife Refuge**: I-5, Exit 115. Salt-marsh habitat wildflws.

4 **Rockford State Pk.**: Beside WA 20. Stately coastal forest.

5 **Mt. Baker**: Access by WA 542. Scenic, flw. meadows, mountain trails.

6 **Mt. Rainier Nat'l Pk.**: Access by several rds. Scenic. Magnificent flw. areas.

7 **Okanogan**: Sagebrush and ponderosa pine. Spring wildflws.

8 **Republic** eastwards: On WA 20. Fine mtn. forest, varied flora.

9 **Mt. Spokane State Pk.**: Pvd. rd. from U.S. 2 to 1792 m (5878'). Good floral variety.

10 **Dry Falls**: Beside WA 17 near U.S. 2. Scenic; sagebrush ecosystem.

11 **Vantage** southwards: Side rd. along W. side of Columbia R. Spring flws., desert.

12 **Vantage–Ellensburg**: Vantage Hwy. (pvd. side rd.) Good variety; sagebrush. Spring–summer.

13 **Yakima Canyon**: Paved side rd. #821. Wide variety of sagebrush–ponderosa pine plants. Spring–early summer.

14 **Columbia Gorge**: 160 scenic kilometres (100 mi.) on WA 14 and on U.S. 30 (Oregon). Rich floral variety.

15 **Steptoe Butte State Pk.**: N. of Colfax by pvd. rd. Rich Palouse flora.

16 **Blue Mtns.**: Mostly gravel rds. in higher mtns. Wide floral variety.

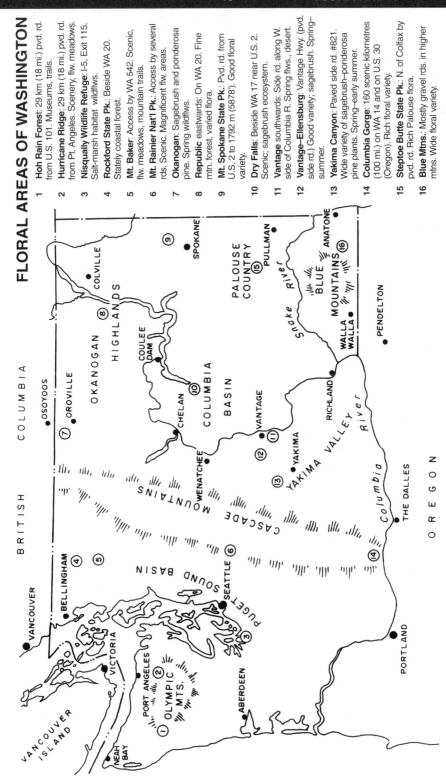

THE ECOSYSTEMS

As more and more studies continue to dissect the landscape and its components, new terminology results. Terms such as 'biotic zone' or 'life zone' formerly used to describe interrelated units have been superseded by the word 'ecosystem.' By definition, an ecosystem is an interacting complex of living organisms (plants, animals, fungi, bacteria) and the physical environment (soil, air, water, bedrock) immediately affecting them. In the broadest sense, a mountain range covered by conifers could be, and is, an ecosystem. Your backyard could be another.

Botanists have used their skills to fine-tune the countryside and its vegetation into a number of classifications: 14 for B.C. and 11 for Washington. An attempt has been made in this book to correlate names and ranges with *Ecosystems of British Columbia,* 1991, Ministry of Forests. Many of the ecosystems in B.C. are also in Washington and cooperative studies are a matter of course. As some of these ecosystems are not easily discernible to the average observer, a simplified system has been followed. It is in part based on earlier, more general classifications and on the author's personal observations.

Despite attempts at classification, it is impossible to mark distinct floral boundaries. For example, western redcedar and mountain hemlock, typical coast species, will be found far east of the Cascades. Conversely, ponderosa pine associated with sagebrush country will stray into the coastal region. However, an observant person will be able to soon identify the surrounding ecosystem.

A very important land division appears on the zone maps. This is the height of land, the mountain crest, that separates the coast from the interior. For simplification, the term 'Cascade Mountains' (or 'Cascade Range,' or 'Cascades') has been applied in both B.C. and Washington, though technically there are two systems in B.C., the Coast Mountains and the Cascade Mountains. Most botanical references frequently and conveniently use the phrase 'east (or west) of the Cascades.' On the ground, the transition is usually both fast and noticeable.

Climactic conditions have the overall greatest effect on the type of vegetation in an area. These weather patterns, together with the topography, create the number of distinct ecosystems in B.C. and Washington. In B.C., the series of mountain ranges running lengthwise through the province intercept moisture-laden westerly winds, thereby forming alternating, parallel wet and dry regions.

Washington's Cascade Mountains, dominated by five great peaks, likewise intercept moisture-laden westerly winds. Thus, for example, the westerly and southern flanks of the Olympic Mountains are saturated with an annual rainfall of 3500 mm (140") while regions in the central interior of the state parch with a meagre 300 mm (12").

The varied topography and climate of this region have produced dramatic contrasts difficult to find elsewhere in North America. Here, high volcanic peaks lie encased in ice hundreds of metres (or feet) thick—yet not far from their feet, almost-barren plateaus stretch away for a hundred kilometres (or miles).

The formidable mountain barrier that divides the west from the east in both B.C. and Washington has been crossed by impressive feats of highway construction. Crossing points on main highways in B.C. are Allison Pass in Manning Park on the Hope–Princeton (Highway 3), the summit of the Coquihalla (Highway 5) and the Duffy Lake section of Highway 99 between Pemberton and Lillooet. In addition, a long way further to the north is Highway 20 in the Chilcotin while still another jump northwards is Highway 16 in the Skeena Valley.

Washington, likewise has a series of west-to-east highways crossing distinct mountain passes. The most northerly of these is the North Cascades Highway (WA 20). Further south is the Everett-to-Wenatchee route (U.S. 2) through Steven's Pass. Paralleling it a little to the south is I-90, which traverses Snoqualmie Pass. Next is WA 410, which passes through Mt. Rainier Park, giving options for easterly travel from the park. U.S. 12, through White Pass, is the most southerly of the series.

Note: Colour ecosystem maps for British Columbia and Washington are located inside the front and back covers of this book.

THE ECOSYSTEMS AND THEIR MOST COMMON FLORA

GULF AND SAN JUAN ISLANDS

Those rocky islands in Georgia Strait south of Quadra Island and an easterly strip of Vancouver Island south of Campbell River make up the B.C. portion of this small ecosystem. It extends eastwards to embrace the San Juan Islands between the southern tip of Vancouver Island and the Washington mainland. South-facing land features open, grassy slopes and rocky knolls, with arbutus and Garry oak providing unique landscaping. In protected draws and on northern slopes, a rich mixture of deciduous and evergreen trees is common.

The climate is world-famous for its mildness and precipitation is low because of protection from mountains to the west and south. Rainfall averages only 750 mm (30") per year, most of which is between October and March. Snow is unusual.

Note: In the following tables,
↓ *indicates a species found in Washington only and* * *indicates an introduced species.*

TREES

Garry oak	arbutus	Douglas-fir
shore pine	bigleaf maple	western flowering dogwood
aspen	bitter cherry	grand fir
Rocky Mountain juniper		

SHRUBS

* Scotch broom	* gorse	hairy manzanita
hairy honeysuckle	evergreen huckleberry	Indian-plum
poison-oak	soopolallie	oceanspray
salal	tall Oregon-grape	gummy gooseberry
red-flowering currant	baldhip rose	kinnikinnick
west'n trumpet honeysuckle	snowberry	trailing snowberry
blue elderberry		

FLOWERS • WHITE

western trillium	miner's lettuce	bunchberry
* oxeye daisy	* corn chamomile	white fawn lily
meadow death-camas	Alaska saxifrage	common whitlow-grass
wild carrot	American wild carrot	field chickweed
yerba buena	fool's onion	Nuttall's pussytoes
baneberry	Pacific water-parsley	white hawkweed
cow-parsnip	Scouler's harebell	

• YELLOW

buttercup species	violet species	stonecrop species
spring gold	barestem desert-parsley	brittle prickly-pear cactus
* common St. John's-wort	Pacific sanicle	Puget Sound gumweed
* wall lettuce	skunk cabbage	yellow waterlily
* hairy cat's-ear		

• PINK–RED

woodland star	vari-leaved collomia	nodding onion
satinflower	prince's pine	* dovefoot geranium
sea blush	red maids	candy stick
peavine species	twinflower	

• PURPLE

Menzies's larkspur	Howell's triteleia	harvest brodiaea
blue-eyed Mary	purple cudweed	Douglas aster
common California aster		

• BLUE

common harebell	lupine species	common camas

• BROWN–GREEN

chocolate lily	pineapple weed	

COAST FOREST

In the broadest sense, this ecosystem comprises the forest extending from ocean beaches to the subalpine elevations of the coastal mountains. More-complex classification schemes usually recognize 2 zones within what is here considered as a single ecosystem. This division is based on the replacement, with increasing elevation, of certain species of evergreen trees by others. Typically, lowland coastal forest species such as Sitka spruce, western hemlock and western redcedar are replaced by mountain hemlock, amabilis fir, alpine fir and yellow-cedar.

The average forest traveller does not note this distinction upon passing into what is often called the Canadian Zone; it is therefore shown on the ecosystem map as part of the coast forest. But with increasing elevation, at a point between 1200 m (4000') in the south of Washington and 1500 m (5000') in northern B.C., there is a noticeable difference as the forest dwarfs and opens. For our purposes, this is the upper limit of the coastal forest and the start of the subalpine forest.

At the ocean-side levels of the coast forest ecosystem, winters verge on freezing temperatures while summers are moderate. Rainfall on coastal cities such as Vancouver, B.C., is annually 1170 mm (46") while Seattle has an average of 850 mm (33"). In the upper part of the coastal forest zone it gets considerably colder; the heavy snowfall encourages the building of ski resorts.

Although the ecosystem range map shows the entire coast forest zone as one colour, separate species lists for lower and upper (Canadian) parts help clarify the floral picture.

LOWER LIMIT—COAST FOREST

TREES

Douglas-fir	western hemlock	western redcedar
grand fir	western white pine	western yew
shore pine	red alder	bigleaf maple
black cottonwood	bitter cherry	western flowering dogwood
vine maple	cascara	Pacific willow

SHRUBS

trailing blackberry	black twinberry	devil's club
red elderberry	stink currant	falsebox
goatsbeard	wild gooseberry	hardhack
beaked hazelnut	red huckleberry	false azalea
Pacific ninebark	oceanspray	black raspberry
salal	red-osier dogwood	salmonberry
thimbleberry	Saskatoon	sweet gale
twinflower	common snowberry	dull Oregon-grape
↓ whipplea		

FLOWERS • WHITE

* oxeye daisy	bunchberry	queen's cup
sweet coltsfoot	false bugbane	false Solomon's-seal
Siberian miner's lettuce	clasping twistedstalk	Hooker's fairybells
pearly everlasting	yarrow	rattlesnake-plantain
white-veined wintergreen	one-sided wintergreen	vanilla leaf
pathfinder	sundew	western trillium
false lily-of-the-valley	foamflower	Indian-pipe
white rein-orchid	fringecup	alumroot species
↓ Oregon oxalis	↓ inside-out flower	fool's onion
cow parsnip	Pacific water-parsley	↓ bigroot

• YELLOW

yellow monkey-flower	* hairy cat's-ear	Canada goldenrod
stream violet	buttercup species	stonecrop species
* common St. John's-wort	curly-cup gumweed	* wall lettuce
large-leaved avens	skunk cabbage	yellow waterlily
silverweed	* common tansy	dune tansy
↓ western yellow oxalis	fern-leaved goldthread	

• ORANGE

tiger lily	spotted coralroot	

- **PINK**

oaks toothwort	fireweed	shootingstar species
* common stork's-bill	pink wintergreen	prince's pine
smooth Douglasii	beach morning-glory	nodding onion
broad-leaved starflower	* foxglove	broad-leaved peavine
giant vetch	sea blush	thrift
rosy pussytoes	fairy slipper	Pacific bleeding heart
candystick	red columbine	spotted coralroot

- **PURPLE**

harvest brodiaea	self-heal	beach pea
purple pea	* purple loosestrife	burdock
Cooley's hedge-nettle	Douglas's aster	

- **BLUE**

American brooklime	wood forget-me-not	camas
Howell's triteleia	blue-eyed Mary	early blue violet
king gentian	Menzies's larkspur	↓ western mertensia
marsh skullcap	lupine species	* self-heal
↓ Columbia kittentail	↓ synthyris species	

- **BROWN–GREEN**

wild ginger	chocolate lily	western mountain bells
stinging nettle	* curly dock	youth-on-age
mitrewort species	pineapple weed	silver bur-weed
twayblade species	western meadowrue	

UPPER LIMIT (CANADIAN ZONE)

TREES

amabilis fir	mountain hemlock	yellow-cedar
subalpine fir	Sitka alder	

SHRUBS

copperbush	false azalea	black huckleberry
oval-leaved huckleberry	Sitka mountain-ash	white-flwr. rhododendron
bog-rosemary	bog-laurel	Labrador tea
Alaska blueberry		

FLOWERS

bunchberry	five-leaved bramble	queen's cup
Indian-pipe	one-sided wintergreen	fern-leaved goldthread
bracted lousewort	single delight	heart-leaved twayblade
Indian hellebore	mitrewort species	fringed grass-of-Parnassus

SUBALPINE FOREST

You reach an entirely different realm when you enter this ecosystem. A backpacker, sweating while bushwhacking up a mountainside, first senses the change as the tangle of blueberries and other shrubs begins to dwarf and open up. Glades appear and mountain valerian, arnica and Indian hellebore promise easier going ahead. Engelmann spruce and subalpine fir stand as picturesque spires. Blue lupine and red paintbrush mingle with yellow butterwed. Fingers of flower meadows widen among the subalpine fir and expand in a riot of colour. To many people, it is the choicest, most beautiful terrain of the entire mountain scene.

In B.C., this zone embraces the superb meadowlands of Garibaldi Park, the high country of Manning and Cathedral Lakes parks, the magnificent floral displays at the end of the road into Mt. Revelstoke National Park, the unbelievable expanse of blooms on the Trophy Mountains of Wells Gray Park. It continues eastwards to the rocks, meadows and lakes of the Bugaboos and then on to Berg Lake in Mt. Robson Park in the heart of the Rockies.

In Washington, the glory of Olympic National Park has been preserved for over 50 years. The snowfields and glaciers of Mt. Olympus serve as a matchless background for meadows of wildflowers. Cliffs, talus slopes and canyon walls have their own floral treasures, with over a dozen species found no other place in the world. Mt. Baker, Mt. Shuksan, Mt. Rainier, Mt. Adams and Mt. St. Helens—that chain of great volcanoes—are accessible to some degree by good roads. Trails lead to a wide variety of terrain, each with its specially adapted plants. The floral display around timberline and beyond into the alpine tundra zone can be beautiful beyond belief. Parts of the Blue Mountains in southeastern Washington reach the subalpine; they have their own particular floral beauty.

Watching amazed summer visitors at the road end in a national park, I thought a simple explanation would answer most questions: 'The red flowers are paintbrush, the spikes of blue flowers are lupines, the pink-purple daisies are mountain daisies, the tall, white heads of small flowers are either mountain valerian or bistort, the large yellow flowers are arnicas and the picturesque trees are subalpine fir.'

Hundreds of other high-country areas, less known and not as accessible, are equally beautiful. This subalpine terrain near the top of the world displays the greatest number and variety of flowers per hectare (or acre) to be found in our region—and always supplemented by superb scenery! The deep snows and meadowlands also create conditions ideal for ski resorts.

Winters are long and severe with freezing temperatures. Snow may reach a depth of 6 m (20') or even more. Vegetation is adapted to withstand these conditions and buffeting winds besides. Heavy spring rains and swirling mists are common. Spring comes late and fast, leaving a short summer season. Generally, flowers are at their best from mid-July to mid-August.

TREES

subalpine fir	limber pine	white bark pine
Engelmann spruce	alpine larch	

SHRUBS

white-flw. rhododendron	dwarf blueberry	blue-leaved huckleberry
crowberry	pink mountain-heather	white mountain-heather
yellow mountain-heather	common juniper	alpine wintergreen
shrubby cinquefoil	falsebox	

FLOWERS • WHITE

cow parsnip	Pacific water-parsley	western springbeauty
fringed grass-of-Parnassus	Sitka valerian	western pasqueflower
American bistort	globe-flower	cotton-grass
bear-grass	yarrow	partridgefoot
tufted saxifrage	leather-leaf saxifrage	thread-leaved sandwort
Parry's campion	white bog-orchid	elmera
alpine pussytoes	woolly pussytoes	Fendler's waterleaf
Olympic aster	Gray's lovage	

• YELLOW

fan-leaved cinquefoil	villous cinquefoil	mountain arnica
broad-leaf arnica	sibbaldia	subalpine buttercup
snow buttercup	golden fleabane	Elmer's butterweed
Lyall's goldenweed	silvery butterweed	other butterweed spp.
sulphur eriogonum	avalanche lily	round-leaved violet
northern goldenrod	alpine goldenrod	arrow-leaved butterweed
Elmer's butterweed	yellow willowherb	yellow mountain avens
bracted lousewort	slender luina	large-flowered agoseris

• ORANGE

tiger lily	

• PINK

pink owl-clover	scarlet paintbrush	moss campion
spreading phlox	shootingstar species	western springbeauty
Columbia lewisia	pink monkey-flower	subalpine daisy
sickletop lousewort	elephant's head	roseroot
paintbrush		

• PURPLE

small-flowered paintbrush   ↓ purple aster

• BLUE

Cusick's speedwell	mountain bog gentian	gentian species
Arctic bluebell	mountain bluebell	rockslide larkspur

• GREEN

Indian hellebore	bog-orchid species

ALPINE TUNDRA

Though the name 'alpine tundra' has replaced the more common name 'alpine,' for most people there is still a problem: How do you know the boundary between subalpine and alpine tundra? An easy and widely accepted answer is to say that it's the timberline, but stunted trees creep high into the rocks. An easily visible transition occurs where the rich growths of meadows and flower beds are replaced with near-barren rocky slopes and ridges. This is where the rockery plants such as heather, campanula, moss campion and saxifrage take over. However, there is some leeway in this transition, with an overlapping of some flowers into, and out of, the subalpine.

This changeover occurs at about 1350 m (4500') on southern coastal mountains and rises to 1800 m (6000') east of the Cascades, but it becomes progressively lower as you go northwards. Weather conditions are severe, with long winters and short summers. More and more people, formerly limited to a short period of summer recreation, now find the alpine tundra spectacular for winter sports.

SHRUBS

pink mountain-heather	yellow mountain-heather	white mountain-heather
crowberry	bog blueberry	alpine-azalea
alpine willows	Davidson's penstemon	

FLOWERS • WHITE

elmera	alpine anemone	American bistort
alpine pussytoes	woolly pussytoes	lance-fruited draba
white mountain-avens	partridgefoot	Tolmie's saxifrage
spotted saxifrage	leather-leaf saxifrage	cordate-leaved saxifrage
rusty-haired saxifrage	wedge-leaved saxifrage	globe-flower

• YELLOW

Lyall's goldenweed	evergreen saxifrage	spreading stonecrop

• PINK

moss campion	spreading phlox	mountain sorrel

• PURPLE

silky phacelia

• BLUE

alpine lupine	mountain harebell	alpine speedwell

• GREEN

mountain sagewort

MOUNTAIN FOREST

For orientation, let us imagine that you are travelling from west to east. You have climbed through the coastal forest to one of the high passes in the Cascade Mountains. Many of these passes are on the lower edge of the subalpine ecosystem. If you chose to climb higher, you would pass through the beautiful flower meadows of true subalpine terrain and on into the alpine tundra ecosystem. Continuing through the pass instead, you would descend into the forest of the eastern slopes, here termed the mountain forest.

This mountain forest system is quite different from the moist westside coastal forest. And as with the coastal forest ecosystem, this arbitrary classification has subdivisions. Continuing from west to east, you begin in an attractive forest almost entirely of white and Engelmann spruce and alpine fir. With descending elevation, there is a change to lodgepole pine, cottonwood and birch. Engelmann spruce continues into these lower elevations, but in hybrid form. In southern B.C., the lower border of this ecosystem fades into the ponderosa pine ecosystem.

In B.C., the mountain forest ecosystem consists in general of the vast mountainside forests east of the Cascades and above the ponderosa pine ecosystem. It is the most widespread area for commercial timber harvest in the province. Several large valley bottoms in eastern B.C. have special status as the interior cedar-hemlock ecosystem (described separately in this section). In the north, the mountain forest terminates rather abruptly where it is replaced by the boreal forest ecosystem.

In Washington, the bulk of the mountain forest ecosystem lies on the easterly flank of the Cascade Mountains. Another large area is found at higher elevations of northeastern Washington, within the Okanogan and Colville National Forests. There is also a small segment on the northern slope of the Blue Mountains.

Only a generalized weather picture can be given. The severe heat of summer in some valleys and on some plains contrasts remarkably with the tempering effect of increased moisture found at higher elevations and on north-facing slopes. Summer can produce anything from temperate to high temperatures while winter brings freezing temperatures and sufficient snow to operate ski resorts.

TREES

Douglas-fir	western redcedar	western hemlock
western white pine	lodgepole pine	grand fir
↓ noble fir	western larch	Engelmann spruce
trembling aspen	black cottonwood	paper birch
bitter cherry	Sitka alder	Douglas maple
amabilis fir (higher elevations)		

SHRUBS

blue clematis	sticky currant	snowbrush
red-osier dogwood	falsebox	black gooseberry
goatsbeard	kinnikinnick	↓ squaw carpet
Labrador tea	trapper's tea	Sitka mountain-ash
white-flw. rhododendron	Saskatoon	scrub birch
soopolallie	highbush-cranberry	sweet gale
alpine-wintergreen	thimbleberry	black twinberry
Utah honeysuckle	mallow ninebark	birch-leaved spirea

FLOWERS • WHITE

Hooker's fairybells	false Solomon's-seal	mountain ladyslipper
pearly everlasting	yarrow	Indian pipe
pussytoes species	long-stalked starwort	queen's cup
rattlesnake-plantain	western trillium	one-leaved foamflower
pathfinder	clasping twistedstalk	Fendler's waterleaf
snow bramble	five-leaved bramble	Canada violet
spotted saxifrage	wild strawberry	heart-leaved springbeauty
anemone species	white shootingstar	Parry's catchfly
silverleaf phacelia	↓ white false hellebore	marsh valerian
night-flowering catchfly	sticky false asphodel	cut-leaf luina
silverback luina	* oxeye daisy	* corn chamomile
cut-leaved daisy	yarrow	pearly everlasting
tufted fleabane	white-veined wintergreen	round-leaved rein-orchid
mountain ladyslipper	bog-orchid species	white rein-orchid

• YELLOW

pinesap	spreading stonecrop	worm-leaved stonecrop
stream violet	prairie cinquefoil	large-leaved avens
yellow willowherb	* common St. John's-wort	* sow thistle species
gumweed	Canada goldenrod	

• ORANGE

orange-flowered agoseris	orange hawkweed	large-flowered collomia

• PINK-RED

coralroot	nodding onion	shootingstar
prince's pine	pink wintergreen	twinflower
broad-leaved star-flower	fairyslipper	red columbine
scarlet gilia		

• PURPLE

purple pea	penstemon species	common butterwort

- **BLUE**

Cusick's speedwell	American brooklime	↓ white-stem frasera
↓ clustered frasera	penstemon species	* wood forget-me-not
showy Jacob's-ladder	tall Jacob's-ladder	early blue violet
bracted vervain	larkspur species	lupine species
* self-heal	great northern aster	

- **BROWN–GREEN**

wild ginger	youth-on-age	rein-orchid species
twayblade species	seabeach sandwort	chocolate lily

PONDEROSA PINE

This ecosystem is readily distinguished by the presence of ponderosa pine, either as widely scattered individuals or in substantial numbers mixed with Douglas-fir, up to an elevation of 900 m (3000'), where it merges with the mountain forest. At lower elevations in B.C., the attractive ponderosa pine forest blends into what has been designated as the sagebrush ecosystem. The intermediate bunchgrass ecosystem, because of limited extent, has been integrated into the sagebrush category on the B.C. ecosystem map. The much more evident bunchgrass ecosystem of Washington reaches south to where Garry oak mingles with ponderosa pine.

Because of the dry climactic conditions in the lee of the Cascade Range, a strip of ponderosa pine forest of varying width extends from the Columbia River northwards through Klickitat, Kittitas and Chelan counties to the hillsides of the Okanogan Valley and on to B.C. Ponderosa pine is also common both north and south of Spokane.

It ranges through the Okanagan and Similkameen valleys of B.C., probes northwards beyond Kamloops into the Cariboo and fingers westwards to Lytton and the Fraser Canyon. In eastern B.C., it occurs again around Kootenay Lake and in the Kootenay River valley. Ground cover is comparatively scarce, often grassland species only, which makes for easy going on foot or by horse. Stream and river bottoms and rocky places have a variety of smaller trees and shrubs as well as ponderosa pine.

The climate is relatively arid, with hot summer days and freezing winter temperatures. Annual rainfall ranges between 250 mm (10") and 500 mm (20").

TREES

ponderosa pine	Douglas-fir	Rocky Mountain juniper
trembling aspen	black cottonwood	black hawthorn
Douglas maple	water birch	choke cherry

SHRUBS

common juniper	antelope-brush	big sagebrush
sumac	tall Oregon-grape	poison-ivy
mock orange	blue elderberry	wild rose species
red-osier dogwood	soopolallie	squaw currant
wolf-willow	thimbleberry	red-raspberry
blue clematis	white clematis	shrubby penstemon

FLOWERS • **WHITE**

strawberry species	yarrow	pearly everlasting
mariposa lily species	death camas	woodland star species
spring whitlow grass	* common watercress	pale evening primrose
velvet lupine	meadow death camas	Hooker's thistle
buckwheat species	rock onion	tufted fleabane
shaggy fleabane	white hawkweed	wild carrot
hemlock water-parsnip	Douglas water-hemlock	* white sweet-clover
western saxifrage	white plectritis	round-leaved alumroot
varileaf phacelia	tufted white prairie aster	

- **YELLOW**

yellow bell	buttercup species	wallflower species
Columbia bladderpod	lemonweed	sulphur cinquefoil
brown-eyed Susan	balsamroot species	mule's-ears
common sunflower	Rocky Mtn. helianthella	* great mullein
blazing star	* common St. John's-wort	* yellow iris

• YELLOW (cont.)
brittle prickly-pear cactus
* Dalmatian toadflax
desert-parsley species
worm-leaved stonecrop

gold star
yellow rattle
* common salsify

* butter and eggs
yellow penstemon
* sow thistle species

• PINK
dagger pod
western springbeauty
old man's whiskers
bitterroot
* knapweed

Holboell's rockcress
smooth woodland star
nodding onion
rosy pussytoes
pink fairies

long-leaved phlox
* common stork's-bill
strict buckwheat
pinedrops

• ORANGE
collomia

• RED
scarlet gilia

• PURPLE
American vetch
edible thistle

* woolly vetch
showy aster

ballhead waterleaf
lupine species

• BLUE
early blue violet
Burke's larkspur
Douglas' brodiaea
* Canada thistle
Howell's triteleia
lupine species
bitter fleabane

western blue flax
common bluebell
* self-heal
showy aster
* chicory
penstemon species
Lindley's aster

Nuttall's larkspur
long-flowered bluebell
* common thistle
bracted vervain
* bachelor's button
↓ blue gilia
great northern aster

• GREEN
pineapple weed

BUNCHGRASS

Normally, the bunchgrass ecosystem lies between the lower elevations of the ponderosa pine forest and the upper part of the sagebrush community. In B.C. the steep terrain and narrow valleys often eliminate the bunchgrass zone, but the plant listings below will apply to much of the attractive valley lowlands. By contrast, although the great basins and rolling hills of Washington provide ample range for bunchgrass, widespread cultivation, mostly under irrigation, has largely eliminated it. The once-great continuity of this ecosystem can only be imagined now.

Bunchgrass flanks the desert-like sagebrush terrain of Douglas, Adams, Yakima, Franklin and Walla Walla counties, with the Columbia and Spokane rivers as a northern boundary. Its elevation range is from 450 m (1500') to 720 m (2400'). Extremely hot summers contrast with the freezing temperatures experienced most of the winter. Snow to a depth of 30–60 cm (1–2') covers the ground for several months.

TREES
ponderosa pine

Douglas-fir

Rocky Mountain juniper

SHRUBS
common juniper
↓ golden currant
poison-ivy

antelope-brush
squaw currant
rabbit-brush

white clematis
mock-orange
Saskatoon

FLOWERS • WHITE
yarrow
cut-leaved daisy
common evening-primrose
pepperpod
little grey aster

pearly everlasting
Hood's phlox
pale evening-primrose
pepperpod

woodland star species
* common watercress
meadow death camas
hoary false yarrow

• YELLOW
common sunflower
* common salsify
woolly sunflower

↓ Cusick's sunflower
Oregon sunshine
balsamroot species

↓ yellow bee plant
line-leaved fleabane
brown-eyed Susan

- **YELLOW** (CONT.)

fern-leaved desert-parsley	yellow sagebrush violet	lemonweed
goldstar	yellow bell	rough wallflower
sagebrush buttercup	prairie cinquefoil	fiddleneck species
* great mullein	blazing-star	common sunflower
buckwheat species		

- **PINK**

pink fairies	shootingstar species	grass widow
rosy pussytoes	bitterroot	long-leaved phlox
* knapweed	Geyer's onion	strict buckwheat
showy phlox		

- **RED**

scarlet gilia	* common hound's-tongue

- **PURPLE**

fern-leaved desert-parsley	sagebrush mariposa lily	ballhead waterleaf
sticky geranium	penstemon species	old man's whiskers
sagebrush violet		

- **BLUE**

Nuttall's larkspur	Burke's larkspur	western blue iris
↓ white-stemmed frasera	Douglas's brodiaea	* teasel
thread-leaved phacelia		

- **GREEN**

western mugwort

SAGEBRUSH

The ecosystems in this book are generally arranged both from west to east and in terms of adjacent elevations. Here we drop to our most arid region to explore the sagebrush ecosystem.

Like ponderosa pine, sagebrush provides obvious boundaries—there are actually a number of 'sagebrushy' plants that fall into the common concept of sagebrush. They cover a large part of the Columbia Plateau of central Washington, with a wide arm branching along the Yakima River towards Cle Elum. More sagebrush is found bordering the Columbia River from Pasco westwards to near The Dalles, Oregon, in the Columbia Gorge. A number of isolated pockets are scattered throughout southern Washington.

From the international boundary, it extends into B.C. through the Similkameen and Okanagan valleys, forming a narrow fringe of arid terrain below the zone of ponderosa pine, except where bunchgrass intervenes. Fingers of sagebrush reach westwards to Lytton and then continue northwards along the Fraser Canyon beyond its confluence with the Chilcotin, to Farwell Canyon. Separate small areas are centred at Grand Forks and Creston. An additional strip lies in the Kootenay plains from the border to Cranbrook.

Although sagebrush appears to typify the most arid terrain, it nevertheless breaks our concepts of where it should grow and climbs high on suitable mountainsides. It ranges up to 540 m (1800') in Washington, and as high as 1200 m (4000') in B.C.

Draws, stream courses, ponds and marshes create various small habitats within the main sagebrush ecosystem, which accounts for the wide variety of plants.

The summer heat is intense, often reaching 38°C (100°F). Winters see freezing temperatures from November to March. Annual rainfall is less than 250 mm (10").

TREES

ponderosa pine	Douglas-fir	Rocky Mountain juniper
trembling aspen	choke cherry	black cottonwood
Columbia hawthorn		

SHRUBS

antelope-brush	greasewood	big sagebrush
threetip sagebrush	↓ rigid sagebrush	pasture sage
rabbit-brush	grey horsebrush	Saskatoon
sumac	common snowberry	tall Oregon-grape
poison-ivy	mock-orange	squaw currant
↓ golden currant	wolf-willow	white clematis
↓ hop sage	↓ winter fat	↓ four-winged salt bush

FLOWERS • WHITE

smooth woodland star	meadow death camas	yarrow
hoary false yarrow	Hood's phlox	field locoweed
fleabane species	pepperpod	

• YELLOW

sagebrush buttercup	yellow bell	brown-eyed Susan
* great mullein	balsamroot species	gold star
Oregon sunshine	woolly sunflower	line-leaved fleabane
buckwheat species	slender hawksbeard	Thompson's paintbrush
cushion fleabane	pale wallflower	desert-parsley species
common sunflower	↓ Simpson's cactus	brittle prickly-pear cactus
many-spined prickly-pear cactus		

• ORANGE

collomia	Munro's globemallow	↓ rigid fiddleneck

• PINK

bitterroot	western springbeauty	shootingstar species
long-leaved phlox	owl-clover	white-leaved globemallow

• PURPLE

↓ Dorr's sage	sagebrush mariposa lily	old man's whiskers

• BLUE

thread-leaved phacelia	↓ elegant penstemon	wild blue flax
silky lupine	small-flw. blue-eyed Mary	long-leaved fleabane

INTERIOR CEDAR-HEMLOCK

Knowing no political boundaries, this ecosystem occupies several parallel valleys in eastern B.C. and extends fingers southwards into northeastern Washington. It was once called the interior wet belt—deservedly so since it has sufficient precipitation to grow redcedar and western hemlock, two important trees of the coastal forest ecosystem. Other well-known coastal plants, such as devil's club and skunk cabbage, occupy swamp areas. Mostly, however, this forest is luxuriant with birches, aspens and cottonwoods, with many shrubs filling every forest glade.

In B.C., this ecosystem occupies the valley of Kootenay Lake and the West Kootenays from the border northwards along the Arrow Lakes and as far west as Mabel and Shuswap lakes. Also the 'Big Bend' of the Columbia River and a narrow strip northwards bordering Kinbasket Lake. An isolated occurrence of this ecosystem in northwestern B.C. includes the Nass Basin and surrounding flanks of the Hazelton and Skeena mountains.

A Washington traveller heading eastwards from Republic on WA 20 will notice the character of the interior-cedar hemlock ecosystem by the lushness of the vegetation and by the most visible clue, the yellowish green fronds of western redcedar. There is a noticeable strip on either side of Sherman Pass. The eastern section is much the wider and continues until the Colville National Forest boundary a few kilometres (miles) west of the Columbia River. Another north-south strip, rather narrow, lies west of the Pend Oreille River, just inside the eastern boundary of this block of the Colville National Forest. A further area occurs east of the Pend Oreille in the extreme northeastern corner of the state.

The annual precipitation, ranging from 500 mm (20") to 1000 mm (40"), is caused by several mountain systems, climaxed by the bulk of the Rocky Mountains to the east.

TREES

Douglas-fir	western redcedar	western hemlock
western white pine	grand fir	western larch
lodgepole pine	Engelmann spruce	western yew
Rocky Mountain juniper	trembling aspen	balsam poplar
paper birch	bitter cherry	mountain alder
Douglas maple	water birch	

SHRUBS

common juniper	black twinberry	devil's club
red elderberry	blue elderberry	false azalea
black gooseberry	hardhack	beaked hazelnut

SHRUBS (cont.)

west'n trumpet honeysuckle	black huckleberry	oval-leaved huckleberry
goatsbeard	kinnikinnick	Labrador tea
tall Oregon-grape	Sitka mountain-ash	mallow ninebark
red raspberry	red-osier dogwood	Utah honeysuckle
white-flw. rhododendron	Saskatoon	scrub birch
birch-leaved spirea	American bush-cranberry	western tea-berry
thimbleberry		

FLOWERS • WHITE

northern bedstraw	bunchberry	cow parsnip
Hooker's fairybells	rough-fruited fairybells	false Solomon's-seal
one-leaved foamflower	pearly everlasting	yarrow
queen's cup	rattlesnake-plantain	pathfinder
clasping twistedstalk		

• YELLOW

skunk cabbage	glacier lily	heart-leaved arnica

• ORANGE

tiger lily

• PINK

striped coralroot	spotted coralroot	fairyslipper
prince's pine	pink wintergreen	

• RED

wood lily

• PURPLE

meadowrue	showy aster

• BLUE

arctic lupine   * self-heal

• BROWN–GREEN

wild ginger	wild sarsaparilla	green wintergreen
Indian hellebore		

CARIBOO PARKLANDS

Cariboo parklands is not recognized as a distinct ecosystem among the 14 in B.C., perhaps because of limited size. However, the term does appear in an early land classification system and it accurately describes the way many people see this attractive area. This zone features a pastoral landscape, log fences bounding hay meadows, ponds and marshes with coots and ducks, and rolling hillsides with clumps of aspen backed by pleasant forests of lodgepole pine and spruce. The zone is bounded on the east by the height of land that has drainage to the Clearwater and North Thompson rivers and on the west by the rising gradient of the Cascade Range. The southern boundary is near Clinton and the northern one about 25 km (15 miles) south of Quesnel. A further small area lies east of Prince George, visible along Highway 16 to McBride. There is no appreciable change in elevation within this zone, which averages about 600 m (2000'). The mountain forest ecosystem almost completely surrounds these parklands.

Spring can be wet but summers are warm and dry. Winters are long, with freezing temperatures but little snowfall.

TREES

Douglas-fir	lodgepole pine	Engelmann spruce
white spruce	subalpine fir	Rocky mountain juniper
trembling aspen	black cottonwood	paper birch
Douglas maple	choke cherry	bitter cherry

SHRUBS

common juniper	creeping juniper	black twinberry
velvet-leaved blueberry	kinnikinnick	tall Oregon-grape
pasture sage	big sagebrush	dwarf blueberry
red-osier dogwood	Saskatoon	scrub birch
wolf-willow	soopolallie	birch-leaved spirea

SHRUBS (cont.)

highbush cranberry	prairie rose	prickly rose
red raspberry	shrubby cinquefoil	

FLOWERS • WHITE

northern bedstraw	bunchberry	meadow death-camas
star-flw. false Solomon's-seal	tufted white prairie aster	creamy peavine
rattlesnake-plantain	Richardson's geranium	baneberry

• YELLOW

pale agoseris	yellow owl-clover	spike-like goldenrod
heart-leaved arnica		

• PINK–RED

nagoonberry	scarlet paintbrush	twinflower
spreading dogbane		

• PURPLE

sticky geranium	western meadowrue	purple peavine
old man's whiskers	small-flowered penstemon	showy aster

• BLUE

big-leaved lupine	wild blue flax	showy Jacob's-ladder

• GREEN

tarragon

PEACE RIVER PARKLANDS

This arbitrary division is based on aspects similar to those of the Cariboo parklands. A view of this area from high in the sky would give the impression of an immense and magnificently landscaped golf course. The irregular shapes of hay and cereal fields are bordered by groves of aspen, poplar and white birch. A darker green indicates forest stands of white spruce and lodgepole pine. Tucked into wetter places, thickets of black spruce and willow add yet more colours and forms. Much of this area is boreal forest (described last) that has been altered through agriculture and extensive forestry operations, a process facilitated by ease of access and nearby processing plants.

You enter this parkland just west of Chetwynd on Highway 97 and leave it west of Ft. Nelson on the Alaska Highway. In the south, it extends westwards to Bennet Dam and Hudson's Hope. In an eastward direction, it reaches into the farmlands of Alberta. A big block of northeastern B.C. east of the Rockies might be related to this zone but the forest there is much denser and has been classified as boreal forest.

Summers are moderately warm throughout and annual precipitation averages about 450 mm (18"). Winters are very cold, with temperatures often below -18°C (0°F).

TREES

white spruce	black spruce	lodgepole pine
Rocky Mountain juniper	subalpine fir	trembling aspen
black cottonwood	paper birch	green alder
mountain alder	choke cherry	balsam poplar

SHRUBS

western snowberry	prickly rose	prairie rose
common juniper	black twinberry	crowberry
velvet-leaved blueberry	dwarf blueberry	black huckleberry
lingonberry	red raspberry	northern gooseberry
black gooseberry	beaked hazelnut	kinnikinnick
Labrador tea	red honeysuckle	western mountain-ash
pasture sage	tall Oregon-grape	red-osier dogwood
Saskatoon	scrub birch	wolf-willow
soopolallie	birch-leaved spirea	highbush cranberry

FLOWERS • WHITE

baneberry	northern bedstraw	bunchberry
cow-parsnip	star-flw. false Solomon's-seal	false Solomon's-seal
tufted white prairie aster	creamy peavine	white geranium
white bog-orchid	field pussytoes	

- **YELLOW**

large-leaved avens	heart-leaved arnica	spike-like goldenrod

- **ORANGE**

orange agoseris	* orange hawkweed

- **PINK–RED**

spotted coralroot	scarlet paintbrush	dwarf nagoonberry
rosy pussytoes	twinflower	spreading dogbane

- **PURPLE**

great northern aster	showy aster	showy fleabane
western meadowrue	small-flowered penstemon	purple peavine

- **BLUE**

showy Jacob's-ladder	bitter fleabane	arctic lupine

- **BROWN–GREEN**

wild sarsaparilla

BOREAL FOREST

'Boreal forest' as used here is a broad catch-all for much of the northern third of B.C. and its vegetation. It occupies the lower elevations along the Stikine and Dean River plateaus and the Atlin–Teslin lakes region in the northwestern corner of B.C. It also extends to take in much of the northeastern part of the province. A fairly dense forest lies below timberline, which is at around 1200 m (4000'). Overall, though logging in B.C.'s boreal forest is comparatively minimal at present, it is on the increase. Large tracts of boreal forest in other Canadian provinces are designated to be logged for pulp; the Russian boreal forest appears headed in the same direction, already with one clearcut that measures 160 by 480 km (100 by 300 miles).

Physiographic and climactic variances produce differing proportions of white and black spruce, willow and birch. Experts recognize within this continuum several separate ecosystems. One, which occupies a large area, has many meadows and limited tree growth. It fits our definition of subalpine; at higher elevations, the alpine-tundra description applies. These 2 regions are shown this way on the B.C. ecosystem map. Washington has no boreal forest.

Winters are cold, with a short growing season. Temperatures are below 0°C (32°F) for 5–7 months per year, with the ground remaining frozen for that period. Annual precipitation is about 430 mm (17"), with half of it falling as snow.

TREES

black spruce	white spruce	subalpine fir
tamarack	aspen	balsam poplar

SHRUBS

Alaska mountain-heather	four-angled mtn.-heather	pink mountain-heather
yellow mountain-heather	white mountain-heather	black huckleberry
shrubby cinquefoil		

FLOWERS • WHITE

white bog-orchid	partridgefoot	Sitka valerian
northern grass-of-Parnassus	leatherleaf saxifrage	

- **YELLOW**

mountain arnica	fan-leaved cinquefoil	arrow-leaved butterweed
bracted lousewort	sibbaldia	

- **PINK–RED**

fireweed

- **PURPLE**

elephant's head

- **BLUE**

arctic lupine

- **BROWN-GREEN**

Indian hellebore

Elevation Ranges of B.C.'s Common Trees, Shrubs and Flowers

(See also plant lists in 'Ecosystem' section, pp. 15–27)

TREES	SHRUBS	FLOWERS
(*above and near timberline*) stunted subalpine fir	white, pink and yellow mountain-heathers, four-angled and Alaska mountain-heathers, black huckleberry, crowberry, alpine willows, common juniper	subalpine daisy, arnicas, gentians, pink and yellow monkey-flowers, red columbine, Sitka valerian, lupines, saxifrages, moss campion, mountain harebell, spreading stonecrop, drabas
mountain hemlock, subalpine and amabalis firs, yellow-cedar, western white pine, Sitka alder	false azalea, copperbush, white-flowered rhododendron, crowberry, mountain-ash, black huckleberry, subalpine spirea, shrubby cinquefoil	Douglas's aster, queen's cup, false Solomon's-seal, twistedstalk, partridgefoot, cotton-grass, Indian hellebore, paintbrushes, arnicas, butterweeds, five-leaved bramble
Douglas-fir, western hemlock, redcedar, amabalis fir, grand fir, western yew, red alder, bigleaf and vine maples, dogwood, cascara, black cottonwood, Pacific willow, Pacific crab apple, black hawthorn, arbutus	falsebox, salal, goatsbeard, devil's club, black twinberry, salmonberry, thimbleberry, mock-orange, oceanspray, red elderberry, hardhack, red huckleberry, western trumpet honeysuckle, hazelnut, Oregon-grape, Indian-plum, red-osier dogwood	bunchberry, vanilla-leaf, wild lily-of-the-valley, starflower, purple pea, rattlesnake plantain, trillium, bleeding heart, fairybells, prince's pine, blue-eyed Mary, shootingstars, miner's lettuce, violets, queen's cup, starflowers, buttercups, skunk cabbage, trillium
(*above timberline*)	white and pink mountain-heathers, alpine willows, black huckleberry, alpine wintergreen, dwarf blueberry, black elderberry, common juniper, crowberry	mountain arnica, yellow columbine, white and yellow mountain-avens, cushion buckwheat, gentians, lupines, saxifrages, moss campion, drabas
Engelmann spruce, subalpine fir, lodgepole pine, Douglas-fir, western larch, whitebark pine, Rocky Mountain juniper, trembling aspen, mountain and sitka alders, paper birch, bitter cherry	black twinberry, Utah honeysuckle, grouseberry, black huckleberry, kinnikinnick, Labrador tea, Sitka mountain-ash, soopolallie, white clematis, common juniper, snowbrush, falsebox, kinnikinnick, mock-orange, snowberry	mountain arnica, bunchberry, red columbine, fairy-slipper, false Solomon's-seal, paintbrushes, lupines, meadowrue, subalpine daisy, Sitka valerian, prince's pine, butterweeds, violets, yarrow, pearly everlasting, common St. John's-wort, Indian pipe
ponderosa pine, Douglas-fir, trembling aspen, black cottonwood, Douglas maple, water birch, mountain alder, choke cherry, hawthorn	antelope-brush, sagebrush, rabbit-brush, Saskatoon, smooth sumac, tall Oregon-grape, blue elderberry, poison-ivy, squaw currant, soopolallie, white clematis	bitterroot, yellow bell, shootingstars, Menzies's larkspur, phloxes, phacelias, scarlet gilia, arrow-leaved balsamroot, fringecup, mariposa lilies, cacti
(*above timberline*)	grouseberry, dwarf blueberry, white and pink mountain-heathers, alpine willows, crowberry, common juniper, black huckleberry	grass-of-Parnassus, gentians, ladies' tresses, moss campion, subalpine daisy, Sitka valerian, western anemone, bracted lousewort, lupines, paintbrushes, elephant's head, white rein-orchid, Sitka valerian, arrow-leaved butterweed, saxifrages, cotton-grass
Engelmann and white spruces, subalpine fir, mountain hemlock, whitebark pine	black elderberry, false azalea, black huckleberry, Sitka mountain-ash, white-flowered rhododendron	Indian-hellebore, tiger lily, grass-of-parnassus, mountain arnica, butterweeds, glacier lily
Douglas-fir, western hemlock, redcedar, western white pine, grand fir, lodgepole pine, western yew, western larch, trembling aspen, black cottonwood, paper birch, bitter cherry, mountain alder, Douglas maple	black twinberry, devil's club, red elderberry, falsebox, hardhack, hazelnut, goatsbeard, kinnikinnick, Utah honeysuckle, thimbleberry, squashberry, soopolallie, redstem ceanothus, false azalea, Oregon-grape	heart-leaved arnica, bunchberry, lupines, red columbine, cow-parsnip, foamflowers, wood lily, yellow glacier lily, queen's cup, prince's pine, wild sarsaparilla, mountain ladyslipper, paintbrushes, tiger lily, self-heal
(*above timberline*)	alpine willows, white, pink and yellow mountain-heathers, black huckleberries, lingonberry, common juniper	moss campion, common harebell, elephant's head, gentians, mountain deathcamas, subalpine daisy, cushion buckwheat, lupines, grass-of-parnassus, columbines, pink monkey-flower, false forget-me-not, white rein orchid, yellow monkey-flower
subalpine fir, lodgepole, whitebark and limber pines, Engelmann spruce, subalpine larch, Rocky Mountain juniper, Douglas-fir, Sitka alder	Columbia clematis, red elderberry, Labrador tea, shrubby cinquefoil, western trumpet honeysuckle, black twinberry, white-flowered rhododendron, Sitka mountain-ash, crowberry, spreading dogbane, soopolallie, scrub birch, grouseberry, black huckleberry, twinflower, snowbrush, red-osier dogwood	cow-parsnip, gentians, fireweed, Sitka valerian, bunchberry, penstemons, shootingstars, blue-eyed Mary, fairy-slipper, wintergreens, yellow mountain-avens, violets, mountain sagewort, ladies' tresses, showy fleabane, arrow-leaved butterweed, arnicas, paintbrushes, fireweed, self-heal
Engelmann spruce, white spruce, larch, junipers, trembling aspen, Douglas maple, cottonwood, paper birch, mountain alder	squashberry, birch-leaved spirea, kinnikinnick, twinflower, false azalea, Saskatoon, snowbrush	mountain ladyslipper, streambank butterweed, bunchberry, yellow columbine, wintergreens, lupines

Elevation Ranges of B.C.'s Common Trees, Shrubs and Flowers

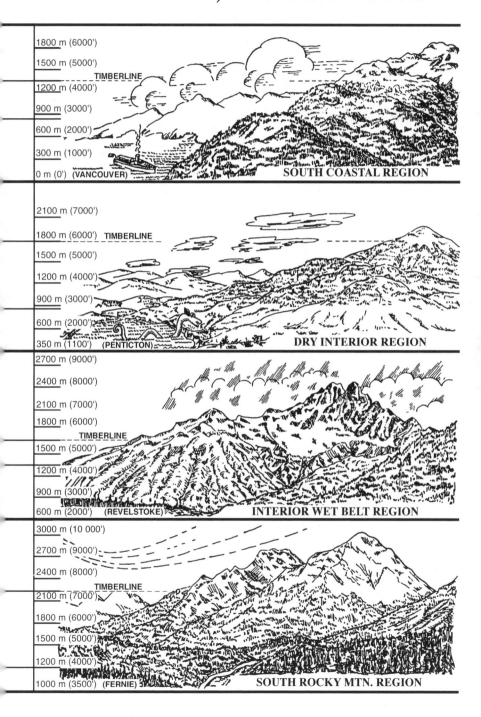

1800 m (6000')
1500 m (5000')
TIMBERLINE
1200 m (4000')
900 m (3000')
600 m (2000')
300 m (1000')
0 m (0') (VANCOUVER) **SOUTH COASTAL REGION**

2100 m (7000')
1800 m (6000') TIMBERLINE
1500 m (5000')
1200 m (4000')
900 m (3000')
600 m (2000')
350 m (1100') (PENTICTON) **DRY INTERIOR REGION**

2700 m (9000')
2400 m (8000')
2100 m (7000')
1800 m (6000')
TIMBERLINE
1500 m (5000')
1200 m (4000')
900 m (3000')
600 m (2000') (REVELSTOKE) **INTERIOR WET BELT REGION**

3000 m (10 000')
2700 m (9000')
2400 m (8000')
TIMBERLINE
2100 m (7000')
1800 m (6000')
1500 m (5000')
1200 m (4000')
1000 m (3500') (FERNIE) **SOUTH ROCKY MTN. REGION**

Elevation Ranges of Washington's Common Trees, Shrubs and Flowers

(See also plant lists in 'Ecosystem' section, pp.15–27)

TREES	SHRUBS	FLOWERS	
(*above and near timberline*) stunted subalpine fir, whitebark pine	white, pink and yellow mountain-heather, alpine willows, shrubby cinquefoil, alpine wintergreen, black huckleberry, crowberry, common juniper	subalpine daisy, arnicas, gentians, Indian-hellebore, pink and yellow monkey-flowers, columbines, Sitka valerian, white and yellow glacier lilies, saxifrages, moss campion, drabas, stonecrops	
mountain hemlock, subalpine and amabalis firs, yellow-cedar, western white and whitebark pine, subalpine larch, Sitka alder	false azalea, copperbush, white-flowered rhododendron, crowberry, mountain-ash, black huckleberry, subalpine spirea, shrubby cinquefoil	western mountainbells, queen's cup, false Solomon's-seal, twistedstalk, partridge-foot, bear-grass, bunchberry, fairy-slipper, lupines, paintbrushes, five-leaved bramble	
Douglas-fir, hemlock, redcedar, amabalis and grand firs, western yew, Rocky Mountain juniper, red alder, bigleaf and vine maples, dogwood, cascara, black cottonwood, Pacific willow, Pacific crab apple, black hawthorn, Oregon ash, arbutus	falsebox, salal, goatsbeard, devil's club, black twinberry, salmonberry, thimbleberry, mock-orange, oceanspray, red elderberry, hardhack, red huckleberry, twinflower, western trumpet honeysuckle, Pacific rhododendron, silk tassel, deerbrush, western wahoo	oxalises, vanilla-leaf, wild lily-of-the-valley, starflower, purple pea, rattlesnake plantain, trillium, bleeding heart, fairybells, prince's pine, blue-eyed Mary, shootingstars, miner's lettuce, violets, skunk cabbage, yerba buena, buttercups, satin-flower, lupines, trillium	
(*above timberline*)	white, pink and yellow mountain-heathers, alpine willows, alpine wintergreen, dwarf blueberry, common juniper, black huckleberry	mountain arnica, yellow columbine, white mountain-avens, cushion buckwheat, elephant's head, yellow glacier lily, gentians, saxifrages, drabas, silky phacelia, mountain harebell, goldenweed	
western white pine, whitebark pine, Engelmann spruce, lodgepole pine, Douglas-fir, amabalis and subalpine fir, western larch, trembling aspen, mountain and sitka alders	black twinberry, Utah honeysuckle, grouseberry, black huckleberry, kinnikinnick, Labrador tea, Sitka mountain-ash, soopolallie, birch-leaved spirea, twinflower, black gooseberry, common juniper	mountain arnica, bunchberry, columbines, fairy-slipper, false Solomon's-seal, paintbrushes, lupines, meadowrue, subalpine daisy, Sitka valerian, prince's pine, larkspurs, butterweeds, yarrow, pearly everlasting, common St. John's-wort	
ponderosa pine, Douglas-fir, black cottonwood, Douglas maple, water birch, mountain alder, choke cherry, hawthorn, white alder	antelope-brush, sagebrush, rabbit-brush, Saskatoon, smooth sumac, tall Oregon-grape, blue elderberry, poison-ivy, squaw currant, soopolallie, white clematis, hazelnut, golden currant	bitterroot, yellow bell, shootingstars, Menzies's larkspur, phloxes, phacelias, scarlet gilia, arrow-leaved balsamroot, fringecup, mariposa lilies, cacti, mule's-ears, buckwheats	
(*above timberline*)	huckleberry, dwarf blueberry, white and pink mountain-heathers, dwarf and alpine willows, crowberry, shrubby cinquefoil, common juniper	grass-of-Parnassus, gentians, ladies' tresses, moss campion, subalpine daisy, lupines, Sitka valerian, western anemone, bracted lousewort, elephant's head, paintbrushes, bear-grass, white marsh-marigold, yellow and white glacier lilies, owl-clovers, butterweeds, saxifrages, drabas, cotton-grass, Sitka valerian	
Engelmann spruce, subalpine and noble firs, mountain hemlock, amabalis fir, yellow-cedar	black elderberry, false azalea, black huckleberry, Sitka mountain-ash, white-flowered rhododendron	white false hellebore, tiger lily, grass-of-parnassus, mountain arnica, western anemone	
Douglas-fir, western hemlock, redcedar, western white and lodgepole pines, grand fir, western yew, trembling aspen, black cottonwood, paper birch, bitter cherry, mountain alder, Douglas, vine and bigleaf maples	black twinberry, devil's club, red elderberry, falsebox, hardhack, goatsbeard, kinnikinnick, salmonberry, thimbleberry, western trumpet honeysuckle, redstem ceanothus, twinflower, snowbrush, red huckleberry, dwarf bramble	heart-leaved arnica, bleeding heart, bunchberry, lupines, red columbine, cow-parsnip, foamflowers, queen's cup, prince's pine, catchflies, paintbrushes, meadowrue, baneberry, wild ginger, alumroots, saxifrages, five-leaved bramble	
(*high mountain ridges*) bitter cherry, mountain mahogany, whitebark pine	snowbrush, red-stem ceanothus, falsebox, grouseberry	white false hellebore, Sitka valerian, subalpine daisy, white-leaved phacelia, old man's whiskers, sulphur buckwheat, yarrow, death-camas, Davidson's and shrubby penstemons, owl-clovers, fraseras	
ponderosa pine, Douglas-fir, grand fir, Douglas maple, choke cherry, Engelmann spruce, western larch, aspen, Sitka alder	blue elderberry, oceanspray, kinnikinnick, western trumpet honeysuckle, Sitka mountain-ash, mallow ninebark, thimbleberry, squaw currant, alder-leaved buckthorn, black huckleberry, Saskatoon	heart-leaved arnica, star-flowered Solomon's-seal, penstemons, paintbrushes, self-heal, bracted lousewort, prince's pine, wintergreens, false bugbane, columbines, stonecrop, collomias, scarlet gilia, nettle-leaved giant-hyssop, pink fairies, sugarbowls, Hooker's balsamroot	
Rocky Mountain juniper, choke cherry, hackberry	blue elderberry, rabbitbrush, sagebrush, white clematis, poison-ivy, golden currant, hop sage	teasel, St. John's-wort, yellow salsify, common sunflower, butterweeds, milkweed, great mullein, grass widow, cattail	

Elevation Ranges of Washington's Common Trees, Shrubs and Flowers

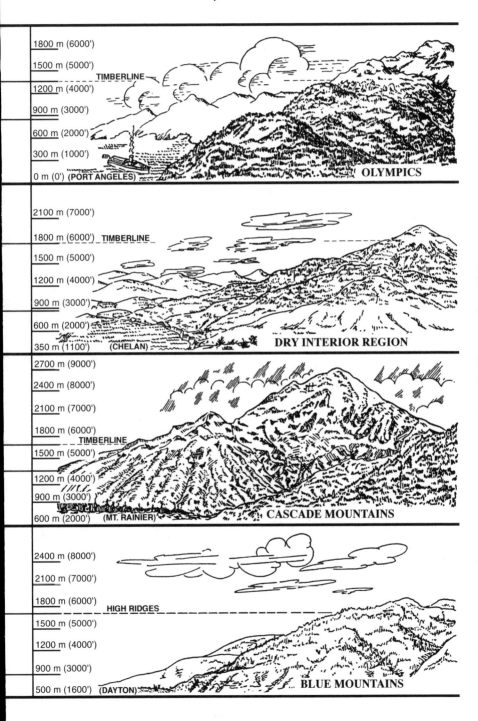

1800 m (6000')
1500 m (5000')
- - - - TIMBERLINE
1200 m (4000')
900 m (3000')
600 m (2000')
300 m (1000')
0 m (0') (PORT ANGELES)

OLYMPICS

2100 m (7000')
1800 m (6000') TIMBERLINE
1500 m (5000')
1200 m (4000')
900 m (3000')
600 m (2000')
350 m (1100') (CHELAN)

DRY INTERIOR REGION

2700 m (9000')
2400 m (8000')
2100 m (7000')
1800 m (6000')
- - TIMBERLINE
1500 m (5000')
1200 m (4000')
900 m (3000')
600 m (2000') (MT. RAINIER)

CASCADE MOUNTAINS

2400 m (8000')
2100 m (7000')
1800 m (6000')
HIGH RIDGES
1500 m (5000')
1200 m (4000')
900 m (3000')
500 m (1600') (DAYTON)

BLUE MOUNTAINS

EARLY BOTANISTS

Most pioneer botanists would also qualify as explorers. Their travels, often into the wilderness, involved considerable personal danger. Clothes, food and equipment were unbelievably scant, yet the lure of new country and undiscovered flora led them forward. Fortunately their names live on, commemorated in the names of plants that become more and more treasured with the passage of time.

William Anderson (1750–1778): Anderson was surgeon/botanist on James Cook's third voyage, which arrived at Nootka Sound in 1778. When he made the first known collections of plants for northwestern America, the science of classifying and naming plants was only 25 years old! Unfortunately, Anderson was quite ill at the time and died shortly, leaving to others the task of collecting and naming our flora.

Archibald Menzies (1754–1842): On Captain George Vancouver's voyage to the northeastern Pacific in 1792, Menzies was the expedition's surgeon and botanist. Included in his voluminous collections were false azalea (*Menziesia ferruginea*) and arbutus (*Arbutus menziesii*), 2 of many species named in his honour. It was Menzies who described the shores of Puget Sound as 'impenetrable stretches of Pinery' and named Douglas-fir after a companion botanist, David Douglas.

Meriwether Lewis (1774–1809) and **William Clark** (1770–1838): The Lewis and Clark Expedition (1804–06) from St. Louis to the mouth of the Columbia River was responsible for the discovery of more than 100 species of 'new' plants. The genera *Lewisia* and *Clarkia*, as well as many species designations (both plant and animal) commemorate their considerable achievement.

David Douglas (1799–1834): Douglas first arrived at the mouth of the Columbia River in April, 1825. His collections of both plants and seeds through Washington, Oregon and British Columbia were impressive. His collections were divided between William Hooker (described below) and the Horticultural Society of London. David Douglas's contribution to the understanding of the botany of the Northwest is unequalled. This botanist from Glasgow, Scotland, is rightly honoured in the naming of Douglas-fir, the genus *Douglasia* and many species designations.

John Scouler, M.D. (1804–1871): In 1825, Scouler was appointed geologist/surgeon/naturalist on the sailing ship *William and Ann*. Thus he accompanied David Douglas on his first trip to the Pacific Northwest. Species that commemorate his name include Scouler's harebell (*Campanula scouleri*) and Scouler's valerian (*Valeriana scouleri*).

Sir William Hooker (1785–1865): It was Hooker who supported and advanced David Douglas's career as an explorer/collector for the Horticultural Society of London. Also, as the recipient of much of the material Douglas collected, it was Hooker who described and published many of the results. Hooker's onion (*Allium acuminatum* Hook.) is among these species. Specimens he distributed to other botanists were occasionally named in his honour. For example, Hooker's (white) thistle (*Cirsium hookerianum*) was named by Thomas Nuttall.

Thomas Nuttall (1786–1859): What greater tribute might be paid to an explorer than to have the western flowering dogwood (*Cornus nuttallii*) named for him? Nuttall was a 'botanical explorer' who ventured into many areas of western North America. In addition, he named many plants collected by other botanists.

Sir Joseph Hooker (1817–1911): The son of Sir William (above), he was renowned as one of the most colourful and talented botanists in Europe. Early in his career he collected plants in the Rocky Mountains. Hooker's fairybells (*Disporum hookeri*) recognises his contribution.

David Lyall, R.N. (1817–1895): During the International Boundary Survey, Lyall reported the occurrence of subalpine larch (*Larix lyalli*) along the 49th parallel. Lyall was a surgeon in the Royal Navy during the coastal surveys of Captain Richards, 1857–59.

John Macoun (1831–1920): John Macoun and his son James made many expeditions into Canada's northwest to collect all manner of natural objects, primarily plants, birds and mammals. At least 20 plant species and varieties were named in his honour. His major contributions to Canadian botany were his large collection and his *Catalogue of Canadian Plants*, published in parts between 1883 and 1892. Macoun's meadowfoam (*Limnanthes macounii*), endemic to southern Vancouver Island, honours John Macoun (rare, not in this book).

This colour key contains 450 photos of flowering shrubs, wildflowers, irregular plants and ferns, selected to include a wide variety of both unusual and similar-looking species (some common species are absent). The order is roughly the same as for the individual write-ups—by colour and flower type—though some species are placed in new contexts. The photo ID numbers are used for cross-referencing from the individual write-ups.

velvet-leaved blueberry 001

bog blueberry 002

grouseberry 003

Alaskan blueberry 004

oval-leaved blueberry 005

evergreen huckleberry 006

oval-leaved blueberry 007

black raspberry 008

highbush cranberry 009

black huckleberry 010

stink currant 011

sticky currant 012

gummy gooseberry 014

red-flowering currant 013

squaw currant 015

golden currant 016

trailing black currant 017

bog cranberry 018

trailing blackberry 019

western teaberry 020

Alaskan mountain-heather 021

crowberry 022

kinnikinnick 023

white mountain-heather 024

pink mountain-heather 025

falsebox 026

bog-laurel 027

bog-rosemary 028

yellow mountain-heather 029

Labrador tea 030

scrub birch 031

sweet gale 032

trapper's tea 033

hardhack 034

western trumpet
honeysuckle 035

big sagebrush 036

antelope-brush 037

poison-oak 038

common rabbit-brush
039

hop sage 040

poison-ivy 041

soopolallie 042

greasewood 043

wolf-willow 044

Davidson's penstemon 045

shrubby penstemon 046

Dorr's sage 047

tall Oregon-grape 048

dull Oregon-grape 049

clustered wild rose 050

baldhip rose 051

devil's club 052

oceanspray 053

common snowberry 054

false azalea 055

Utah honeysuckle 056

mock-orange 057

Pacific ninebark 058

tree lupine 059

black twinberry 060

salal 061

redstem ceanothus 062

snowbrush 063

hairy manzanita 064

Indian-plum 065

Sitka mountain-ash 066

western mountain-ash 067

Pacific rhododendron 068

subalpine spirea 069

enchanter's nightshade 070   Lyall's mariposa lily 071   bunchberry 072

white trillium 073   water-plantain 074   wapato 075   pepperpod 076

northern bedstraw 077   Alaska saxifrage 078   grassland saxifrage 079

pale evening-primrose 080   tufted saxifrage 081   Tolmie's saxifrage 082

common watercress 083   Mertens's saxifrage 084   western saxifrage 085

field chickweed 086

white water-buttercup 087

yerba buena 088

thread-leaved
sandwort 089

wild strawberry 090

five-leaved bramble
091

cloudberry 092

long-stalked starwort
093

Scouler's popcorn
flower 094

three-leaved
anemone 095

round-leaved sundew
096

small-flowered
woodland star 097

miner's lettuce 098

globeflower 099

white shootingstar
102

western springbeauty 100

dwarf hesperochiron 101

Oregon oxalis 103

Canada violet 104

one-sided wintergreen 105

fringecup 106

three-leaved foamflower 107

fringed grass-of-Parnassus 108

Indian-pipe 109

single delight 110

buckbean 111

Sitka mistmaiden 112

round-leaved alumroot 113

small-flowered alumroot 114

desert thelopody 115

silverleaf phacelia 116

white fawn lily 119

queen's cup 117

bigroot 118

western anemone 120

Howell's triteleia 121

clasping twistedstalk 122

white mountain-avens 123

false Solomon's seal 124

Hooker's fairybells 125

inside-out flower 126

white false hellebore 127

Parry's catchfly 128

Sitka valerian 129

white marsh-marigold 130

bladder campion 131

longleaf milk-vetch 132

velvet lupine 133

false lily-of-the-valley 134

white sweet-clover 135

woolly pussytoes 136

racemose pussytoes 137

partridgefoot 138

cotton-grass 139

narrow-leaved
bur-reed 140

meadow
death-camas 141

American glehnia 142

vanilla-leaf 143

pathfinder 144

American bistort
145

baneberry 146

star-flowered false
Solomon's-seal 147

palmate coltsfoot 148

goatsbeard 149

false bugbane 150

pearly everlasting 151

large-fruited
desert-parsely 152

strict buckwheat 153

creamy buckwheat
154

bear-grass 155

hemlock
water-parsnip 156

poison-hemlock 157

wild carrot 158

white hawkweed 159

cow-parsnip 160

English daisy 161

silverback luina 162

shaggy fleabane 163

white layia 164

cut-leaved daisy 165

oxeye daisy 166

corn chamomile 167

round-leaved
rein-orchid 168

rattlesnake-plantain
169

white-veined
wintergreen 170

white rein-orchid 171

mountain ladyslipper
172

phantom orchid 173

Columbia bladder pod 174

pinesap 175

rough wallflower 176

yellow willow-herb 177

rape 180

creeping buttercup 181

small yellow water-buttercup 178

sagebrush buttercup 179

villous cinquefoil 182

sibbaldia 183

sticky cinquefoil 186

silvery cinquefoil 187

fan-leaved cinquefoil 184

silverweed 185

worm-leaved stonecrop 188

spreading stonecrop 189

large-leaved avens 190

yellow montane violet 191

round-leaved violet 192

western yellow oxalis 193

yellow bell 194

common
St. John's-wort 195

great mullein 196

lemonweed 197

wall lettuce 198

tufted loosestrife 199

fringed loosestrife
200

blazing-star 201

yellow
mountain-avens 202

yellow glacier lily 203

yellow iris 204

many-spined prickly
pear cactus 205

brittle prickly pear
cactus 206

yellow waterlily 207

woodland tarweed
208

yellow monkey-flower
209
Inset: chickweed
monkey-flower 210

Dalmatian toadflax
211

gold star 212

yellow rattle 213

common touch-me-not 214

golden corydalis 215

Pacific sanicle 216

greater bladderwort 217

bird's-foot trefoil 218

yellow sand-verbena 219

Thompson's paintbrush 220

bracted lousewort 221

Canada goldenrod 222

spring gold 223

common tansy 224

skunk cabbage 225

barestem desert-parsley 226

fern-leaved desert-parsley 227

narrow-leaved desert-parsley 228

cushion buckwheat 229

sulphur buckwheat 230

heart-leaved buckwheat 231

short-beaked
agoseris 232

yellow salsify
233

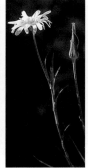
meadow salsify
234

smooth
hawksbeard 235

silvercrown 236

large-flowered agoseris 237

common dandelion 238

woolly butterweed 239

western hawkweed
240

Canadian butterweed
241

Elmer's butterweed
242

arrow-leaved
butterweed 243

prickly
sow-thistle 244

brown-eyed Susan
245

tansy butterweed 246

Oregon sunshine 247

golden fleabane 248

heart-leaved arnica 249

Puget sound gumweed 250

nodding beggarticks 251

arrow-leaved balsamroot 252

Hooker's balsamroot 253

groundcone 254

mule's-ears 255

mountain sneezeweed 256

common sunflower 257

Rocky Mountain helianthella 258

yellow ladyslipper 259

California poppy 260

orange hawkweed 264

Munro's globemallow 261

rigid fiddleneck 262

spotted coralroot 263

orange agoseris 265

American searocket
265

oaks toothwort 266

pink fairies 267

fireweed 268

Henderson's
shootingstar 269

beach
morning-glory 270

smooth douglasia 271

old man's whiskers
272

desert shootingstar
273

nagoonberry 274

smooth woodland
star 275

common
stork's-bill 276

dovefoot geranium
277

small-leaved montia
278

spreading phlox 280

long-leaved phlox 281

Robert geranium 282

moss campion 279

pink wintergreen 283

prince's pine 284

Menzies's pipsissewa 285

pink fawn lily 286

satin-flower 287

grass widow 288

nodding onion 289

Hooker's onion 290

Columbia lewisia 291

broad-leaved starflower 292

bitterroot 293

scalloped onion 294

bicoloured linanthus 295

wild bergamot 296

pink monkey-flower 297

foxglove 298

purple dead-nettle 299

big-head clover 300

springbank clover 301

giant vetch 302

woolly pod
milk-vetch 303

broad-leaved peavine
304

gnome-plant 305

water smartweed 306

seablush 307

spotted knapweed 308

mountain owl-clover
309

rosy pussytoes 310

mountain sorrel 311

elephant's head 312

hemp dogbane 313

subalpine daisy 314

fairy-slipper 315

candystick 316

spreading dogbane
317

steer's head 318

Pacific bleeding heart 319

red maids 320

scarlet gilia 321

common
hound's-tongue 322

pinedrops 323

wood lily 324

Brown's peony 325

scarlet paintbrush
326

magenta paintbrush
327

round-leaved
trillium 328

blue mustard 329

strawberry-blite 332

red columbine 333

Sagebrush mariposa
lily 330

sagebrush violet 331

alpine speedwell 334

naked broomrape 335

clustered broomrape 336

groundcone 337

purple mountain saxifrage 338

European bittersweet 339

marsh cinquefoil 340

silky phacelia 341

field mint 342

Columbia monkshood 343

purple loosestrife 344

pasqueflower 345

harvest brodiaea 346

American vetch 347

Simpson's cactus 348

common butterwort 349

blue-eyed grass 350

common salsify 351

Cooley's hedge-nettle 352

small-flowered penstemon 353

coast penstemon 354

woolly vetch 355

beach pea 356

showy locoweed 357

showy fleabane 358

Philadelphia
fleabane 359

roseroot 360

ballhead waterleaf 361

self-heal 362

burdock 363

common thistle
364

edible thistle
365

wavy-leaved
thistle 366

showy aster 367

fern-leaved desert-parsley
368

cushion fleabane 369

Cusick's speedwell 370

western meadowrue 371

Douglas's aster 372

American brooklime 373   clustered frasera 374   common harebell 375   blue stickseed 376

early blue violet 377   Olympic harebell 378   bachelors button 379

showy Jacob's-ladder 380   Menzies's larkspur 381   common camas 382   chicory 383

large-flowered blue-eyed Mary 384   king gentian 385   long-flowered bluebell 386

viper's bugloss 387   blue skullcap 388   marsh skullcap 389   tall bluebell 390

spurred lupine 391

silky lupine 392

bicoloured lupine 393

Columbia kittentails 394

teasel 395

wild ginger 396

common plantain 397

chocolate lily 398

western mountainbells 399

western mugwort 400

tarragon 401

curly dock 402

sheep sorrel 403

youth-on-age 404

seabeach sandwort 405

Brewer's mitrewort 406

three-toothed mitrewort 407

Indian hellebore 408

pineapple weed 409

mountain sagewort 410

stinging nettle 411

Alaska rein-orchid 412

green-flowered bog-orchid 413

wild sarsaparilla 414

Eurasian water-milfoil 415

watershield 416

heart-leaved twayblade 417

pondweed 418

giant horsetail 419

scouring-rush 420

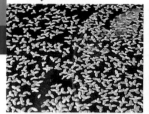
duckweed 421

licorice fern 422

maidenhair fern 423

American glasswort 424

common cattail 425

leathery polypody 426

deer fern 427

TREES

KEY TO EVERGREEN TREES

needle or scale-like

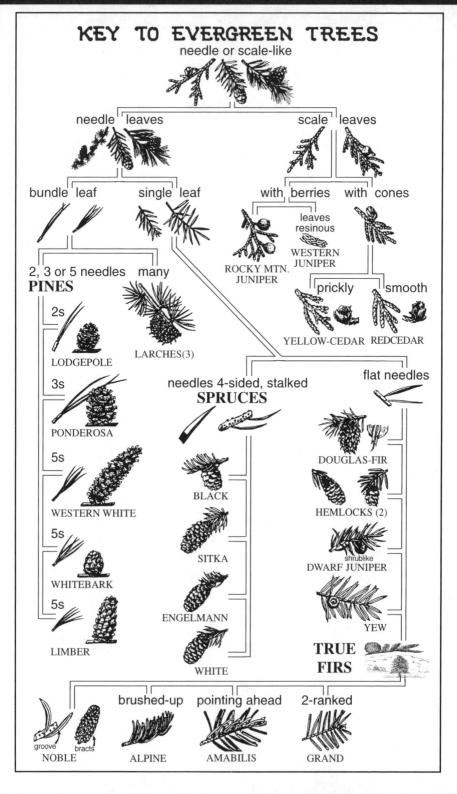

needle leaves

scale leaves

bundle leaf

single leaf

with berries

with cones

leaves resinous

WESTERN JUNIPER

ROCKY MTN. JUNIPER

prickly

smooth

YELLOW-CEDAR

REDCEDAR

2, 3 or 5 needles
PINES

many

2s

LODGEPOLE

LARCHES(3)

3s

PONDEROSA

5s

WESTERN WHITE

5s

WHITEBARK

5s

LIMBER

needles 4-sided, stalked
SPRUCES

BLACK

SITKA

ENGELMANN

WHITE

flat needles

DOUGLAS-FIR

HEMLOCKS (2)

shrublike
DWARF JUNIPER

YEW

TRUE FIRS

groove   bracts
NOBLE

brushed-up

ALPINE

pointing ahead

AMABILIS

2-ranked

GRAND

KEY TO BROADLEAF TREES

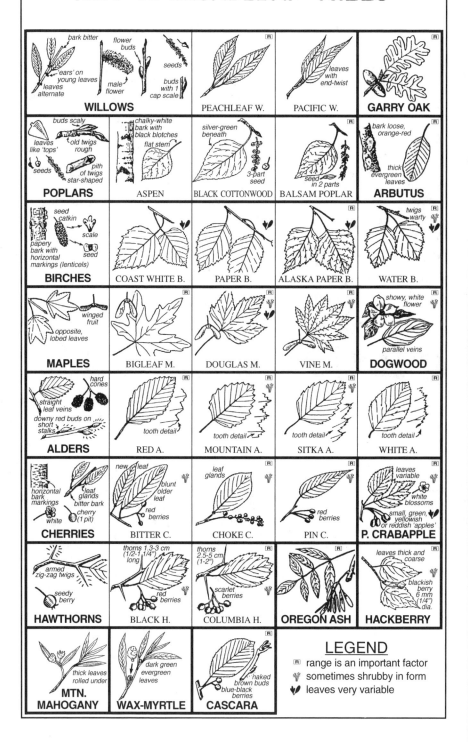

WILLOWS — bark bitter; flower buds; 'ears' on young leaves; leaves alternate; male flower; seeds; buds with 1 cap scale

PEACHLEAF W. [R]

PACIFIC W. — leaves with end-twist [R]

GARRY OAK [R]

POPLARS — buds scaly; leaves like 'tops'; old twigs rough; seeds; pith of twigs star-shaped

ASPEN — chalky-white bark with black blotches; flat stem

BLACK COTTONWOOD — silver-green beneath; 3-part seed

BALSAM POPLAR — seed in 2 parts [R]

ARBUTUS — bark loose, orange-red; thick evergreen leaves [R]

BIRCHES — seed catkin; scale; seed; papery bark with horizontal markings (lenticels)

COAST WHITE B.

PAPER B.

ALASKA PAPER B.

WATER B. — twigs warty [R]

MAPLES — winged fruit; opposite, lobed leaves

BIGLEAF M. [R]

DOUGLAS M.

VINE M. [R]

DOGWOOD — showy, white flower; parallel veins

ALDERS — hard cones; straight leaf veins; downy red buds on short stalks

RED A. — tooth detail [R]

MOUNTAIN A. — tooth detail

SITKA A. — tooth detail [R]

WHITE A. — tooth detail [R]

CHERRIES — horizontal bark markings; leaf glands; bitter bark; white; cherry (1 pit)

BITTER C. — new leaf; blunt older leaf; red berries [R]

CHOKE C. — leaf glands [R]

PIN C. — red berries [R]

P. CRABAPPLE — leaves variable; white blossoms; small, green, yellowish or reddish 'apples' [R]

HAWTHORNS — armed zig-zag twigs; seedy berry

BLACK H. — thorns 1.3-3 cm (1/2-1 1/4") long; red berries

COLUMBIA H. — thorns 2.5-5 cm (1-2"); scarlet berries [R]

OREGON ASH [R]

HACKBERRY — leaves thick and coarse; blackish berry 6 mm (1/4") dia.

MTN. MAHOGANY — thick leaves rolled under

WAX-MYRTLE — dark green evergreen leaves

CASCARA — naked brown buds; blue-black berries [R]

LEGEND

[R] range is an important factor

sometimes shrubby in form

leaves very variable

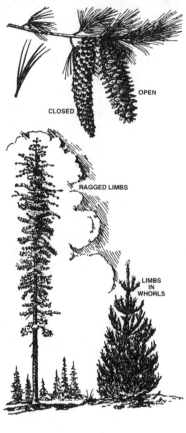

OPEN

CLOSED

RAGGED LIMBS

LIMBS IN WHORLS

Pine Family, *Pinaceae*

WESTERN WHITE PINE
Pinus monticola
White Pine, Silver Pine

FORM: When growing under fair conditions, it has a remarkably straight trunk, like a sturdy flagpole. Trees 30–90 cm (1–3') in diameter and 36 m (120') high are most common. Regular whorls of horizontal limbs are characteristic and these form a narrow crown on the top ⅔ of the tree. Towards the crown on mature trees, long, protruding limbs make a design recognizable from afar.

BARK: Silvery grey on trees up to 15 cm (6") in diameter. On larger trees, becoming dark and deeply fissured to form a regular pattern of small, thick plates that distinguishes white pine from any other native tree.

LEAVES: Bundles of 5 needles to a sheath. Needles 5–12.5 cm (2–5") long, thin and bluish green. Other 5-needle pines have needles about 5 cm (2") long.

FRUIT: An unusual cone 12.5–23 cm (5–9") long, slightly curved and with the tips of the scales a darker brown. The cones are so large that the clusters can be seen plainly on the topmost limbs during summer and fall months.

WOOD: A light, fine-grained wood, easily worked and a high proportion free from defects. Prized for special construction purposes and for match blocks. An ideal wood for carving.

DID YOU KNOW that there is usually a thick carpet of brown needles under this pine? In heavy timber, its presence is often noticed this way. White pine blister rust has decimated this tree across B.C. This disease requires the showy red currant, or another currant or gooseberry species, as an alternate host in order to spread.

QUICK CHECK: From afar, a few protruding limbs and clusters of large cones in upper crown. A 5-needle pine with needles 5–12.5 cm (2–5") long. Curved cones 12.5–23 cm (5–9") long.

RANGE BC: Vancouver Island southwards from Port Hardy. Adjacent Coast Mountains. Common in the interior cedar–hemlock ecosystem: Kootenays, Arrow Lakes, Shuswap and northwards to Quesnel Lake. On the coast, often growing to near subalpine elevations. In the interior, drops out near 1050 m (3500').

RANGE WA: Generally at middle elevations as characterized by the mountain forest ecosystem in the Olympics, Cascades, mountains of northeastern Washington. Sporadic in coastal and subalpine regions. Usually well scattered among other forest zones.

Pine Family, *Pinaceae*

WHITEBARK PINE
Pinus albicaulis

FORM: A crooked, lop-sided tree seldom over 9 m (30') high and 50 cm (20") in diameter. Long, limber branches drooping to ground on old trees. Sometimes shrublike from severe exposure.

BARK: Thin, smooth and light-coloured on young trees, but becoming grey and brown with reddish tones and loosely scaly on old trees.

LEAVES: 5 needles to a sheaf. Needles are stout, slightly curved and 4–7.5 cm (1 ½–3") long. They tend to cluster thickly at twig ends and thus give a view of whitish-barked limbs throughout the tree.

FRUIT: A heavy, soft, purplish cone to 7.5 cm (3") long. Often very pitchy.

WOOD: A soft, light wood only used in cases of necessity.

RANGE BC: A tree of subalpine elevations where it grows on rocky, exposed places. Mountains above 1050 m (3500'). Cascades and eastwards into Rockies at 1500 m (5000') and above. Occurs only in the southern half of B.C. Subalpine meadows of Manning Park.

RANGE WA: A timberline tree growing in rocky, exposed situations—the only pine in Washington found at elevations of 1500 m (5000') and over. High places in the Cascade Mountains. Mt. Baker, Mt. Rainier, Mt. Adams, Olympics.

LIMBER PINE, *P. flexilis*: Closely resembles whitebark pine in its twisted, irregular form and high-mountain habitat. However, its cones are 8–20 cm (3–8") long, whereas those of whitebark pine do not exceed 7.5 cm (3"). In general, its range is limited to high slopes of the southern Rockies north of the Crow's Nest Pass. Numerous trees at an elevation of 990 m (3300') along the highway between Golden and Field, B.C.

Pine Family, *Pinaceae*

PONDEROSA PINE
Pinus ponderosa
Yellow Pine, Bull Pine

FORM: Distinctive: a straight trunk topped by a loose mass of heavy branches with tufts of bushy foliage. Large, twisted branches stick out here and there without plan, but nevertheless produce a tousled, narrow crown. Few trees are over 90 cm (3') in diameter, but fine specimens about 60 cm (2') across and 24–27 m (80–90') high are quite common. Young trees have a distinct whorl to their branches, a characteristic voluptuousness of glossy green foliage.

BARK: The flaky, terra-cotta red bark of this stately pine stands out boldly on distant hillsides, making recognition possible from afar. Young trees have very dark bark; as they age, reddish furrows begin to show. Trunks over 30 cm (12") in diameter produce thin, flaking scales, like jigsaw pieces. On warm days, the bark furrows emit a vanilla aroma.

LEAVES: Ponderosa pine has the longest needles of any evergreen in our area. The brushy tufts are made up of needles about 15–23 cm (6–9") long with 3 needles to the sheaf. Sometimes 4 or 5 to the bundle are found here and there on young trees.

FRUIT: Roundish, shiny, light-brown cone ripens on tree; usually 7.5–10 cm (3–4") long, 5–7.5 cm (2–3") in diameter. Plump, heavy seed; stout wing whirls it through the air during September or later.

WOOD: Light yellow with contrasting dark brown knots when freshly cut. Soft and light. Used for interior finishing (knotty pine), boxes.

DID YOU KNOW that this pine has a flammable chemical in the needles? Quick ground fires of flaking scales and ground needles do not harm the tree but do destroy competing vegetation. Aboriginals used bark scales to make small, hot, smokeless fires that cooled rapidly, not revealing their movements. Pine cones are also excellent for quick, hot fires.

QUICK CHECK: From afar, orange-red bark. 3 long needles to a sheaf is positive proof. Do not confuse with larch (more, shorter needles per cluster).

RANGE BC: Common in valley bottoms and on mountain slopes, to 900 m (3000') in Okanagan and Similkameen valleys. Extends just north of Clinton in the Cariboo. Also at Lower Arrow Lake, Kootenay Lake and Kootenay River valley; as far north as Columbia Lake. Isolated stand in lower Skagit valley.

RANGE WA: Common in lowest forested part of central and eastern Washington. Usually at 450–1050 m (1500–3500') elevation. Klickitat and Yakima counties. Often in bunchgrass ecosystem—scattered individuals, small groves. Steptoe Butte. Very local, meagre occurrences in dry, gravelly places west of Cascades. Lake Crescent, Shelton, Sequim and Hillhurst. With Garry oak in Columbia Gorge.

Pine Family, *Pinaceae*
LODGEPOLE PINE
Pinus contorta var. *latifolia*
Jack Pine

NOTE: 'Jack pine,' used rather loosely for this species, is correctly used for *P. banksiana*, a close relative east of the Rockies. True jack pine extends into B.C. in local areas north and east of Fort Nelson.

FORM: Under normal conditions, a tall, slender tree to 45 cm (18") in diameter and 30 m (100') high. The crown is narrow and rounded, with the thin limbs often occurring only on the top third of the tree. Young trees are narrowly conical with regular whorls of bushy, up-pointing limbs.

BARK: Mottled, dark grey with some trees showing light brown areas. Light covering of small, loose scales. Coarse plates on very old and coastal trees.

LEAVES: 2 needles to a bundle. 4–6 cm (1 ½–2 ½") long and often tinged yellowish green.

FRUIT: A hard, oval cone, spiny and up to 5 cm (2") long. Usually clustered and hanging unopened on the tree for several years.

WOOD: A straight-grained, light wood, pale in colour. Once regarded as a weed species and used largely for railway ties, mine props and fuel, it has assumed high importance for small-dimension construction material (studs).

DID YOU KNOW that most burned-over forests east of the Cascades grow up in a very dense stand of lodgepole pine? This is because the cones activate with high heat and open to release their seed.

QUICK CHECK: A 2-needle pine, the only one in our area.

RANGE BC: A tree likely to be found almost any place in B.C., from middle-mountain to subalpine elevations.

RANGE WA: A tree of widespread range likely to occur almost anywhere in the mountain forest ecosystem and up to subalpine elevations. Extends here and there into ponderosa pine forest. Whidbey Island.

SHORE PINE, *Pinus contorta* var. *contorta*: Recognized as the coastal version; it grows in poor soil or swampy ground close to the sea. Generally twisted and crooked in form with bushy, irregular limbs. In other features, it very closely resembles lodgepole pine. Shore pine grows in boggy regions along the coast and in dry, gravelly places of eastern Vancouver Island and Puget Sound.

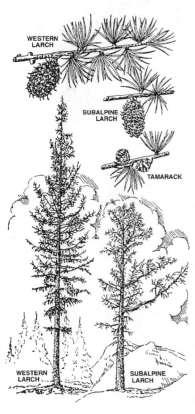

Pine Family, *Pinaceae*

WESTERN LARCH
Larix occidentalis
Larch

FORM: A straight, tapering trunk possibly 90 cm (3') thick and 45 m (150') high. Most trees 30–60 cm (1–2') in diameter. In the narrow, open crown every one of the short, horizontal limbs can be seen. Limbs near tree top have a distinctive up-curve to their tips; lower ones turn downwards.

BARK: Changing from thin, scaly, light brown bark on trees up to 25 cm (10") in diameter to deeply furrowed, orange-red and loosely scaly bark on old trees. May resemble ponderosa pine from a distance.

LEAVES: Needle-like, in clusters of 1–2 dozen arising from knobs on the twigs. About 2.5 cm (1") long and yellow-green. Turn light golden yellow in fall before dropping to the ground.

FRUIT: A light-coloured cone 2.5–4 cm (1–1 ½") long with protruding bracts. Old cones often fail to drop from limbs.

WOOD: Heavy, reddish wood; very durable, even in contact with the ground. Used for ties, pit props and general construction. Commercially important.

DID YOU KNOW that in the fall, larches on distant mountains can be identified by their yellow-gold colour? Grouse often eat the fallen needles.

QUICK CHECK: Loose, open crown displays all limbs; ends twist up or down. Needles in bunches.

RANGE BC: Eastwards from Okanagan Lake to flank of the Rockies. Northwards to Shuswap Lake and Columbia Lake. Many fringes or pockets here and there. Altitudinal range approximately 600–1200 m (2000–4000').

RANGE WA: Generally east of the Cascades and occurring most frequently in the mountain forest ecosystem of northeastern Washington. Spotty occurrence along eastern slope of Cascades at 800 m (3000') and higher. Also in Blue Mountains. Altitude range approximately 600–1350 m (2000–4500').

TAMARACK, *L. laricina*: Like western larch but only to 24 m (80') high and 50 cm (20") in diameter. Prefers bog habitat. Has small cones about 1.3 cm (½") long. Northeastern B.C. to Cassiar Mountains. Spotty at Cluculz, Aleza Lake, Chilako River and Liard River. Isolated trees west of Quesnel and Prince George. Not in Washington.

SUBALPINE LARCH, *L. lyallii*: Tends to develop a craggy, windswept form with irregular, heavy branches. Other features are quite similar to those of western larch. In B.C., it is confined to high mountains along the international border: Cascades—including Manning and Cathedral Lakes parks—Selkirks, Monashees, Purcells, Rockies. In Washington it adjoins the above-mentioned parks; prominent in higher mountains of Pasayton Wilderness.

Pine Family, *Pinaceae*

WHITE SPRUCE
Picea glauca

NEAT, SYMMETRICAL CONE

NOTE: See Engelmann spruce, p. 66, for notes on the problem of differentiating these 2 trees. The ranges overlap in central B.C. However, white spruce and black spruce have the northern half of B.C. to themselves and are easily told apart. White spruce does not extend into Washington.

FORM: No clear distinction from the many varieties of Engelmann spruce. Limbs have a tendency to be more widespreading and bushy with a triangular effect caused by tassel-like side branchlets. Limbs often extend to the ground.

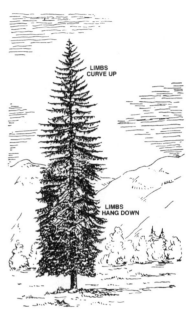

LIMBS CURVE UP

BARK: Greyish and scaly with reddish brown tinges showing between and underneath the thin scales.

LEAVES: 4-sided, sharp-pointed leaves to 2.5 cm (1") long. Tendency to twist to upper side of limb. Pungent smell when bruised.

FRUIT: When ripe, a brown cone 3–5 cm (1 ¼–2") long. Scales very smooth, thicker and neatly rounded on their margins; trim and symmetrical in comparison to Engelmann spruce.

LIMBS HANG DOWN

WOOD: Soft, light, clear-grained and light coloured; Generally not distinguishable from Engelmann spruce. Important tree in northern B.C., where it is the main timber species.

DID YOU KNOW that squirrels often gather hundreds of spruce cones into a cache? Piles of scales left after the cones are broken apart may be half a metre or more (several feet) deep.

QUICK CHECK: Clearly a spruce by its pointed, squarish or stalked needles. A neat, symmetrical cone to 5 cm (2") long.

RANGE BC: Generally in low to middle-mountain forests. Where these spruces overlap, Engelmann spruce will continue into the subalpine ecosystem. The forest spruce tree of northern B.C., found as far south as Lillooet, Cache Creek and Vernon, but not in the Coast Mountains. Also in the Big Bend area of the Columbia River and scattered southwards along the Rockies to Sparwood–Crowsnest Pass region.

RANGE WA: White spruce does not extend into Washington.

RAGGED CONE

Pine Family, *Pinaceae*

ENGELMANN SPRUCE

Picea engelmannii

There has been much confusion among white spruce, Engelmann spruce and various hybrids. The latest B.C. authority, *The Vascular Plants of British Columbia*, 1989, recognizes only Engelmann and white spruce. How do you tell the difference in a wide band where they overlap, in central B.C.? Some clues: young twigs of Engelmann are sparsely hairy; generally smooth for white spruce. Leaves of Engelmann are often 2–3 cm (1") long; white spruce's often under 1.3 cm (½"). Also see 'Fruit,' below.

You may find the hybrid white spruce, *P. glauca* x *engelmannii*, given in some references as the common spruce of central B.C.

FORM: In thick stands, narrowly conical with straight, clean trunk. Topmost branches twist upwards, middle ones point outwards and lower ones drop strongly. Particularly in open or at lower elevations, branchlets hang like tassels from main boughs. Typically 30–90 cm (1–3') in diameter and 30–42 m (100–140') high.

BARK: Trunk covered by loose, greyish scales between which shows a brownish or rusty red tinge.

LEAVES: 4-sided, dull-pointed, blue-green needles to 2 cm (¾") long. Tend to curve towards top side of twig. Each needle fastens to a tiny spur that remains on the limb after the needle has fallen. Pungent smell when crushed, as with white spruce.

FRUIT: Light brown cone 2.5–6 cm (1–2 ½") long. Scales thin, finely ridged on back; tip slightly wrinkled. Cones do not have the neat symmetrical look of white-spruce cones. Purplish when immature.

WOOD: Soft, straight-grained and creamy white. Very important to interior forest industry. Cut extensively for general construction and pulp.

DID YOU KNOW that a quick way to tell a spruce needle, except Sitka, is to roll it between thumb and finger? The 4 edges allow it to roll easily whereas flat needles (fir, hemlock) do not.

QUICK CHECK: A spruce by its stalked, prickly needles. Cone scales with flexible, finely-waved margin. Tree untidy in appearance.

RANGE BC: Widespread in southern half. Extends across to Rockies. Sometimes with white spruce in wetter valley bottoms. The only spruce of higher elevations, 1150 m (3800') to near timberline. Replaced by white spruce in northern half of province, but present in Skeena and Omineca mountains and southwards. Westerly flank of Cascades.

RANGE WA: Starts high on westerly slopes of Cascades; most common on easterly side, from 1200 m (4000') to near timberline. Also in wet places in low valley bottoms in bunchgrass and ponderosa pine ecosystems. In Blue Mountains and Olympics.

Pine Family, *Pinaceae*

SITKA SPRUCE
Picea sitchensis
Tideland Spruce, Coast Spruce

FORM: Varies considerably depending on whether a forest tree or growing in the open. In humid coastal forests, it produces a long, clean trunk 0.9–1.8 m (3–6') in diameter topped by a thin crown of short branches reaching 45 m (150') in height. In the open, limbs are strongly out-thrust and carry a triangular fringe of drooping branchlets. Limbs extend almost to the ground. This latter form is easily distinguishable from a considerable distance.

A small stand of magnificent mature Sitka spruce in B.C. has achieved worldwide attention because of attempts to save it from logging. It lies in the valley of the Carmanah River, remote although only about 85 km (55 miles) north of Victoria. Some trees are over 800 years old. Nurtured by heavy coastal rains, remaining groves of giant trees are a vestige of similar forests once a major part of the west-coast flora in our area.

BARK: Covered with thin, loose, crisp scales of rusty brown hue.

LEAVES: Bristling out in all directions from the twig. Flattish rather than 4-angled. 2 white lines showing on upper surface, 2 fainter ones on lower. Very sharp to the touch.

FRUIT: An easily recognized cone by reason of its disorderly array of thin, irregular, wavy-edged scales. Most cones about 5 cm (2") long.

WOOD: Fairly light and soft and varying from creamy to pale buff in colour. Forest trees have a large proportion of clear, straight-grained wood that is excellent for fine construction. Also very important as a pulp tree.

DID YOU KNOW that Sitka spruce from the Queen Charlotte Islands was chosen during World War II as the most desirable wood for aircraft construction? Howard Hughes's 'Spruce Goose' was built from this tree.

QUICK CHECK: Coastal habitat, needles bristling out all around twig and light, disorderly cone.

RANGE BC: A narrow strip along the coast of B.C. and on adjacent islands. Confined to elevations of less than 300 m (1000') and seldom found more than 80 km (50 miles) from tidewater. Most common along immediate coastline. Scattered trees in Fraser Valley as far east as Hope and then eastwards along Hope-Princeton Highway to Sumallo Grove. As far north as Garibaldi Station, Cheakamus River valley.

RANGE WA: Occupies a narrow strip along the coast. Seldom more than 80 km (50 miles) inland or at elevations over 600 m (2000'). Although frequent along the coast, its occurrence around Puget Sound is very sporadic.

THICK, KNOBBY TOP

Pine Family, *Pinaceae*

BLACK SPRUCE
Picea mariana

FORM: Usually quite distinctive with its narrow, irregular crown that has a tendency to form thick clumps and bulges. The straight trunk has very little taper. Most trees are 12.5–25 cm (5–10") in diameter and up to 15 m (50') high. In northern limits, it tends to become twisted and shrublike. Most limbs are short and horizontal.

BARK: A dirty grey, scaly bark quite similar to that of lodgepole pine. Twigs blackish and hairy.

LEAVES: Stiff, shortish needles about 1.3 cm (½") long.

FRUIT: Almost round, grey-brown cone 1.3–2.5 cm (½–1") long—the smallest of all the spruce. Usually several in a cluster. They often remain on the tree for several years.

WOOD: Light, soft and very fine grained. Pale in colour. Important for pulp in eastern Canada, but not much used in B.C.

DID YOU KNOW that black spruce grows almost to the Arctic Ocean and ranges across all the northern regions of Canada? On the East Coast, it dips as far south as Maryland, a clue to its botanical name.

QUICK CHECK: From afar, a spire-like tree with knobby top, in a swamp habitat. Small cones with many old ones remaining on tree.

RANGE BC: In broad terms, the northern half of the province lying east of the Cascades. Characteristically found in swamps and bogs but less limited to this habitat in more northerly regions. Common along the Alaska Highway. Found southwards to Smithers, Fraser Lake and Quesnel, down along the North Thompson River to Blue River and in an isolated occurrence in the Chilcotin. Noticeable in the Jasper region of the Rockies.

RANGE WA: Black spruce does not extend into Washington.

Pine Family, *Pinaceae*

WESTERN HEMLOCK

Tsuga heterophylla

FORM: A large tree thriving in dense shade. From 0.6 to 1.2 m (2 to 4') in diameter and 35–50 m (120–160') high. Limbs long and irregularly spaced on trunk. The topmost twig (leader) droops in graceful fashion. Foliage on young trees is drooping, feathery and very attractive.

BARK: About 2.5 cm (1") thick with flat, scaly ridges and deep furrows on mature trees. Dark, rich brown in colour. Young trees have thin, fine-scaled bark of lighter colour.

LEAVES: About 1.3 cm (½") long, flat and blunt. More or less 2-ranked. 2 fine white lines on undersurface. Each needle has a short stalk and a twist at base.

FRUIT: A light cone, seldom over 2.5 cm (1") long, that ripens in fall and drops during winter.

WOOD: A tree that more than 20 years ago was considered of low value but now ranks high in importance as a pulpwood species. Also finds wide use in lumber industry. Bark is high in tannin.

DID YOU KNOW that hemlock is a prized ornamental in Great Britain? It can grow and reproduce in dense shade. Young trees often start on top of stumps or fallen logs. It usually grows with Douglas-fir, redcedar and Sitka spruce. Aboriginals prized the wood and bark for many uses; for example, the soft springtime cambium layer as food.

QUICK CHECK: The drooping tree tip identifies hemlock from afar. Leaves and cones are good points of reference, too.

RANGE BC: Coastal forests up to 850 m (2800') and interior cedar–hemlock ecosystem up to 1500 m (5000').

RANGE WA: Common in coastal forests extending to elevations of 1200 m (4000'). Also occurs on moist eastern slopes of Cascades from 600 to 1350 m (2000 to 4500') and in northeastern Washington.

MOUNTAIN HEMLOCK, *T. mertensiana*: Common along coast of B.C. (but not Washington) in proximity to western hemlock but at higher elevations from 780 m (2600') up to timberline. Scattered occurrences east of the Cascades, ranging from 1200 to 1700 m (4000 to 5500'). The dark green needles grow in disorderly array around the twigs, giving a thick, tufted appearance. Branches tend to have an upward sweep at their tips. The stout leader has the characteristic 'hemlock droop.' Cones are 2.5–5 cm (1–2") long, twice the length of western hemlock cones.

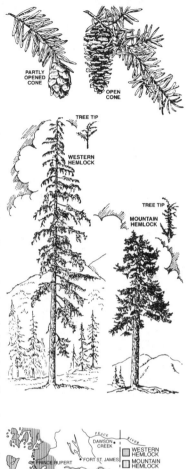

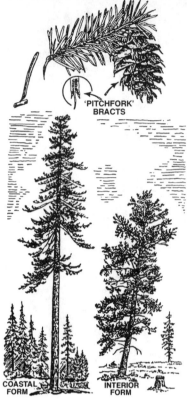

'PITCHFORK' BRACTS

COASTAL FORM

INTERIOR FORM

Pine Family, *Pinaceae*
DOUGLAS-FIR
Pseudotsuga menziesii
Fir, Douglas Spruce, Oregon Pine

Coastal (*P. menziesii* var. *menziesii*) and interior (*P. menziesii* var. *glauca*) populations of Douglas-fir are slightly different: e.g., coastal has larger cones.

FORM: Trees over 60 m (200') high and 1.8 m (6') in diameter seldom seen now. Most mature trees on coast are 0.9–1.2 m (3–4') in diameter and 60 m (200') high; in the interior, few are more than 75 cm (30") in diameter. Young trees form broad-sloping pyramid: lower branches straight or drooping, upper branches curving upwards. Old trees develop heavy, crooked limbs and flattened or irregular tops. In shade, lower limbs drop off, leave long, clear trunk.

BARK: Smooth grey-brown with resin blisters on young trees up to 15 cm (6") in diameter. Becomes thick and deeply fissured into reddish brown ridges, sometimes up to 30 cm (1') thick, as tree ages. Bark prevents damage from fires but makes excellent fuel.

LEAVES: Flat, sharp-pointed needles about 2.5 cm (1") long. Not prickly to touch like spruce.

FRUIT: Cone ripens in fall, hangs downwards, drops to ground; 5–10 cm (2–4") long for coastal form or 4–7.5 cm (1 ½–3") long for interior variety. Unmistakable 3-pronged bracts between cone scales protrude so as to be easily visible.

WOOD: Generally reddish but sometimes yellowish, with prominent annual rings. Splits cleanly, very strong. Important for heavy construction and interior and exterior finishing.

DID YOU KNOW that David Douglas was a famous botanist at age 27? The British Royal Horticultural Society sent him on 2 trips to western North America, in 1825 and 1830. He discovered many plants; some bear the specific name *douglassi*. Menzies named this majestic tree after him.

At coast, in association with hemlock, redcedar, gand fir; in interior, ponderosa pine; at higher elevations, lodgepole pine, white pine, larch.

Canada's largest Douglas-fir, nearly 1000 years old, is near Port Renfrew, Vancouver Island; diameter at breast height is 4.0 m (13'1"). Though its broken top now reaches only to 89 m (291'), it is believed that this tree once stood over 98 m (320') tall!

QUICK CHECK: 'Pitchfork' bracts on cone.

RANGE BC: Most of southern half. Biggest on Pacific coast; variety of soils, to approximately 850 m (2800'); in southern interior to 1050 m (3500'). May grow at elevations of 1800 m (6000') in Rockies. Northwards as far as Stuart and McLeod lakes.

RANGE WA: Most widespread conifer in state, likely to be found wherever there are evergreens below subalpine elevations. Largest in coastal forests. Altitudinal limit approximately 1350 m (4500').

Pine Family, *Pinaceae*

SUBALPINE FIR

Abies lasiocarpa

Alpine Fir

FORM: Distinguished by its symmetrical, narrow, spire-like form. Branches in whorls, very short and stiff towards the top. Most slope downwards. Trees may reach 1.2 m (4') in diameter and 45 m (150') high but are usually 25–40 cm (10–16") in diameter and 15–23 m (50–75') high. At timberline, it becomes stunted and sprawls with limbs to the ground.

BARK: On young trees, thin, smooth and grey with conspicuous resin blisters. On old trees 30 cm (12") or more in diameter, irregular, shallow furrows and reddish scales. Remaining smooth on upper section.

LEAVES: Flat needles about 2.5 cm (1") long, twisting upwards from underside of twigs to bush densely around twig. Pungent smell. Normally blunt, but sharp-pointed on cone limbs. Blue-green, with silvery tinge to new growth. Twig-ends orange-brown; older growth hairy.

FRUIT: A group of heavy, hard, purple wooden cones 5–10 cm (2–4") long, standing erect near the top of the tree. Often blotched with sticky pitch and quickly disintegrating.

WOOD: A light-coloured soft wood of little commercial importance.

DID YOU KNOW that the picturesque, symmetrical trees in most subalpine and timberline pictures are subalpine fir? The short, stiff branches usually slope downwards and are adapted to withstand heavy loads of ice and snow that may completely encase them in midwinter. The pitch from the bark formed the basis for many aboriginal medicines.

QUICK CHECK: Flat needles leave circular scars on twigs. Stiff, erect cones. Altitude range (though at low elevations in some coastal valleys of northern B.C.)

RANGE BC: A tree of subalpine elevations, it is found throughout B.C. except for on the Queen Charlotte Islands. Locally frequent at higher elevations on Vancouver Island. It is most common east of the Cascades, where it appears at moderate elevations around 600 m (2000') on the interior plateau and continues to timberline.

RANGE WA: A tree of subalpine elevations. Occurs on all the higher mountain systems of Washington. Usually at elevations of 1350–2100 m (4500–7000'). Chinook Pass, Snoqualmie Pass, Stevens Pass. Isolated groves in the Olympic Mountains.

CONE BRANCH

LOWER BRANCH

CONES IN TREE TIP

POSSIBLE VARIATION IN LOWER BRANCHES

TOP NEEDLES POINTING AHEAD

WHITE BLOTCHES ON BARK

Pine Family, *Pinaceae*

AMABILIS FIR

Abies amabilis

Pacific Silver Fir, Balsam

FORM: A straight tree up to 30 m (100') high and 75 cm (2 ½') in diameter. Spire-pointed, but with rounded conical shape for older trees. Thickly foliated and very symmetrical. Branches, except those on the top third, curve downwards and away from the trunk, often extending to the ground. Remarkably beautiful in form and colour. At high elevations it resembles subalpine fir in general shape.

BARK: Smooth, unbroken, ash-grey bark splotched with chalky white patches and resin blisters. Old trees usually seamed and much like mountain hemlock.

LEAVES: Flat needles about 4 cm (1 ¼") long, grooved on upper side and ridged with 2 white lines below. Needles 3-ranked and blunt; most with small notch on end. Notice how they point forwards along the top of the twigs and how those near the underside twist upwards to produce a flattish effect when a bough is seen from beneath. These characteristics should limit confusion with grand fir, which has 2-ranked needles. Twigs tend to be hairy.

FRUIT: An erect, dark purple cone 10–12.5 cm (4–5") long. Falls apart in early fall leaving spike-core standing on bough for several months.

WOOD: Soft, light and yellowish brown. Generally cut for pulp.

DID YOU KNOW that the silvery colour is given by 2 white lines on the lower side of the needles? Trees likely to be associated with it at lower levels are grand fir, Douglas-fir, hemlock and white pine. At higher elevations, mountain hemlock and subalpine fir are common companions. Many old-time wilderness campers preferred the springy boughs of amabilis fir to those of any other tree for making a 'bough bed.' However, those days are past!

QUICK CHECK: Circular leaf scars on twigs identify true firs. A ridge of needles pointing forwards along top sides of twigs. Young trees and those in good growing sites have smooth bark with white patches.

RANGE BC: Western slopes of the Coast Mountains and on Vancouver Island. Varies from sea level to 1350 m (4500'). Usually on moist, shady bottomlands above the general level of Douglas-fir–hemlock forest. Prominent at 850–1200 m (2800–4000') on Coast Mountains.

RANGE WA: A fairly common tree of the Olympic and Cascade mountain systems. Wide tolerance to elevation, from near the coast to elevations of 1800 m (6000').

Pine Family, *Pinaceae*
GRAND FIR
Abies grandis
Lowland Fir, Western Balsam

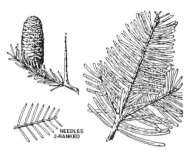

NOTE: The true firs, *Abies* spp., are often referred to as 'balsam fir' or simply 'balsam,' names derived from the true balsam fir, *A. balsamea*, a species of northern and eastern North America.

FORM: A tall, straight tree averaging 60–90 cm (2–3') in diameter and up to 40 m (125') high. A broad, pyramidal shape when young, with a distinct horizontal pattern to branch ends. Older trees have rounded tops and conical form when open grown but are free of lower branches and irregular of crown in shade. Lower branches have a graceful downward swing, forming a broad skirt.

NEEDLES 2-RANKED

BARK: Young trees up to 20 cm (8") in diameter have thin, smooth bark with numerous resin blisters. White mottling may lead to confusion with the bark of amabilis fir or young Douglas-fir. Older trees have hard, irregular furrows and ridges.

LEAVES: Flat, strongly 2-ranked needles 3–5 cm (1 ¼–2") long. Blunt or slightly notched and in 1 flat rank from each side of twig. Upper surface is dark green and has a central groove while underside is made silvery by 2 whitish stripes.

TOP OF OLD TREE

FRUIT: An erect cone 5–10 cm (2–4") long and to 4 cm (1 ½") in diameter. Very noticeable on tree tops in late August. As the cones ripen, the scales and seeds start to fall, leaving an untidy brown mass of spikes. Most of the cones break apart by October.

WOOD: A soft, light wood with a faint brownish colour. Grand fir cut as part of a logging operation is usually used for pulp or as lumber for cheap wood products.

DID YOU KNOW that some folks placed great faith in the healing properties of gum from resin blisters? It makes a handy glue, too. Grand fir is usually found in valley bottoms growing as part of the Douglas-fir–hemlock–redcedar forest. Named 'grand' by botanist David Douglas because of its great height.

QUICK CHECK: From afar, heavy, conical crown is a reliable indicator. If close, make sure it is a true fir—either circular leaf scars on twigs or erect cones. 2-ranked needles make it grand fir.

RANGE BC: Lower slopes and valleys of southern coastal region and Vancouver Island. From Butte Inlet southwards to Victoria. Probably not general above 900 m (3000') elevation. Northwards along the Fraser River to Boston Bar. Also valley bottoms of Kootenay and Arrow lakes region.

RANGE WA: Grows best in moist situations such as valley bottoms but extends upwards to 1500 m (5000'). Coastal forests and slopes of Olympics, Cascades and Blue Mountains. Also mountain systems of northeastern Washington.

Pine Family, *Pinaceae*

NOBLE FIR

Abies procera

FORM: In common with subalpine, amabilis and grand fir, noble fir has beautiful symmetry of form. The straight trunk with its slight taper is complemented by a thin, conical outline formed by stiff out-pointing branches. Lower branches often droop. The crown is characteristically narrow and rounded but quite open, in marked contrast to the very dense top of most grand firs. Mature trees range from 0.6 to 1.5 m (2 to 5') in diameter and may reach 60 m (200') in height.

BARK: Relatively smooth with resin blisters on trees less than 35 cm (14") in diameter, then starting to seam into ridges that break crossways to form long, shingly plates that easily flake off with increasing age. Grey when young, with a tinge of brown when older. New, exposed bark a dark rusty shade.

LEAVES: A 4-sided needle 2.5–4 cm (1–1 ½") long with conspicuous groove along the upper surface. Blue-green in colour, given a silvery tinge by rows of tiny white markings. Needles on lower limbs more flattened and sometimes notched at the tip. They twist upwards from the branches and are more densely clustered near the top of the tree.

FRUIT: An erect, hard cone 10–15 cm (4–6") long, shingled with sharp-pointed bracts. Light brown when mature in early fall. They fall apart in October.

WOOD: A light brown wood with reddish tinges. Hard yet lightweight with a medium-fine grain, it finds a place in airplane manufacture, interior finish and general construction purposes.

DID YOU KNOW that noble fir is an exceptionally long-lived tree with some favoured specimens possibly reaching close to 1000 years?

QUICK CHECK: Range limited to middle-mountain slopes of Washington Cascades. Leaf-scars on twigs and channeled needles. Shingled, spiny cone.

RANGE BC: Noble fir does not extend into B.C.

RANGE WA: A tree of comparatively high mountain slopes, generally between 750 and 1500 m (2500 and 5000') elevation. Occurs on both sides of the Cascade Mountains. Common in Mt. Rainier region. Reported in Olympic Mountains but if so is extremely limited in range. Does not extend into the mountain forest ecosystem.

Cypress Family, *Cupressaceae*

WESTERN REDCEDAR
Thuja plicata

FORM: A giant tree on the coast, often over 45 m (150') high and 1.8 m (6') in diameter. The trunk tapers from a fluted base to a long spike-like top that is often dead. On mature trees, the branches are long, irregular and usually point very distinctly downwards. The frond-like branchlets impart a feathery or lacy appearance to the form. Regular, spreading branches give young trees a conical outline. A yellowish green colour distinguishes this tree from other, dark green conifers.

BARK: Thin and stringy, it can be pulled off in long strips. No other native tree except yellow-cedar has similar bark. Aboriginals valued it highly for making baskets, clothes and mats.

LEAVES: Scaly, blunt, paired leaves pressed tightly to the twig. Branchlets hang like fronds or sprays from main boughs.

FRUIT: A small, erect cone 1.3 cm (½") long that hangs on over winter.

WOOD: A reddish, fragrant wood that splits into thin boards with remarkable ease. It is very light and free from pitch or resin. Used at one time by aboriginals for massive war canoes and lodges. Now valuable for shingles, siding and posts because of its resistance to decay. Fallen trees may remain sound for more than 100 years.

DID YOU KNOW that western redcedar is the provincial tree of B.C.? The largest redcedars grow on Vancouver Island. One 4.1 m (13'6") in diameter and 875 years old was cut in 1948 near Comox. Aboriginals hollowed canoes 18 m (60') long from a single trunk. Much of their outstanding culture can be attributed to the many uses of this magnificent tree.

QUICK CHECK: Bark stringy and dark, trunk fluted at base. Twigs smooth; not prickly to the touch like yellow-cedar when stroked against the grain.

RANGE BC: Common forest tree west of the Cascades, where it grows up to elevations of 850 m (2800'). Also abundant as far north as Prince George in the interior cedar–hemlock ecosystem, with an upper altitudinal limit of 1280 m (4200'). Scattered pockets and riverside fringes in the Similkameen, Okanagan, Columbia and Kootenay valleys. Botanie Creek.

RANGE WA: Achieves best growth and greatest abundance in moist soils west of the Cascades at elevations below 1050 m (3500'). Elsewhere, though usually confined to creek bottoms and similar wet places, it does occur up to nearly 1500 m (5000') elevation.

PRICKLY SCALES

Cypress Family, *Cupressaceae*

YELLOW-CEDAR

Chamaecyparis nootkatensis

Yellow Cypress, Alaska Cedar

FORM: A shaggy tree usually less than 24 m (80') high. The trunk is often slightly twisted and tapers quickly towards the top. Well-formed trees up to 60 cm (2') in diameter are common, but growth is slow and breakage becomes heavy with age. The limbs sweep outwards and downwards, with fernlike fronds hanging from them. The slender tip of the tree droops very much like hemlock. Young trees are shrubby and warped but gradually straighten as they grow taller.

BARK: On trees over 20 cm (8") in diameter, the bark appears a distinctive dirty white from a distance. It is stringy and brittle and hangs in loose, rough pieces. Unlike redcedar, it will not pull off in long strips, but breaks in stiff sections instead. Young trees have a fairly smooth, reddish bark with elongated scales.

LEAVES: Scaly, overlapping leaves very similar to redcedar but prickly to the touch when stroked against the grain.

FRUIT: A knobby, round berry, greenish white, that ripens into a reddish brown cone in late September or October, which then falls during the winter.

WOOD: Soft, light yellow, with a very noticeable sharp fragrance when freshly cut. Easily worked and resistant to rot and insects, it is a favourite of boat builders. Also one of the most popular woods for arrow-making and carving.

DID YOU KNOW that 2 gnarled monarchs, 1.5 m (5') and 2.5 m (8') in diameter, may be the oldest living things in Canada, with ages over 1000 years?

QUICK CHECK: Prickly cedar-like foliage and stringy silvery bark. Coastal range and elevation.

RANGE BC: On coastal slopes and islands from Alaska southwards. Spotty occurrence from sea level to alpine in northern half of range. Common on upland plateaus and mountains of Vancouver Island. Mingles with mountain hemlock and amabilis fir on North Shore mountains above 850 m (2800') elevation. Patches in Slocan Valley.

RANGE WA: Ranges from near sea level on eastern slopes of Olympics to customary 900–2100 m (3000–7000') elevations in Cascades. Most plentiful in northern Cascades. Southern limit near Mt. Adams.

Cypress Family, *Cupressaceae*

ROCKY MOUNTAIN JUNIPER
Juniperus scopulorum

NOTE: 2 junipers have been classified as 'Shrubs,' see p. 136.

FORM: Generally a bushy, shrublike tree, but sometimes conical or even sprawling. Commonly 1.8–6 m (6–20') high and up to 30 cm (1') in diameter.

BARK: Thin, stringy; reddish brown tinges.

LEAVES: Often 2 distinct kinds on same tree. Young shoots with sharp, needle-like leaves about 1.3 cm (½") long. Older branches with smooth, scaly leaves like western redcedar. Variable foliage colour and a pleasing form make it exceedingly attractive.

FRUIT: Lumpy, smooth, bluish purple berries require two years to ripen and contain 1–2 large, grooved seeds.

WOOD: A reddish heart-wood with a wide ring of white sapwood. Light and soft. Used occasionally for small ornamental work. In eastern Canada and the U.S.A., a closely related species is much sought after for pencil wood.

DID YOU KNOW that juniper berries are used in the flavouring of gin?

QUICK CHECK: Often a small tree. Leaves without a glandular pit on the back; scaly on older branches. Smooth, bluish purple berries.

RANGE BC: South half of the province, east of the Coast Mountains to the Rocky Mountain Trench. Sporadic occurrences in north-central B.C. Sparsely scattered along eastern side of Vancouver Island. Sometimes abundant in arid interior and Cariboo parklands ecosystems. Mostly found in dry places at low- to middle-mountain elevations; sometimes on the edges of swamps and river courses. Spotty occurrences in Peace River area, Babine Lake, North Tweedsmuir Park.

RANGE WA: Sporadic enough to be considered scarce. High northern ridges of Olympics, northern San Juans. Sporadic in bunchgrass and sagebrush ecosystems of southeastern Washington but common on steep slopes in eastern section of Columbia Gorge.

WESTERN JUNIPER, *J. occidentalis*: Very similar to the juniper above in that it may be either a conical or bushy tree or a twisted shrub. Old but vigorous specimens—e.g., south of Sunnyside and on road to Bickelton—can be up to 45 cm (18") in diameter and to 9 m (30') tall. Trunk usually branches into several stout limbs. Strong bark furrowed with brown-cinnamon tones. Leaves scale-like, with a shallow glandular pit on back—may be sticky with a drop of resin. Generally a juniper of great size will be this one. Ranges widely but sporadically on dry foothills and low mountains from Yakima County southwards and eastwards. Klickitat and Asotin counties.

KNOBBY, BLUE BERRIES

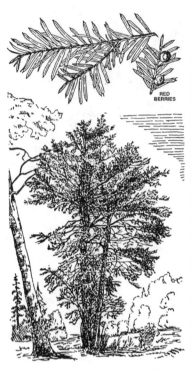

RED BERRIES

Yew Family, *Taxaceae*

WESTERN YEW

Taxus brevifolia

Pacific Yew

FORM: Small, bushy, untidy tree averaging 4.5–9 m (15–30') high. Seamed, twisting trunk is seldom over 30 cm (12") in diameter. Prefers deep shade beneath larger trees; develops an ungainly limb pattern. This characteristic and dull green colour distinguish it from other conifers.

BARK: Rich red tints of rough, scaly bark usually draw attention to this hide-away tree. Often, thin bark is fluted and seamed as if tree suffered agonies in its slow growth. Yew, a 'trash tree' in logging operations, has achieved worldwide fame for taxol, a chemical produced from its bark. Taxol is successfully used to treat some cancers. Since one treatment requires 2 grams of taxol—from the bark of 10 good-sized trees—yew now rates intensive management.

LEAVES: Flat, sharp-pointed, dull green needles; short stems. Most needles are 1.3–2 cm (½–¾") long, but some twigs have shorter ones. Although new shoots have a brushy needle effect, older limbs appear roughly 2-ranked, much like hemlock.

FRUIT: Female trees have eye-catching peanut-sized, greenish berries, with horny seed, that turn red in September. Birds carry them to new locations. Male trees have small 3 mm (⅛") round clusters of stamens in early spring.

WOOD: The contrast between yellow sapwood and rich red heartwood makes it perhaps the most attractively coloured native wood. Extremely hard and durable, highly valued by aboriginals and fashioned into a wide range of small implements, but of negligible commercial importance because of small size and rarity. Attempts are being made to extract taxol from the wood in addition to the bark.

DID YOU KNOW that various tough and springy yew species supplied fighting bows for ancient armies and are still prized by do-it-yourself archers?

QUICK CHECK: Scaly, red trunk. Sharp-pointed needles with 2-tone colour beneath.

RANGE BC: From sea level to 350 m (1000') on Vancouver Island and in coast forest ecosystem, but up to elevations of 1200 m (4000') in interior cedar-hemlock ecosystem. Generally a tree of river banks and damp canyons, and in shade of other trees. Cariboo, Monashee, Selkirk and Purcell mountains. One small grove near Kelowna. Other trees in Sparwood–Fernie area.

RANGE WA: River banks and deep canyons; prefers shade of other trees. From sea level to near 1500 m (5000') in Cascades and in mountains of northeast. Very widespread as scattered individuals; optimum range is west of Cascades. Western slopes of Stevens, Snoqualmie and Chinook Passes.

Willow Family, *Salicaceae*

TREMBLING ASPEN

Populus tremuloides

Quaking Aspen

FLAT
LEAF STEM

FORM: East of the Cascades, it grows 6–9 m (20–30') high in low, spreading groves where there is evidence of moisture. In wetter regions, graceful trees to 24 m (80') high are common. These have straight trunks to 40 cm (16") in diameter; the top half bears a loose, rounded crown of brittle branches. The characteristic groves of aspen result when new trees clone from spreading roots.

BARK: Mostly smooth and white with black 'horseshoe' markings here and there. A chalk-like substance can be rubbed off. Older trees have fissured and blackened bark near their bases. Young cottonwood trees, with which aspen might be confused, have smooth, green-white bark without black patches.

LEAVES: An abruptly tipped, rounded or heart-shaped leaf to 7.5 cm (3") long. Leaf stems long and flattened at right angles to leaf blade. Foliage is an attractive fresh green colour.

FRUIT: Appears early in spring, before the first leaves, as inconspicuous catkins of small, greenish capsules. Male and female catkins on separate trees.

WOOD: A weak, soft wood, almost white. Brittle and fast-rotting. Used for pulp and crating in eastern North America and favoured in parts of the United States for excelsior (shavings for packing) and match stock. In B.C. it is used for pulp, waferboard and chopsticks.

DID YOU KNOW that the leaves tremble with the slightest breeze because of the long, flattened leaf stems? The chalky substance on the bark is supposed to be heaviest on the south side of the tree. Thus, aspen may act as a direction guide.

QUICK CHECK: Whitish bark marked with black splotches. Flat leaf stems.

RANGE BC: Abundant throughout the province east of the Cascades. Extends from valley bottom to 1200 m (4000') in elevation. Occasional clumps on eastern part of Vancouver Island south of Campbell River. Victoria region, Gulf Islands.

RANGE WA: Found in most of the state except for the Olympic Peninsula, and the sagebrush and subalpine ecosystems. Very sporadic in occurrence, at usual elevations of 300 m (1000') to 1800 m (6000'). Ellensburg, Bickelton, Blue Mountains.

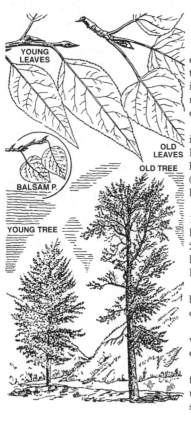

Willow Family, *Salicaceae*

BLACK COTTONWOOD
Populus balsamifera ssp. *trichocarpa*

FORM: Young and old trees often mistaken for distinct species: young trees, to 12 or 15 m (40 or 50') tall, have symmetrical conical form; stout, up-pointing branches. With age, limbs become very thick, irregular and crooked, turn to point outwards, twist downwards. Top flattens as higher limbs break.

BARK: Smooth and green on trees to 15 cm (6") in diameter (aspen has white bark, black blotches). Bark-furrowing increases with age; mature trees have hard, dark grey ridges several centimetres (inches) thick. Sweet-flavoured inner bark, cambium layer; eaten by aboriginals in early spring.

LEAVES: In general, widely triangular and 6–12.5 cm (2 ½–5") long. Yellowish green on top; much lighter beneath. Leaves on young trees or sprouts to 20 cm (8") long, narrow, tapering. Teeth on new leaves quite regular; ragged and wavy on old leaves. Leaf stems round, not flattened like aspen.

FRUIT: In spring, long string of rough, greenish beads; splits into 3 segments. They ripen and burst open in a cottony mass that blows far and wide.

WOOD: Soft, lightweight, drab; used largely for veneer and plywood. Perhaps the most important local broadleaf tree. Rots quickly on ground.

DID YOU KNOW that the thick, sticky buds are fragrant in springtime? The substance exuded can be used as ointment on cuts or as a glue. Spring 'snowstorms' of cottony seeds give cottonwood its name.

QUICK CHECK: Yellowish green foliage. Smooth, whitish green bark on young trees; thick, furrowed bark on older trees. Riverine and bottomland habitat.

RANGE BC: Possibly most widespread of larger broadleaf trees. Low–middle elevations throughout. North of Prince George and east of Rockies in northeast. On river banks, gravel bars, low-lying land. Generally not on outer coast or offshore islands.

RANGE WA: Very extensive range; usually only on river banks, gravel bars, low-lying land. Most common east of Cascades; even along sagebrush region watercourses. Sea level to 1350 m (4500').

BALSAM POPLAR, *P. balsamifera* ssp. *balsamifera*: Like black cottonwood, but balsam poplar generally east of Rockies and into far north-central B.C., black cottonwood to south. Leaves paler green beneath, not silver. Fruits (seeds) in 2 sections, not 3; hairy, not smooth. Across northern B.C.; Alberta to Atlin and Stikine River. Not in Washington.

SOUTHERN COTTONWOOD, *P. deltoides*: Rare introduction into southeastern B.C., e.g., Osoyoos. Leaves green on both sides, distinctly serrated along basal margins. Not in Washington.

LOMBARDY P. *P. nigra* var. *italica*.: See p.102.

Willow Family, *Salicaceae*

PACIFIC WILLOW

Salix lucida ssp. *lasiandra*

Shining Willow, Black Willow

FORM: A crooked trunk branching into a number of upright limbs that give a ragged, rounded outline to the crown. Seldom more than 9 m (30') high or 40 cm (16") in diameter. The thin leaves and long slender twigs impart a graceful appearance. Probably the largest native willow.

BARK: Blackish. Channeled into irregular, rough plates by many furrows and cross-seams. Thick, smooth branchlets are orange to brown, while new twigs and suckers are yellow.

LEAVES: Distinctive long, thin point with a sideways twist. From 5 to 12.5 cm (2 to 5") long and very finely toothed. A shiny, dark green above, but smooth and whitish below. Stomata (pores) only on the lower surface. Obvious stipules (small growths at leaf-stem bases).

FRUIT: Thick catkins about 5 cm (2") long appear with leaves. Bright yellow at maturity, followed by fuzzy, white cotton that often remains until July.

WOOD: Pale brown and brittle. Very soft and not used for any specific purpose.

DID YOU KNOW that willow's liking for water often results in long rows of these trees outlining the margins of sloughs and streams? Sections of limbs stuck in wet ground will root easily and quickly. Most willows have two small 'ears,' or wings, at the base of each leaf stem when the leaves are young.

QUICK CHECK: A black-barked, ragged tree growing near water. Shiny, dark green leaves with a long, thin point, usually twisted to the side.

RANGE BC: Widespread throughout B.C. along streams, rivers and wetlands at low to medium elevations. In addition to the 3 willows described here, under favourable conditions the following may also reach tree proportions: *S. bebbiana, S. scouleriana* and *S. sitchensis*.

RANGE WA: A tree of stream banks and lowlands generally confined to the coast forest ecosystem. Occasional occurrence east of the Cascades. Lake Chelan, Almota.

TAIL-LEAVED WILLOW, *S. lucida* ssp. *caudata*: Differs by having leaves slightly rough beneath, with numerous stomata (pores) on both surfaces instead of only on undersurface. Young twigs smooth. Confined to south-central B.C. along stream banks and wetlands; possibly in central Washington.

PEACH-LEAF WILLOW, *S. amygdaloides:* Up to 12 m (40') high. Thin, drooping yellow branchlets, narrow leaves but no sideways twist at tip. Leaves finely toothed, pale yellow-green. Lakesides, river banks. Scattered throughout southern Okanagan Valley of B.C., southwards into central Washington.

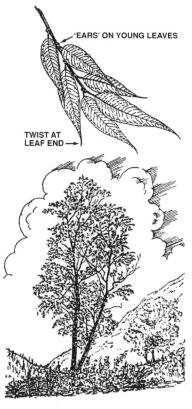

'EARS' ON YOUNG LEAVES

TWIST AT LEAF END →

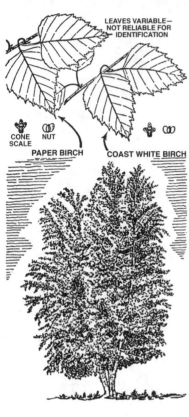

LEAVES VARIABLE—
NOT RELIABLE FOR
IDENTIFICATION

CONE NUT
SCALE

PAPER BIRCH   COAST WHITE BIRCH

Birch Family, *Betulaceae*

THE WHITE or PAPER BIRCHES
Betula spp.

Forty years ago I wrote, 'These birches are most confusing....' But those classifications were based on the information of the day. The birches have since been juggled around unmercifully, and can (for the meantime) be segregated as described below (correlated to *The Vascular Plants of British Columbia*, 1989).

PAPER BIRCH, *Betula papyrifera* var. *papyrifera*: Often a tall tree reaching 30 m (100') or more in height and 40 cm (16") in diameter. Leaves to 7.5 cm (3") long, many wedge-shaped at base and sharp-pointed. Mostly doubly-serrate. The bark is white on old trees and peels easily, but doing so disfigures the tree. Possibly the most variable tree species in Canada. Interior aboriginals found many uses for the bark, among them were the construction of baskets and canoes.

RANGE BC: Common throughout most of B.C. Absent on Vancouver and Queen Charlotte islands and along mainland coastal strip with heavy rainfall.

RANGE WA: Most common regions of occurrence are shown on range map. Seldom found over any extensive area. San Juan Islands eastwards at low to medium elevations.

COAST WHITE BIRCH, *B. papyrifera* var. *commutata*: An arbitrary common name, since no other could be found for this variety.

It is a relatively small tree to 20 m (60') high. Leaves are to 7.5 cm (3") long, squarish or slightly heart-shaped at the base. The bark is in thin layers, creamy white to light orange on old trees but often reddish brown on young trees. The botanist's differentiation from paper birch is that the fruiting bracts have lateral lobes that spread at right angles while for paper birch the lobes are ascending.

Found in Fraser Valley in extreme southwestern B.C.; very likely extends southwards along the coast.

ALASKA PAPER BIRCH, *B. neoalaskana*: A small tree to 9 m (30') high with smallish, triangular leaves. Twigs are covered with resin glands. Bark varies from white to red. Northeastern B.C., Peace River area and northwards to Fort Nelson.

EUROPEAN BIRCH, *B. pendula*: A horticultural escape. Now a brushy shrub, to 12 m (40') tall, of bogs. Leaves about 2.5 cm (1") long, rounded. Bark a chestnut brown. Lower Fraser Valley. Southeastern Vancouver Island. Likely in similar habitats in Washington.

SCRUB BIRCH, *B. glandulosa*: Described in the 'Shrubs' section, p. 125.

Birch Family, *Betulaceae*

WATER BIRCH
Betula occidentalis
Black Birch, Red Birch, Mountain Birch

FORM: Most commonly a wide-spreading, graceful shrub to 6 m (20') high, with stems rising from one main clump. In rich soils beside creeks or in meadows, it occasionally becomes a tree to 15 m (50') high and 30 cm (12") in diameter. Branches are very slender and willowy.

BARK: Rich reddish brown with prominent light-coloured markings. Not peeling except on larger trees, where there is some curling. Young twigs greenish and very warty. Mature twigs shiny red.

LEAVES: A roundish leaf with an abrupt, sharp point. Mostly 2.5–5 cm (1–2") long. Light green undersurface is often finely dotted.

FRUIT: Catkins showing prominently by April. Thickish, about 2.5 cm (1") long.

WOOD: Fine grained, soft and light, splitting easily. Larger trees used for firewood and farm purposes.

DID YOU KNOW that all birch bark is very durable? The wood of fallen trees will rot away, leaving a shell of bark. The attractive, richly coloured bark was used by the aboriginals in their basket weaving to create decorative patterns.

QUICK CHECK: Wet places along creeks and meadows within drier regions. Copper-brown bark, new twigs heavily warty.

RANGE BC: Extending eastwards from the Cascades to the Rockies. Most abundant in drier regions of south-central B.C. Limited occurrences to northeastern B.C., but generally absent in interior cedar–hemlock ecosystem.

RANGE WA: East of the Cascades, mostly in bunchgrass and ponderosa pine ecosystems. Oroville, Tonasket, Ferry County, Dayton.

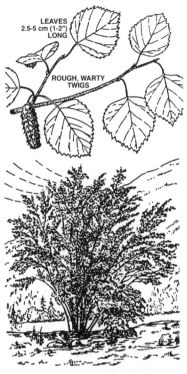

LEAVES
2.5-5 cm (1-2")
LONG

ROUGH, WARTY
TWIGS

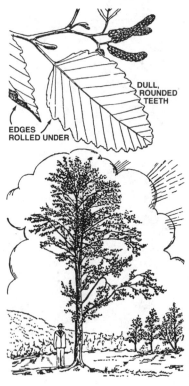

DULL, ROUNDED TEETH

EDGES ROLLED UNDER

Birch Family, *Betulaceae*

RED ALDER

Alnus rubra

The alders of B.C. and Washington are known as 'red,' 'white' and 'mountain'; 'green' and 'Sitka' are regarded as subspecies. Male and female catkins both form in the fall and are visible throughout winter. Catkins are fully developed when leaves appear. They hang in pollen-producing tassels; thus ornamented, alder resembles a flowering tree. Alders produce naked buds in summer. Small, woody cones form in fall, remaining until the following spring.

FORM: Varies greatly with age and density of stand. In open, young trees to 9 m (30') high have irregularly spaced, long, straight limbs that point outwards and upwards, making a broadly conical form. With increased age, lower limbs disappear, leaving a long, clear trunk and a narrowly conical crown. Older trees commonly 20–35 cm (8–14") in diameter, up to 24 m (80') high—but much larger trees do occur.

BARK: Slightly roughened bark, dirty grey in colour. Older trees with clear trunks have blotchy white markings, making them quite distinctive. Bases are sometimes seamed or thickly scaly.

LEAVES: Most leaves are 7.5–12.5 cm (3–5") long, dull green above and sometimes rusty-tinged below. The dull teeth form rounded lobes quite unlike the other 3 native alders. The extreme outer leaf edge curls under on most leaves. Note that the straight veins run to the leaf margin.

FRUIT: A cluster of green cones that turn brownish, each slightly more than 1.3 cm (½") long, remaining on the tree until late spring. The nut is bordered by very narrow wings. Striking attractive catkins to 10 cm (4") long shed pollen in early spring before the leaves appear.

WOOD: Light, fine grained and easily worked without splintering. Does not check or warp and is valuable in furniture construction. Serves as a base for expensive veneers, but finishes well itself. The dry wood makes excellent fuel and burns leaving very little ash. Native peoples have used it for centuries in barbecuing salmon.

DID YOU KNOW that freshly peeled alder bark turns bright orange on the undersurface? The aboriginals used this inner bark to make a dye. Catkins and buds are a source of food for ruffed grouse. Well known for its ability to 'fix' nitrogen from the air.

QUICK CHECK: Round-lobed leaves with edges rolled under. Splotchy bark.

RANGE BC: Very common west of the Cascades. Northwards to Alaska.

RANGE WA: Very common west of the Cascades but probably appearing infrequently at low elevations on eastern slopes. Streams, valley bottoms and slopes with rich soils.

Birch Family, *Betulaceae*
WHITE ALDER
Alnus rhombifolia

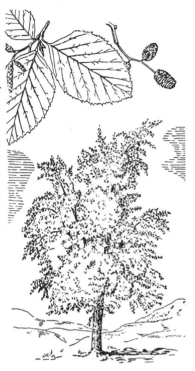

FORM: Resembles red and mountain alders in having a broad, open crown formed by strong branches. Large trees can reach around 50 cm (20") in diameter and 18 m (60') in height but the usual size is less than 30 cm (1') in diameter and about 9 m (30') high. Often the trunks are clear of limbs for half of their length if in dense growth. Otherwise, long, ungainly limbs branch from near the base.

BARK: Light grey and smooth on young trees up to 15 cm (6") in diameter. Then turning greyish brown and breaking into rectangular scales that continue well up on the trunk. Contrast with red alder's mottled bark.

LEAVES: Light green on top surface and 5–9 cm (2–3 ½") long, half as long as for red alder. Undersides and stems of leaves carry very fine, downy hairs. Prominent yellow midrib. Margins irregularly toothed with gland-tipped teeth (dark spots) on old leaves. Best seen with a hand lens.

FRUIT: Clusters of brown cones, each 0.8–2 cm (⅓–¾") long. Seeds shed in late fall or winter months. These are very small and surrounded by a narrow, hard wing of extreme thinness. Thick clumps of catkins show in January and February, giving the tree a rusty appearance.

WOOD: Sapwood dull white and heartwood light brown. Soft, brittle and comparatively light. Limited use because of meagre supply but suitable for furniture and wooden ware. Good fuel wood.

DID YOU KNOW that birch and alder are close relatives, each bearing seed as a small, winged nut held in a cone? In birch the cones fall apart to release the seed; in alders the seeds fall from a hard cone.

QUICK CHECK: Limited range is a good guide. Also, small size of leaves and smooth grey bark.

RANGE BC: Does not extend into British Columbia

RANGE WA: Along stream banks in the sagebrush ecosystem, particularly in southeastern Washington. For example, along Rock Creek in Klickitat County to elevations over 1200 m (4000'). Because of its confusing resemblance to both red and mountain alder, the westerly and northerly extents of white alder are not well known. However, it does not usually extend west of the Cascades.

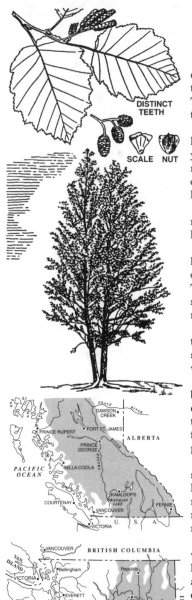

DISTINCT TEETH

SCALE   NUT

PEACE RIVER
DAWSON CREEK
PRINCE RUPERT   FORT ST. JAMES
ALBERTA
PRINCE GEORGE
PACIFIC OCEAN   BELLA COOLA
FRASER
KAMLOOPS
COURTENAY   OKANAGAN LAKE   FERNIE
VANCOUVER
VICTORIA   U.   S.   A.

VANCOUVER   BRITISH COLUMBIA
VAN. ISLAND
Bellingham   Republic
VICTORIA
EVERETT
IDAHO
SEATTLE   Grand Coulee   SPOKANE
WENATCHEE   Cle Elum
TACOMA
YAKIMA
Goldendale
PORTLAND   COLUMBIA RIVER   OREGON

Birch Family, *Betulaceae*

MOUNTAIN ALDER
Alnus tenuifolia
Thinleaf Alder

FORM: At higher elevations, much like Sitka alder: sprawling and shrublike. In valleys, it is a small tree to 12 m (40') high and 30 cm (12") in diameter. The branches are fairly straight and point upwards to form a loose, conical crown.

BARK: Thin, smooth, dirty, green-grey with lighter horizontal markings especially prominent on younger trees. Older trees tend to flake and scale near their bases. In the wetter parts of its range, it is often partially covered by a scaly, whitish green lichen.

LEAVES: Distinctly double-toothed with definite sharp teeth. Leaves dark green, 5–10 cm (2–4") long, with yellow central vein.

FRUIT: Male catkins are shed soon after the pollen is released. The female catkins become a cluster of 3–9 hard green cones about 1.3 cm (½") long. These hang on the trees throughout summer and are still prominent the following spring. The nut is narrowly winged.

WOOD: A soft, light wood similar to Sitka alder that, because of mountain alder's larger size, does see some use as fuel. Rots quickly when in contact with the ground.

DID YOU KNOW that alder leaves do not turn brown in the fall, but remain green until they drop? A decoction of alder bark was used by the aboriginals in the treatment of rheumatic fever. The bark was found to contain salicin, now a standard medicinal prescription for this disease.

QUICK CHECK: Recognize as alder by its small, woody cones, naked buds or straight-veined leaves. Distinctly double-toothed leaves confirm it as mountain alder. Note that leaves of red alder are rolled inwards on their margins and that its general range is west of the Cascades.

RANGE BC: East of the Cascades and northwards to the Yukon. Occurs from valley bottoms to high mountain elevations. Often growing with Sitka alder at higher elevations. Prefers wet ground, creek edges and borders where there is good exposure. The common alder of the southern interior, but in the central and northern interior, Sitka alder may be more common.

RANGE WA: East of the Cascades, where it is the most common alder, largely in the mountain forest ecosystem. Habitats as for B.C.

Birch Family, *Betulaceae*

SITKA ALDER

Alnus crispa ssp. *sinuata*

Mountain Alder

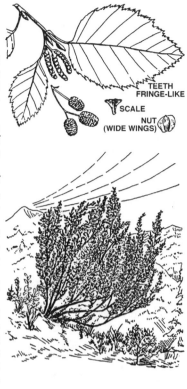

TEETH FRINGE-LIKE

SCALE

NUT (WIDE WINGS)

FORM: Often a sprawling shrub to 3 m (10') high, with crooked, upward-curving limbs. Along creek courses and wet areas an erect tree to 9 m (30') high. Branches stick outwards to form a loose, ragged outline.

BARK: Smooth, greyish green, marked with light-coloured, warty lenticels. In the wetter parts of its range it is often mottled by a scaly, greenish white lichen.

LEAVES: Readily distinguished from those of other alders by broad lobes or wavy margins. Fine, sharp teeth give a fringe-like appearance to the margin. Leaves are from 1.3 to 7.5 cm (½ to 3") long.

FRUIT: A cluster of 3–6 hard, brownish cones about 1.3 cm (½") long. The small nuts, which can be shaken from the cones, have wide wings on either side.

WOOD: A soft wood with little strength. Used only in case of necessity.

DID YOU KNOW that the crooked, sprawling limbs are protection against the deep snows of winter? They bend to the ground when weighted and rise back as they are released. Many avalanche paths are covered with this alder. It is a horror to the hiker who tries to push through a thicket of it, where stiff limbs project in every direction.

QUICK CHECK: An alder by reason of cones, naked buds or leaves with straight veins running to the margin. Fringe-like margins make it Sitka alder.

RANGE BC: Westwards from the Rocky Mountains throughout southern B.C. on mountain slopes of 900 m (3000') and upwards. Occasionally reaches to low levels. Ranges northwards to overlap range of green alder and fades out. Almost always in damp places such as along streams or bordering swampy meadows.

RANGE WA: Throughout Washington on sites as for B.C. Will stand partial shade. Range map includes regions where Sitka alder may be found at low levels.

GREEN ALDER, *A. crispa* ssp. *crispa*: The northern subspecies of Sitka alder is very similar but leaves do not have wavy-lobed edges. Green alder flowers as the leaves come out. In the fall a good clue would be the wide-winged nutlets (seeds). A common plant of the boreal forest. Peace River and to the north and east of the Rocky Mountain Trench.

Beech Family, *Fagaceae*
GARRY OAK
Quercus garryana
White Oak

FORM: Most picturesque when mature, because of its huge, gnarled limbs and massive, shaggy crown. Heavy trunk may be up to 90 cm (3') in diameter. Short branches become stout limbs. Form not mistakable for any other native tree. Can be scrubby, low tree—forms dense thickets on mountain slopes to 1050 m (3500'); largely in Klickitat County.

BARK: Light grey. Soon fissuring to produce narrow, horny ridges with occasional stout scales.

LEAVES: Typical lobed oak leaf 7.5–15 cm (3–6") long, thick and dark green.

FRUIT: Smooth, brown acorn 1.3–2.5 cm (½–1") long, dropping in fall, with knurled or roughened cup often attached to base. Acorns were a good food for natives, if soaked in water to remove bitter tannins. Sought by band-tailed pigeons, Steller's jays.

WOOD: Hard, strong and heavy, like other oaks. Wood checks and warps upon drying, which, together with the short trunks, gives Garry oak little commercial value. It makes excellent fuel when thoroughly dry. Rots quickly when in contact with the ground.

DID YOU KNOW that David Douglas first collected this oak in 1825, on lower Columbia River? The oaks were 'generally low and scrubby...[and] interspersed over the country in an open manner, forming belts or clumps along the tributaries of the larger streams....' In naming it, Douglas wrote 'I have great pleasure in dedicating this species to N. Garry, Esq., Deputy Governor of the Hudson's Bay Company....'

The shiny acorns are used in making novelty items and costume jewelry.

QUICK CHECK: A massive, shaggy tree of limited range. Characteristic oak leaves and acorns.

RANGE BC: B.C.'s only wild oak. Very limited range. Vancouver Island: southern tip, extending northwards in patches along eastern coast in open, rocky locations as far north as Comox. Common on Gulf Islands. Small grove on south side of Fraser River 2.4 km (1 ½ miles) above Yale, and on Sumas Mountain. Low elevations. Often with arbutus.

RANGE WA: State's only wild oak, but with considerable range. On San Juan Islands, southwards in patches along Puget Sound; open, rocky locations. East of Cascades in several localities: Tampico and Tieton rivers in Yakima County. Vicinity Cle Elum. Mixes with ponderosa pine throughout Columbia Gorge. From Goldendale both northwards to Toppenish and northeastwards to Bickelton.

Impressive grove—size and beautiful specimen trees—at historic Fort Simcoe east of Toppenish. Another grove 50 km (30 miles) to north, beside U.S. 12, may be most northerly inland occurrence in state.

Rose Family, *Rosaceae*
PACIFIC CRAB APPLE
Malus fusca

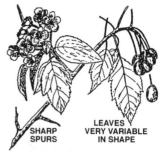

FORM: Usually a small, scraggly tree to 9 m (30') high and 30 cm (12") in diameter, but often shrublike with a number of straight, smooth stems a couple of centimeters or so (an inch or two) in diameter. Very bushy when growing in the open. Numerous stout, blunt spurs a few centimeters (an inch or two) long give a realistic imitation of thorns.

SHARP SPURS

LEAVES VERY VARIABLE IN SHAPE

BARK: Very fissured, scaly and patchy on old trunks and branches.

LEAVES: Much like those of domesticated apple trees but often with irregular lobes and in a variety of shapes. Thick, sharply toothed and with prominent veins. Mostly 5–9 cm (2–3 ½") long.

FRUIT: Clusters of white, fragrant, 'apple blossoms' appear in early spring, to be followed by bunches of little, oblong apples about 1.3 cm (½") long. At first greenish, they become yellowish or blushed with red. They are edible, but rather acidic. Coastal natives used large quantities for food either fresh or after storage in water.

WOOD: Very compact and fine grained. Sometimes used in small ornamental turnery because of toughness and brownish hue of wood.

DID YOU KNOW that crabapple leaves turn beautiful shades of yellow and russet in autumn? In winter, the fruit left on the trees is a preferred food of purple finches.

QUICK CHECK: Wetland habitat, some lobed 'apple' leaves and fragrant flowers or clusters of tiny apples.

RANGE BC: Coastal strip of entire province, including adjacent islands. Low, damp places, such as stream banks and swamp edges, where it often forms an impenetrable thicket. Common on low ocean frontage.

RANGE WA: Coast forest ecosystem and San Juan Islands. Sites as for B.C.

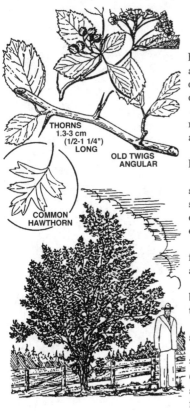

THORNS
1.3-3 cm
(1/2-1 1/4")
LONG

OLD TWIGS
ANGULAR

COMMON
HAWTHORN

Rose Family, *Rosaceae*

BLACK HAWTHORN
Crataegus douglasii

FORM: Either a small, bushy tree to 6 m (20') high or a tangle of low, shrublike growth that forms an impenetrable thicket. Twigs have alternating crooks. Spines are 1.3–3 cm (½–1 ¼") long and needle-sharp, not always too noticeable.

BARK: Dirty grey, very rough and scaly. Often mottled with patches of lichen. Young shoots smooth and shiny. Very similar to Pacific crabapple.

LEAVES: Thickish, oval leaves to 7.5 cm (3") long with 5–9 small lobes at top end.

FRUIT: Showy clusters of smelly white blossoms during April and May, followed by bunches of small, black-purple 'apples' ripe by late July. Though quite edible, rough seeds make them unpopular for eating. The 'apples' wither quickly once ripe.

WOOD: Tough and close-grained, and therefore favoured by native peoples for small tools. The spines also have practical uses.

DID YOU KNOW that the legendary giant lumberjack Paul Bunyan always used a big hawthorn tree as a back-scratcher?

QUICK CHECK: A bushy tree with sharp spines 1.3–3 cm (½–1 ¼") long. Zig-zag twigs.

RANGE BC: Wide range throughout southern ⅔ of the province. Scattered northwards to Dawson Creek. Low to middle-mountain elevations. Edges of streams and meadows. Common along roadsides and fields.

RANGE WA: Wide range from low to middle-mountain elevations. Mostly coast forest and ponderosa pine ecosystems. Tonasket, Republic, Stevens County, Whitman County, Spokane. Sites as for B.C.

RED HAWTHORN, *C. columbiana*: Has slender thorns 3–5 cm (1 ³⁄₁₆–2") long and dark scarlet berries. Found east of Cascades, from stream banks to hillsides, but less common than black hawthorn.

COMMON HAWTHORN, *C. monogyna*: A widespread escape on southern portion of Vancouver Island and Lower Mainland, on Gulf and San Juan islands, and southwards. Spiny limbs, deeply lobed leaves and colourful bunches of scarlet berries that persist over winter. Introduced from Europe.

BLACK HAWTHORN
RED HAWTHORN
ALBERTA

BLACK HAWTHORN
RED HAWTHORN

Rose Family, *Rosaceae*

BITTER CHERRY
Prunus emarginata

FORM: Wide variation, depending on soil type and climactic factors. At the coast it is a slender tree up to 25 cm (10") in diameter and 18 m (60') high. The branches are straight and point upwards. On drier sites and at higher elevations, a low, crooked shrub up to 3 m (10') may be expected. Leaves grow along branches rather than on side twigs, thereby quite accurately outlining the limb framework.

BARK: On young trees, shiny reddish brown with prominent raised lenticels in horizontal rows. On older trees, a dirty greyish brown, lightly roughened and marked with greyish lenticels to 5 cm (2") long. Used for decorative purposes by native peoples.

LEAVES: Varying from 4 to 7.5 cm (1 ½ to 3") long on shrubby trees, but may be 10 cm (4") long on tall trees. Blunt leaf points on older trees, sharper on new growth. Fine, rounded teeth. 2 small knobs or glands at the base of the stem.

FRUIT: Flat-topped clusters of 5–10 fragrant white or pinkish flowers toward limb ends, during April or May, are replaced by pea-sized, bright red cherries that taste extremely bitter. Each fruit is on a short stem about 1.3 cm (½") long that branches off a stouter central stem 1.3–2.5 cm (½–1") in length.

WOOD: A brittle, quick-rotting wood sometimes cut for fuel or used for ornamental purposes.

DID YOU KNOW that cherry bark can be peeled from the tree and polished to a rich red? Aboriginals used strips of this bark in their basket weaving to give colour to their work.

QUICK CHECK: Small, narrow leaves outline limb framework, making it visible from a distance. Glands on a dull-pointed leaf make it the bitter cherry. Flat-topped clusters of white flowers (5–10) or bright red cherries on stems about 1.3 cm (½") long. Do not confuse with choke cherry, p. 92, which has similar clusters, but with darker fruits.

RANGE BC: Lower elevations on Vancouver Island and southern coastal zone, to 900 m (3000'). Throughout the interior cedar–hemlock ecosystem to the subalpine. Common on eastern slopes of mountains between Grand Forks and Nelson. Ootsa Lake.

RANGE WA: Coastal zone to 900 m (3000') and east of Cascades—most common in ponderosa pine habitat. Kamiak Butte, Yakima, White Pass, Blue Mountains.

PIN CHERRY, *P. pensylvanica*: Small tree, wideranging, as Latin name suggests. Leaves slender, tapering to sharp point. Red fruits on stems about 2.5 cm (1") long. Poor soils and forest borders across Canada, westwards to Cascades in southern half of B.C. Probably in northern Washington.

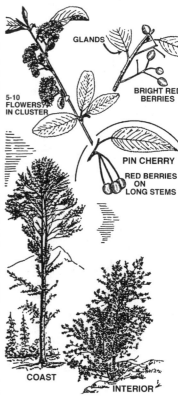

GLANDS

5-10 FLOWERS IN CLUSTER

BRIGHT RED BERRIES

PIN CHERRY

RED BERRIES ON LONG STEMS

COAST

INTERIOR

WHITE FLOWERS

GLANDS

DARK PURPLE BERRIES

Rose Family, *Rosaceae*
CHOKE CHERRY
Prunus virginiana

For the amateur, the one name 'choke cherry' will suffice to cover both recognized subspecies: *Prunus virginiana* ssp. *melanocarpa*, with berries red to purple or black, west of the Cascades and *Prunus virginiana* ssp. *demissa*, with black fruits, east of the Cascades.

Three other species not described here are introductions from Europe and have very limited ranges.

FORM: Varying from heavy, crooked shrubs less than 3 m (10') high to small trees 6 m (20') high. Groups often form a sprawling, irregular mass. Note the abundance of dark green leaves and the drooping pencils of flowers or berries.

BARK: Greyish brown, roughened by numerous small lenticels.

LEAVES: Dark green. Sharply pointed and finely saw-toothed; wider above the middle. Leaf stems about 1.3–2 cm (½–¾") long.

FRUIT: White flowers, 1.3 cm (½") across, very noticeable in May; in dense, cylindrical clusters near the ends of the limbs. Dark berries are 0.6–1.3 cm (¼–½") in diameter. Although very puckery to taste, they make fine jams and jellies. Aboriginals blended them with meat to make that important food, pemmican.

WOOD: Brittle, fine grained and of no economic importance.

DID YOU KNOW that the bruised bark gives off a pungent smell? It has a very bitter taste, like the bark of all cherries.

QUICK CHECK: From a distance, note the crooked form and heavy, dark green foliage. In spring, long clusters of white flowers; in fall, masses of dark purplish berries.

RANGE BC: Generally east of the Coast Mountains but also in the Fraser delta and on southeastern Vancouver Island. Common in dry interior regions across the province. Spotty in Peace River and Stikine River valleys. Although often found in exposed locations, it prefers damp ground of fair richness. White clematis and poison ivy are likely to be growing nearby.

RANGE WA: Generally east of the Cascade Range. Sagebrush, bunchgrass and ponderosa pine ecosystems. Sites and companions as for B.C.

Maple Family, *Aceraceae*

BIGLEAF MAPLE

Acer macrophyllum

Broadleaf Maple, Oregon Maple

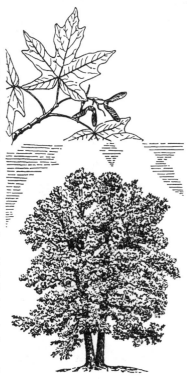

FORM: A massive, bushy tree. In the open, it grows to 24 m (80') high; the short trunk, to 60 cm (2') in diameter, branches into numerous upright limbs. Among forest trees, it grows straight, a loose crown of up-pointing branches surmounting a clear trunk. Often forked or several trunks close together.

BARK: Green on young trunks becoming finely roughened on trees to 15 cm (6") in diameter. Then becoming a drab grey-brown, furrowed into narrow, horny ridges.

LEAVES: The largest tree leaf in our area, occasionally to 40 cm (16") long, but more commonly 15–25 cm (6–10"). Divided into 5 prominent lobes. Pale yellow fall colour does not match brilliant reds of our other 2 maples but is beautiful nevertheless.

FRUIT: Prominent pale yellow flowers hanging in clusters appear in April, before the leaves. Seeds are paired maple wings about 5 cm (2") long and at 45 to 60 degrees to one another; late summer to early winter.

WOOD: Fine grained and fairly dense. Very valuable for furniture, interior finishing and other specialty uses.

DID YOU KNOW that a thick mat of moss and ferns often make their home on maple trunks and lower limbs? The maple syrup of the bigleaf lacks the high quality of syrup from eastern maples.

QUICK CHECK: Large, 5-lobed maple leaves.

RANGE BC: West of the Coast Mountains at elevations under 300 m (1000'). Bella Coola southwards to Vancouver. Follows valleys inland from west, e.g., up the Fraser Canyon as far as Boston Bar and to 16 km (10 miles) east of Hope along Hope–Princeton Highway. The several other tree maples in B.C. not described in this book are garden escapes with very limited range.

RANGE WA: Mostly west of the Cascade Mountains at elevations under 300 m (1000')—it is a significant, moss-draped tree in the Olympic rainforest. Also found east of the Cascades along some watercourses in Chelan and Klickitat counties. Near Peshastin and Entiat. Columbia Gorge eastwards as far as The Dalles, Oregon. A few good specimens along Rock Creek in Klickitat County, far from the coast.

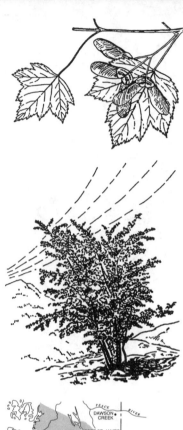

Maple Family, *Aceraceae*

DOUGLAS MAPLE

Acer glabrum

Rocky Mountain Maple

FORM: Sometimes a number of spreading stems up to 7.5 m (25') high and forming a loose, wide-spreading crown. Very often shrublike and under 3 m (10') high.

BARK: Smooth and grey-brown on main trunks. New twigs bright red.

LEAVES: A relatively small leaf 2.5–7.5 cm (1–3") across, with 3–5 toothed lobes. Dark green. but often blotched with vivid red dabs. The leaf stem is grooved and usually bright red. Very colourful red and crimson autumn foliage.

FRUIT: Pairs of maple 'wings' about 2.5 cm (1") long, joined together at almost a right angle. Pinkish tinge during late summer.

WOOD: Not used because of tree's small size. Was a favourite wood of aboriginal peoples because of its toughness; it was used for snowshoe frames and other items.

DID YOU KNOW that this maple leaf is very much like the one on the Canadian flag? A note from the past: The forked limbs made very fine sling-shot crutches. None of our maples produce maple syrup of the quality of the sugar maple of eastern North America.

QUICK CHECK: Small 'maple' leaf with 3–5 toothed lobes.

RANGE BC: Very abundant and widespread east of the Cascades and in southern ⅔ of the province. Spotty occurrence along coast southwards from Alaska. Grows to over 1200 m (4000') in elevation and extends considerably northwards of the Canadian National Railway tracks.

RANGE WA: Very abundant and widespread at middle-mountain elevations east of the Cascades but might be found in any treed region of the state. Ranges to elevations over 1200 m (4000'). Okanogan county and southwards to the Columbia River.

Maple Family, *Aceraceae*
VINE MAPLE
Acer circinatum

FORM: Typically a bushy mass to 9 m (30') high supported by a number of stems 5–12.5 cm (2–5") in diameter, with a straggly, crooked outline. In the open it can develop a recognizable tree form, however; seldom with a single trunk.

BARK: Smooth, pale green but occasionally becoming dull brown.

LEAVES: Opposite circular leaves with 7–9 short, sharply toothed lobes that look like spread, blunt fingers. Leaves 5–15 cm (2–6") across. Dark green until the middle of summer, when shades of yellow and red often start to show.

FRUIT: Maple 'wings' so widely spread as to be almost in a straight line. From 2 to 4 cm (¾ to 1 ½") long and quite red when ripe.

WOOD: A surprisingly heavy wood of fine grain. Rots quickly when in contact with earth. Seldom used today because of its small size and crooked form. Because it is flexible when green, natives used it for snowshoe frames and various implements.

DID YOU KNOW that camp cooks once used green vine maple for pot hooks and reflectors around campfires because it is almost impossible to burn? Vine maple has the most vividly coloured autumn foliage of any coastal tree or shrub.

IN SHADE   OPEN-GROWN

QUICK CHECK: Leaf with 7–9 blunt, spreading fingers or lobes. Widely spread seed 'wings.' Limited range.

RANGE BC: Lower and middle elevations of coastal forest from Knight Inlet southwards. Very rare on Vancouver Island. Probably sporadic occurrence in wetter places of interior, since it grows in Wells Gray Park, 240 km (150 miles) north of Kamloops. A tree of damp places along creeks or meadows where soils are fairly good. Tolerant of shade but usually found along forest borders.

RANGE WA: Lower and middle elevations from B.C. to California. Very noticeable on western sides of main mountain passes, especially when showing fall colours. Sporadic occurrence in wetter places east of Cascades, along creeks or meadows where soils are fairly good. Very tolerant of shade and usually found growing under other trees. Head of Lake Chelan and valley bottoms north of Peshastin. Leavenworth, Mt. Rainier, Olympics.

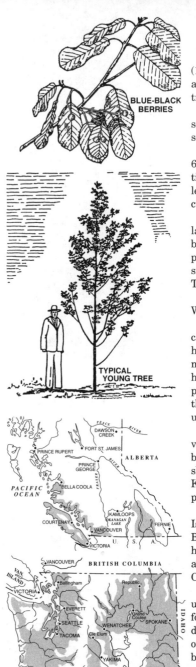

BLUE-BLACK BERRIES

TYPICAL YOUNG TREE

Buckthorn Family, *Rhamnaceae*

CASCARA

Rhamnus purshiana

FORM: Older trees to 9 m (30') high and 25 cm (10") in diameter are seldom seen now. Often twisted and irregularly limbed if fighting for light. Young trees straight. Limbs relatively few and upright.

BARK: Smooth, slightly mottled grey bark resembling young red alder. Older trees tend to be scaly near their bases.

LEAVES: Large 'cherry-like' alternate leaves 6–15 cm (2 ½–6") long, finely toothed, with distinctive shape and clustering tendency. Prominent parallel veins. Paler green beneath. Leaf buds naked. Uncurling leaves coppery in early spring.

FRUIT: Clusters of small, greenish flowers in late spring usually go unnoticed. The plump, blue-black berries, about 8 mm (⅓") across, are very prominent in late summer. With short stems, in small clusters at the ends of longer main branches. The seedy berries are edible but not highly rated.

WOOD: Light, brittle and of no value; during World War II tests were made on its drug content.

DID YOU KNOW that the drug cascara sagrada came from the bark? It was used as a laxative and held in high regard by the medical profession for many decades. During World War II the bark was harvested in great quantities, bringing up to 44 cents per kilogram (20 cents per pound), big money in those days. Natives also had a number of medicinal uses for the bark.

QUICK CHECK: Large, ovalish, strongly veined leaves in loose whorls. Clusters of blue-black berries. Alder-leaved buckthorn, a closely related shrub that overlaps the range of cascara in the Kootenays, has singly borne black berries (see p. 125).

RANGE BC: At low elevations on Vancouver Island and Gulf Islands. Coastal strip of mainland to Bella Coola. Southern part of the interior cedar–hemlock ecosystem, e.g., Arrow and Kootenay lakes and Adams River. Also from 16 km (10 miles) east of Creston to Yahk.

RANGE WA: An infrequent tree that often goes unnoticed. Wide range but most abundant in low foothills of coast below 750 m (2500'); prefers shady, damp places. Often among new growth in logged-over areas. Sporadic along stream banks in sagebrush, bunchgrass and ponderosa pine ecosystems.

Dogwood Family, *Cornaceae*

WESTERN FLOWERING DOGWOOD
Cornus nuttallii
Pacific Dogwood, Flowering Dogwood

4-6 SHOWY BRACTS

RED BERRIES

FORM: Sometimes a bushy, lop-sided tree to 30 cm (12") in diameter and 9 m (30') high, often with several short trunks that divide into heavy, upward-pointing limbs. However, quite common as a very bushy shrub less than 3 m (10') high. Twigs are symmetrically branched opposite to one another and at right angles to the preceding pair.

BARK: Blackish brown and smooth, but finely ridged on older trees.

LEAVES: Glossy dark green above and much lighter below. Opposite, 7.5–10 cm (3–4") long; characteristic 'dogwood veins' curving parallel to leaf edges. In the fall, leaves are generally tinged with red.

FRUIT: Dogwood blooms from April to June and sometimes again in September, but the later blooming does not produce fruit. The blossom is 6–12.5 cm (2 ½–5") across and may have 4–6 white, showy bracts surrounding a rounded knob of greenish flowers. This central cluster turns into a compact group of red, bead-like berries, thus adding gay colour to the tree during August and September.

WOOD: A hard, fine-grained wood that had value for making small, durable items, but can no longer legally be used in B.C.

DID YOU KNOW that the dogwood is the official floral emblem of B.C.? It and the cascara are the only trees protected by law in B.C. Skewers or 'dags' were once made from its wood, hence the name 'dagwood,' later popularized as 'dogwood.' Kousa dogwood is a popular small horticultural tree also having large, white flowers. Pileated woodpeckers, flickers and band-tailed pigeons are some of the birds that enjoy dogwood berries.

QUICK CHECK: Unmistakable 'dogwood' leaves, flowers and fruit. In winter, the symmetrical 'scalloped' branching is sufficient.

RANGE BC: Common along eastern coast of Vancouver Island and adjacent mainland coast and in Port Alberni region. Extends into Fraser Canyon at least 50 km (30 miles) northwards beyond Hope. Upper elevation limit approximately 300 m (1000').

RANGE WA: Fairly common in coastal region on a variety of soils: rich valley bottoms and rocky places. Upper elevation limit approximately 450 m (1500'). Abundant around Mt. Vernon. Extends up Columbia Gorge to Hood River, Oregon. Sporadic occurrence in mountain forest ecosystem north of the Columbia Gorge.

SHRUB FORM

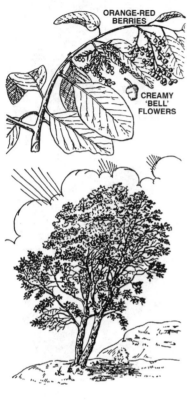

ORANGE-RED BERRIES

CREAMY 'BELL' FLOWERS

Heath Family, *Ericaceae*

ARBUTUS
Arbutus menziesii
Madrone, Pacific Madrone

FORM: Seldom more than 40 cm (16") in diameter with twisting trunk and heavy, irregular branches. Open grown, assumes irregular, rounded outline; shape varies in shady surroundings.

BARK: On older trees, an unusual reddish orange peeling bark. Large, loose scales formed by a drying and shrinking process. If young or if scales have fallen, trunk is quite smooth and green. The only native tree with this type and colour of bark.

LEAVES: Alternate, evergreen leaves 7.5–15 cm (3–6") long with a thick leathery texture. Glossy dark green above and whitish green beneath.

FRUIT: Clusters of fragrant, creamy-white, bell-shaped flowers as early as mid-April are followed in late summer by irregular masses of small, orange-red berries each 8 mm (⅓") across. Though seedy and inedible to people, they are much sought after by birds, particularly by robins and varied thrushes during the winter months.

WOOD: Soft and easily worked when green, but becomes extremely hard when dry. Its tendency to warp and check limits its use to a few novelty items. The bright colour of the bark is not found in the wood, which is brownish.

DID YOU KNOW that arbutus is the only native broadleaf evergreen in Canada? Usually a good indicator of rock or a hard subsoil lying not far beneath the surface. Its glossy leaves are adept at shedding water, thereby protecting themselves from a damaging ice coating in winter.

The scientific name *Arbutus* means 'strawberry tree' in Latin, referring to red berries. Commonly known as madrone in California, a name that comes from early Spanish explorers, who related it to the *madrono* or strawberry tree of the Mediterranean. Archibald Menzies, a botanist and surgeon in the Royal Navy, collected it here in the 1880s and 1890s.

QUICK CHECK: A tree of limited range with orange-red bark that peels in scales and thick, glossy, evergreen leaves.

RANGE BC: A prominent tree of southeastern Vancouver Island and Gulf Islands region. Often associated with Garry oak. Very sparse west of Alberni summit; scattered trees at Great Central Lake, Buttle Lake and Gold River. Scattered along coast of mainland and islands from Bute Inlet southwards.

RANGE WA: A prominent tree of San Juan Islands, often associated with Garry oak. Scattered along coast of Puget Sound. Not found along exposed outer coast. Isolated grove on bluffs along south fork of Snoqualmie River 50 km (30 miles) east of Seattle.

Ash Family, *Oleaceae*
OREGON ASH
Fraxinus latifolia

FORM: If open grown, a rounded, heavily foliated tree to 15 m (50') high. Limbs are stout and widespreading. When competing for light with other trees, it grows to 23 m (75') in height and has a short and narrow top of rather compact branches. Most trees seen now are under 60 cm (2') in diameter.

BARK: On young trees, greyish, blotchy, lightly roughened by welts. Old trees carry prominent crisscrossing fissures.

LEAVES: A particular gracefulness of foliage arises from the symmetry of opposite leaves. The main stems, to 30 cm (12") long, are prominently grooved and carry 5–7 leaflets ranging from 7.5 to 15 cm (3 to 6") long. Light green and generally quite woolly beneath, though a hairless variety has been found.

FRUIT: Small flowers precede the leaves, blooming during March and April. Fruits are single wings 2.5–5 cm (1–2") long. On female trees in conspicuous large clusters, ripening by late summer.

WOOD: Light brown, with grain varying from medium to coarse. Sees limited use for tool handles and in the construction of barrels, boxes and furniture. Also used for fuel and as a shade tree.

DID YOU KNOW that the ashes belong to the olive family, which includes such popular garden shrubs as the lilacs, forsythias and sweet jasmines? All have opposite leaves. Legend says that poisonous snakes will not be found where this tree grows.

QUICK CHECK: Compound leaf with 5–7 leaflets, hairy below. Single winged seeds 2.5–5 cm (1–2") long.

RANGE BC: Only recently discovered in B.C.: a few specimens on coastal Vancouver Island. Nahmint Valley, Pacific Rim National Park.

RANGE WA: West of the Cascades from sea level to near 900 m (3000') elevation. Prefers rich bottomland soils but grows well on poorer soils bordering streams. Usually scattered among other deciduous trees. A tree of the westerly part of the coastal forest area. Abundant in King and Snohomish counties.

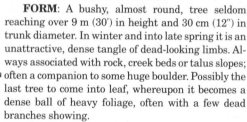

LEAVES TO 7.5 cm (3") LONG
STIFF TO COARSE

PEBBLY TWIGS

RUSTY BROWN
TO PURPLE
BERRY
6 mm (1/4")
LONG

HEAVY RIBS BENEATH

Elm Family, *Ulmaceae*

HACKBERRY

Celtis reticulata

FORM: A bushy, almost round, tree seldom reaching over 9 m (30') in height and 30 cm (12") in trunk diameter. In winter and into late spring it is an unattractive, dense tangle of dead-looking limbs. Always associated with rock, creek beds or talus slopes; often a companion to some huge boulder. Possibly the last tree to come into leaf, whereupon it becomes a dense ball of heavy foliage, often with a few dead branches showing.

BARK: Numerous thin fissures produce a fine weave of grey and black tones on the limbs. Trunks are more deeply ridged.

LEAVES: Leaves have a lifeless appearance, being heavy, thick and dull green. To 7.5 cm (3") in length, with very heavy ribs on the undersurface; rough and unpleasant to the touch. Note that the sides are of uneven lengths, like hackberry's close relatives, the elms.

FRUIT: Tiny, inconspicuous flowers are produced in spring. Those at the bases of the leaves near the ends of the branchlets develop into single cherry-like fruits. These are about 6 mm (¼") in diameter and range from dusty brown to purple or black when ripe. Most of the berry is filled with a large, hard seed. The edible fruit is dryish and sweet but not prized.

WOOD: Heavy and brittle with low-to-medium hardness. Sapwood is slightly paler than the light yellow-brown heartwood. Because of its small size and limited range, it is of no economic importance.

DID YOU KNOW that this tree, though shrublike in Washington, may reach 27 m (90') in height and over 60 cm (2') in diameter when grown several hundred kilometers to the east?

QUICK CHECK: Very limited range. Uneven elm-like leaves and single cherry-like berries.

RANGE BC: Not in B.C.

RANGE WA: Common occurrence in southeastern Washington along the dry, rocky banks of the Snake and Columbia rivers and their tributaries. Sometimes on the edges of streams and also on adjacent rocky places. Noticeable in the vicinity of Clarkston, Pomeroy and Dayton. Common on talus slopes below Dry Falls. Extends westwards to about The Dalles in the Columbia Gorge, where it is often the only tree on steep, rocky slopes. Far removed from its usual habitat, there is a small tree along Rock Creek in Klickitat County.

Rose Family, *Rosaceae*

MOUNTAIN MAHOGANY
Cercocarpus ledifolius
Curl-leaf Mahogany

This is the only mountain mahogany in Washington, although there are a number to the south. They take their name from the hard, mahogany-coloured wood. Generally it is a much-branched shrub to 4.5 m (15') high but described here because it is technically a tree. Occasionally this mountain mahogany reaches 6 m (20') in height and forms a crooked, squat tree with a trunk to 25 cm (10") in diameter. Older twigs are marked by raised leaf scars. The trunk, measuring 7.5–10 cm (3–4") in diameter, is covered with irregular, small, greyish brown scales. Very spotty in occurrence. Usually found on dry, open ridges as individuals or small groves.

The tiny leaves hardly clothe the limb framework. They vary from 1 to 3 cm (½ to 1 ¼") long, with margins tightly rolled over. The upper surface appears grooved because of the heavy midrib while beneath there is a fine mat of yellowish hairs. From 1 to 4 hairy flowers form in the axils of the leaves and later change into spectacular seeds, each carrying a silky tail to 7.5 cm (3") long.

QUICK CHECK: Limited range and unmistakable rolled-over leaves.

RANGE BC: Not in B.C.

RANGE WA: High, open ridges of Blue Mountains.

THICK, SHINY EVERGREEN LEAVES TO 2.5 cm (1") LONG

EDGES ROLLED UNDER

HAIRY FLWS. TO 1.3 cm (1/2") ACROSS

SEED WITH SILKY TAIL TO 7.5 cm (3") LONG

INTRODUCED TREES AND SHRUBS

RUSSIAN OLIVE, *Elaeagnus angustifolia*: This is an introduced relative of wolf-willow, *E. commutata*. Both have sage-green leaves, but while the latter is usually under 1.8 m (6') high, Russian olive forms a bushy-leaved tree perhaps 10 m (33') high. Many trees are almost round in shape with lower branches sweeping the ground. The trunk and main limbs have a dark, fissured bark. Leaves are narrow, 10 cm (4") long, with the underside silvery and the upper sage-green. Olives the size of a small peanut and matching the leaves in colour grow singly here and there along the branches. They are considered inedible, but much used by wildlife. It is very noticeable fringing the South Thompson River east of Kamloops, B.C. Some trees are used in landscaping nearby motels. Old trees are a picturesque feature of the southern Okanagan, where they grow close to the lake and on bordering clay banks. In Washington, this tree is used quite extensively in landscape planting in very dry areas. Also found sporadically along roadsides. Quite common on lands bordering the Columbia and Snake rivers. Wenatchee, Sunnyside, Lyon's Ferry.

ENGLISH HOLLY, *Ilex aquifolium*: A cultivated holly tree with its masses of red berries looks quite at home and unmistakable in a garden. But as an ill-shaped, berryless shrub on a coastal island or adjacent mainland, it is less familiar. A foreign visitor prowling the woods could easily assume he had found a native shrub, for holly is abundant and widespread. Only Oregon grape, low and shrubby, has a similar leaf. 'Misplaced' holly can be found at low elevations in the Gulf and San Juan islands.

EUROPEAN MOUNTAIN-ASH, *Sorbus aucuparia:* The most conspicuous mountain-ash because of its height, up to 5 m (16'), and showy plumes of creamy flowers followed by large clusters of red berries. Leaflets usually numbering more than 13 will be definite identification. Meadows, fields and roadsides in western B.C. and Washington.

LOMBARDY POPLAR, *Populus nigra* var. *italica*: The distinctive columnar shape allows identification without detailed checking of leaf shape, flowers and fruit. French landscape paintings feature this poplar as rows of tall, thin, conical trees bordering a road or meadow. In our region, it is a favourite for similar plantings. It grows quickly, up to 30 m (100') high. Not easily confused with any other tree. Common at lower elevations across B.C. and Washington.

BLACK LOCUST, *Robinia pseudo-acacia*: Native to eastern U.S.A. but widely introduced locally because it will grow in inhospitable areas and because the durable wood was once valuable for firewood and fence posts. A stout, erect tree to 25 m (80') high, with fissured, black bark. In winter, the top third is heavily draped with brown masses of pea-pod-like seed cases. In May, clusters of whitish-to-pink blossoms cover the tree. Black locust has spiny twigs. Be sure not to confuse with Oregon ash. Common around farms, silted areas and stream banks. Most common east of the Cascades.

MANITOBA MAPLE or **BOX-ELDER**, *Acer negundo*: A large tree native to Ontario and Manitoba. It differs from other maples in having leaves divided into 3–7 leaflets. The fruit is a pair of long, wrinkled, winged seeds. Valley bottoms of southern B.C. and throughout western and eastern Washington, usually close to habitations.

COMMON HAWTHORN, *Crataegus monogyna*: See p. 90.

EUROPEAN BIRCH, *Betula pendula*: See p. 82.

SPURGE-LAUREL, *Daphne laureola*: A confusing shrub reminiscent of rhododendron with its dark evergreen leaves in rough whorls. Stems are too rubbery to break. Small, yellow-green flowers develop into a cluster of poisonous purple-black berries. Sporadic from backyards to forest openings. Southern Vancouver Island, Gulf and San Juan islands and adjacent mainland. In places, threatening to become a pest.

SHRUBS

The world is so full
of a number of things
I'm sure we should all
be as happy as kings.

—Robert Louis Stevenson

LIST OF SHRUBS

This is not a key! It is a list of shrubs in the order in which they appear in this book. Each is listed in only one category and is thus omitted from other likely classifications. However, this simplified grouping should serve as a general guide. Note that some small shrubs, listed here as such, may appear as flowers to most people (e.g., a few pentemons, a few buckwheats and Dorr's sage). Goatsbeard is with the shrubs while dogbane is with the flowers, as are some of the smaller *Rubuses* and twinflower, for example.

↑ *found only in B.C.*   ↓ *found only in Washington*

WILLOWS

PALATABLE BERRIES
blueberry,
↑ velvet-leaved
bog
dwarf
grouseberry
bilberry, low
blueberry,
oval-leaved
Alaskan
↑ highbush
huckleberry,
red
evergreen
blue-leaved
blue
black
raspberry,
red
black
blackberry,
Himalayan
evergreen
cranberry, highbush
bush-cranberry, Amer.
↓ viburnum, oval-lvd.

UNPALATABLE BERRIES
currant,
stink
northern black
red-flowering
sticky
squaw
↑ skunk
trailing black
↑ red swamp
maple-leaved
↓ golden
gooseberry,
black
white-stemmed
wild
gummy
Idaho
northern

CREEPERS
cranberry, bog
snowberry, creeping
tea-berry, western
wintergreen, alpine
blackberry, trailing
kinnikinnick
↑ bearberry, alpine
squaw carpet
↓ whipplea

SMALL SHRUBS (under 90 cm (3'))
mountain-heather,
white
four-angled
Alaskan
pink
yellow
crowberry
lingonberry
cinquefoil, shrubby
falsebox
azalea, alpine

BOG OR MARSH SPECIES
bog-laurel
western small-lvd.
↑ bog-rosemary
Labrador tea
trapper's tea
northern Labrador tea
birch, scrub
buckthorn, alder-lved.
hardhack
sweet gale
wax-myrtle, Calif.

CLIMBERS
clematis, white
traveller's joy
clematis,
golden
Columbia
sugarbowls
honeysuckle,
western trumpet
red
hairy or purple

SAGEBRUSH- PONDEROSA PINE
sagebrush,
big
threetip
↓ rigid
antelope-brush
greasewood
rabbit-brush,
common
green
horsebrush, grey
sagewort, prairie
↓ hop-sage
↓ saltbrush, fourwing
↓ winter fat
↓ sage, Dorr's
penstemon,
Davidson's
shrubby
Scouler's
buckwheat,
↓ Douglas
↓ thyme-leaved
poison-ivy
poison-oak
sumac, smooth
soopolallie
buffalo-berry, thorny
↑ wolf-willow
juniper,
common
creeping

WIDESPREAD SPECIES
elderberry,
blue
red
black
rose,
Nootka
baldhip
clustered
prickly
prairie
dog
Oregon-grape,
creeping
dull
tall

WIDESPREAD SPECIES (cont.)
devil's club
thimbleberry
hazelnut,
beaked
California
Saskatoon
mock-orange
oceanspray
goatsbeard
dogwood, red-osier
snowberry,
common
western
trailing
azalea, false
twinberry, black
honeysuckle, Utah
ninebark,
Pacific
mallow
↓ deerbrush
snowbrush
ceanothus, redstem
spirea,
birch-leaved
pyramid

COASTAL SPECIES
broom, scotch
gorse
lupine, tree
manzanita,
hairy
green-leaf
salal
salmonberry
Indian-plum
↓ silk-tassel
wahoo, western

MOUNTAIN SPECIES
copperbush
rhododendron,
Pacific
white-flowered
↑ rosebay, Lapland
spirea, subalpine
mountain-ash,
Sitka
western

Willow Family, *Salicaceae*

WILLOWS
Salix spp.

Willows are one of the most familiar and widespread groups of shrubs in our area. Although most grow along creeks or rivers, certain species are found high on mountain slopes, where they form a shrubby mat only a few centimeters (inches) tall. All willows like sunlight and prefer open spaces.

Most willow leaves are long and graceful, with smooth or slightly toothed edges. In spring and early summer, each leaf stem has 2 shiny false leaves growing at its base. Winter buds have a single, hood-like scale. 'Pussy willows,' the white, fluffy catkins common to most willows, are very noticeable in spring. Willow bark is exceptionally bitter.

With more than 50 willows in our area, only a trained botanist can cope with their identification, because many species flower before their leaves appear, male and female flowers are on different shrubs, leaf and twig characteristics often vary greatly with age and hybridization is common. 2 willows reaching tree size are peachleaf willow and Pacific willow. They are described under 'Trees' (see p. 81). 4 others listed below vary from shrubs to small trees. Here is a selection of the more common and recognizable willows:

ALPINE WILLOWS
• to 15 cm (6") tall; shrubby, mat-like

CASCADE W., *S. cascadensis*
Leaves 2.5–5 cm (1–2"), pea-green. Coast to Cascades.

NETTED W., *S. reticulata*
Leaves 0.3–1.3 cm (⅛–½"), silvery beneath. Widespread.

ARCTIC W., *S. arctica*
A dwarf shrub, often prostrate. Leaves 1.3–4 cm (½–1 ½") long, sharp-pointed, tuft of hairs at tip; stout, brown stems. Subalpine and alpine areas throughout B.C. and into north Cascades of Washington. Wallowa Mountains, Oregon.

SHRUB TO TREE WILLOWS
• leaves wide, roundish

SCOULER'S W., *S. scouleriana*
Leaves 5–10 cm (2–4") and a third this wide, rounded or broad tip. Smooth on both sides. Widespread. First to bloom.

SITKA W., *S. sitchensis*
Leaves 5–10 cm (2–4") long and a third this wide, rounded tip. Velvety with fine hairs beneath. Uncommon except on coast.

HOOKER'S W., *S. hookeriana*
Leaves 5–15 cm (2–6") long and half as wide; dull-pointed. Woolly hairs beneath. Wet or dry land. Southern Vancouver Island and southwards along coast. Puget Sound.

BEBB'S W., *S. bebbiana*
Leaves to 5 cm (2") long and half as wide, round-pointed. Common across B.C. east of Cascades. Eastern border of Washington.

'SANDBAR' WILLOWS
• to 4.5 m (15') tall; slender limbs
• narrow leaves

SOFT-LEAVED W., *S. sessilifolia*
Leaves 0.6–2 cm (¼–¾"), glossy, pointed both ends, hairy beneath. Rare in B.C.'s Fraser Valley, but in damp areas southwards along coast, into Columbia Gorge.

COYOTE or **SANDBAR W.**, *S. exigua*
Leaves 5–10 cm (2–4") long, 6–10 mm (¼–⅜") wide. Silvery green. A pioneer on gravel bars, east of Cascades in southeastern, south-central and northern B.C. Same habitat in Washington, also east of Cascades.

MISCELLANEOUS WILLOWS

BARCLAY'S W., *S. barclayi*
Variable leaves 5–10 cm (2–4") long, ovalish, sharp-pointed, hairy above, bloom beneath. Common willow across B.C., low elevations–subalpine. Ranging southwards across Washington to Mt. Adams.

DRUMMOND'S W., *S. drummondiana*
Leaves 2.5–7.5 cm (1–3") long, ovalish, white velvety beneath with mat of hairs. Common in B.C. in moist areas east of Cascades. Middle-mountain to subalpine. In Washington from the Palouse to Spokane.

BOG W., *S. pedicellaris*
Low shrub to 1.5 m (5'). Leaves to 5 cm (2"), edges rolled, short stems. Bogs and wet places east of Cascades in B.C. Mostly in and west of Cascades in Washington.

MACKENZIE'S W., *S. prolixa*
Shrub or small tree. Leaves 4–9 cm (1 ½–3 ½") long, sharp-pointed, rounded at base, 'powdery' beneath. Stems hairy. Low to high elevations. Widespread in B.C.

WILLOWS

ALPINE ZONE WILLOWS

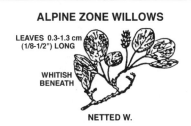

LEAVES 0.3-1.3 cm
(1/8-1/2") LONG

WHITISH
BENEATH

NETTED W.

'SANDBAR' WILLOWS

6–20 mm
(1/4-3/4")

5-10 cm
(2-4")

SOFT-LEAVED   COYOTE

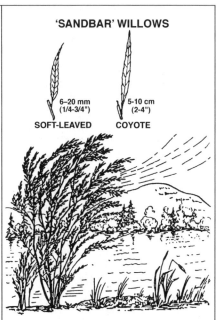

SHRUB TO TREE WILLOWS

5-10 cm
(2-4")

SCOULER'S W.

5-10 cm
(2-4")

SITKA W.

5-15 cm
(2-6")

HOOKER'S W.

2.5-5 cm
(1-2")

BEBB'S W.

MISCELLANEOUS WILLOWS

SILKY HAIR

WAXY GREEN
LEAVES

5-10 cm
(2-4")

2.5-7.5 cm
(1-3")

BARCLAY'S W.   DRUMMOND'S W.

EDGES SLIGHTLY
ROLLED

2.5-5 cm
(1-2")

BOG W.

4-9 cm
(1 1/2-3 1/2")

MACKENZIE'S W.

Heather Family, *Ericaceae*
BLUEBERRIES, HUCKLEBERRIES AND WHORTLEBERRIES
Vaccinium spp.

Blueberry, huckleberry, whortleberry, bilberry and cranberry are all used as common names in the genus *Vaccinium*, thus providing a great opportunity for confusion.

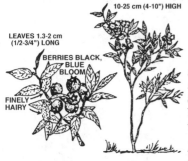

10-25 cm (4-10") HIGH

LEAVES 1.3-2 cm (1/2-3/4") LONG

BERRIES BLACK, BLUE BLOOM

FINELY HAIRY

VELVET-LEAVED BLUEBERRY [001]
Vaccinium myrtilloides
Canada Blueberry

Plants tend to bunch together; open form when alone. Leaves sharp-pointed, no teeth. Young stems and leaves soft-hairy. White-to-pinkish 'bell' flowers. Sweet black-to-blue berries; heavy, lighter blue bloom; ripe by August. Abundant in north, where they carpet vast areas of forest floor.

QUICK CHECK: Range. Less than 30 cm (12") tall. Flowers or berries in clusters.

RANGE BC: Common throughout southeastern, central and northeastern B.C.
RANGE WA: Not known in Washington.

BOG BLUEBERRY or BOG BILBERRY [002], *V. uliginosum*: To 60 cm (2') high, but dwarfed in subalpine and alpine. Leaves alternate, leathery, broadest above middle, rolled edges; veins prominent on underside. Sepals stay with mature fruit. Flowers on long stem, petals often reflexed. Blue fruit, waxy bloom; edible. Muskeg, peatmoss bogs at low elevations. Beautiful purple-red in fall. Wide-ranging: coast and mountains of far north.

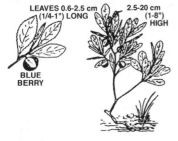

LEAVES 0.6-2.5 cm (1/4-1") LONG

2.5-20 cm (1-8") HIGH

BLUE BERRY

DWARF BLUEBERRY
Vaccinium caespitosum
Dwarf Huckleberry

Sometimes tiny. Many thin, round twigs; heavy mass of finely toothed leaves. Vein network on underside. Pinkish flowers narrowly urn-shaped; blue berries, pale grey bloom. Abundant but solitary and smallish; prized for sweet flavour.

QUICK CHECK: Usually 10–20 cm (4–8") tall. Bushy. Leaves 0.6–2.5 cm (¼–1") long, widest above centre; green both sides. Single flowers, berries.

RANGE BC: Widespread; higher elevations. Common on edges of swamps, in mountain meadows. Also rocky ridges at high elevation. Alaska Highway and southwards throughout.

RANGE WA: Throughout, in habitats as for B.C.

7.5-15 cm (3-6") HIGH

LEAVES 0.6-1.3 cm (1/4-1/2") LONG

RED BERRIES

GREEN, ANGLED TWIGS

GROUSEBERRY [003]
Vaccinium scoparium
Grouse Huckleberry

Small! May form extensive lacy green carpet on dry, gravelly forest soil. Sour odour. Generally green or yellowish in fall; changes to sombre red-purple in harsh habitat. Usually 1050 m (3500') and above.

QUICK CHECK: Bushy. Angled green twigs; oval leaves to 1.3 cm (½") long. Small, red berries.

RANGE BC: Mountains east of Cascades. Very widespread south of Prince George. Manning Park.

RANGE WA: Mountains east of Cascades. Blue and Wenatchee mountains. Chinook Pass.

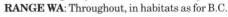

LOW BILBERRY, *V. myrtillus*: To 25 cm (10") tall; twiggy ground cover. Younger branches angled, greenish yellow. Leaves 1.3–2.5 cm (½–1") long, toothed, prominent mid-rib on lower surface. Flowers pink, single. Berry deep red to blue-black. Southeastern B.C. and east of Washington's Cascades.

OVAL-LEAVED BLUEBERRY [005 007]
Vaccinium ovalifolium
**Tall Huckleberry, Oval-Leaf Whortleberry,
Tall Blue Huckleberry**

Tallest huckleberry in area, commonly to 1.8 m (6') high under ideal conditions. Usually in shade, scraggly form. Single pinkish 'bell' blossoms from leaf axils, usually before leaves. Berries ripe by July; welcomed by coastal peoples. Cool, shady forests.

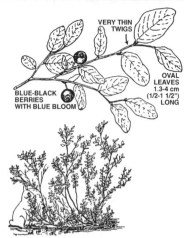

QUICK CHECK: Egg-shaped leaf with smooth edges. Large berries with bluish bloom. Style does not extend beyond flower tube.

RANGE BC: Widespread along coast; sea level to subalpine. Also as far north as Prince George in and near interior cedar-hemlock ecosystem.

RANGE WA: Widespread west of Cascades. Most abundant in mountain forest; sea level to subalpine.

ALASKAN BLUEBERRY [004]
Vaccinium alaskaense

Fine eating quality; northern Oregon to Alaska, but west of Cascades only. Height, to 1.2 m (4'), and oval-shaped leaves are similar to oval-leaved (above). Alaskan may show scattered glandular hairs along midrib of leaf; bend a leaf in half lengthwise to check.

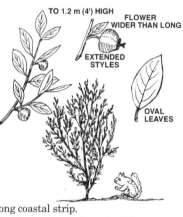

With luck, found with flowers—generally appearing with emerging leaves. Pinkish 'bell' flowers. Berries dark bluish black, without bloom. Fruit stalk enlarged at base of berry.

QUICK CHECK: Single flowers, broader than long, appear as leaves form. Style extends beyond flower. Enlarged fruit stalk.

RANGE BC: West of Cascades. Fairly common along coastal strip.

RANGE WA: West of Cascades. Coastal strip of Washington to northwestern Oregon.

HIGHBUSH BLUEBERRY, *V. corymbosum*: Introduced cultivated species; limited range in boggy areas of southwestern B.C.

RED HUCKLEBERRY
Vaccinium parvifolium

Unmistakable lacy, bright green bush; edible, shiny red berries. Sometimes 1.8 m (6') in height; open, upright form with mass of small, oval leaves each less than 2.5 cm (1") long; on rare occasions toothed. Some may remain over winter. Twigs as green as leaves, sharply angled. Sometimes in fairly shady places; more often at edges of forest openings and along roadsides. Old stumps a favourite perch. Cool coastal forests.

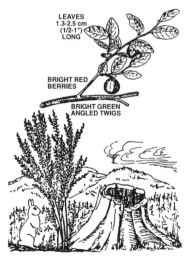

QUICK CHECK: Note range. Angled, green twigs and tart, red berries.

RANGE BC: Below 300 m (1000'). Common on Vancouver Island, Gulf Islands; northwards to Queen Charlottes and Alaska. Spotty occurrences in interior. Sicamous to Revelstoke. Kootenay Lake.

RANGE WA: Most common below 450 m (1500'). Olympics.

FLOWERS PINK-WHITE

THICK, WAXY GREEN LEAVES, 1.3-2.5 cm (1/2-1") LONG

HAIRY, REDDISH TWIGS

EVERGREEN HUCKLEBERRY [006]
Vaccinium ovatum

Mostly bushy and dense, often 1.8 m (6') high. Fine-toothed, leathery leaves; mostly under 2.5 cm (1") long; very short stems. Many small flowers; groups along or near branch ends. Abundant berries. Sweet flavour—very desirable for pies, other culinary uses. *Shot-oolalie* or 'shotberry' to aboriginals because of berry size and shape. Rocky, gravelly soils.

QUICK CHECK: Limited range. Small evergreen leaves; toothed, egg-shaped, sharp-pointed. Woolly twigs. Clusters of pink 'bell' flowers during May; shiny black berries ripen late summer—may remain to end of November.

RANGE BC: Scattered on Gulf Islands, more plentiful in south, and southeastern and southwestern coast of Vancouver Island. More abundant and vigorous towards Pacific coast. Sooke, Alberni. Occasionally on mainland. Pemberton.

RANGE WA: Scattered throughout San Juans and coast forest ecosystem. Olympia, Tacoma and low elevations of Olympics.

BLUE-LEAVED HUCKLEBERRY, *V. deliciosum*: Small, bushy; to 30 cm (12") high. Leaves 0.6–2.5 cm (¼–1") long, rounded tips, whitish bloom beneath. Berries single, deep blue; bloom. Excellent eating. B.C.: near timberline; Coast Mountains, Forbidden Plateau, Mt. Arrowsmith, Garibaldi Park. Washington: subalpine meadows; Olympics, Cascades.

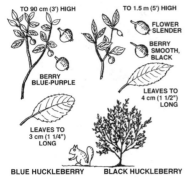

TO 90 cm (3') HIGH

TO 1.5 m (5') HIGH

FLOWER SLENDER

BERRY SMOOTH, BLACK

BERRY BLUE-PURPLE

LEAVES TO 4 cm (1 1/2") LONG

LEAVES TO 3 cm (1 1/4") LONG

BLUE HUCKLEBERRY   BLACK HUCKLEBERRY

BLUE HUCKLEBERRY
Vaccinium globulare
Globe Huckleberry

Identifying *Vaccinium*s can be puzzling unless you examine flower parts in detail. However, both blue huckleberry and low bilberry (p. 108) are most common from southeastern B.C. through eastern Washington to Oregon. Recognize as huckleberry by small leaves, 'bell' flowers and characteristic berry.

Much branched; oval leaves to 3 cm (1 ¼") long, finely toothed, undersides lighter green. 'Globe' flower not unique. Berries blue-purple, very tasty.

QUICK CHECK: Range important. Flowers and berries borne singly in leaf axils.

RANGE BC: Extreme southeastern B.C.

RANGE WA: Eastern Washington to Rockies.

BLACK HUCKLEBERRY [010]
Vaccinium membranaceum
Black Blueberry, Black Mountain Huckleberry

Mountain shrub. Elevations over 750 m (2500') on coast, over 1200 m (4000') in interior; often the most common shrub. All elevations in central/northern interior of B.C.

At times with pink mountain-heather, oval-leaved blueberry (different leaves) or dwarf form of grouseberry (p. 108). May be 15 cm (6") tall near timberline. Often an extensive low, shrubby growth prominent late summer–early fall, as a beautiful red-purple carpet. Old branches have shredding bark. Crooked plant with slightly angled twigs. Sometimes a few twisting stems; sparse form, but often a bushy shrub. Leaves 1.3–4 cm (½–1 ½") long, pointed both ends, with fine teeth. Small, whitish flowers appear with or after leaves—May or June. Reddish black berries without bloom ripe by September; popular with berry pickers.

QUICK CHECK: Thin leaves, fine teeth. Smooth, black berries. Note elevation ranges.

RANGE BC: Common throughout; mountain slopes; most abundant on burnt land.

RANGE WA: Habitats as for B.C.

Rose Family, *Rosaceae*
BLACK RASPBERRY [008]
Rubus leucodermis
Black Cap

Note: Raspberries come off a central core while blackberries are picked with the core.

Not all raspberries are red—this flavourful fruit is black when fully ripe. Most of the summer, its 'raspberry' look will serve for rough identification. Many of the shoots are quite straight and may reach 1.5 m (5') in height. Other stems bend over and droop to the ground. Stout, flattened thorns, with their points recurved, bristle along the branches and even up onto the leaf stems. The sharp-pointed, sharply toothed, crinkly leaflets are in 3s or 5s; their contrasting silvery undersurfaces provide a fairly reliable identity check. Small, dense clusters of 5-petalled white flowers bloom from April to June. The bristly fruit looks like a common raspberry until it turns almost black later in the summer.

QUICK CHECK: A mostly upright shrub, thorny, 3–5 leaflets with very silvery undersurface. Flower petals are shorter than the hairy, reflexed (folded back) sepals. Blackish fruit with light bloom when ripe.

RANGE BC: West of the Selkirk Mountains, to elevations of 600 m (2000') at the coast and 1200 m (4000') in the interior.

RANGE WA: Wide distribution throughout Cascades and Blue Mountains. Sagebrush to mountain forest ecosystem. Seattle, Snake River.

HIMALAYAN BLACKBERRY, *R. discolor*: Himalayan and evergreen blackberries, though with fruit more like trailing blackberry (p. 118), are more like these raspberries in habit, but more robust and invasive. A foreign introduction and now a dominant part of the vegetation along road edges and in waste places west of the Cascades. Thick, arching branches to 3 m (10') in length are armoured with heavy recurved spines. Great masses of pale-pink-to-white flowers appear in June, followed by abundant crops of tasty blackberries in late summer.

EVERGREEN BLACKBERRY, *R. laciniatus*: Another introduced plant almost as sturdy as the one above. Its evergreen leaflets are so dissected as to appear fringed. Berries are black and tasty; firmer than on the Himalayan. Found west of the Cascades along roadsides and waste places. In western Washington, dense thickets are a common sight along roadsides.

Rose Family, *Rosaceae*
RED RASPBERRY
Rubus idaeus

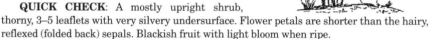

A red-fruited raspberry otherwise quite similar in form and leaf to black raspberry. However, a range primarily east of the Cascades and a preference for very dry places, such as rockslides, help identify red raspberry. Many plants have a very noticeable blue bloom over the brownish stalks. The leaves are hairy beneath and do not have the high silvery sheen of black raspberry.

QUICK CHECK: Calyx folded back. Blue bloom on many stalks. Thorns not flattened or hooked. Fruit red. Watch range limitations.

RANGE BC: Widely distributed east of Cascades and sporadic to west. Northern B.C.

RANGE WA: Widely distributed east of Cascade summit in bunchgrass and ponderosa pine ecosystems.

VARIABLE LEAVES

DARK RED BERRIES

Honeysuckle Family, *Caprifoliaceae*
HIGHBUSH-CRANBERRY [009]
Viburnum edule
Squashberry

This plant usually attracts attention through its bunches of bright red berries or when its leaves turn crimson in fall, accurately marking its position in the floral picture. Perhaps its straggling growth and relatively low height of 0.6–1.8 m (2–6') keep it from otherwise being noticed. Once identified in its damp, shady habitat, it will be recognized throughout much of B.C. and Washington. At higher elevations it grows in more open situations, frequently among rocks.

The long-stemmed, crinkled leaves are sharply toothed and grow in opposing pairs. Most are 3-lobed but young ones are often variable in shape. Fine hairs and prominent ribs are found on the underside. The bark is reddish. Small, white flowers about a centimetre (half an inch) or so across form a cluster. The 2 to 5 yellow berries ripen to a dark red. They are bitter tasting and have a large, flat seed.

QUICK CHECK: Rounded, 3-lobed leaves, opposite, usually hairy beneath. Flower cluster approximately 2.5 cm (1") across. Several smooth, red berries. Ninebarks have somewhat similar leaves but they are alternate.

RANGE BC: Damp, shady locations at lower levels, more exposed at higher elevations. Throughout B.C. except for sagebrush and subalpine ecosystems. Common in northern B.C., extending into Yukon and Alaska.

RANGE WA: Damp, shady locations in mountain forest ecosystem. More exposed at higher elevations in the subalpine ecosystem. Drops to low coastal elevations on western side of Olympics. Lake Chelan, Stevens Pass, Mt. Adams, Simcoe Mountains.

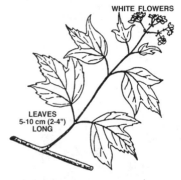

WHITE FLOWERS

LEAVES 5-10 cm (2-4") LONG

Honeysuckle Family, *Caprifoliaceae*
AMERICAN BUSH-CRANBERRY
Viburnum opulus
High-bush Cranberry

May be fairly abundant in certain limited localities, but otherwise so scarce as to be missed by most people. General form closely resembles *V. edule*, but leaves more deeply 3-lobed, hairy beneath; flowers creamy white and very showy, in broad clusters. Central flowers small; outer, sterile ones large and showy. Fruit is bright red and very tart. When cooked, it closely resembles cranberry sauce in flavour. The seeds, like high-bush cranberry, are particularly large. Shady, brushy thickets along streams are preferred habitat.

QUICK CHECK: Leaves strongly 3-lobed. Unusual design of white flower clusters 5–10 cm (2–4") across.

RANGE BC: Edges of interior cedar-hemlock ecosystem, but only as far north as Prince George. Quesnel, Salmon Arm, Sicamous, Cranbrook.

RANGE WA: Moist woods southwards to Columbia Gorge.

OVAL-LEAF VIBURNUM

OVAL-LEAF VIBURNUM, *Viburnum ellipticum*: To 3 m (10') high with a white flower cluster. Opposite, sharply toothed leaves shaped like miniature badminton paddles are sure identification. West of Cascades at lower elevations in southern Washington. Columbia Gorge. Not found in B.C.

Currant and Gooseberry Family, *Grossulariaceae*
CURRANTS AND GOOSEBERRIES
Ribes spp.

Although classified as 'unpalatable,' some of these berries are used for various purposes. Currants and gooseberries have a close resemblance in leaf and fruit to the cultivated varieties. The leaves are 3–5 lobed, usually toothed and have prominent veins radiating from the base to all the lobes. The pungent smell of the crushed leaves is a help if in doubt. **Note that currants are unarmed, while gooseberries carry spines.**

STINK CURRANT [011]
Ribes bracteosum
Blue Currant

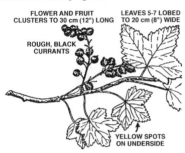

FLOWER AND FRUIT CLUSTERS TO 30 cm (12") LONG   LEAVES 5-7 LOBED TO 20 cm (8") WIDE

ROUGH, BLACK CURRANTS

YELLOW SPOTS ON UNDERSIDE

A more or less erect shrub to 2.5 m (8') high with a few sprawling branches. The large 'currant' leaves up to 20 cm (8") in width are wider than long, with 5–7 sharp-tipped lobes. Yellow resin ducts are scattered on the undersurface. The erect, green-to-pink, saucer-shaped flowers and rough, black fruit stand in long, loose clusters of 2 or 3 dozen; these clusters are quite different from the less bountiful ones of other *Ribes*. The smell of the leaves is very pronounced and accounts for the common name. While the berries are not generally popular, they are used by some northwest coast peoples.

QUICK CHECK: Large, blue-green leaves, 5–7 lobed. Flowers and fruit in 15–30 cm (6–12") long cluster. Strong currant smell to leaves.

RANGE BC: Usually confined to low elevations in coast forest ecosystem and on Vancouver Island.

RANGE WA: Most common at low elevations in coast forest ecosystem, but extending up mountain slopes to near subalpine elevations.

NORTHERN BLACK CURRANT
Ribes hudsonianum
Hudson Bay Currant

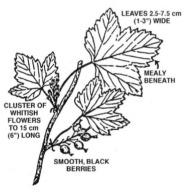

LEAVES 2.5-7.5 cm (1-3") WIDE

MEALY BENEATH

CLUSTER OF WHITISH FLOWERS TO 15 cm (6") LONG

SMOOTH, BLACK BERRIES

This shrub resembles stink currant in leaf shape and in the clusters of flowers and berries but it differs in that leaves are seldom over 7.5 cm (3") wide. Flowers and fruit, 8–16 in number, form clusters 4–15 cm (1 ½–6") long.

This shrub often has sturdy shoots reaching 1.5 m (5') into the air. It may catch the eye in the spring, when numerous erect clusters of white flowers show in advance of the full leaf growth. The flower and berry cluster is only a few centimeters or so long but from this main stalk 8–16 short stems branch. The leaves are slightly broader than long. There are 3 definite rounded lobes, usually with a minor division into 5. Although smooth-textured above, they are mealy beneath with yellow dots. The berry is rather small, smooth, black and of poor flavour.

QUICK CHECK: Upright flower and berry stalk. 8–16 whitish flowers with long petals and hairy sepals. Smooth black berries.

RANGE BC: Generally east of the Coast Mountains and along shady watercourses.

RANGE WA: Southwards from south-central B.C. into Okanogan County.

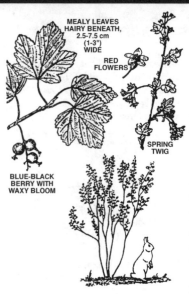

MEALY LEAVES
HAIRY BENEATH,
2.5-7.5 cm
(1-3")
WIDE

RED
FLOWERS

SPRING
TWIG

BLUE-BLACK
BERRY WITH
WAXY BLOOM

RED-FLOWERING CURRANT [013]
Ribes sanguineum
Winter Currant, Red Flower Currant

To persons living at the coast, this currant surely will be the one best known. True enough, it has only a few weeks of glory, but it is at a time in early spring when colour is much appreciated. Small, red flowers in drooping clusters attract the first migrant hummingbirds about April 1, but blooms may be seen into May.

Red-flowering currant consists of several crooked stems forming a loose bush. Cultivated shrubs reach 3 m (10') in height, but in the woods most are not over 1.5 m (5'). The mealy, dull green leaves are 2.5–7.5 cm (1–3") across and matted on the undersurface with almost invisible hair. Flowers, usually a dozen or more, are in erect or drooping clusters. The fruit is a globular, blue-black berry with a whitish waxy bloom. It has an unpleasant taste.

David Douglas, enthusiastic 'old-country' botanist, introduced this plant to European botanical gardens.

QUICK CHECK: Roundish, 3–5 lobed currant leaves with fine, white hairs beneath. Red flower clusters.

RANGE BC: Low elevations in southern part of coastal forest and Gulf Islands regions. Dry, open woods. Logged areas or roadsides are favoured areas. Sporadic occurrences at Arrow Lake and Slocan Lake. Along Fraser Canyon to near Lytton.

RANGE WA: Low elevations in coast forest ecosystem. San Juan Islands, Seattle, Tacoma, Silverton.

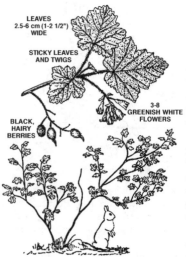

LEAVES
2.5-6 cm (1-2 1/2")
WIDE

STICKY LEAVES
AND TWIGS

3-8
GREENISH WHITE
FLOWERS

BLACK,
HAIRY
BERRIES

STICKY CURRANT [012]
Ribes viscosissimum

Most bushes are 0.6–1.2 m (2–4') high, the few stems twisted and stout at their bases. They are thornless but with rather shreddy bark on older branches. Sticky currant grows in semi-open forests on mountains east of the Cascades. Although individual shrubs are well separated, they are often distributed across extensive mountain slopes.

Sticky pores and hairs on the leaves, twigs and fruit are the most noticeable feature of this currant. The greenish white flowers with pinkish tinge are in clusters of 3 to 8. The black berries, which are covered with short, stiff, sticky hairs, are very few in number. Note the unusual odour of this plant, especially on sunny days.

According to David Douglas, an early explorer, 2 or 3 berries will cause vomiting.

QUICK CHECK: 'Currant' leaves. Twigs, leaves and fruit sticky to the touch.

RANGE BC: Semi-open mountain slopes from 500 m (1700') to 1800 m (6000') from Cascades across southern B.C. to Rocky Mountains. Manning Park, Hedley Mountains.

RANGE WA: Semi-open mountain slopes from 540 m (1800') to 1800 m (6000') from Cascades eastwards. Ponderosa pine and mountain forest ecosystems.

SQUAW CURRANT [015]

Ribes cereum

Wax Currant

Only on drier slopes of sagebrush and ponderosa pine country. Here, along the foot of rocky slopes or on almost barren benches, squaw currant can be recognized by its compact, rounded outline to 1.2 m (4') high and small, drab olive-green leaves. By August, most shrubs are bright with hundreds of small, red currants.

Crooked, upright stems are brushy towards top; bark is greyish with reddish tones and dull white markings. Leaves vary: dime-sized to quarter-sized.

BRIGHT RED	DARK RED, GLANDULAR	PURPLISH BLACK BLOOM	SMOOTH, RED
SQUAW	SKUNK	TRAILING	RED SWAMP

QUICK CHECK: Bushy shrub with small, olive-green, paddle-shaped leaves. Smooth, red currants in August and September. Grows with common rabbit-brush and cactus.

RANGE BC: Wide range in south-central area. Most abundant in Similkameen and southern Okanagan valleys. Scattered from Kamloops to Lytton.

RANGE WA: Sagebrush and bunchgrass ecosystems. Tonasket, Republic, Yakima and Chelan counties. Blue Mountains.

SKUNK CURRANT, *R. glandulosum*: Sprawling, spineless shrub to 1 m (40") high. Leaves give off strong odour when crushed. 5–7 deeply cut, toothed lobes give the leaves a 'maple' look. Lobes sharp-pointed, as are those on trailing black currant, below. Flowers greenish white to purple; berries dark red and bristly. Widespread northwards from Williams Lake, B.C., in open valley forests to subalpine heights. Not known to occur in Washington.

TRAILING BLACK CURRANT [017], *R. laxiflorum*: Somewhat like skunk currant (leaves similar), but usually under 90 cm (3') tall. Tends to be trailing; sometimes a climber in Washington. Flowers reddish to purplish, inedible berries purplish black with bloom. Widely distributed along coast from Alaska to California. Locally common in Rockies.

RED SWAMP CURRANT, *R. triste*: Spineless. If erect, usually to 90 cm (3') tall. Typically broad leaves; 3 rounded, toothed main lobes—sometimes 5. Drooping clusters of 6 to 15 red-to-purplish flowers. Berries bright red; edibility a matter of personal taste. Moist to wet ground. Low–middle elevations in northern half of B.C. Also in Cascades to northern Oregon.

MAPLE-LEAVED CURRANT, *R. howellii.*: To 90 cm (3') tall. Typical currant leaves, 3–5-lobed 'maple' shape, but slightly longer than wide. Plant finely hairy. Red petals, recurved. Berry globe-shaped, smooth, bluish black. Wet places in subalpine ecosystem in southwestern B.C. and Washington in the Cascades and Olympics.

GOLDEN CURRANT [016]

Ribes aureum

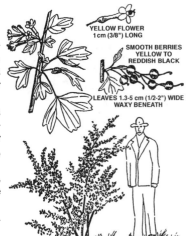

YELLOW FLOWER 1 cm (3/8") LONG

SMOOTH BERRIES YELLOW TO REDDISH BLACK

LEAVES 1.3-5 cm (1/2-2") WIDE WAXY BENEATH

Eye-catching during April and May, when clusters of bright yellow flowers adorn slender branches. Fortunately for the amateur botanist, yellowish green, 3-lobed leaves are unlike those of any other native shrub.

Majority are 1.2–1.8 m (4–6') tall. Commonly in lowland thickets and damp places, some in rocky surroundings. Yellow flowers, 5 petals, may have red or purple tinge. Slightly less than 1.3 cm (½") long, clusters of 6 to 15; much favoured by insects.

Smooth berry, 6 mm (¼") in diameter, with long, chaffy protuberance. Berries usually yellow, can be red to almost black; considered edible.

QUICK CHECK: Distinctive 3-lobed leaves. Yellow flowers and berries.

RANGE BC: Not in B.C.

RANGE WA: Relatively common in brushy, damp places or on rocky sidehills in bunchgrass ecosystem. Wenatchee, Ellensburg, Yakima, Spokane, Almota.

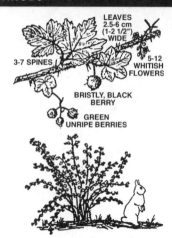

LEAVES 2.5-6 cm (1-2 1/2") WIDE

3-7 SPINES

5-12 WHITISH FLOWERS

BRISTLY, BLACK BERRY

GREEN UNRIPE BERRIES

BLACK GOOSEBERRY
Ribes lacustre
Black Swamp Gooseberry

Very plentiful in damp, shady places from sea level to the higher mountains, where it may form an extensive, low thicket of weakly upright, spiny stems about a metre (several feet) high. Stems are richly endowed with many small, golden spines. Young stems are bristly with 3–7 heavier spines at the nodes. Older branches are almost smooth. Maple-like, alternate leaves are deeply cut into 5 lobes. The saucer-shaped flowers, tinged with maroon and red, are about 6 mm (¼") across and hang 5–12 per cluster. The purple-black berry is bristly and insipid.

QUICK CHECK: Young stems bristly. Flowers and berries as described.

RANGE BC: Sea level to mountain slopes to 1800 m (6000') throughout, and on interior plateau.

RANGE WA: Mountain slopes from coast to 1800 m (6000') throughout Washington. Port Ludlow, Olympics, Wenatchee, Mt. Adams and Blue Mountains.

WHITE-STEMMED GOOSEBERRY, *R. inerme*: The name of this 0.9–1.5 m (3–5') high shrub comes from its light grey bark (reddish beneath). Clusters of 2–4 greenish purplish flowers produce smooth, reddish purple berries. This gooseberry ranges from moist areas in valley bottoms to open high mountain areas. Sporadic in southern B.C. and southwards throughout Washington east of the Cascades.

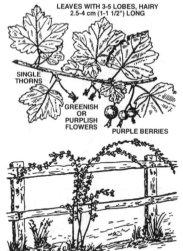

LEAVES WITH 3-5 LOBES, HAIRY 2.5-4 cm (1-1 1/2") LONG

SINGLE THORNS

GREENISH OR PURPLISH FLOWERS

PURPLE BERRIES

WILD GOOSEBERRY
Ribes divaricatum
Common Gooseberry

Sends up several sturdy 'whip' limbs that may reach 1.8 m (6') in the air before bending over. Most of the stem is unarmed but at every joint there is 1 spine (occasionally 2 or 3). Flowers, 4 or fewer per cluster, are greenish or purplish; 5 petal-like lobes hang down. Smooth berry is wine-coloured.

QUICK CHECK: Single thorn at joints. Long, conspicuous flowers with reflexed petal-like lobes.

RANGE BC: Coast forest and Gulf Island ecosystems below 600 m (2000') elevation.

RANGE WA: Coast forest generally below 600 m (2000') elevation. Hoquiam, Puget Sound, Seattle, Port Ludlow.

GUMMY GOOSEBERRY [014], *R. lobbii*: Noticeable in spring because of handsome red-petalled flowers, 4 or fewer flowers per cluster. Spiny stems to 1.2 m (4') high; 3 large spines at nodes. Deeply 3-lobed leaf to 2.5 cm (1") across. Leaf and stem sticky. Berry large, hairy and purple. Shady forests of southern Vancouver Island and Lower Mainland. Coast forest area in Washington and along eastern side of Cascades.

IDAHO GOOSEBERRY or **MOUNTAIN GOOSEBERRY**, *R. irriguum*: Resembles *R. divaricatum* (above) except that it has fine thorns all along its branches, with 1–3 large ones at joints. 1–3 greenish white flowers. Berry smooth, purple to black. In southeastern B.C., along mountain streams; in Washington, along streams in bunchgrass and ponderosa pine ecosystems. Spokane, Blue Mountains.

NORTHERN GOOSEBERRY or **SMOOTH GOOSEBERRY**, *R. oxyacanthoides*: Low shrub. Stems lightly armed, several stout thorns at nodes. Leaves seldom over 2.5 cm (1") across, 3–5 lobed, blunt teeth. Small, greenish flowers in pairs. Paired berries smooth, blue-black. Along streams east of Cascades from southern B.C. northwards, but not in Washington.

Heather Family, *Ericaceae*
BOG CRANBERRY [018]

Oxycoccus oxycoccos
Vaccinium oxycoccos

WHITE-TO-RED BERRY 6 mm (1/4")

EVERGREEN LEAVES ± 6 mm (1/4") LONG

THIN VINE TO 1.2 m (4') LONG

Although it belongs to the blueberry family, this cranberry bears little resemblance to the other members. It is found in peaty, mossy bogs—a very thin vine perhaps 1.2 m (4') long, almost hidden in the moss. Tiny evergreen leaves branch out from this slender stem. A hunt is often needed to find one of the pink-to-red flowers with its protruding stamens. The berry changes colour from white to red as it ripens.

Labrador tea and swamp laurel often hide this dainty cranberry from view.

Cranberries were much sought after by the first Europeans in America. They called them 'crane berries' because of a fanciful resemblance of the flower to the head and neck of a crane. Long before, coastal peoples prized them for food.

QUICK CHECK: Very slender vine growing in sphagnum bogs. Alternate evergreen leaves, sharp-pointed and almost stemless. Flowers pink and berries white or red.

RANGE BC: Probably quite common in low coastal forest sphagnum bogs. Westwards from km 12 (7 ½ miles east of Hope) on Hope–Princeton Highway. In the north, fairly common along the Alaska Highway and wherever there are sphagnum bogs.

RANGE WA: Boggy situations in the coast forest ecosystem. Vicinity of Seattle and Tacoma.

Heather Family, *Ericaceae*
CREEPING-SNOWBERRY

Gaultheria hispidula

THICK LEAVES 1.5-6 mm (1/16-1/4") LONG BRISTLY BENEATH

WHITE BERRY WITH BLACK BRISTLES

REDDISH HAIR ON GREENISH TWIGS

RUNNERS TO 20 cm (8") LONG

A tiny, creeping evergreen shrub limited in both range and habitat. What it lacks in size is made up in daintiness, for its numerous slender runners are only a few centimeters (an inch or so) long. A person's hat might cover it completely. Usually found in a swampy or damp location perched on some hummock or rotting log. Superficially it resembles twinflower or bog cranberry, but the stem, leaves and fruit are quite different. Small, pinkish flowers are usually borne singly in the leaf axils.

The fine, greenish stems are covered with short, reddish hairs, while the thick, leathery leaves are alternate, less than 6 mm (¼") long, and dotted on the underside with small, black hairs. The white berries, not to be outdone, bristle with similar short spines. Few in number, they are usually hidden from view by the mat of leafy runners. While not poisonous, they are of poor flavour. Growing in the vicinity might be black spruce, Labrador tea, scrub birch, twinflower and pyrola.

QUICK CHECK: Thick leaves up to 6 mm (¼") long, hairs on stem and leaves. White berries.

RANGE BC: Southern half of B.C. interior in wet forests and bogs from valleys to middle-mountain heights. Babine Lake, Mt. Robson and Sicamous to Nelson.

RANGE WA: Limited to northeastern sphagnum bogs and deep coniferous woods.

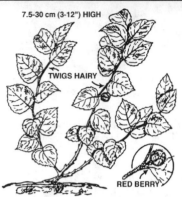

7.5-30 cm (3-12") HIGH

TWIGS HAIRY

RED BERRY

Heather Family, *Ericaceae*
WESTERN TEA-BERRY [020]
Gaultheria ovatifolia
Bush Wintergreen, Oregon Wintergreen

Small, sprawling, with bright red berries that go unnoticed beneath trim little leaves. Several thin, reddish, kinked branches. Heart-shaped alternate evergreen leaves, 1.3–4 cm (½–1 ½") long, artistically arranged on hairy twigs. Minutely toothed; shiny, waxy green above but dull beneath. Small, single white or pinkish flowers. Sepals hairy, reddish brown; smooth for alpine wintergreen (below). Edible berries quite unusual: grooved into segments.

Prefers semi-open forest land above 600 m (2000') elevation. Commonly with lodgepole pine, white pine, Douglas-fir, with sparse undergrowth of falsebox and twin-flower—but also in bogs and to subalpine elevations.

QUICK CHECK: Low shrub less than 30 cm (12") high. Alternate, glossy green, heart-shaped leaves on hairy, red twigs. Single red berries.

RANGE BC: Mt. Arrowsmith and Forbidden Plateau, Vancouver Island, km 40 (25 miles east of Hope) Hope–Princeton Highway, Mt. Revelstoke. Garibaldi Park at 840 m (2800') elevation. Selkirk and Monashee mountains from Revelstoke southwards.

RANGE WA: Olympics, Cascades; 600 m (2000')–1800 m (6000'). Mts. Rainier, Adams.

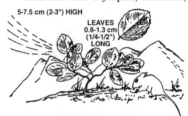

5-7.5 cm (2-3") HIGH

LEAVES
0.6-1.3 cm
(1/4-1/2")
LONG

Heather Family, *Ericaceae*
ALPINE WINTERGREEN
Gaultheria humifusa
Mountain Tea-berry

Miniature edition of western tea-berry (above). Usually in damp locations, subalpine–alpine. Seldom reaches 7.5 cm (3") in height. Small, rounded, very finely toothed leaves. Flowers like western tea-berry, but sepals smooth, not hairy.

QUICK CHECK: Note range limitation. Low shrub; several short, crooked stems. Alternate, rounded leaves finely toothed. Red berry.

RANGE BC: Timberline throughout southern B.C. Vancouver Island. Manning Park, Garibaldi Park, Mt. Revelstoke.

RANGE WA: Near timberline in Cascades and Olympics. Mt. Rainier, Mt. Angeles.

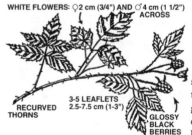

WHITE FLOWERS: ♀2 cm (3/4") AND ♂4 cm (1 1/2") ACROSS

RECURVED THORNS

3-5 LEAFLETS
2.5-7.5 cm (1-3")

GLOSSY BLACK BERRIES

Rose Family, *Rosaceae*
TRAILING BLACKBERRY [019]
Rubus ursinus
Dewberry

Long, thorny, blue-stemmed creeper; tough and sinewy. Its foot-catching and thorn-sticking capabilities might be twined over 4.5 or 6 m (15 or 20') of ground or draped artistically over logs or rock outcrops. Most abundant and widely distributed blackberry on coastal slopes (2 common roadside blackberries on p. 111). Pioneers with bracken and fireweed on recently logged or burned-over forests.

Male flower larger and showier than female, about 4 cm (1 ½") versus 2 cm (¾") across; on separate plants, berries only on female. Blooms in May. Usually 3 leaflets, occasionally 5. Thorns slender, curved back at tip. Fruit—many small drupelets, like raspberry (see 'Note,' p. 111), but more robust—ripens by August; much sought by humans, bears, birds, even deer.

QUICK CHECK: Slender, thorny crawler; alternate, 3-leaflet leaves. Glossy 'blackberry.'

RANGE BC: Most abundant in burns and logging areas of Douglas-fir coastal forests. To 900 m (3000') elevation at coast, to 1050 m (3500') in Cascades.

RANGE WA: Habitats as for B.C.

Heather Family, *Ericaceae*
KINNIKINNICK [023]
Arctostaphylos uva-ursi
Common Bearberry

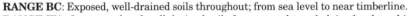

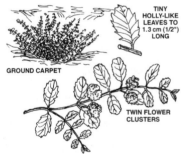

5-10 cm (2-4") HIGH

THICK, LEATHERY
EVERGREEN LEAVES
1.3-2.5 cm
(1/2-1") LONG

BRIGHT RED
BERRIES

PINKISH
'BELL' FLOWERS

Common, widely distributed. On coarse gravel soil, a low green mat; or extends exploring arms over rotten logs and rocks or down road-cut faces. Roots pursue twisting course through rocky soil. Can form large, unbroken ground cover in open, dry forest. Quickly attracts attention. Compelling bright red berries dotted among glossy evergreen leaves. Prominent August–late winter; berries favoured by grouse and bear. Alternate, leathery, evergreen leaves. Larger stems have reddish bark.

Kinnikinnick is aboriginal for 'a smoking mixture.' Both leaves and berries were used by natives.

QUICK CHECK: Trailing plant, alternate evergreen leaves 1.3–2.5 cm (½–1") long. Small pink 'bell' flowers in terminal cluster or mealy red berries.

RANGE BC: Exposed, well-drained soils throughout; from sea level to near timberline.

RANGE WA: On exposed and well-drained soils from coastal to subalpine levels and in ponderosa pine ecosystem. Leavenworth. Olympics.

ALPINE BEARBERRY, *A. alpina*: Leaves deciduous—bright red in autumn. Berries may be purplish black to red. Mountainsides to alpine heights in northern B.C. but scarcer in southeastern B.C. Not in Washington.

Buckthorn Family, *Rhamnaceae*
SQUAW CARPET
Ceanothus prostratus

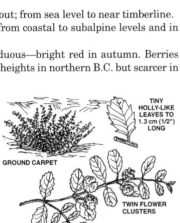

TINY
HOLLY-LIKE
LEAVES TO
1.3 cm (1/2")
LONG

GROUND CARPET

TWIN FLOWER
CLUSTERS

Not easily confused with anything else. Kinnikinnick may be closest, but squaw carpet has non-alternate leaves half the size and no red berries.

Can spread in a low carpet over dry forest floor, sometimes to 3 m (10') across. Opposite evergreen leaves to 1.3 cm (½") long, comparatively thick. Coarsely toothed, giving leaf a 'holly' look. Very tiny flowers grouped in twin bell-like clusters, each about 2 cm (¾") across. Flower colour varies, white–blue, as for other *Ceanothus* spp., (pp. 147–48.)

QUICK CHECK: Sprawling mat of small, 'holly-like' leaves.

RANGE BC: Not in B.C.

RANGE WA: Dry forests; eastern slope of Cascades, Yakima southwards throughout.

Hydrangea Family, *Hydrangeaceae*
WHIPPLEA
Whipplea modesta
Whipplevine

LEAVES
1.3-2.5 cm (1/2-1")
LONG

SMALL, WHITE
FLOWERS

SLIGHT
TOOTHING

Name relates to a Lt. A.W. Whipple who led U.S. Railroad Expedition of 1850s. First collected in California while he was there.

Can be difficult to find, but may be a very common ground cover. Slender vine roots like strawberry runner, to 0.9 m (3') long. Numerous small, opposite, stemless leaves 1.3–2.5 cm (½–1") long; suggestion of teeth, 3 veins. Short stem about 7.5 cm (3") high from each rooted node; terminates in white cluster of tiny flowers (5–6 petals) in late spring. Flower chock-full of bulbous stamens—later replaced by number of large seeds.

RANGE BC: Not in B.C.

RANGE WA: Dry, open forests along Hood Canal and southwards west of Cascades.

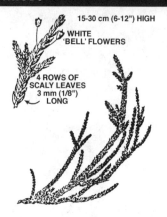

15-30 cm (6-12") HIGH

WHITE 'BELL' FLOWERS

4 ROWS OF SCALY LEAVES 3 mm (1/8") LONG

Heather Family, *Ericaceae*
WHITE MOUNTAIN-HEATHER [024]
Cassiope mertensiana
Club-Moss Mountain-heather
White Moss Heather

A common plant of subalpine and alpine areas, it may grow in neat clumps a few metres (yards) across and about 30 cm (1') high, or in ragged masses sprawling over extensive areas of rock and ground. The small, white flowers are bell-shaped. Opposite, stemless leaves are arranged in 4 rows and hold tightly to the stem.

QUICK CHECK: A low, matted plant with scales pressed to twigs. Small, white 'bell' flower.

RANGE BC: Subalpine and alpine mountains across B.C.

RANGE WA: Subalpine and alpine heights of the Cascades and Olympics.

FOUR-ANGLED MOUNTAIN-HEATHER, *C. tetragona*: Closely resembles *C. mertensiana,* (above) except that its leaves are deeply grooved on their backs. Common at and above the timberline east of the Cascades in B.C. Only in northern half of Washington.

ALASKAN MOUNTAIN-HEATHER or **MOSS HEATHER** [021], *C. stelleriana*: Alternate, flat leaves spread from the stem. Do not confuse with crowberry or pink mountain-heather, which have leaves grooved underneath. Flowers single, white with orange sepals. Subalpine and alpine ecosystems northwards from Mt. Rainier.

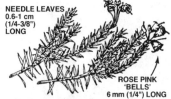

NEEDLE LEAVES 0.6-1 cm (1/4-3/8") LONG

ROSE PINK 'BELLS' 6 mm (1/4") LONG

Heather Family, *Ericaceae*
PINK MOUNTAIN-HEATHER [025]
Phyllodoce empetriformis
Red Heather

A common evergreen plant of the high mountains, low and matted, to 40 cm tall. Sometimes almost the only ground cover over extensive alpine slopes. When in bloom, it is topped with hundreds of rose-pink 'bell' blossoms—an ample reward of floral beauty for the mountain climber. The alternate leaves are relatively short and needle-like, not scale-like as for *Cassiope* spp. Do not confuse with crowberry (p. 121) which has shorter leaves and inconspicuous smaller flowers.

QUICK CHECK: A 'heather' plant with short needle leaves. Rose-pink flowers.

RANGE BC: Subalpine and alpine slopes of most of B.C.

RANGE WA: Cascades and Olympics at subalpine and alpine heights.

YELLOW MOUNTAIN-HEATHER [029], *P. glanduliflora*: Pink and yellow mountain-heather are similar except that the latter has creamy flowers. Often the 2 mingle in a spreading mat with the 2 colours in vivid contrast. Their ranges are the same. Hybrids between these 2 species have pale pink flowers.

Crowberry Family, *Empetraceae*
CROWBERRY [022]
Empetrum nigrum

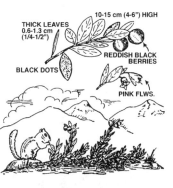

THICK, SHINY LEAVES 0.3-1.3 cm (1/8-1/2") LONG

5-15 cm (2-6") HIGH

SMOOTH, BLACK BERRY

Crowberry differs from the mountain-heathers in having inconspicuous small, purplish flowers or conspicuous round berries as black as a crow. Rather sour, they are tasty in jelly or pies.

Whereas mountain-heathers usually grow in dry, exposed places near timberline or above it, crowberry generally prefers moist, shady places beneath trees. Needle-like leaves about 6 mm (¼") long, thick and needle-like.

QUICK CHECK: Thick, short 'needle' leaves. Berries smooth and black.

RANGE BC: Bogs and bluffs along our coast. Also, to timberline in moist, shaded forests, above 900 m (3000') at the coast and above 1200 m (4000') in the interior. Alaska Highway.

RANGE WA: Moist, shaded areas between 900 m (3000') and timberline in Cascades. Sea coast to alpine slopes of Olympics.

Heather Family, *Ericaceae*
LINGONBERRY
Vaccinium vitis-idaea
Rock Cranberry, Rock Bilberry

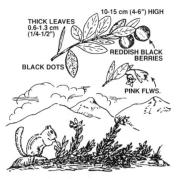

THICK LEAVES 0.6-1.3 cm (1/4-1/2")

10-15 cm (4-6") HIGH

REDDISH BLACK BERRIES

BLACK DOTS

PINK FLWS.

Little-known dwarf shrub, about 15 cm (6") tall, with bright evergreen leaves and large, red berries. Might be mistaken for a puny kinnikinnick, but leaves are smaller, usually less than 1.3 cm (½") long; black dots beneath. Sometimes a small mat from which protrude a few weak stems. Several small, pink flowers cluster near top of twigs. Large, reddish black berries were an important food for northern peoples. Unlike most mountain blueberries, prefers fairly damp habitat, such as muskeg, but also in rocky places, coniferous woods.

QUICK CHECK: Note range limits. Alternate, oval evergreen leaves less than 1.3 cm (½") long. Black dots beneath. Ripe berries soft and reddish black.

RANGE BC: Northern B.C. Dease and Liard rivers. Prince Rupert and Peace River. Abundant in muskeg and damp, mossy forests along Alaska Highway. Selkirks and Rocky Mountains above 1500 m (5000'). Queen Charlottes. Occasional on northern and western coasts of Vancouver Island.

RANGE WA: Not in Washington.

Rose Family, *Rosaceae*
SHRUBBY CINQUEFOIL
Potentilla fruticosa

VELVETY LEAVES 3-7 FINGERS

YELLOW FLOWER 2 cm (3/4") ACROSS

Note: 'Flower' cinquefoils on pp. 221–23 & 297.

Seen often once recognized, otherwise may be a total stranger. A fine-limbed, sprawling shrub up to 90 cm (3') high, or even higher in the north. This plant is widely used as a garden shrub; many flower-colour variants have been developed. Among the shrubs of our region, the leaves are very distinctly shaped, with 3–7 segments.

QUICK CHECK: Velvety leaves with 3–7 'fingers' and yellow 'buttercup' flowers.

RANGE BC: Cascades–Rockies. Sporadic; mostly high, exposed places. Common on Windermere Valley roadsides. Fairly abundant north-central B.C.–Yukon. Cathedral Lakes.

RANGE WA: Cascades, Olympics. Exposed places up to subalpine and alpine elevations.

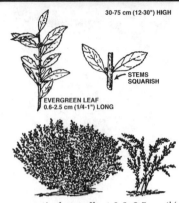

30-75 cm (12-30") HIGH

STEMS SQUARISH

EVERGREEN LEAF
0.6-2.5 cm (1/4-1") LONG

Staff-tree Family, *Celastraceae*
FALSEBOX [026]
Paxistima myrsinites
Mountain Boxwood

In certain forest and subalpine regions, falsebox makes up for its lack of conspicuous flowers or leaves by its sheer abundance. This evergreen plant is characteristic of Douglas-fir, lodgepole pine or Engelmann spruce forests. Adaptable, it also mingles with subalpine flowers on sunny slopes. In the shade it grows loose and sprawling, or even flat on the ground, rarely reaching a height of 60 cm (2'). On rocky slopes it forms attractive, compact balls of greenery. The opposite, thick evergreen leaves are comparatively small, at 0.6–2.5 cm (¼–1") long. Branches are thin and angled.

The small, 4-petal flower clusters in the leaf axils are greenish or reddish and usually go unnoticed. No berries are formed. The fresh, bright green look and compact form make it a fine shrub for landscaping and floral arrangements. It is a favourite browse of mule deer.

QUICK CHECK: Small, opposite evergreen leaves on thin, angled twigs. Comparatively abundant in its range.

RANGE BC: Widespread throughout B.C. in damper coniferous forests from sea level to subalpine heights, northwards as far as Ft. St. John.

RANGE WA: Widespread throughout in damper coniferous forests from sea level to timberline. Coast, mountain and subalpine ecosystems. Spotty in ponderosa pine ecosystem.

2.5-7.5 cm (1-3") HIGH

PINK FLOWERS

LEAF EDGE ROLLED UNDER

LEAVES (3-6 cm)
1/8-1/4" LONG

Heather Family, *Ericaceae*
ALPINE-AZALEA
Loiseleuria procumbens
Trailing Azalea

Perhaps alpine-azalea deserves some special attention, for it may be the smallest native shrub. Its twiggy, matted form may be only a couple of centimeters (an inch or so) high and is thus easily overlooked on the rocky slopes of the high mountains where it grows. In keeping with the dwarf stature of this shrub, its leaves are only 3–6 mm (⅛–¼") long. Clustering abundantly on the twigs, they are thickish, with rolled-over edges; dense hair makes them white beneath. The small, pink 'bell' flowers are in twig-end clusters. Dry seed husks form after the flowers.

QUICK CHECK: Tiny subalpine or alpine shrub. Leaves less than 6 mm (¼") long with rolled edges.

RANGE BC: Bogs of lowland slopes along the coast. Exposed places in subalpine and alpine terrain. Most common west of Cascades and northwards to Alaska. Strathcona Park, Mt. Arrowsmith. Prince Rupert, Queen Charlotte Islands.

RANGE WA: Considered rare in Washington.

Heather Family, *Ericaceae*
BOG-LAUREL [027]

Kalmia microphylla ssp. *occidentalis*
Kalmia polifolia

American Laurel, Alpine Kalmia, Bog Kalmia

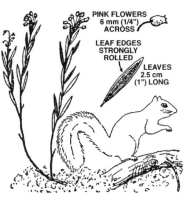

LEAVES TO 2.5 cm (1") LONG

ROSE FLOWERS 1.3 cm (1/2") ACROSS

Watch for this dainty shrub if you are crossing a bog, though the small, narrow leaves and thin, twisting stems up to 60 cm (2') high easily lose their identity in the thicker tangle of Labrador tea. Of course, if bog-laurel is displaying its beautiful reddish purple blossoms, you will find it quickly. While the opposite, leathery leaves with their edges strongly rolled over look much like those of Labrador tea from above, they are less than 2.5 cm (1") long and have beneath a velvety whiteness rather than brownish wool.

The saucer-shaped flowers are worth examining—some will have their stamens bent back and held by their tips in pits in the petals. Should an insect disturb one, the elastic stalk will straighten, spraying the intruder with pollen.

QUICK CHECK: A bog shrub with opposite, pointed leaves, tough and shiny, to 2.5 cm (1") long. Edges rolled under, velvety white beneath. Pink saucer flowers. Contrast with bog rosemary, which has alternate leaves.

RANGE BC: Generally given as 'common in bogs throughout B.C.,' but also in wet mountain meadows. Widespread in northern B.C.

RANGE WA: Boggy situations and wet meadows. Often occurring with Labrador tea. Most common in lowlands of coast forest ecosystem.

WESTERN SMALL-LEAVED BOG-LAUREL, *K. microphylla* ssp. *microphylla*: A close copy of bog-laurel that grows in wet meadows at subalpine elevations. Flowers may be only 1.3 cm (½") across. Leaves are 1–2 cm (⅜–¾") long. Frequent in Cascades of Washington and northwards throughout B.C. to Alaska, but favours western side of Cascades.

Heather Family, *Ericaceae*
BOG-ROSEMARY [028]

Andromeda polifolia

Moorwort

PINK FLOWERS 6 mm (1/4") ACROSS

LEAF EDGES STRONGLY ROLLED

LEAVES 2.5 cm (1") LONG

Bog-rosemary closely resembles bog-laurel in form and habitat. Its crooked, slight stems seldom reach 60 cm (2') above the peaty bogs it lives in. The dull green leaves with white undersurfaces are its main distinguishing feature; so strongly are they rolled over that they are almost tubes. Although approximately 2.5 cm (1") long, often they are only 3 mm (⅛") wide on the top surface. The pink flowers are bell-shaped and form in loose clusters at the ends of the twigs. Although relatively widespread, it is not a particularly common plant.

QUICK CHECK: Bog habitat. Slender shrub with alternate, narrow, dull leaves strongly rolled over and white beneath. Pink 'bell' flowers with distinctive long, pink stems.

RANGE BC: Peat bogs northwards to Alaska. Alice Arm, Prince Rupert, north-central B.C.

RANGE WA: Presence in Washington is doubtful.

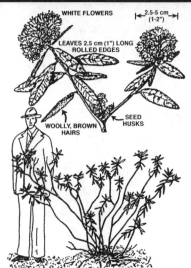

WHITE FLOWERS |← 2.5-5 cm →| (1-2")

LEAVES 2.5 cm (1") LONG ROLLED EDGES

SEED HUSKS

WOOLLY, BROWN HAIRS

Heather Family, *Ericaceae*
LABRADOR TEA [030]
Ledum groenlandicum

Labrador tea is widely distributed across Canada and the United States. Eastern aboriginals used it for making a tea—a practice that was copied by early explorers and settlers—but there is little record of this use in the West. The leaves are picked and left to dry, and then crushed and steeped like tea. Note the comments about trapper's tea (below) and how to easily tell it from Labrador tea.

Found generally on mossy, spongy bogs and sometimes on wet, rocky sidehills. Its thin, twisting stems, seldom over 90 cm (3') high, and thick, narrow leaves with their woolly mats of reddish hairs on the undersurface, are quite distinctive. This wool, and the rolled-over edges of the leathery evergreen leaves, are nature's way of preventing water loss. The leaves are grouped to appear roughly whorled.

During June and July, depending on altitude, this twisted little shrub raises a showy head of white, star-like blossoms with protruding stamens. These are replaced by clusters of dry husks that often hang on until the following spring. Common associates are bog-laurel, scrub birch and willow.

QUICK CHECK: Leaves to 6 cm (2 ½") long, with thick, rusty wool on undersides.

RANGE BC: Bogs throughout B.C. from sea level to subalpine terrain. Common ground cover in parts of northern B.C.

RANGE WA: Bogs of coast and mountain forest ecosystems. Sometimes at higher elevations. Vicinity Seattle and Tacoma. Humptulips, Mt. Rainier.

NORTHERN LABRADOR TEA, *L. palustre*: A low shrub to 60 cm (2') high. Check for reddish hairs on underside of leaf. White flowers have 10 stamens versus 5–7 for Labrador tea. North of Fort Nelson, ranging from lowland bogs to alpine tundra.

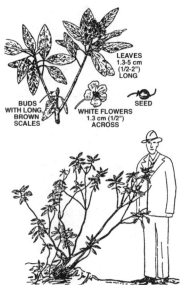

LEAVES 1.3-5 cm (1/2-2") LONG

BUDS WITH LONG BROWN SCALES

WHITE FLOWERS 1.3 cm (1/2") ACROSS

SEED

Heather Family, *Ericaceae*
TRAPPER'S TEA [033]
Ledum glandulosum
Mountain Labrador Tea

Although slightly larger than Labrador tea, this shrub is quite similar in its crooked branching, white flowers and leaf-whorl characteristics. On the undersides of its leaves, instead of Labrador tea's reddish wool, it has whitish wool more like that of bog-laurel and bog-rosemary. Trapper's tea grows on shady, damp mountainsides where it often forms a bushy tangle. The leaves are deciduous. The common name 'trapper's tea' must be a misnomer, for the leaves of this plant contain a poisonous alkaloid.

QUICK CHECK: Leaf whorls and slightly rolled-over edges. White beneath.

RANGE BC: Most abundant east of the Cascades on moist mountain slopes. Found up to 1500 m (5000'). Associated with swamp gooseberry and Engelmann spruce. Manning Park.

RANGE WA: Habitats and associates as for B.C.

Birch Family, *Betulaceae*
SCRUB BIRCH [031]
Betula glandulosa
Bog Birch, Dwarf Birch

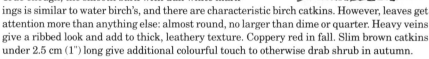

In 1943, 'Mick' Trew and I, the Provincial Parks Branch at that time, were making a reconnaissance of large and virtually unknown Tweedsmuir Park. We rode horses through untold kilometres (miles) of trackless meadow and muskeg covered by an unknown shrub. It resisted identification using the meagre technical botanical guide we carried. The frustration lived with me for years and probably led to my effort to produce a floral guidebook for the amateur, *Trees, Shrubs and Flowers to Know in British Columbia*, which came out in 1952.

'Any resemblance to any birch living or dead is purely coincidental' might be the descriptive theme! True enough, the smooth bark with dull white markings is similar to water birch's, and there are characteristic birch catkins. However, leaves get attention more than anything else: almost round, no larger than dime or quarter. Heavy veins give a ribbed look and add to thick, leathery texture. Coppery red in fall. Slim brown catkins under 2.5 cm (1") long give additional colourful touch to otherwise drab shrub in autumn.

Wet meadows, swamps and occasionally damp hillsides can be covered almost completely by this shrub. Seldom as a lone individual. From a rather compact base, a dozen or more crooked branches rise and spread outwards to form a loose, sprawling top, often to 4.5 m (15') tall. Height decreases northwards; in central B.C., it may be only 1.2 or 1.5 m (4 or 5') high.

Labrador tea and black spruce often keep it company.

QUICK CHECK: A shrub of swamps and low ground. Leaves thick and round, 1.3–2 cm (½–¾") across. Twigs rough and sticky.

RANGE BC: Usually damp ground, but occasionally high open slopes. Throughout, east of Cascades, valley bottom–subalpine levels. Far more common in north; continues to Alaska.

RANGE WA: Sporadic occurrences in coast ecosystem and along edges of mountain forest ecosystem. Also in Klickitat County in ponderosa pine ecosystem.

Buckthorn Family, *Rhamnaceae*
ALDER-LEAVED BUCKTHORN
Rhamnus alnifolius
Buckthorn

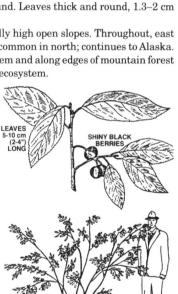

No thorns—probably takes buckthorn name from related species to south. Confined to marshy places such as wet meadows, where it grows with scrub birch, red-osier dogwood and twinflower, and may be quite plentiful.

Most alder-leaved buckthorns are under 1.5 m (5') high and have a loose limb structure, something like red-osier dogwood. Symmetrical, dark green leaves, finely toothed, 5–10 cm (2–4") long; more noticeable than small, greenish flowers or scattered black berries. Plump, 3-seeded berries usually single; juicy but very bitter.

QUICK CHECK: Marsh habitats. Distinctive symmetrical leaf, heavily veined beneath. Small, greenish flowers or single black berries in leaf axils.

RANGE BC: Very limited in range but plentiful in Kootenay River valley, east of Creston and in the vicinities of Salmo and Kimberley. Extends to Rocky Mountains.

RANGE WA: Range very limited: wet, brushy places in eastern Washington. Ponderosa pine and mountain forest ecosystems.

FLUFFY, PINKISH FLOWERS

LEAVES 5-7.5 cm (2-3") LONG

WINTER TWIG

Rose Family, *Rosaceae*
HARDHACK [034]
Spiraea douglasii
Steeple Bush, Douglas Spirea
Note: Mountain-slope spireas on pp. 148 & 153.

Hardhack usually masses in dense clumps around wet places and always favours open places where there is plenty of light. It is crowned with such a showy display of fluffy, pink plumes that it can not help but be noticed. Its slender, reddish brown stems, some 0.9–1.2 m (3–4') high, show up like a miniature forest during the winter months. At this time, the brownish pyramidal husks are very noticeable—possibly they give rise to the alternative common name of 'steeple bush.' In summer, the twigs are covered by narrow, oblong leaves. Coarse notches on the upper part of the leaf are characteristic of the spireas. Flowers are pink to deep rose and form a compact cluster several times longer than broad.

QUICK CHECK: In winter, thin reddish brown stems with dry pyramidal husk. In summer, upright pink plumes of tiny flowers. Spirea leaf, toothed on upper half, woolly beneath.

RANGE BC: Margins of ponds and meadows at low elevations throughout coastal forest and interior cedar-hemlock ecosystems. Damp places in central B.C. Quesnel, Burns Lake.

RANGE WA: Margins of ponds and meadows at low elevations throughout coast forest ecosystem. Mountains of Stevens County.

GREENISH CATKINS

LEAVES 1.3-5 cm (1/2-2") LONG

Sweet Gale Family, *Myricaceae*
SWEET GALE [032]
Myrica gale
Sweet gale might not be noticed unless a special effort is made to find it. Because it prefers a moist, humid climate and a damp habitat, sweet gale is found along swamps and lake margins. It closely resembles hardhack in form and often seems to vanish when with the showier shrub. When crushed, the leaves have a strong fragrance.

The slender, grey-brown stems grow to 1.2 m (4') high. The thin, wedge-like leaves, 1.3–5 cm (½–2") long, are coarsely notched along their upper thirds. They are dotted with yellow glands above and below, and have a whitish tinge on the undersurface. Before the leaves appear, clumps of greenish yellow catkins form along the ends of the branches. These later change to brown, cone-like husks that become prominent in winter.

QUICK CHECK: See habitat limitations. Unmistakable catkins and 'cone' seeds.

RANGE BC: Margins of ponds and lakes. Gulf Islands and coastal forest areas. Cowichan Lake, Bella Coola, Peace River, Alaska Highway.

RANGE WA: Margins of ponds, bogs and lakes in coast and mountain forest ecosystems. Ocean City, Lake Quinault.

CALIFORNIA WAX-MYRTLE or **BAYBERRY**, *M. californica*: Extremely limited range, thus a little-known native shrub in B.C., even though a sturdy shrub or small tree in coastal Washington. Evergreen leaves shaped like those of sweet gale. Along extreme western coast of Vancouver Island between Tofino and Ucluelet—common on Chesterman's Beach and at Amphitrite Point. Also coastal Washington; low hills and valley bottoms. Most abundant in southern region, particularly so along Oregon coast. In open or in fairly heavy shade.

Buttercup Family, *Ranunculaceae*
WHITE CLEMATIS
Clematis ligusticifolia
Old Man's Beard, White Virgin's Bower

Vigorous climber—artistically decorates land-scape east of Cascades. Often chooses fence, tree or shrub to climb on. Slender vines may be 12 m (40') in length and bunch together to form mat-like festoons, but dwarfed in its northern range; will form clumps on a bare hillside. In spring, owes its beauty to numerous clusters of small, white flowers, gradually replaced by soft, fluffy masses of silver fleece. New shoots and flower bunches prolong blooming season from mid-June to mid-August, when both flowers and fleece decorate same vine. Fluffy seed masses until late October; perhaps over winter. Prefers most parched regions of our area, where most of them have their roots in fair soil, such as waste places along edges of farms. Ponderosa pine, black cottonwood and chokecherry usually grow in vicinity.

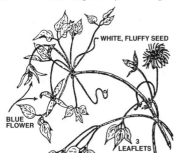

QUICK CHECK: Vigorous climber; leaflets in 3s; clusters of small flowers or fluffy seed masses.

RANGE BC: Dry interior area; associated with sagebrush and ponderosa pine. Abundant from Princeton to Osoyoos. Lytton to Kamloops. Lillooet, Cranbrook. Sporadic in Cariboo parklands ecosystem and lower Fraser Valley.

RANGE WA: East of Cascades. Republic, Wenatchee, Spokane, Pullman, Yakima, Pomeroy, Dayton, Pasco.

TRAVELLER'S JOY, *C. vitalba*, and **GOLDEN CLEMATIS**, *C. tangutica*: Garden escapes, both with yellowish green flowers. Traveller's joy has leaves coarsely toothed; golden clematis with smooth margins. Traveller's joy is on southeastern Vancouver Island and Gulf Islands. Sporadic in Puget Sound region. Golden clematis rare in B.C., possibly Washington.

Buttercup Family, *Ranunculaceae*
COLUMBIA CLEMATIS
Clematis occidentalis
Blue Clematis, Virgin's Bower

Columbia clematis is a rather uncommon shrub. Its slender, twining form is all but lost on the bushes upon which it climbs. Were it not for the large, showy blue flowers or white, fluffy seedheads, it might never be noticed by the casual observer. The shoots branch from the main stems in a very symmetrical fashion and carry the tendrils by which it clings to its support. Leaflets are in 3s. Unlike the sturdy white-flowering clematis, this one rises only 1 m (3') into the air. The flowers, set off by a yellow centre, may be 5–10 cm (2–4") across. Blooms during May and into June. Seedheads are white and fluffy, like miniature dust mops.

QUICK CHECK: Short, slender climber; large, blue flowers; fluffy, white 'dust mop' seedheads.

RANGE BC: East of the Cascades, with wide altitudinal range, on shady mountain slopes and in valley bottoms. From Peace River and Ft. St. James southwards; Adams Lake, Windermere valley, Castlegar, Creston, Penticton Creek.

RANGE WA: Shady mountain slopes or valley bottoms east of the Cascades. Upper edge of the ponderosa pine ecosystem and extending into the mountain forest ecosystem. Ferry County, Spokane, Mt. Carlton. Roadsides west of Sherman Pass.

SUGARBOWLS, *C. hirsutissima*: A vine so slender and delicate that many people would not think of it as a shrub. And since it has showy blue flowers besides, it is described in that section (see p. 312).

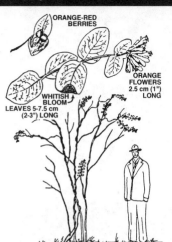

Honeysuckle Family, *Caprifoliaceae*
WESTERN TRUMPET
HONEYSUCKLE [035]
Lonicera ciliosa
Orange Honeysuckle

Any native shrub west of the Cascades or in the interior cedar-hemlock ecosystem that rises from the ground in several thin stems and climbs up shrubs and trees by spiralling tightly around the branches will be western trumpet honeysuckle. Sometimes the stems extend 9 m (30') in the air.

Opposite, oval, entire leaves, as in all honeysuckles; whitish bloom beneath, which rubs off. Twin end-leaves join to make irregular disk, stem through centre. Breaks into leaf early in season, closely following Indian-plum. During May, its clusters of orange-red flowers, thin and tube-like, attract attention, but then fade into background. In September, tidy bundles of transclucent orange-red berries become prominent. Usually, 3 or 4 berries develop at expense of others. Inedible; filled with reddish pulp and number of large, yellow seeds. Hummingbirds especially attracted to flowers; many bird species enjoy berries.

QUICK CHECK: Supple vine tightly twisted around shrubs or trees. Flowers and berries orange-red.

RANGE BC: Below 450 m (1500') elevation in southern B.C. From Vancouver Island to the Kootenay River valley.

RANGE WA: Particularly abundant below 600 m (2000') elevation west of the Cascades. Shady places at upper elevations of ponderosa pine ecosystem. Spokane, Blue Mountains.

Honeysuckle Family, *Caprifoliaceae*
RED HONEYSUCKLE
Lonicera dioica
Smooth-leaved Honeysuckle

Thin stem and opposite leaves of the honeysuckles, but often stretches uncharacteristically vine-like along ground for several metres (yards). Sometimes these thin, straight vines literally cover the ground or other low vegetation, giving little clue as to where they originate. Variable leaves 5–9 cm (2–3 ½") long, whitish bloom beneath. Leaf edges finely hairy. Flowers are yellow to red (not orange or purple, as for other *Lonicera* species) and berries likewise. Semi-open, dry forests above 600 m (2000') are favourable. Associates with Douglas-fir, white pine, falsebox, western tea-berry.

QUICK CHECK: Thin, straight vine on ground. Opposite leaves with bloom below and hairs along edges.

RANGE BC: Most common in southeastern region in low to medium elevation forest. Field, Cariboo, Peace River. Rocky Mountain regions.

RANGE WA: Possibly in northeast.

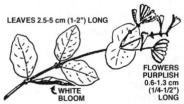

HAIRY or **PURPLE HONEYSUCKLE**, *L. hispidula*: Slender, trailing plant to 3.6 m (12') long that favours dry places. Young stems hairy. Hairy, oval leaves 2.5–5 cm (1–2") long, with bloom beneath. Pink-purple flowers, red berries. Common in dry, open situations on southern Vancouver Island, Gulf Islands and San Juan Islands. Sporadic occurrences west of Cascades southwards to California.

Sunflower Family, *Asteraceae (Compositae)*
BIG SAGEBRUSH [036]
Artemisia tridentata
Sagebrush

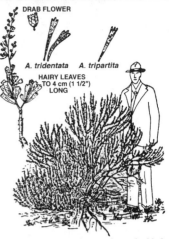

Note: Another *Artemisia* is also a shrub (p. 131); some are with the green flowers (pp. 332–33 & 337).

Big sagebrush can easily be identified by one feature: thin, triple (trident-like) notched leaves. But be careful, because sometimes new spring leaves are not notched. From afar, gnarled form of shrub appears so characteristically massed over barren, arid terrain that it just has to be big sagebrush. 'Sage' smell of foliage and twisted, loose bark are other features. It is a ritual for me, every time I enter sagebrush habitat, to crush a handful of leaves and enjoy that distinctive smell. Extremely small and drab, flowers bloom from mid-September to mid-October.

Under favourable conditions often to 2.5 m (8') in height, but averages about half that high. Considered an intruding weed plant in overgrazed or otherwise impoverished soils. Interestingly, almost always on soils of volcanic origin, seldom on those derived from granite.

QUICK CHECK: Gnarled, grey-green shrub 0.6–2.1 m (2–7') high; thin 'trident' leaves.

RANGE BC: Dry interior regions. Prominent Ashcroft to Kamloops, Similkameen and Okanagan valleys. Usually valley bottoms and lower slopes, but sometimes to 1500 m (5000') on southern exposures.

RANGE WA: Westerly portion of sagebrush ecosystem and scattered in bunchgrass ecosystem. Not in Snake River Canyon or in channelled scablands, (flood-scoured area between Moses Lake and Spokane). Cusick southwards. Columbia River east of The Dalles, Oregon.

THREETIP SAGEBRUSH, *A. tripartita*: Can be difficult to find. Differs from big sagebrush principally in having leaves quite deeply notched. Smaller and generally more bushy. Range is similar to that of big sagebrush—and possibly wider, because it is adaptable to moist conditions—but field research is required to verify its precise limits.

Sunflower Family, *Asteraceae (Compositae)*
RIGID SAGEBRUSH
Artemisia rigida
Stiff Sagebrush

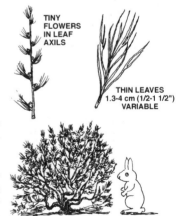

You could have difficulty trying to identify this sagebrush amidst others. One side of the road, sloping and rocky, might carry a growth of rigid sagebrush while the other side, a benchland, may be predominantly big sagebrush. A quick visual clue is that big sagebrush only grows to 60 cm (2') high and has a more rounded, compressed outline. But on some dry, rocky slopes it is low and ragged, the only shrub.

Look closer and you will see a small copy of the larger sagebrush. Those thick, twisted lower stems look as if they have been in agony for 100 years. After a short distance, they branch into short, stout spurs, hence 'rigid.' This 'pitchfork' framework is hidden and softened by a dense growth of olive-green leaf tufts. Leaves appear as thin singles, but many are forked in a variety of shapes. They have a pungent sage smell. Tiny flowers form in leaf axils—generally go unnoticed.

QUICK CHECK: Low, stiff sagebrush with thin leaves.

RANGE BC: Not found in B.C.

RANGE WA: From very dry benchlands to rocky slopes in eastern and central areas of the state. Extends northwards to vicinity of Grand Coulee. David Douglas, the explorer botanist, passed this way in 1826; did he notice the difference between the 2 sagebrushes? Vantage–Ellensburg (back road). Columbia River benches and slopes east of The Dalles, Oregon.

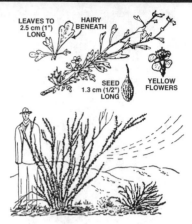

LEAVES TO 2.5 cm (1") LONG

HAIRY BENEATH

SEED 1.3 cm (1/2") LONG

YELLOW FLOWERS

Rose Family, *Rosaceae*
ANTELOPE-BRUSH [037]
Purshia tridentata
Bitterbrush

The dull green of this shrub blends very well with the neighbouring sagebrush. However, it is quite different in form, having a spreading structure of stiff, awkward branches. The thin, notched leaves are grey-hairy beneath. They are so tiny that the outline of the fibrous limbs is quite distinctive. In the southern Okanagan it may grow 2.4 m (8') high but in the Kootenay River valley it seldom reaches 1.2 m (4') and is more brushy. In places, possibly because of browsing by wildlife, it may be reduced to a low twiggy shrub or even to a mat.

In early spring, it is dotted with small, yellow blossoms that give this ungainly shrub strong prominence in the desert-like landscape.

QUICK CHECK: Note limited range. Short leaves up to 2.5 cm (1") long, triply notched like big sagebrush. Yellow flowers or 'teardrop' seeds on limbs.

RANGE BC: Most arid benches of southern Okanagan and Similkameen valleys. Prominent from Osoyoos northwards to Kaleden. Below approximately 450 m (1500'). Kootenay River valley, vicinity Fort Steele and northwards to Radium Hot Springs.

RANGE WA: Sagebrush and bunchgrass ecosystems. Tonasket to Brewster, Leavenworth, Wenatchee, Ellensburg, Yakima, Pasco, Walla Walla County, Gifford, Entiat, Peshastin, The Dalles, Oregon.

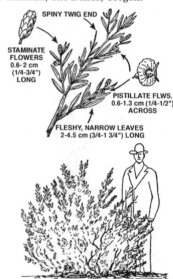

SPINY TWIG END

STAMINATE FLOWERS 0.6- 2 cm (1/4-3/4") LONG

PISTILLATE FLWS. 0.6-1.3 cm (1/4-1/2") ACROSS

FLESHY, NARROW LEAVES 2-4.5 cm (3/4-1 3/4") LONG

Goosefoot Family, *Chenopodiaceae*
GREASEWOOD [043]
Sarcobatus vermiculatus

Green, green, green! That is the key factor in identifying this dark green shrub intermingled with the more-common, similarly shaped, but grey sagebrush. Note, however, that green rabbit-brush (facing page) is a similar green. A further clue to watch for is the presence of alkali, as shown by a white, pasty substance on the ground, especially around drying ponds, for greasewood is very partial to this situation. Spiny twig ends are a further feature in clinching its identity.

Growing from 0.6 to 2.4 m (2 to 8') high, greasewood forms a stout, erect shrub; it is spiny and densely leaved. Unlike sagebrush, it has a smooth, light grey bark. Leaves are narrow and thickish, tapering inwards at their bases.

The same shrub carries 2 types of flowers, around the middle of summer. The male flower looks like a small catkin because its parts are hidden by numerous small scales. The female flower usually occurs as a single bloom; as it goes to fruit, it develops a heavily veined wing with a ragged edge.

QUICK CHECK: Dark green shrub usually with or near sagebrush. Alkaline habitat. Spiny twigs and distinctive flowers.

RANGE BC: Not found in B.C.

RANGE WA: Widely scattered in suitable localities throughout the sagebrush ecosystem. Prominent from Coulee City to Ephrata. Vantage to Othello. Yakima, Pasco, Washtucna.

Sunflower Family, *Asteraceae (Compositae)*
COMMON RABBIT-BRUSH [039]
Chrysothamnus nauseosus

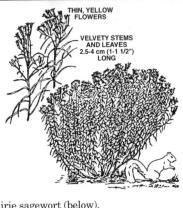

THIN, YELLOW FLOWERS

VELVETY STEMS AND LEAVES 2.5-4 cm (1-1 1/2") LONG

Lovely, compact, olive-green shrub complements sagebrush scene. Colour comes from very fine wool on stems and leaves. Averages 0.6–1.2 m (2–4') in height. Many thin, branching stems almost hidden by long, string-thin leaves. Masses of small, yellow flowers, occasionally by early August, more commonly late summer–early fall.

Might be confused with almost-identical grey horsebrush or sagebrushes—similar form and habitat. Tousled yellow blooms in June probably indicate grey horsebrush (below). Comparatively wide leaves, tips notched, indicate sagebrush. Also see prairie sagewort (below).

QUICK CHECK: Soft sage colour. Leaves thin, velvety. Massed yellow flowerhead, usually late summer.

RANGE BC: Common in drier valleys east of Cascades. Often associated with sagebrush but more adaptable to higher elevations.

RANGE WA: East of Cascades, favours upper elevations of sagebrush community. Chelan, Lyon's Ferry, Vantage; The Dalles, Oregon.

GREEN RABBIT-BRUSH, *C. viscidiflorus*: Often with common rabbit-brush; similar size and shape—key word is 'green.' Abundant narrow leaves, bright green distinct from rest of sagebrush community. (Greasewood, also green, usually over 90 cm (3') high, with spiny twig ends, facing page.) Retains dried flower clusters even until spring. In B.C., mostly in south-central/southeastern areas; Kootenay River valley, Fort Steele. In Washington, dry open places east of Cascades, valleys to foothills. Vantage.

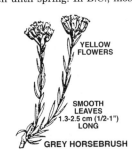

YELLOW FLOWERS

SMOOTH LEAVES 1.3-2.5 cm (1/2-1") LONG

GREY HORSEBRUSH

GREY HORSEBRUSH, *Tetradyma canescens*: Very closely resembles common rabbit-brush in colour and form. However, horsebrush blooms in June, not in late summer. 4 oblong bracts surround each flowerhead. Without flowers in bloom, identify by narrow, twisted leaves about 1.3 cm (½") long. Common in sagebrush zone of south-central B.C.; similar habitat in Washington. Tonasket, Dry Falls.

Sunflower Family, *Asteraceae (Compositae)*
PRAIRIE SAGEWORT
Artemisia frigida
Fringed Sagebrush, Pasture Wormwood

FLOWERS

FRINGED LEAVES

Note: Other *Artemisia*s are also shrubs (p.129), while some are with the green flowers (pp. 332–33 & 337).

Beautiful little shrub; mingles with big sagebrush and common rabbit-brush; pick out by small size and softer colour. Very pungent 'sage' smell to crushed leaves. Leaves differ from rabbit-brush in being finely hairy, divided into 2 or 3 thin, fringe-like leaflets. Small, dull flowers cluster along thin twigs.

Ranges beyond big sagebrush, common rabbit-brush. Favours dry, open slopes, southerly exposure. Perhaps with other wormwoods. *Frigida* reflects Siberian specimen collected 150 years ago.

QUICK CHECK: Low, olive-green; 10–50 cm (4–20") high; finely divided, hairy leaves.

RANGE BC: Widely distributed throughout dry interior region and Cariboo parklands. Into Cariboo to 32 km (20 miles) south of Quesnel; up North Thompson River to Blue River, Kootenay River valley, vicinity of Fort Steele. Also Peace River, Dawson Creek, Ft. St. James.

RANGE WA: Infrequent. Okanogan and Ferry counties have greatest abundance.

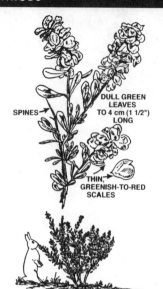

SPINES →

DULL GREEN
LEAVES
TO 4 cm (1 1/2")
LONG

THIN,
GREENISH-TO-RED
SCALES

Goosefoot Family, *Chenopodiaceae*
HOP SAGE [040]
Grayia spinosa

A bushy shrub associated with the sagebrush desert country. Although ranging from 0.3 to 1.2 m (1 to 4') high, it most commonly forms a tousled, much-branched shrub around 60 cm (2') in height. In spring it is quite conspicuous as it displays on ends of its twigs dense clusters of fruiting bracts or scales, often decoratively tinged with red.

The leaves, like those on most plants of dry areas, are relatively small, being seldom over 4 cm (1 ½") long and more commonly around 1.3 cm (½"). They are thick and tend to be rough to the touch. Hidden among them on twig ends are sharp spines. Older branches have a dark brown, shredding bark.

QUICK CHECK: Seldom found anywhere but in the sagebrush ecosystem, where it is the only 'sagebrush' shrub that has spines. Fruiting scales distinctive.

RANGE BC: Not in B.C.

RANGE WA: Mostly in south-central sagebrush community in Douglas, Yakima, and Franklin counties. Also found near Grand Coulee.

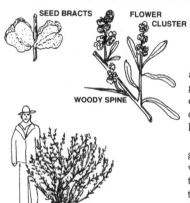

SEED BRACTS   FLOWER
                CLUSTER

WOODY SPINE

Goosefoot Family, *Chenopodiaceae*
FOURWING SALTBRUSH
Atriplex canescens

Generally nondescript, a dull green shrub associated with such dryland plants as sagebrush, greasewood, antelope-brush and rabbit-brush. It is easily overlooked, for it does not have any special character until early summer, when some of the leaves take on a bronze tinge.

In general, it fits the outline of sagebrush and greasewood. Some old shrubs tend to sprawl and develop an open center, exposing a mass of twisted, thick stems. Saltbrush does not have true spines but the many sharp twig ends make a good substitute. It is difficult to see how this shrub would rate as a valuable range browse food, but it does.

Small, upright leaves are paddle-shaped and vary from 1.3 to 2.5 cm (½ to 1") in length. On some twigs they are well spaced and alternate, on others much smaller and clustered.

Not all saltbrush shrubs carry flowers. Those that do, do so in April and May, while long-leaved phlox is in bloom nearby and Munroe's globemallow is in bud. However, saltbrush produces no floral display, for the flowers are very small and greenish. They form a small, vertical cluster at each twig end.

As seeds ripen, they are protected by 2 large, greenish bracts or wings. The way these are formed and join together gives a roughly 4-winged appearance. In late summer they take on several colour tones, which gives the shrub somewhat greater prominence in the landscape.

QUICK CHECK: Part of sagebrush community. Paddle-shaped leaves. Winged seeds.

RANGE BC: Not in B.C.

RANGE WA: Generally favouring saline soils east of the Cascades. Othello to Vantage. Sunnyside. Some steep slopes in Yakima Canyon.

Goosefoot Family, *Chenopodiaceae*
WINTER FAT
Eurotia lanata
White Sage, Winter Sage

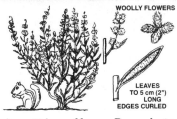

WOOLLY FLOWERS

LEAVES
TO 5 cm (2")
LONG
EDGES CURLED

Important rangeland nutrition during winter months, particularly for domestic sheep, but difficult to find. Look for extensive patches on low alkali ground. Erect stems 30–60 cm (1–2') high. Overall whitish appearance comes from abundance of white hairs on twigs and leaves. Dense clusters of very soft, woolly flowers in leaf axils make top half of plant fuzzy. Male flowers and very attractive female ones (4 dense tufts of white-to-brown, silky hairs) on same plant.

QUICK CHECK: Shrubby; sparse number of small, flat leaves, curled-over edges.
RANGE BC: Not in B.C.
RANGE WA: Low elevations east of Cascades. Douglas and Yakima counties.

Mint Family, *Lamiaceae*
DORR'S SAGE [047]
Salvia dorii

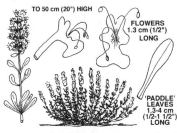

TO 50 cm (20") HIGH

FLOWERS
1.3 cm (1/2")
LONG

'PADDLE'
LEAVES
1.3-4 cm
(1/2-1 1/2")
LONG

Thick, broad, shrubby. Not always present, but quite abundant in various sagebrush areas. Stout, contorted framework; like miniature sagebrush. Lower twiggy limbs often covered with crusty orange substance. Crush leaves for unusual sour, spicy odour. Leaves opposite, thick, silvery.

Can be mass of bright bloom April–June; many dozen purplish flower stems per shrub. Tiny flowers in whorls, each supported by 2 small leaflets and broad, purplish bracts. 2 long projecting stamens, even longer style. Flower's lower lip is fine landing platform for a bee; arching style and stamens brush its back as it enters. David Douglas, pioneer botanist, collected seeds 'on the plains of the Columbia, near the Priest's Rapid.'

QUICK CHECK: Like a large, shrubby penstemon. Distinctive flowers.
RANGE BC: Not in B.C.
RANGE WA: Dry, rocky soils; usually with rigid sagebrush. Valleys and hillsides around Vantage. In small Sandhills Park, south edge of Adams County. Southwards in sagebrush.

Figwort Family, *Scrophulariaceae*
DAVIDSON'S PENSTEMON [045]
Penstemon davidsonii
Menzies's Penstemon

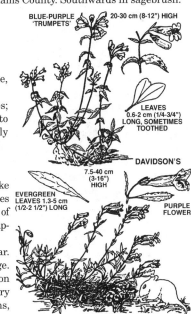

BLUE-PURPLE
'TRUMPETS'

20-30 cm (8-12") HIGH

LEAVES
0.6-2 cm (1/4-3/4")
LONG, SOMETIMES
TOOTHED

DAVIDSON'S

7.5-40 cm
(3-16")
HIGH

EVERGREEN
LEAVES 1.3-5 cm
(1/2-2 1/2") LONG

PURPLE
FLOWER

SCOULER'S

Note: White/yellow, p. 230; purple, p. 302; blue, p. 322.

Low, mat-like. Thick, ovalish, evergreen leaves; entire or short-toothed; bract-like on flower stems, to 1 cm (⅜") long. Flower throat bristly; anthers woolly tipped. Blooms July–August; high, rocky places.

RANGE BC: Coast Mountains, Cascades.
RANGE WA: Coast Mountains, Cascades.
SHRUBBY PENSTEMON [046], *P. fruticosa*: Like Davidson's, but leaves broader, lower elevations. Masses of purple May blooms. Anthers woolly tipped. East of Cascades to Rockies; Washington, southern B.C. Adaptable: foothills–subalpine; rocky cliffs.

SCOULER'S PENSTEMON, *P. fruticosa* var. *scouleri*: Unlike shrubby penstemon, but same range. Narrow, finely serrated leaves. As compact clumps on dry, rocky terrain. Flowers to 4 cm (1 ½") long, very showy clusters. Blooms in April at lower elevations, June near subalpine heights. Not in B.C. Okanogan, Ferry and Klickitat counties in Washington.

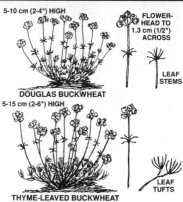

5-10 cm (2-4") HIGH

FLOWER-HEAD TO 1.3 cm (1/2") ACROSS

LEAF STEMS

DOUGLAS BUCKWHEAT

5-15 cm (2-6") HIGH

LEAF TUFTS

THYME-LEAVED BUCKWHEAT

Buckwheat Family, *Polygonaceae*
DOUGLAS BUCKWHEAT
Eriogonum douglasii

A dwarf, shrubby dryland gem, often with large plants such as rigid sagebrush; presses close to ground. A wonder of nature, raised mass of flower-heads can be seen from afar, mostly in May.

Deep root, a number of miniature, twisting, shrub-like stems. Tiny olive-grey leaves about 6 mm (¼") long. Thin flower stems to 5 cm (2") above leaf level. Flowerhead to 1.3 cm (½") across.

QUICK CHECK: Tuft of leaves at stem top. Whorl of leaves near centre of stem, topped by cluster of pale cream-to-yellow flowers.

RANGE BC: Not in B.C.

RANGE WA: Sagebrush and ponderosa pine terrain of south-central Washington and southwards into Oregon. Brewster, Yakima Canyon, Vantage.

THYME-LEAVED BUCKWHEAT, *E. thymoides*: Low, tightly compact, shrubby growth very similar to Douglas buckwheat, but flowers may be white, yellow, pink or red. Tiny leaves in tufts, not on stalks. Distribution similar, but wider range into foothills. Vantage.

30-50 cm (12-20") HIGH

DULL WHITE BERRY

3 GLOSSY GREEN LEAFLETS

Sumach Family, *Anacardiaceae*
POISON-IVY [041]
Rhus radicans

Regional abundant, with well-known poisonous properties, yet few can identify it. A sun lover with specialized habitat: low patches brighten rock slides, stony places and road edges in warm interior; e.g., Okanagan, southern Kootenay and Columbia river valleys. Identify by large, wavy-edged leaves in 3s, scarlet in fall and soon lost, and by whitish berries in loose clusters part way up (may overwinter).

All parts contain oil that causes severe skin irritation, even via contact with shoes or clothes that have touched them, or smoke from burning. If exposure is suspected, wash in several changes of water. Chamomile lotion is standard remedy, else try a coating of either soft soap or a strong solution of Epsom salts over affected parts. In any case, irritation usually clears up in several days. Strangely enough, sheep, goats and cattle graze on poison-ivy without harm and may help eradicate it.

QUICK CHECK: Glossy green leaves in 3s. Dull white berries close to stem.

RANGE BC: Sagebrush, ponderosa pine ecosystems. Quesnel, Seton Lake. Southern Kootenay and Columbia river valleys. Radium Hot Springs, Trail.

RANGE WA: Sagebrush and bunchgrass areas and edge of ponderosa pine ecosystem. Ferry County, Oroville, Wenatchee, White Pass, Yakima, Snake River.

GREENISH WHITE BERRIES

GLOSSY, LEATHERY LEAFLETS VARIABLE IN SHAPE

POISON-OAK [038], *R. diversiloba*: Fairly rare in B.C. and Washington. May be a straggly, erect shrub to 1.5 m (5') high, a climber or a scrubby ground cover. 3 roughly round leaflets, sometimes irregularly lobed or toothed, often resemble oak leaf. Occasionally in dry places on southern half of Vancouver Island and Gulf Islands. Coastal forest areas. Vicinity of Seattle, Tacoma, Hoodsport. Common in central region of Columbia River Gorge, eastwards to the Dalles.

LEAVES IN 3, LET IT BE!

Sumach Family, *Anacardiaceae*
SMOOTH SUMAC
Rhus glabra

13-21 LEAFLETS

REDDISH SEED CONE

A picturesque, many-branched shrub found in drier regions of our area. Often a loose thicket develops as new plants shoot up from long, rambling roots. Usually about 90 cm (3') high, but along Okanagan Lake and in areas of Washington, bright green masses 1.8 m (6') in height are quite common; in some places to 3 m (10'). The crooked, twisting limb structure makes it look like an Oriental dwarf shrub with characteristic 'plume' foliage.

Some plants have conspicuous conical mass of round seeds near top. In September and October, plush-like seed covering is rich red. Also during October, leaflets and stem turn bright crimson, bringing a vivid red rash of autumn colour to lower mountain slopes. Sumac exudes a milky juice when bruised. Some species are used in preparing waxes, dyes and varnishes. 'Indian lemonade' is made from the velvety seeds: husks are steeped in hot water, then water is strained, sweetened and left to cool.

QUICK CHECK: Stout leaf bearing 13–21 toothed leaflets. A similar relative, poison sumac, occurs in eastern U.S.A. but not in our area.

RANGE BC: Dry interior at lower elevations. Grand Forks, Okanagan Valley, Princeton to Keremeos, Lillooet.

RANGE WA: Sagebrush and bunchgrass areas, occasionally on edges of ponderosa pine forest. Okanogan River, Snake River and tributaries. Ferry, Chelan and Yakima counties.

Oleaster Family, *Elaeagnaceae*
SOOPOLALLIE [042]
Shepherdia canadensis
Soap Berry, Canada Buffalo-berry

COPPERY DOTS

DARK GREEN LEAVES 2.5-6 cm (1-2 1/2") LONG

ORANGE-TO-RED BERRIES

Soopolallie berries, rubbed between the hands, make a soapy froth (aboriginal *soop* = 'soap' plus = *olallie* 'berry'). A more attractive name was 'Indian ice-cream.'

In semi-open, bushy and upright, 0.9–1.2 m (3–4') tall. On westerly slopes of Rockies, a higher, more sprawling form is common. Abundant in semi-open forests at medium elevations. Large areas of lodgepole pine have an undergrowth of soopolallie.

Leaves opposite, entire, roughly egg-shaped and 2.5–6 cm (1–2 ½") long. Although dark green above, undersurfaces are a combination of silvery hairs and rusty brown or coppery spots. Twigs pebbled with same spots. Small, yellowish brown male and female flowers are borne on separate bushes. Orange-to-red, almost transparent fruit, in small clusters along stems and twigs. Most who taste them would agree that 'soop' could be significant.

QUICK CHECK: Opposite, dark green leaf with silvery hairs and rusty brown spots on undersurface. Twigs pebbly with rust. Berries orange-red, translucent.

RANGE BC: Throughout B.C., except most of coast forest. Limited on southeastern Vancouver Island. Below 1350 m (4500') in interior, 1650 m (5500') in Rockies. Alaska Highway.

RANGE WA: Generally east of the Cascades, but isolated occurrences in Olympics and San Juan Islands. Okanogan and Stevens counties. Rockport.

THORNY BUFFALO-BERRY, *S. argentea*: Very rare shrub. Leaves silvery on both sides; twigs thorny. Dry places in B.C.'s ponderosa pine ecosystem and adjacent slopes. Extends into northeastern B.C. Not known in Washington.

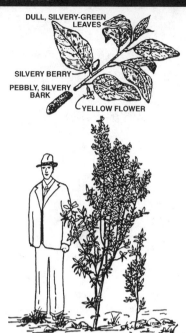

DULL, SILVERY-GREEN LEAVES

SILVERY BERRY

PEBBLY, SILVERY BARK

YELLOW FLOWER

Oleaster Family, *Elaeagnaceae*
WOLF-WILLOW [044]
Elaeagnus commutata
Silverberry

East of the Cascades, one sometimes sees a group of shrubs with leaves so silvery that they appear diseased or as if covered with a fine, glistening silt—wolf-willow. An erect shrub sometimes 3.6 m (12') high, more often 0.9–1.5 m (3–5'). Usually as small groves near stream edges or back channels of larger creeks and rivers. Occasionally on high banks or mountain slopes, usually on a seepage area.

Leaves alternate, 5–10 cm (2–4") long; silvery sheen caused by very small, overlapping scales. While smaller limbs are same colour, often both very young twigs and main leaf vein are finely pebbled a beautiful copper. Pleasantly perfumed small, yellow flowers are out in June; dry, olive-like, silvery berry in August and September. Interior natives once strung berries on necklaces; tough bark provided material for weaving and rope.

QUICK CHECK: Silvery twigs; alternate, entire, silvery leaves. Yellow flowers or silvery berries.

RANGE BC: Main watercourses of dry interior. Nicola, Similkameen and Okanagan rivers. Seton Lake, northwards along Fraser River to Quesnel, Kootenay River valley, Cranbrook, Peace River parklands and northwards to Yukon.

RANGE WA: Not known in Washington.

WHITISH CHANNEL

LEAVES IN 3s

BLUE BERRIES WITH BLOOM

Pine Family, *Pinaceae*
COMMON JUNIPER
Juniperus communis
Ground Juniper

Sprawling; holds close to rocks at higher elevations but often a bushy, upright mat to 1 m (1 yard) high, 1.2–3 m (4–10') across, in valley bottoms.

The bark is thin, reddish grey and rough with scales. Narrow, sharp-pointed needles are 6 mm (¼") long, with a whitish channel on the underside. Note that the needles are arranged in whorls of 3. The fruit is a fleshy, knobby berry, dark blue to black and covered with a whitish bloom. It is stemless, unlike the fruit of the similar creeping juniper (noted below). None of the junipers has edible berries, but they are used to flavour beer and gin. A favoured winter food of Townsend's solitaires.

QUICK CHECK: Evergreen shrub with leaves in whorls of 3. Stemless berries.

RANGE BC: Widely ranging throughout. Common on rocky places in the interior and with wide altitudinal tolerance. Alaska Highway. The most widely distributed shrub in the northern half of the world, it is circumpolar.

RANGE WA: High timberline and alpine regions of Olympics and Cascades. Stevens Pass, Mt. Rainier, Mt. Adams.

CREEPING JUNIPER, *J. horizontalis*: A low, creeping shrub, as name implies. Needles are scale-like and sheath the stems. Not prickly like those of common juniper. Berries are bluish purple with a grey bloom. They have short stems. Ranges across northern half of B.C., favouring valley bottoms of major rivers and extending through Rocky Mountain Trench to East Kootenays. Does not reach Washington.

Honeysuckle Family, *Caprifoliaceae*
BLUE ELDERBERRY
Sambucus cerulea

As their names show, the elders can be distinguished by their berries. Otherwise, the individual habitats, ranges and times of blooming can be definite aids. All elder stems have a large, soft pith.

Blue elderberry partly overlaps red elderberry in range, but grows in dry, open situations. Leaflets are generally in 9s. although 5s and 7s often occur. The white, flat flowerhead is 12.5–20 cm (5–8") across. Blooms throughout the summer; may continue into late August. The berries, in flat-topped clusters, are tinged with a lighter bluish bloom.

QUICK CHECK: Large, flat-topped blooms in June, July and August. Masses of small, blue berries August to September.

RANGE BC: Dry places, Gulf Islands ecosystem, Malahat, Duncan, Alberni. Occasional shrubs in upper Fraser Valley. Very abundant from Princeton to Penticton in dry interior areas. Nelson, Trail.

RANGE WA: Widespread across the state.

Honeysuckle Family, *Caprifoliaceae*
RED ELDERBERRY
Sambucus racemosa

Red elderberry is a common shrub of the coast and in the wetter regions of the interior, where it raises a spreading mass of stout stems 3–6 m (10–20') high. The opposite leaves consist of 5–7 sharp-pointed, toothed leaflets. Vigorous young shoots sometimes have 9 leaflets. The rounded heads of small, yellowish white flowers, in bloom throughout May, have an unpleasant smell. The smooth berries tinge with red in early June. They are edible only if cooked, as was done by aboriginal peoples. Generally little used today.

QUICK CHECK: Compound leaf with toothed, pointed leaflets 5–12.5 cm (2–5") long. Yellowish white flowers in a pyramidal head. Berries bright red, but occasionally yellow.

RANGE BC: Coastal forest and Gulf Island ecosystem below 600 m (2000'). Interior cedar-hemlock ecosystem below 1050 m (3500'). North-central B.C. Mt. Robson.

RANGE WA: Generally west of Cascades.

Honeysuckle Family, *Caprifoliaceae*
BLACK ELDERBERRY
Sambucus racemosa var. *melanocarpa*

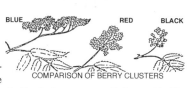

A variety of the red species above, black elderberry is seldom over 2.4 m (8') high. It has the same typical ascending spread and similarity in leaf and bark. Confusion can usually be avoided by noting the high mountain range limitation. The flower cluster differs from that of the coastal red elderberry in being more rounded and looser.

QUICK CHECK: Loose flowerhead. Shiny black berries. 5–7 leaflets.

RANGE BC: Mountain slopes east of the Cascades, from 1200 m (4000') to timberline. Aberdeen Mountain (near Vernon). Mt. Revelstoke Park, Kokanee Glacier Park, Nelson, Salmo–Creston Summit.

RANGE WA: Mostly on eastern slopes of the Cascades from 900 m (3000') to timberline. Mountain forest ecosystem. Blue Mountains.

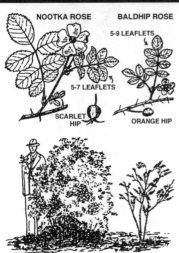

NOOTKA ROSE
BALDHIP ROSE
5-9 LEAFLETS
5-7 LEAFLETS
SCARLET HIP
ORANGE HIP

Rose Family, *Rosaceae*
WILD ROSES
Rosa spp.

Wild roses are easily recognized the country over, but it may take an expert to identify the species of a particular specimen. Over 100 different types grow in North America, with at least 8 in B.C. and Washington. Each features white-to-deep-rose flowers with fragrant perfume. Almost as well known as the flowers are the fruits, or 'hips,' which hang on all winter; the mealy peel is the part eaten. The leaves have an odd number of leaflets; a wing-like sheaf clasps the base of each leaf. Most roses have thorny stems. Note that native roses have straight spines while introduced roses have curved spines. Here are some of the more common roses in our region:

NOOTKA or **COMMON WILD ROSE**, *R. nutkana*: Probably the most common bush rose. A bushy shrub, to 3 m (10') high, armed with stout, straight prickles, usually paired, beneath each leaf axil. Leaves have 5–7 toothed leaflets. Showy flowers are often 5 cm (2") across. Mostly borne singly, but sometimes with 1 or 2 others. Blooms from May to July. Fruit is a large, showy, scarlet hip. Prefers fairly rich soils. Ranges throughout B.C. and Washington at lower elevations.

2 varieties are recognized: *R. nutkana* var. *nutkana*, which ranges west of the Cascades, and *R. nutkana* var. *hispida*, which is found mostly east of the Cascades. The former has leaflets doubly serrate; the latter, singly serrate.

BALDHIP or **DWARF ROSE** [051], *R. gymnocarpa*: A spindly shrub, to 1.2 m (4') high, with slender stems bristly with weak, straight spines. 5–9 leaflets. Pale pink flowers 1.3–2 cm (½–¾") across borne singly at the ends of twigs. Orange fruit with sepals not attached. Found in rocky, exposed situations at lower elevations across B.C. south of Quesnel. At lower elevations throughout Washington.

CLUSTERED WILD or **SWAMP ROSE** [050], *R. pisocarpa*: Weakly armed with straight spines and a pair of larger spines at each leaf base. 5–7 sharp-pointed leaflets. Several small-ish, pale flowers in twig-end clusters. Fruit pea-sized. Usually in swampy locations, southern Vancouver Island and Lower Mainland. Throughout Washington at lower elevations.

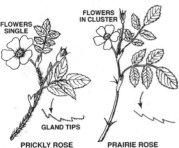

FLOWERS SINGLE
FLOWERS IN CLUSTER
GLAND TIPS
PRICKLY ROSE
PRAIRIE ROSE

PRICKLY ROSE, *R. acicularis*: This rose might be the most abundant overall, yet rate as the most inconspicuous. You will see it as part of the low ground cover in partly open forest stands. Often without a blossom or hip, it tells you it is a rose by its toothed 5–7 leaflets and uniform, weak, straight, bristly spines. Often it is only 30 cm (1') high, but under ideal conditions it will form a loose shrub of slender stems to 1.2 m (4'). Borne on short branches, its showy, solitary blooms are pink to deep rose. The hips become dark red and are high in vitamin C. Common across Canada east of the Cascades. Under shady forests or on open ground northwards to Alaska, but not extending southwards into Washington.

PRAIRIE or **WOOD'S ROSE**, *R. woodsii*: Generally distinguished, with difficulty, from prickly rose by supposedly having a pair of spines at the nodes of the stems and by having clustered flowers. There may be a half dozen flowers or red hips in each close cluster. A spindly shrub to 90 cm (3') high. Variable in habitat, preferring damp places at low elevations and drier sites at higher elevations. Found east of Cascades from U.S. border northwards beyond Dawson Creek, where it seeks dry valley slopes. May occur in northeastern Washington.

DOG ROSE, *R. canina*: This is presumably the white rose you will see occasionally. Usually it is a dense roadside thicket to 1.5 m (5') high. Southern Washington. Snake River, Steptoe Butte. Southwestern B.C.

Barberry Family, *Berberidaceae*
OREGON-GRAPE
Mahonia spp.
Mahonia, Holly Grape, Barberry

Oregon-grape is immediately identified by its evergreen, holly-like leaflets, for no other shrub native to B.C. or Washington bears any resemblance to it. There may be considerable variation in 2 of the 3 species, but the main characteristics remain: 'holly' leaf, bright yellow flowers and green 'grape' berries ripening to a waxy-blue colour. Berries can be used to make jellies and jams.

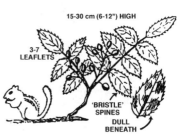

15-30 cm (6-12") HIGH

3-7 LEAFLETS

'BRISTLE' SPINES

DULL BENEATH

CREEPING OREGON-GRAPE, *M. repens*: Has 3–7 leaflets per leaf, dull beneath but possibly with a whitish tinge. Main central vein. Very similar to tall Oregon-grape, but less than 60 cm (2') high and with teeth on the leaves very short and weaker. In dry places to 1800 m (6000') in southeastern B.C. Occasionally in central and southern interior and on eastern side of Vancouver Island. Widely scattered throughout Washington in dry places to 1500 m (5000') elevation. Bunchgrass and ponderosa pine ecosystems and higher into mountains. Ellensburg, Wenatchee, Colville, Spokane, Pullman.

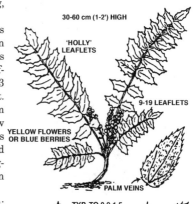

30-60 cm (1-2') HIGH

'HOLLY' LEAFLETS

9-19 LEAFLETS

YELLOW FLOWERS OR BLUE BERRIES

PALM VEINS

DULL OREGON-GRAPE [049], *M. nervosa*: This plant produces sprays of 'holly' to 45 cm (18") in length, each spray with 9–19 leaflets. Young stems carry leaves that are dull green on both sides. Leaflets on older stems are dark green and shiny. The 3 veins, often given as a key feature, are indistinct. The spray of leaves gives an average height to 60 cm (2'). Under the usual forest cover, there may be few yellow flowers or green-to-blue berries. In B.C., it is abundant at lower elevations in the coast forest and the interior cedar-hemlock ecosystems. In Washington, it is generally confined to lower elevations in coastal forests.

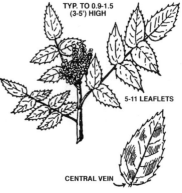

TYP. TO 0.9-1.5 (3-5') HIGH

5-11 LEAFLETS

CENTRAL VEIN

TALL OREGON-GRAPE [048], *M. aquifolium*: Has leaflets shiny on both sides, possibly 5–11 per leaf but generally 7. Main central vein. Favours exposed situations such as road edges and stony clearings. This species is the one most commonly noticed. Sometimes grows as thick, irregular stems reaching 0.9–3 m (3–5') in height (on occasion, even 5 m (16')), but more often a low, sprawling shrub. Bright yellow flowers are prominent in clusters along the stem during May and June. Dark blue berries with whitish bloom are well formed by August and hang on until late fall. They make good jelly when dead ripe or touched with frost. Aboriginals used the bright yellow wood in making a yellow dye. The attractive foliage, flowers and berries together with the brilliant autumn colouring make it a desirable ornamental. Wide distribution throughout southern B.C. in exposed situations with poor rocky soils. Altitudinal range to 1200 m (4000') in the interior. Victoria, Okanagan Valley, Grand Forks. Wide distribution throughout Washington in exposed situations with poor rocky soils. Tonasket, Republic, Spokane, Chelan, Wenatchee Mountains, Tacoma, White Salmon.

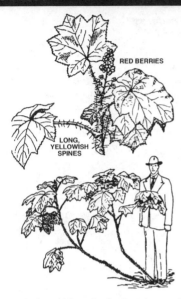

RED BERRIES

LONG, YELLOWISH SPINES

Ginseng Family, *Araliaceae*
DEVIL'S CLUB [052]
Oplopanax horridus

Devil's club stands out because of its large maple-like leaves and thick, spiny stems. In the mottled shadows of cedar swamps, devil's club grows abundantly wherever the ground is black, soft and damp. Its light brown stems rise in crooks and twists to support large, exotic leaves that spread like green platters to catch the sun's filtered rays. Long, yellowish spines bristle from the stems while sparse, thin thorns project from the undersides of the leaves. Terminal clusters of white flowers appear in June. They later change into a pyramid of bright red berries very noticeable during August. They are not edible.

QUICK CHECK: Large 'maple' leaves. Coarse stems about 2.5 cm (1") in diameter, bristling with light brown, needle-like spines.

RANGE BC: Vancouver Island and coastal forests northwards to Alaska. Interior cedar-hemlock ecosystem and northwards to Smithers. Often in association with redcedar but more widespread. Elevation range from sea level to approximately 1350 m (4500').

RANGE WA: From sea level to approximately 1350 m (4500') in coast and mountain forest ecosystems.

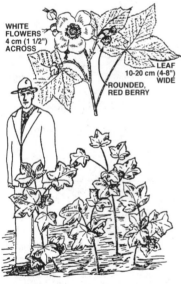

WHITE FLOWERS 4 cm (1 1/2") ACROSS

LEAF 10-20 cm (4-8") WIDE

ROUNDED, RED BERRY

Rose Family, *Rosaceae*
THIMBLEBERRY
Rubus parviflorus

Thimbleberry, a widespread shrub, is particularly noticeable because of its large 'maple' leaves often 20 cm (8") across. In the coast or interior cedar-hemlock ecosystem, thimbleberry masses in damp places along the edges of roads and forest openings. East of the Cascades, although creek bottoms are preferred, the same shrub will also be found in quite dry places.

Ranging from 2.1 m (7') high at the coast to 60 cm (2') in the interior, thimbleberry is an erect, unarmed shrub with short, branching limbs having a dull bloom. The stark white, tissuey flowers, about 4 cm (1 ½") across, show up very dramatically against the background of interlocked, alternate leaves from May to July, serving as a ready means of identification. The berries are not 'thimble' shaped, but round (except where attached) and much like a raspberry. The 'thimble' name may come from the way a large core remains on the plant when it is picked. Berries are soft and have very small drupelets and tiny seeds. They are bright red. Though not much eaten by humans (they mush easily and are not especially tasty) they are sought after by birds and bears during July and August.

QUICK CHECK: Large white flowers or soft, red, rounded berries. 'Blotting paper' 'maple' leaf.

RANGE BC: Fairly common in damp, shady places throughout B.C. northwards as far as Ft. St. John. To 840 m (2800') at the coast and to 1200 m (4000') in the interior.

RANGE WA: Widespread in damp, shady places. Coast, mountain and ponderosa pine ecosystems.

Birch Family, *Betulaceae*
BEAKED HAZELNUT
Corylus cornuta var. *cornuta*

Rather wideranging in interior east of Cascades; grows abundantly, equally at home in exposed rocky places and shady glades. Numerous, slender stems spread in graceful fashion to form a rounded, bushy shrub with heavy foliage. Hazelnut may reach 6 m (20') but is generally 1.5–3.6 m (5–12') tall.

During February and March, conspicuous slender, yellow catkins hang from slight twigs. By July, the nut is taking shape, enclosed within a green, stocking-like husk that blends perfectly with leaves. Edible, ripe by fall, and very popular with squirrels and Steller's jays. Nuts from domesticated hazelnuts (or filberts), also *Corylus* spp., tend to be larger; may be elongated and only partially enclosed by husk.

QUICK CHECK: Yellow catkins in early spring, green 'stocking' husk in summer and fall. 'Stocking' twice length of nut; if shorter, var. *californica*.

RANGE BC: Common east of Cascades and northwards to Fort St. John.

RANGE WA: Lowlands and lower mountain slopes throughout the state.

CALIFORNIA HAZELNUT, *C. cornuta* var. *californica*: Like beaked hazelnut except as described in 'Quick Check,' above. Limited range on southern Vancouver Island, Gulf and San Juan islands and Lower Mainland. Washington's coast forest, sea level to 750 m (2500'). Upper part of ponderosa pine ecosystem. Tieton River, Chinook Pass, Kittitas and King counties.

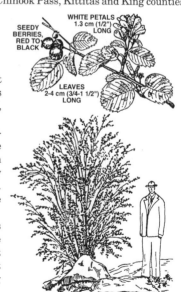

Rose Family, *Rosaceae*
SASKATOON
Amelanchier alnifolia
June Berry, Shadbush, Service Berry

Easily recognized when in bloom or in fruit, but may be so dwarfed or misshapen from typical form as to cause confusion otherwise. Check for small, rounded leaf with big notches along top half.

A preferred browse by cattle and other herbivores. Often under 60 cm (2') high. Variations in shape likely, but when left undisturbed, forms a loose bush with upright limbs fanning outwards at crown. May reach 4.5 m (15') tall, but 1.8–3.6 m (6–12') is typical. Loose, spreading framework, made more noticeable by small leaves—positive identification from afar.

Fragrant clusters of white, 5-petalled blossoms to 2.5 cm (1") across are very abundant and dot the bush from top to bottom during April and May; do not confuse with mock-orange (see p. 142). Berries start to form soon after flowers; a dull red by July, ripening black in early August—a favourite delicacy for bears. Not currently popular, though Natives mixed them with pounded meat to make pemmican.

QUICK CHECK: Clusters of fragrant blossoms during April and May. Many red or half-black, seedy berries during summer and fall. Typically round leaf, notched on top section.

RANGE BC: Most common in southern and central B.C. Sparse around Lower Mainland and on Vancouver Island and Queen Charlottes. Northwards to Alaska. Widespread throughout North America.

RANGE WA: Widespread in Washington. Generally below 600 m (2000') west of Cascades. Very abundant on rocky slopes in all drier regions of eastern Washington.

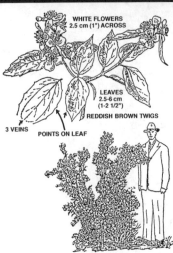

Hydrangea Family, *Hydrangeaceae*
MOCK-ORANGE [057]
Philadelphus lewisii
Syringa, Bridal Wreath

Showy white blossoms invariably arouse curiosity during June. Outshines other shrubs on valley bottoms and rocky slopes with its mass of blooms that continues for at least 2 months. In the interior, seldom grows over 1.8 m (6') high; often a small, ragged shrub niched into clefts on dry, rocky sidehills. At coast, becomes much more robust, to 4.5 m (15') high.

Profuse blossoms can hardly be confused with any other shrub's. Coastal version is very fragrant; interior one almost without perfume. Light green leaves opposite, short-stalked and quite distinctive, with a few points or teeth on each side and a peculiar arrangement of 3 main veins (see drawing). Bright chestnut-brown bark on new twigs, often broken and loose on older stems. Native peoples used straight stems for arrows. The hard, durable wood was also suitable for bows and daily implements.

QUICK CHECK: White 'orange-blossom' flowers from late May to July. Points on leaves and leaf-vein arrangement.

RANGE BC: Low elevations. Sporadic occurrences on southern Vancouver Island. More abundant throughout Fraser Valley, into Fraser Canyon. Frequent occurrences in Okanagan and Similkameen valleys. Also at Shuswap Lake, Christina Lake and the Kootenays.

RANGE WA: Generally at lower elevations. Coast forest, sagebrush, bunchgrass and ponderosa pine ecosystems. Coulee Dam, Dry Falls, Yakima, Spokane, Blue Mountains.

Rose Family, *Rosaceae*
OCEANSPRAY [053]
Holodiscus discolor
Creambush, Arrow-wood

Dogwood has a showier bloom, but oceanspray is more widespread—possibly the most abundant flowering shrub in southern B.C. Nearly every dry forest opening or roadside in this area and northern Washington is softened by masses of loose, creamy plumes from May to July. With erect, spreading stems to 4.5 m (15') in height almost completely covered with bloom; impossible to miss. Not so abundant in interior, and only 1.8–2.4 m (6–8') in height. Like mock-orange, very high in decorative value, being in bloom for several months and having a very visible range along roadsides.

In late summer and throughout winter months, sparse limb framework retains a loose cluster of dried husks that replicates flower plume, a good identifying feature. Attractive even in seed. Prior to blooming, use unusual leaf for identification: severely wedge-shaped base and coarse teeth.

Straight young limbs were a source of arrow shafts, digging sticks and harpoons for aboriginals, for the wood is extremely hard and strong.

QUICK CHECK: Whitish plumes of small flowers or brown plumes of seeds. 'Wedge' leaves with coarse teeth.

RANGE BC: A plant of southern B.C. In dry places from Vancouver Island across B.C. to Columbia Valley. Northwards as far as Cache Creek.

RANGE WA: Dryland shrub of forested areas at lower elevations. Mountain and ponderosa pine ecosystems. Ferry, Stevens and Spokane counties. Steptoe Butte.

Rose Family, *Rosaceae*
GOATSBEARD [149]
Aruncus dioicus

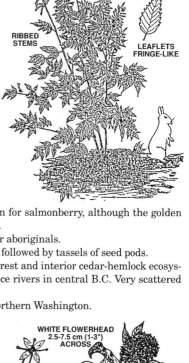

'PENCILS' OF WHITE FLOWERS OR BROWN SEEDS

RIBBED STEMS

LEAFLETS FRINGE-LIKE

Goatsbeard dies to the ground each year and is not a true shrub, but its vigorous growth and shrublike appearance warrant including it here. Found in shady forest borders and seepage areas where the damp soil also supports salmonberry, thimbleberry and elderberry. Its common name probably arises from long, white plumes that make such a showy display during May and June. Much of the top of this 0.9–2.1 m (3–7') high plant often supports many unmistakable, pencil-like strings of very small flowers. Male and female flowers on separate plants.

Thin flower plumes with tiny flowers gradually fade from white to dirty brown as plants form tassels of seed pods. These brown strings last until late summer, and although not very conspicuous, do serve as a means of rapid identification from a distance.

The sharply toothed leaves, with their short side branches often carrying 3 leaflets, might be mistaken for salmonberry, although the golden satiny colour of salmonberry bark is quite distinctive.

This abundant plant had many medicinal uses for aboriginals.

QUICK CHECK: Pencils of small, white flowers followed by tassels of seed pods.

RANGE BC: Low elevations in southern coast forest and interior cedar-hemlock ecosystems. Northwards to Prince George, Skeena and Peace rivers in central B.C. Very scattered on Vancouver Island.

RANGE WA: Very common along roadsides in northern Washington.

Dogwood Family, *Cornaceae*
RED-OSIER DOGWOOD
Cornus stolonifera

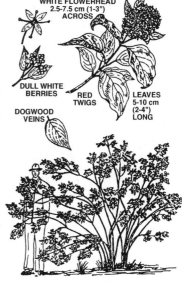

WHITE FLOWERHEAD 2.5-7.5 cm (1-3") ACROSS

DULL WHITE BERRIES

RED TWIGS

DOGWOOD VEINS

LEAVES 5-10 cm (2-4") LONG

Red-osier dogwood has three eye-catching features. In winter and spring, the shiny, bright red bark of the thin stems provides welcome colour. During June, the round heads of dainty flowers are startlingly white against the rich green background of the leaves. In August, it attracts attention with its unusual lead-white berries. The graceful network of 5–7 parallel veins is characteristic of dogwoods, as is the opposite branching.

Usually red-osier dogwood is massed with other shrubs and trees and so in summer loses prominence in the floral picture. In the interior, look along shady creek courses or in damp lowlands with poplar, water birch and willow. Fall weather brings a reddish tint to some of the leaves, which are similar to those of western flowering dogwood (see p. 97). Under ideal conditions along the coast, this dogwood may grow to small tree size and lose its red bark colour.

As with most abundant shrubs, red-osier had various uses with native peoples. In addition, it is an important winter food with the larger browsers, such as deer, elk and moose

QUICK CHECK: In winter, red twig ends and characteristic opposite branching. In summer, 'dogwood' leaf.

RANGE BC: Widespread across B.C. Valley bottoms to near timberline. Alaska Highway.

RANGE WA: Widespread east and west of Cascades. Valley bottoms to near timberline.

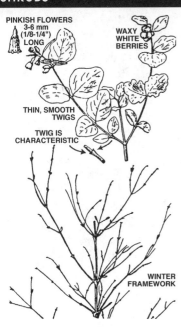

PINKISH FLOWERS
3-6 mm
(1/8-1/4")
LONG

WAXY
WHITE
BERRIES

THIN, SMOOTH
TWIGS

TWIG IS
CHARACTERISTIC

WINTER
FRAMEWORK

Honeysuckle Family, *Caprifoliaceae*
COMMON SNOWBERRY [054]
Symphoricarpos albus
Waxberry

Everyone will remember snowberry for its clumps of waxy, white berries, late summer throughout winter. Forms low, bushy thicket 60–90 cm (2–3') high. Leaves toothless, roughly oval—irregularly lobed on some young plants. Flowers usually clustered near twig ends; in all stages of growth, bud to berry, June–July. Clusters of white, soft berries, irregular in shape, waxy to touch. Wide habitat range, except extreme shade or arid places.

QUICK CHECK: Slender-stemmed shrub usually less than 90 cm (3') high, locally abundant. Thin, opposite leaves, hairy beneath. Clumps of pinkish 'bell' flowers or white, waxy berries.

RANGE BC: Throughout, usually below 600 m (2000') at coast, 900 m (3000') in interior. Perhaps the most widespread low-elevation shrub, but often unnoticed.

RANGE WA: Usually below 600 m (2000') in coastal forests and 1200 m (4000') in east.

WESTERN SNOWBERRY, *S. occidentalis*: Leaves usually not hairy beneath. Stamen and styles longer than petals. Moist places at low to middle elevations east of Cascades. Sporadic occurrences in Washington and B.C.

TRAILING SNOWBERRY, *S. mollis*: Stems trailing and rooting, sometimes the main ground cover. To 20 cm (8") high. Leaves, flowers and fruit smaller than for common snowberry. Open forest areas on southeastern Vancouver Island, Gulf and San Juan islands and adjacent mainland. Generally west of Cascades in Washington.

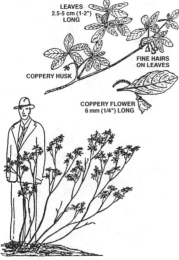

LEAVES
2.5-5 cm (1-2")
LONG

FINE HAIRS
ON LEAVES

COPPERY HUSK

COPPERY FLOWER
6 mm (1/4") LONG

Heather Family, *Ericaceae*
FALSE AZALEA [055]
Menziesia ferruginea
Fool's Huckleberry

The name 'fool's huckleberry' may reflect that while leaves and even flowers closely resemble those of huckleberries, the fruit is just a dry capsule.

Much larger than cultivated azalea (also in *Ericaceae*), with similar loose whorls of small leaves. Common on coast mountains, and where moist conditions prevail; often dominates undergrowth. Usually 0.6–1.8 m (2–6') high; slender, ascending limbs support an artistic pattern of blue-green leaves. Young twigs often covered in fine hairs. Twigs coppery. Leaf bases taper to clasp stems. Unusual pale copper 'bell' flowers on long stems. Often with copperbush, white-flowered rhododendron and black huckleberry. False azalea adds its glow to autumn colours with its distinctive crimson-orange.

QUICK CHECK: Blue-green leaves 2.5–5 cm (1–2") long, in rough whorls; bronze bell-shaped flowers.

RANGE BC: Moist habitats from near sea level to mountain slopes; across B.C. Mt. Revelstoke, Mt. Robson. Central B.C. northwards to Ft. St. John and beyond.

RANGE WA: Moist slopes in mountain forest ecosystem. Olympics, Mason County, Hoquiam, and in the Cascades: Mt. Rainier, Stevens Pass, Skamania County, Mt. Adams.

Honeysuckle Family, *Caprifoliaceae*
BLACK TWINBERRY [060]
Lonicera involucrata
Bearberry Honeysuckle

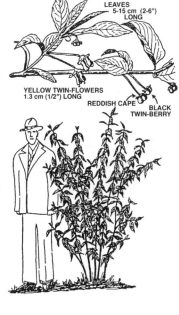

LEAVES
5–15 cm (2–6")
LONG

YELLOW TWIN-FLOWERS
1.3 cm (1/2") LONG

REDDISH CAPE

BLACK
TWIN-BERRY

Black twinberry is a shrub 0.9–3 m (3–10') high that grows on damp ground. On coastal slopes, it may be overshadowed by the luxuriant growth of red-osier dogwood and Pacific crab-apple. East of the Cascades, it grows in more open situations with red-osier dogwood, water birch and cow parsnip.

Yellow twin-flowers 1.3 cm (½") long appear April–June. The unpalatable, black twin-berries with their reddish capes ripen during July and August. Long, light green leaves are tapering and opposite. Flowers, berries and leaves are all excellent identification features, but note that black twinberry is somewhat similar to Utah honeysuckle (below).

QUICK CHECK: Yellow twin-flowers or shiny black twin-berries with contrasting reddish cape.

RANGE BC: Damp places throughout B.C., but especially abundant in coast and interior cedar-hemlock ecosystems. Ranges across B.C. to the Rockies, from sea level to 1500 m (5000'). North-central B.C.: Bella Coola, Quesnel, Chilcotin.

RANGE WA: Damp places throughout Washington, but especially abundant in coastal regions. Ranges from sea level to mountain forests. Coast, mountain and ponderosa pine ecosystems.

Honeysuckle Family, *Caprifoliaceae*
UTAH HONEYSUCKLE [056]
Lonicera utahensis
Red Twinberry

CREAMY
TWIN-FLOWERS
1.3 cm (1/2")
LONG

THIN LEAVES
2.5–5 cm (1–2")
LONG

'JELLY-RED'
BERRIES

The thin stems and opposite leaves of Utah honeysuckle indicate that it belongs to the honeysuckle family. Most shrubs are from 0.6 to 1.5 m (2 to 5') high, with a number of irregular 'spray' branches. Its straggly limbs, small leaves and dirty, dead-looking bark give it a nondescript appearance enlivened only by the twin, yellowish white flowers or jelly-like red berries. Very often 1 flower grows at the expense of the other, resulting in 1 berry being much larger. Flowers are out as early as May at low elevations but might be delayed a month or more on mountain heights. The 2.5–5 cm (1–2") long leaves are very thin, variable in shape and without teeth.

QUICK CHECK: Thin, opposite, ovalish leaves with smooth edges. Twin cream-coloured flowers or pulpy, red 'siamese twin' berries, 1 usually underdeveloped.

RANGE BC: Widely distributed east of Cascades southwards from Anahim Lake, from valley bottom to timberline. Also occurring at subalpine elevations at the coast.

RANGE WA: Widely distributed in mountains east of Cascades, from valley bottom to timberline, but also at subalpine elevations in Olympics. Spokane and Blue Mountains.

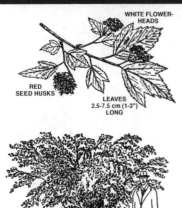

WHITE FLOWER-HEADS

RED SEED HUSKS

LEAVES 2.5-7.5 cm (1-3") LONG

SHREDDING BARK

Rose Family, *Rosaceae*
PACIFIC NINEBARK [058]
Physocarpus capitatus

Ninebarks receive their name from the supposedly 9 thin layers of shreddy bark on the main stems. Older branches do have brown, shredding bark. In coastal regions, Pacific ninebark is a rather dense, upright shrub 2–3.6 m (6–12') high with long, arching branches. Usually it is well scattered and growing with thimbleberry, red-osier dogwood and other eye-catching shrubs.

Distinctive alternate, 3–5 lobed (lobe-toothed) leaves, quite similar to squashberry leaves, often bring ninebark to notice. During May and into June, Pacific ninebark is artistically dappled with white, rounded balls consisting of masses of tiny 5-petalled flowers with many pink stamens. By July 15, these have made a startling change to reddish clumps of rough seed husks. Colour lasts until August, when the shiny yellow seeds are ripe.

QUICK CHECK: 3-lobed, deeply veined, alternate leaves with fine hairs on undersurface. Shreddy, loose bark. Flowers develop into 2 capsules.

RANGE BC: Damp open places, coastal region and in interior cedar-hemlock ecosystem. Creston. Northwards to Alaska.

RANGE WA: Damp, open places, coastal region and near upper edge of ponderosa pine ecosystem. Blue Mountains.

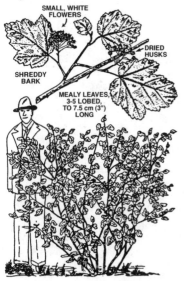

SMALL, WHITE FLOWERS

DRIED HUSKS

SHREDDY BARK

MEALY LEAVES, 3-5 LOBED, TO 7.5 cm (3") LONG

Rose Family, *Rosaceae*
MALLOW NINEBARK
Physocarpus malvaceus

This shrub, because of its rather limited range in the south-central interior of B.C., will not be very familiar to many people in the province. However, it is more widespread in Washington. In September, when the foliage turns a bright russet red, it becomes very attractive and noticeable on semi-open, rocky sidehills. There is considerable similarity to highbush-cranberry at this time of the year.

Mostly stout and bushy, 0.6–1.2 m (2–4') high. With their leaves 3–5 lobed and palmately veined, they might easily be taken for currants. Leaves are mealy. Main stems are light grey with hanging shreds of loose bark. Rounded masses of small, white flowers appear at the twig ends in June; by August they have changed to clusters of brown seed husks.

Mallow ninebark is often associated with Douglas-fir, ponderosa pine, larch, Douglas maple, Saskatoon, Oregon-grape and waxberry.

QUICK CHECK: Shreddy bark on stems; leaves alternate, 3–5 lobed; white flowerhead or mass of brown seed husks. Flowers develop into 3–5 capsules.

RANGE BC: Very abundant on mountains east of Christina Lake and scattered in drier locations around Kootenay Lake. Elko, Fruitvale.

RANGE WA: Very abundant on dry hillsides east of the Cascades. Bunchgrass and ponderosa pine ecosystems. Ferry and Stevens counties. Spokane and Kamiak Butte. Blue Mountains.

Buckthorn Family, *Rhamnaceae*
DEERBRUSH
Ceanothus integerrimus
Entire-leaved Ceanothus

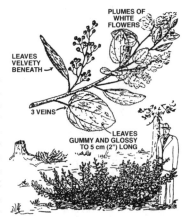

SMALL, BLUISH FLOWERS, IN CLUSTER 5-15 cm (2-6") LONG

LEAVES ALTERNATE, 3-RIBBED TO 5 cm (2") LONG

SEED POD, 6 mm (1/4") ACROSS, HAS 3 DIVISIONS, EACH WITH 1 HARD SEED

This loosely branched shrub with its slender, spreading, greenish limbs might be almost lost in the general floral picture were it not for the deliciously scented clusters of pale-blue-to-white flowers. These are at their best in early May when their misty quality complements the wonders of spring.

The characteristic 3 veins of *Ceanothus* species are evident here in the rather narrow, oval leaves, which seldom reach 5 cm (2") in length. The small flowers bunch in a short plume much like that of a miniature lilac. Usually, each limb branches into twigs, each twig carrying its own cluster of 5-petalled flowers.

The seeds are formed in dry capsules, each having 3 sections bearing a seed apiece.

QUICK CHECK: Tall shrub with narrow, oval, 3-veined leaves. Limited range.

RANGE BC: Not in B.C.

RANGE WA: Generally in exposed places, such as along road edges or rockslides, but tolerant of a fair amount of shade. Lower elevations of easterly slopes of Cascades. Klickitat County. Trout Lake.

Buckthorn Family, *Rhamnaceae*
SNOWBRUSH 063
Ceanothus velutinus
Buckbrush, Sticky Laurel

PLUMES OF WHITE FLOWERS

LEAVES VELVETY BENEATH

3 VEINS

LEAVES GUMMY AND GLOSSY TO 5 cm (2") LONG

Snowbrush grows very abundantly on certain sites. Generally on semi-barren gravel flats or on slopes where sunlight is at a maximum and this bushy, sprawling shrub may form extensive, irregular mats. The smooth, forked limbs twist upwards and carry thick, evergreen leaves about 5 cm (2") long. Their glossy but gummy top surface gives rise to the name 'sticky laurel.' In contrast, the undersurface is soft and velvety. The fine-toothed leaf has 3 main veins, like central fingers, as do all the *Ceanothus* species. When the summer's heat becomes intense, the leaf curls lengthways along its centre, thus preventing water loss as well as minimizing surface exposure.

During June, soft, white heads of tiny, heavily scented flowers rest among the glossy green leaves; they later become small, hard husks. Seeds can lie on the ground for many years until a forest fire stimulates them to germinate. The flowers will make a soapy lather with water and some aboriginal peoples once used the leaves like tobacco.

QUICK CHECK: Glossy evergreen leaves with gummy surface and 3 main veins.

RANGE BC: Upper border of the ponderosa pine ecosystem and into mountain forests. Confined to southern B.C. Abundant near Princeton and on roadside flats between Trail and Castlegar. Cranbrook, Lillooet. Ranges to 900 m (3000') in elevation. Some locations on Vancouver Island.

RANGE WA: Infrequent west of the Cascades but very common in ponderosa pine ecosystem and at higher elevations. Spotty occurrences on lower slopes of Olympics. Ferry, Stevens and Chelan counties. Kettle Falls, Colville, Leavenworth. Blue Mountains.

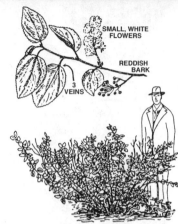

Buckthorn Family, *Rhamnaceae*
REDSTEM CEANOTHUS [062]
Ceanothus sanguineus

Nondescript, bushy; with dead limbs and twigs poking out here and there, seems damaged and unhealthy. Prefers semi-shade and well-drained soils; edges of road clearings, forest openings, rock slides.

Loose, ragged shrub 0.6–1.8 m (2–6') high. Deciduous, oval leaves have characteristic 3 main veins also evident in evergreen relative, snowbrush, and in deerbrush. Leaves very finely toothed. Soft, fluffy masses of small, heavily scented, white flowers near twig ends, out in June. Hard seed pods persist until following year.

In spring, new twigs have a spicy flavour, becomes reddish later in season. With Douglas maple, hazel, oceanspray, Saskatoon, snowbrush. Redstem ceanothus is heavily browsed by deer.

QUICK CHECK: Deciduous, toothed leaves with 3 main veins.

RANGE BC: Upper edges of ponderosa pine ecosystem and higher into mountains. Vernon, Lillooet, Francois Lake. Sparse north of Burns Lake. Some locations on Vancouver Island: Alberni, Buttle Lake.

RANGE WA: Scattered occurrences west of Cascades in coast forests, and bunchgrass ecosystem to upper limits of ponderosa pine ecosystem. Ferry, Spokane and Chelan counties.

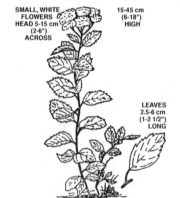

Rose Family, *Rosaceae*
BIRCH-LEAVED SPIREA
Spiraea betulifolia
Flat-top Spirea

Note: For a spirea of bogs and marshes, see p. 126.

Under 60 cm (2') tall, and very common throughout drier, more open forests. Slender stem, often several branches. Leaves rounded like a birch leaf, coarsely toothed along upper ⅔. Dense, flat-topped crown of small, white flowers, occasionally tinged pink, but not all shrubs bloom. Commonly with Douglas-fir, larch, lodgepole pine, soopolallie and kinnikinnick.

QUICK CHECK: Low. Rounded spirea leaf. Flat, white flowerhead on some stems. Do not confuse with scrubby Saskatoon berry.

RANGE BC: East of Cascades in dry, open mountain forests, upper limits of ponderosa pine ecosystem. Cariboo parklands, Stuart Lake, Vanderhoof and northwards to Ft. St. John.

RANGE WA: Mostly east of Cascades in dry, open mountain forests. Bunchgrass and ponderosa pine ecosystems. Chelan, Wenatchee and Pullman.

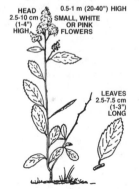

Rose Family, *Rosaceae*
PYRAMID SPIREA
Spiraea pyramidata

The 2.5–10 cm (1–4") long pyramidal head of white or pinkish flowers and large leaves irregularly toothed on the top ⅔ will identify this shrub, which is 0.5–1 m (20–40") in height. Widely distributed.

RANGE BC: Uncommon; middle coastal elevations to 1350 m (4500') east of Cascades. Found in Cariboo parklands ecosystem and as far north as Stuart Lake, at lower elevations.

RANGE WA: East of Cascades in ponderosa pine zone. Kittitas County, Ellensburg, Yakima.

Pea Family, *Fabaceae (Leguminosae)*
SCOTCH BROOM
Cytisus scoparius

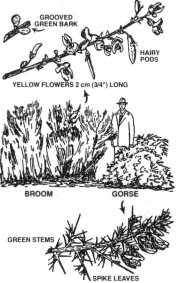

Scotch broom is so widespread on Vancouver Island and southwards throughout the San Juans and Puget Sound that it is included here, although it is not a native shrub. It was introduced to Vancouver Island by Capt. Walter Calhoun Grant in the 1850s. He brought it from Hawaii and planted it on his farm near Sooke, B.C.

The brilliant yellow bloom displayed on dry slopes, rocky knolls and road edges overshadows the efforts of any native plant. Broom is a spindly, ragged mass of slender, unarmed, greenish branches with short, narrow leaves pressed close to them. The flowers stick out from the stem at all angles. Some specimens have blooms that show red and purple shades. Broom is at its most colourful during May, but it does flower into July. However, by this time most bushes are hung thickly with small, black 'pea' pods having many hairs along their edges. The toxic alkaloids contained in these pods pose a risk of poisoning, especially to children. On hot days, pods crack open with easily heard snaps. Scotch broom is often confused with gorse (below).

The ability of this shrub to spread rapidly and crowd out native vegetation places it in the category of a serious weed pest. It is very difficult to eradicate.

QUICK CHECK: Squarish, green twigs with small leaves, yellow flowers or small, hairy 'pea' pods. Note range limitations.

RANGE BC: Roadsides, waste places around Victoria and northwards beyond Campbell River. Gulf Islands. Sporadic occurrences in lower Fraser Valley and at Kootenay Lake.

RANGE WA: Roadsides and waste places throughout western Washington.

GORSE, *Ulex europaeus*: Gorse, an introduced shrub, somewhat resembles Scotch broom by reason of its bushy green form and yellow flowers, which are often out as early as January. It is sprawling, with sharp-pointed needle leaves. Its range and habitat are similar to those of Scotch broom, but it is generally less common. Although very common around Victoria, range is more limited in Washington. Along Puget Sound and western side of Olympic Peninsula.

Pea Family, *Fabaceae (Leguminosae)* [059]
TREE LUPINE
Lupinus arboreus

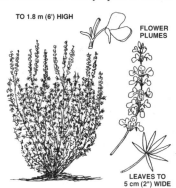

In lieu of a common name for this lupine, a derivation of the botanical name has been used. This name helps to point out the confusion I experienced at discovering a sturdy, thick-stemmed shrublike plant 1.8 m (6') tall covered with a dense growth of small lupine leaves, a 'loner' on top of a rocky bluff above Anacortes. As it was early spring, it as yet gave no evidence of what kind of flowers might appear. Bill Merilees solved my puzzle. I later found it growing abundantly along a stretch of Victoria's waterfront.

QUICK CHECK: A shrub-like plant to 1.8 m (6') tall that bears small lupine leaves with 5–11 leaflets and produces great spikes of yellow 'pea' flowers in late spring.

RANGE BC: Sporadically on southern Vancouver Island and on the Gulf Islands. Powell River.

RANGE WA: Sporadically on the San Juan Islands and Whidbey Island and southwards to California, where it is a native. Anacortes, Port Townsend.

OLIVE-GREEN LEAVES 1.3-4 cm (1/2-1 1/2") LONG

VERY HAIRY TWIGS — WHITE FLOWERS

Heather Family, *Ericaceae*
HAIRY MANZANITA [064]
Arctostaphylos columbiana

Generally limited to Gulf and San Juan islands ecosystem—only on stony slopes or rocky bluffs where sun's heat is at a maximum. Arbutus and kinnikinnick are common companions. Sometimes in coastal Washington several bushes can be on isolated rocky knolls or bluffs high on mountainsides. May be mistaken for a young or dwarfed arbutus (p. 98).

CROOKED, RED LIMBS

Rounded, bushy evergreen shrub, seldom reaches 1.8 m (6') in height. Dull green foliage. Strikingly crooked smooth branches, rich reddish brown colour. Young twigs and leaves very hairy. Urn-shaped, pinkish white flowers, in clusters at branch ends, March–April. They resemble those of salal and arbutus. Later they develop into blackish red, mealy berries. They were gathered by aboriginal peoples and eaten either raw or cooked, but their edibility is not widely known. *Manzanita* is Spanish for 'little apples,' referring to fruit.

QUICK CHECK: Note limited range. Limbs crooked, reddish brown. Thick evergreen leaves. Twigs and young leaves very hairy.

RANGE BC: Rocky, southern exposures in Gulf Islands. Sooke Hills, Malahat Drive, Great Central Lake, Horne Lake.

RANGE WA: Generally west of Cascades in coast forest ecosystem and San Juan Islands. Olympics, Seattle and ranging southwards to Columbia River. Sometimes on isolated knolls on forested mountainsides (but how do they get there?).

GREEN-LEAF MANZANITA, *A. patula*: A 'twin' to the above but more erect and with leaves yellow-green rather than blue-green. Primarily in Klickitat County and in Wasco County, Oregon. Not in B.C.

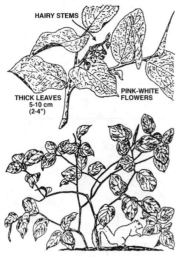

HAIRY STEMS

THICK LEAVES 5-10 cm (2-4") — PINK-WHITE FLOWERS

Heather Family, *Ericaceae*
SALAL [061]
Gaultheria shallon

Named *salal* by coast aboriginals; was highly important because dark, mealy berries were made into a syrup or dried in cakes. Some later residents also know that the edible berries make excellent jams and jellies. Deer also eat them. Possibly the coast forest's most abundant shrub. May vary from low, sparse growth to an impenetrable tangle 3 m (10') or more in height. Thick, tough evergreen leaves egg-shaped, pointed and finely toothed. Stems strong and flexible, to survive heavy, wet snows that may flatten plant to ground. I found the feared dense jungle growth of Yucatan, Mexico, a golf course compared to our own impenetrable coastal salal thickets!

Bell- or urn-shaped flowers hang beadlike along twigs. Blooms mid-May to July, blackish purple berries ripe by August; very long season for both. Dark green, long-lasting leaves favoured by florists; large quantities shipped to eastern Canada and U.S.A. Popular near coast for replanting around construction and in other damaged areas. Grows from seed, but faster from cuttings. Some growers now raise it commercially.

QUICK CHECK: Abundant shrub with leathery evergreen leaves 5–10 cm (2–4") long. White or pink 'bell' flowers; hairy, blackish purple berries.

RANGE BC: Coastal forests to about 750 m (2500') elevation; sporadic at Kootenay Lake.

RANGE WA: Coastal forests to approximately 750 m (2500') elevation.

Rose Family, *Rosaceae*
SALMONBERRY
Rubus spectabilis

SALMON-TO-RED BERRY

LEAFLETS TO 7.5 cm (3") LONG

SATINY BROWN STEMS

RED FLOWERS 2.5 cm (1") WIDE

Common in coastal forest, usually bottomlands or edges of marshes and creeks. Golden, satiny bark and scattered weak spines distinctive in winter. Small bundles of fresh green, toothed leaves appear early April. Blooms an amazingly long time: early April–June. Soft 'raspberry' may be judged insipid, often ripe by late May. Tender young shoots were eaten raw by aboriginals and explorers.

Usually erect, branching, to 1.8–2.4 m (6–8') high, but with favourable conditions, can produce a thicket that would stop Paul Bunyan in his tracks.

Blackberries, raspberries, poison-ivy and poison-oak are only other native shrubs with leaflets in 3s.

QUICK CHECK: Satiny brown stems, very sparse thorns. Leaflets in 3s, pink-red 'tissue' flower 2.5 cm (1") across. Rounded berry, amber or reddish.

RANGE BC: Wet places below 750 m (2500') in coast forest and Gulf Islands ecosystems.
RANGE WA: Wet places below 900 m (3000') in coast forest and Gulf Islands ecosystems.

Rose Family, *Rosaceae*
INDIAN-PLUM [065]
Oemleria cerasiformis
Oso-berry, Bird Cherry, Skunk Bush

LEAVES 7.5-12.5 cm (3-5") LONG

SPRING TWIG

WHITE FLOWERS

BLUISH BLACK BERRIES

First shrub at coast to herald spring as it comes into leaf, often in late February. From each separate cluster of leaves hang 4–9 small, white flowers in various stages of opening. Flowers have peculiar odour, like watermelon rind; torn bark like wood alcohol, crushed leaves also pungent. Prominent, raised veins on leaf undersides. Scattered string of plum-like berries ripen yellowish red to bluish black; large seed, flat flavour. Bushy, erect in open; sprawling if shaded by trees. Often at forest edges.

QUICK CHECK: Upright leaf clusters, strings of white flowers in early spring. Peculiar smell to flower, leaf and bark. Long, tapering leaves and oval berries in summer. Yellow leaves by July.

RANGE BC: Southern Gulf Islands. Victoria–Qualicum. Vancouver–Chilliwack. Squamish.

RANGE WA: Coastal plains, San Juan Islands and low mountain forests west of Cascades. Bellingham, Seattle, Montesano.

SILK TASSEL (Silk Tassel Family, *Garryaceae*), *Garrya fremontii*: Evergreen; opposite leaves dark green. Attractive long, silky tassels of small flowers during winter, slowly replaced by string of purple berries. Limited to brushy borders, western part of Columbia Gorge. Not in B.C.

SILK TASSEL

WESTERN WAHOO (Staff-tree Family, *Celastraceae*), *Euonymus occidentalis*: Straggling, 2.1–5 m (7–16') high; inconspicuous unless in bloom (May–June) or fruit. Small, greenish-to-purple flow-

WESTERN WAHOO

ers 1.3 cm (½") across; neat, cashew-like berries in 3s. Leaves, 5–10 cm (2–4") long, sharp-pointed, with minute teeth. 1 location on Vancouver Island, near Courtenay. Woods west of Cascades, Lewis County to central California. Columbia Gorge.

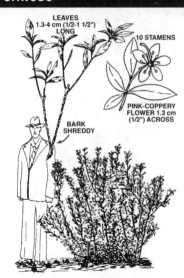

Heather Family, *Ericaceae*
COPPERBUSH
Cladothamnus pyroliflorus

This is a particularly common shrub of coastal forests above 750 m (2500') elevation. It mingles with white-flowered rhododendron and false azalea to form dense thickets. Being 0.9–1.5 m (3–5') high, bushy and with its leaves in rough whorls, it might be passed up for one of its neighbours, especially false azalea. Look closely to see stout stems with loose, coppery bark. Numerous, branching twigs; smooth leaves 1.3–4 cm (½–1 ½") long, covered with a fine, waxy powder when young. Veins show up very prominently on young leaves while older ones have a very noticeable main rib on underside.

If in flower or fruit, you will know it immediately by its round, pale pink-to-coppery flower with curved, protruding stigma or by its bumpy, green seed, which retains this curved protrusion.

QUICK CHECK: Loose, coppery bark on shoots. Old leaves not hairy like false azalea or white-flowered rhododendron. Copper flower with curved style. Fruiting capsule similar in general shape.

RANGE BC: Moist areas in coastal forests, from sea level to subalpine. Most abundant in areas with mountain hemlock and yellow-cedar. Queen Charlotte Islands, north-central B.C. and northwards to Alaska.

RANGE WA: Most abundant in forests with mountain hemlock and yellow-cedar. Subalpine area in Olympics and northerly portion of Cascades.

Heather Family, *Ericaceae*
PACIFIC RHODODENDRON 068
Rhododendron macrophyllum
California or Red Rhododendron

With about 1000 rhododendrons, including azaleas, known in the world, it is amazing that only 27 occur in North America. Of these, only 5 are found in Canada; 3 are described in this book.

Pacific rhododendron, state flower of Washington, is very similar to the cultivated shrub except for its more sprawling form, to 3 m (10') into the air. Generally as extensive patches growing under trees rather than as widely spaced individuals.

Pink blossoms form in June, as round masses strikingly set off by loose rosettes of long, shiny evergreen leaves. Sometimes flower bunches are 15 cm (6") across—a display of colour and beauty surpassing all other native shrubs. Like dogwood, Pacific rhododendron is protected by law in B.C.; in Washington it is only protected from roadside cutting.

QUICK CHECK: Note range. Evergreen leaves 10–15 cm (4–6") long with whitish bloom beneath. Large masses of pink-rose flowers.

RANGE BC: Low to middle elevations. Most abundant in Skagit Valley bordering Washington. Adjacent to Hope–Princeton Highway from 36 to 40 km (22 to 25 miles) east of Hope. On Vancouver Island, a few shrubs in a logged-over area west of Cowichan Lake and at Rhododendron Lake (near Nanoose); also near Quinsam Lake (near Campbell River). An early geologist, G.M. Dawson, found it on mountains above lower Fraser Valley.

RANGE WA: Coast forest ecosystem. Abundant in Skagit Valley bordering B.C. Southeastern flank of Olympics. Sequim, Whidbey Island, Coupeville, Seattle.

Heather Family, *Ericaceae*
WHITE-FLOWERED RHODODENDRON
Rhododendron albiflorum
White Rhododendron

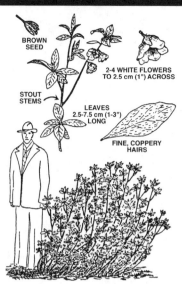

BROWN SEED

2-4 WHITE FLOWERS TO 2.5 cm (1") ACROSS

STOUT STEMS

LEAVES 2.5-7.5 cm (1-3") LONG

FINE, COPPERY HAIRS

Hikers are often so overwhelmed by this shrub's tangle that the descriptive name of 'mountain misery' is hardly out of place. This 0.9–1.8 m (3–6') high shrub favours the coast mountains and the interior cedar-hemlock ecosystem. The conspicuous rough whorls of 5–7 thin, deciduous leaves might be confused with false azalea or copperbush. To verify as rhodendron, turn a leaf so that sunlight glances across the upper surface and look closely for fine, coppery hairs glinting in the sun. The large, white-to-cream flowers are almost 2.5 cm (1") across and are in clusters of 2 to 4. Many shrubs have yellow mottling on the leaves, particularly on mature foliage. Associated shrubs are copperbush, false azalea and mountain-ash.

QUICK CHECK: Leaves 2.5–7.5 cm (1–3") long in rough whorls. Copper-coloured hairs on top of leaf. White flowers in late spring.

RANGE BC: From 750 m (2500') elevation to timberline on coastal slopes. From 1200 m (4000') to timberline in the interior cedar-hemlock ecosystem. On shady, moist mountains throughout the Rockies and central-northern B.C.

RANGE WA: From 1050 m (3500') elevation to near timberline on coastal slopes. On shady, moist mountains throughout the Cascades and Olympic Mountains. Clallam County. Stevens Pass, Mt. Rainier, Mt. Adams.

LAPLAND ROSEBAY, *R. lapponicum*: A low bush with purple flowers. Subalpine and alpine elevations. Northern Rockies. Alaska Highway north of Ft. St. John.

Rose Family, *Rosaceae*
SUBALPINE SPIREA [069]
Spiraea densiflora
Pink Spirea, Mountain Spirea

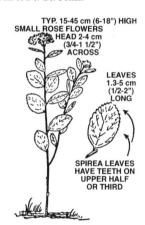

TYP. 15-45 cm (6-18") HIGH
SMALL ROSE FLOWERS
HEAD 2-4 cm (3/4-1 1/2") ACROSS

LEAVES 1.3-5 cm (1/2-2") LONG

SPIREA LEAVES HAVE TEETH ON UPPER HALF OR THIRD

As the name suggests, this trim little shrub is one of high mountain slopes. It is a typical spirea, with a slender stem rarely over 60 cm (2') high, very leafy with oval leaves to 5 cm (2") long. Tiny, pink-to-red flowers, which bloom in August, are massed in small, dense heads 2–4 cm (¾–1 ½") across.

RANGE BC: Subalpine meadows throughout southern B.C.

RANGE WA: Subalpine meadows. Olympics, Cascades (common around Paradise Visitor Center at Mt. Rainier) and Blue Mountains.

FLOWERS WHITE

BERRIES CORAL TO RED

7-11 LEAFLETS

TEETH TO HERE

WESTERN   SITKA

Rose Family, *Rosaceae*
SITKA MOUNTAIN-ASH [066]
Sorbus sitchensis

Most people are familiar with mountain-ash, thanks to the ornamental use of an imported tree species, *S. aucuparia*, with bright red berries (see p. 102). The native mountain-ashes, however, are shrubby growths of the higher mountains. Regardless, the large bunches of red berries that hang on after the leaves have fallen distinguish mountain-ashes from any other tree or shrub.

Sitka mountain-ash, the most common native mountain-ash, comes into fair abundance on high mountain slopes where the trees begin to thin out. Sometimes there are only a few thin stems up to 2.4 m (8') high, with a sparse display of leaves and berries near the top. Under more favourable conditions, a brushy thicket forms. In either case, found in the open.

The compound leaf has 7–11 leaflets with coarse teeth ⅔ of the way to the base. Young twigs have rusty hairs. Small, white-to-cream flowers are carried in round-topped clusters 5–10 cm (2–4") across. These masses of blooms are quite prominent in the high mountains during June.

September sees the bright red berries at their peak, but some hang on until October. Migrating birds sometimes pick off astounding quantities. Berries are not considered edible—they have a sour, mealy taste when raw—but are suitable for making jelly.

QUICK CHECK: Compound leaf with 7–11 leaflets toothed only part way to base. Flowers in round-topped terminal clusters. Berries red with whitish bloom.

RANGE BC: Most abundant above 900 m (3000') at coast, 1200 m (4000') in interior. General throughout. Very common on high mountains between Grand Forks and Rossland. Northwards beyond McLeod Lake.

RANGE WA: Most abundant above 1050 m (3500') in Cascades. General throughout in the mountain and subalpine ecosystems. Spotty in the Olympics and Blue Mountains.

WESTERN MOUNTAIN-ASH [067], *S. scopulina*: Very similar to Sitka mountain-ash but has 9–13 yellowish green, sharp-pointed leaflets with teeth almost their entire length. The flower clusters are usually less than 5 cm (2") across and the berries have a purplish rather than a red colour. The 2 species occasionally grow together. In Washington, this shrub is usually found on the western slopes of the Cascades, but it has been recorded on the eastern side. Quite common in the Olympics. Stevens Pass, Mt. Rainier, Mt. Adams. In B.C., more common east of the Coast Mountains.

FLOWERS

KEY TO THE FLOWERS

Note: In general, but not consistently, each section (for a given colour and number of petals) goes from short to tall. Plants are also grouped according to related species or look-alikes—some secondary species have flower colours different than those they are listed under.

COLOUR: Decide on the basic colour: white, yellow, orange, pink, red, purple, blue, brown or green. A cream-coloured flower would be a variation on white or yellow, while mauve may resolve into either blue or purple on nearby plants. If there is a problem, try a promising alternative. Note: occasionally coded according to fruit colour, if flowers are insignificant.

NUMBER OF PETALS: For flowers with distinct petals, but also see 'Various Shapes' (below). Some plants have bracts or sepals that look like petals. If they look like petals, count them as petals. Some petals are deeply divided, so make sure you are counting whole ones. Now look in the appropriate section for colour and number of petals (2, 3, 4, etc. and then various shapes). E.g., 'Flowers • White,' '4 Petals.'

VARIOUS SHAPES: If the design does not lend itself to an easy petal count, then match the flower with the following designs. Examples are listed in order of colour as they appear in the book, starting with white. Species with several colour possibilities are listed only once, so be sure to cross-reference if necessary.

 TUBULAR: Tubular shape is the overriding key. Many also have a confusing petal design.

Examples: catchflies, campions, valerians, monkey-flower, musk flower, yellow penstemon, butter-and-eggs, Dalmation toadflax, yellow rattle, linanthus, wild bergamot, foxglove, butterwort, monardella, hedge-nettle, penstemons, blue-eyed Mary, gentians, bluebells, mertensia, viper's bugloss, skullcaps.

 'PEA': Typical 'sweet pea' flower. Some are very small and may be grouped into heads; check 'Clustered' as an alternative.

Examples: locoweeds, peavines, milk-vetch, sweet clover, trefoils, hop-clover, sulphur lupine, tree lupine, vetches, clovers, lupines, alfalfa.

CLUSTERED: Clusters, spikes, plumes or heads of flowers usually too small to see their individual structure with the naked eye.

Examples: false lily-of-the-valley, wild lily-of-the-valley, pussytoes, partridgefoot, cotton-grass, white plectrites, meadow death camas, false asphodel, vanilla-leaf, pathfinder, coltsfoot, yarrow, pearly everlasting, burnet, false Solomon's-seal, bistort, baneberry, false bugbane, beargrass, sand-verbena, fiddleneck, skunk cabbage,

SPRING FLOWERS

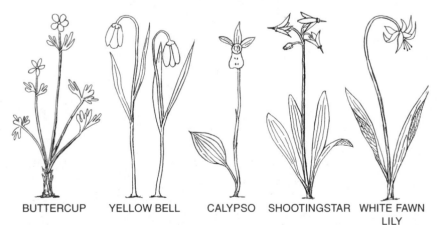

BUTTERCUP   YELLOW BELL   CALYPSO   SHOOTINGSTAR   WHITE FAWN LILY

corydalis, hedysarum, paintbrushes, owl-clover, louseworts, sanicle, goldenrods, tansy, sea blush, thrift, mountain sorrel, elephant's head, roseroot, purple cudweed, ballhead waterleaf, self-heal, Dorr's sage, kittentails, synthyris, teasel, plantain, wormwood, tarragon, dock, pineapple weed, wild sarsaparilla, stinging nettle.

UMBEL: The 'umbrella' design, either single or compound, overrides a possible petal count in very small flowers. Flower stems branch from a common point. Individual flowers usually small.

Examples: buckwheats, desert-parsleys, water-parsnip, water hemlocks, poison-hemlock, wild carrot, cow-parsnip, spring gold.

'DANDELION': Flowerhead of similar narrow ray flowers (like petals). Most plants yellow. Seedhead usually a white ball.

Examples: luina, cat's ear, agoserises, dandelions, silvercrown, microseris, salsify, hawkweeds, hawksbeards, sow thistles, groundsels, ragworts, knapweeds, thistles.

'DAISY': Flowers to 5 cm (2") across, with ray flowers (like petals) and 'button' centres. Usually yellow, white or purple.

Examples: daisies, corn chamomile, fleabanes, butterweeds, goldenweed, haplopappus, brown-eyed Susan, woolly sunflower, Oregon sunshine, arnicas, fleabanes, gumweeds, beggarticks, asters.

'SUNFLOWER': Usually a yellow flower more than 5 cm (2") across. Wide, showy petals often pleated or notched. Contrasting 'button' centre of tiny disk flowers.

Examples: balsamroots, mule's-ears, Cusick's sunflower, common sunflower, mountain sneezeweed, Rocky Mountain helianthella.

'ORCHID': Most people recognize the attractive design of an orchid. Some flowers will be very small and require a close look. The leaves have parallel veins.

Examples: rattlesnake-plantain, ladyslippers, phantom orchid, reinorchids, fairy-slipper, bog-orchid, twayblades, giant helleborine.

UNUSUAL: Plants with unique floral designs.

Examples: groundcones, bleeding heart, steer's head, columbines, meadowrue.

IRREGULAR: Miscellaneous plants of interest.

Examples: pondweeds, water-milfoil, watershield, duckweed, eel-grass, horsetails, scouring-rushes, glassworts, cattail, bulrush.

SUMMER FLOWERS

MARIPOSA LILY   BALSAMROOT   ARNICA   BUTTERWEED   TIGER LILY

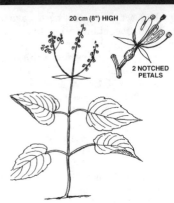

20 cm (8") HIGH

2 NOTCHED PETALS

Evening Primrose Family, *Onagraceae*

ENCHANTER'S NIGHTSHADE [070]

Circaea alpina

The theories for the unusual name are too long and complicated to include; it is enough to explain that Circe was a Greek goddess, the enchantress.

For the amateur botanist, it is interesting to note that this plant stands alone, one of a kind with no close relatives. So what makes it so special besides the name? It might be the clusters of white-to-pink flowers, but **each flower is only the size of a fly**. Look very closely at one. Each of its **2 petals is strongly notched**. There are 2 matching sepals and stamens. Where will you see another flower with such a unique design done in pairs?

This plant is small, averaging about 20 cm (8") tall. Loose clusters of these very small flowers grow on thin stems rising above the opposite, large, heart-shaped, light green leaves. If there are any seeds, wiggle one between your fingers to feel the soft bristles. A preference for damp, shady woods and moist areas. Low to middle elevations.

RANGE BC: Widespread throughout southern B.C.

RANGE WA: Suitable habitats in Washington.

3 PETALS

TO 10 cm (4") HIGH

1-3 FLOWERS 2.5 cm (1") WIDE

Lily Family, *Liliaceae*

ELEGANT MARIPOSA LILY

Calochortus elegans

There are 3 beautiful mariposa lilies in B.C. and a further 3 in Washington. (A large, purple mariposa lily is on p. 293.) Note that they all have **3 broad petals** and **3 shorter, narrower sepals**. All have a purple splotch near each petal base, below which lies a green or yellow nectar gland. 6 symmetrically arranged stamens are easy to see.

Elegant mariposa lily's flower is about 10 cm (4") above ground but the single leaf may extend twice that high. Each very hairy petal has a dark purple horseshoe ring on lower half. With age, green bracts take on a touch of purple and suggest that there are 6 petals. Blooms May through June.

RANGE BC: Not found in B.C.

RANGE WA: Grasslands and nearby ponderosa pine forests in eastern Washington. Pullman. (Also Skyline Drive in Idaho.)

10-30 cm (4-12") HIGH

THREE-SPOT or **BAKER'S MARIPOSA LILY**, *C. apiculatus*: Quickly told from elegant mariposa lily (above) by its greater height, 10–30 cm (4–12"), and a **flower stem that is longer than the leaf**. The several flowers may have a yellow tinge. Each of the 3 petals has a finely hairy lower section with 1 oval-shaped dark mark. These spots show through the petals, giving the plant its name. Blooms late June and into July.

Extends from southeastern B.C.—Canal Flats, Creston, Crowsnest Pass—into northeastern Washington, especially in the ponderosa pine ecosystem, and into Idaho.

Lily Family, *Liliaceae*
LYALL'S MARIPOSA LILY [071]
Calochortus lyallii

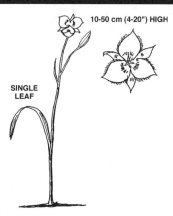

10-50 cm (4-20") HIGH

Calochortus is a Greek word meaning 'beautiful grass,' but the lovely flowers are dominant. Possibly this is the most striking of the white-flowering mariposa lilies. It is 10–50 cm (4–20") tall, with a **single stem leaf about halfway up**; its top might reach the flowerhead. Flowers white but sometimes lavender-tinged. **Above the nectar gland on each petal there is a purple crescent**; on each sepal there is a similar mark, almost hidden. **Petals are sharp-pointed and fringed with slender hairs.** The pointed sepals are about half as long as the petals. Blooms in May.

SINGLE LEAF

RANGE BC: Very rare now; in the high sagebrush–ponderosa pine habitat of the southern Okanagan. It went unrecorded for years; even now its limited habitat is known to only a few people. Sporadic in southeastern part of this ecosystem.

RANGE WA: Very abundant on the eastern slope of the Cascades in open ponderosa pine and Douglas-fir forests. In places, 10 plants per square meter (yard) are found at between 600 m (2000') and 900 m (3000') elevation. Loomis, Winthrop, Leavenworth, Yakima County.

Lily Family, *Liliaceae*
BIG-POD MARIPOSA LILY
Calochortus eurycarpus

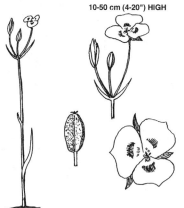

10-50 cm (4-20") HIGH

A tall, erect mariposa lily, 10–50 cm (4–20") in height, with **1 ribbed leaf from the base**, which reaches about a third of the way (or more) up the stem. Usually there is an additional very small leaf about halfway up the stem. The **flowerhead may carry several blooms**, commonly 4, each supported by 3 pale green bracts. Flowers are creamy white to pale lavender. In the middle of each petal is a conspicuous red-purple blotch that shows right through. Petals are as broad as long and are rounded at the tip. A **loose, disorderly pattern of purple hairs is found on the lower inside of each petal.** Blooming season June 15–July 15. False to its name, it bears seed pods that are large neither in comparison to the size of the plant nor in comparison to those of other mariposa lilies.

RANGE BC: Not in B.C.

RANGE WA: Blue Mountains west of Anatone, at an elevation of 1200 m (4000'), in grasslands and open forest of Douglas-fir and ponderosa pine; extends eastwards across central Idaho. Very abundant in certain localities.

SUBALPINE MARIPOSA LILY, *C. subalpinus*: A small plant, 5–30 cm (2–12") tall, with a **single leaf about halfway up the stem**. The flower is creamy, sometimes with a lavender blush. There is a narrow, purple crescent near the base of each petal. There is usually **1 purple dot near the base of each sepal**. Found in loose volcanic soils at medium and higher elevations in the Cascades—the limited range identifies the subalpine species. Southern Washington. Mt. Adams and Mt. St. Helens. Also Mt. Hood, Oregon. Not in B.C.

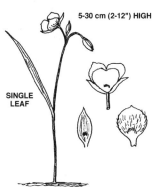

5-30 cm (2-12") HIGH

SINGLE LEAF

WHITE OR PINK FLOWERS PETALS 2.5-5 cm (1-2") LONG

15-60 cm (6-24") HIGH

Lily Family, *Liliaceae*
WHITE TRILLIUM 073
Trillium ovatum
Wake Robin, Western Trillium

Note: A purple-flowered trillium is on p. 293. Unmistakable. A stout stem usually with a whorl of 3 large, net-veined leaves that cradle the short-stemmed white flower, which ages to purple or pink. There are 3 petals 2.5–5 cm (1–2") long and 6 dark, fuzzy stamens. In moist, shady woods, from lowlands to 1200 m (4000'). Blooms April to late May, depending on elevation, often with yellow violets.

RANGE BC: Across southern B.C.

RANGE WA: Widespread.

SESSILE TRILLIUM, *T. chloropetalum*: Moist, shady forests, in the Olympics and southwards into northern Oregon. Distinctive: mottled green leaves and stemless, greenish white flowers, which may become yellow or purplish.

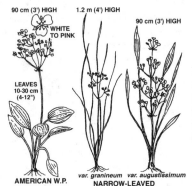

90 cm (3') HIGH   1.2 m (4') HIGH

WHITE TO PINK

90 cm (3') HIGH

LEAVES 10-30 cm (4-12")

AMERICAN W.P.   var. *granineum*   var. *augustissimum*
NARROW-LEAVED

Water Plantain Family, *Alismataceae*
WATER-PLANTAIN 074
Alisma plantago-aquatica

A European **water plant**, now spread across the U.S.A. and Canada. In deep ditches and sloughs and along lake edges. Dark green, pointed leaves, 10–30 cm (4–12") long, are at the ends of long stems to 60 cm (2') in length. Longer, leafless flower stems form a typically whorled, branching top and bear **tiny, 3-petalled flowers**; white but pink-tinged.

RANGE BC: Mostly south of Williams Lake.

RANGE WA: Southwards to California.

NARROW-LEAVED WATER-PLANTAIN, *A. granineum*: **Narrow, grass-like leaves**. Comparatively scarce in south-central B.C. but more abundant further south; extends into California.

A. granineum var. *granineum*, east of the Cascades, has narrow leaves usually floating or submerged. Small flowers white to pink. Shallow lakes and rivers.

A. granineum var. *augustissimum,* west of the Cascades, has wider leaves that are held erect above the water. Marshes and tidal flats.

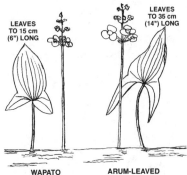

LEAVES TO 15 cm (6") LONG

LEAVES TO 35 cm (14") LONG

WAPATO   ARUM-LEAVED

Water Plantain Family, *Alismataceae*
WAPATO 075
Sagittaria latifolia
Broadleaf Arrowhead, Duck Potato

Shiny green 'arrowheads' rise from shallow waters of ponds, bogs and stream borders. Narrow submerged leaves go unnoticed. Whorls of 3 white waxy flowers on leafless stems. Blooms June–August.

Wapato is Chinook for 'tuberous plant.' A valuable food, the tubers were usually dug out of the mud in water to 90 cm (3') deep and boiled or roasted. Also eaten by diving ducks and muskrats.

RANGE BC: West of the Cascades. Southern Vancouver Island and adjacent mainland.

RANGE WA: West of the Cascades. From B.C. to California. Columbia Gorge.

ARUM-LEAVED ARROWHEAD, *S. cuneata*: More-dramatic leaf shape than wapato. To 35 cm (14") long, **leaves have sharp-pointed spurs**. White flowers arranged in whorls at the top of stiff stalks. East of the Cascades, central B.C. through Washington and Oregon.

Mustard Family, *Brassicaceae (Cruciferae)*
SPRING WHITLOW-GRASS
Draba verna
Vernal Whitlow-Grass

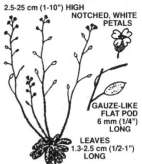

2.5-25 cm (1-10") HIGH
NOTCHED, WHITE PETALS

Often tiny—perhaps not 2.5 cm (1") tall! May be just a white dusting on the ground, but to 25 cm (10") under best conditions.

The 30+ whitlow-grasses, sea level to mountain heights, have either white or yellow flowers (see p. 213). 4 short petals surround 6 stamens, which form a yellow pinhead. Use a magnifying glass. Spring willow-grass has **each petal so deeply divided as to look like 2**—the **only draba with split petals**. In some wonderful fashion, a flat, gauze-like pod materializes from the centre of the fading flower. Only 3 mm (⅛") across, the **round-to-narrow pods alternate from opposite sides of the stem**.

GAUZE-LIKE FLAT POD
6 mm (1/4") LONG

LEAVES
1.3-2.5 cm (1/2-1") LONG

Blooms early, with gold stars, satin-flowers, yellow bells and woodland stars.

RANGE BC: Abundant in low, open areas of southern Gulf and Vancouver islands and adjacent mainland. Occasional in south-central B.C.

RANGE WA: Southern Washington, east of the Cascades. Very common at lower elevations on drier ground. Columbia Gorge.

Mustard Family, *Brassicaceae (Cruciferae)*
LANCE-FRUITED DRABA
Draba lonchocarpa
Twisted Whitlow-Grass

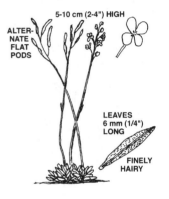

5-10 cm (2-4") HIGH

ALTER-NATE FLAT PODS

Forms a compact bunch of **tiny, narrow leaves**, about 6 mm (¼") long, with **prominent midribs**. Flower stems about 7.5 cm (3") tall, loose heads of small, white flowers. Note the **alternate seed pods**. Minor leaf and seed differences distinguish 3 varieties. Dry meadows and dry, rocky places; subalpine and often above timberline.

LEAVES
6 mm (1/4") LONG

FINELY HAIRY

RANGE BC: Alaska southwards, mostly east of Cascades.

RANGE WA: Throughout in habitats as for B.C.

SNOW DRABA, *D. nivalis*: Whitish hairs make leaves pale grey. Rarer than lance-fruited draba and distinguished partly by diameter of the leaf hairs; same general range.

Mustard Family, *Brassicaceae (Cruciferae)*
PEPPERPOD [076]
Idahoe scapigera
Scale Pod

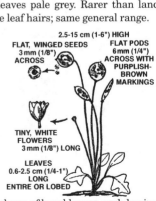

2.5-15 cm (1-6") HIGH
FLAT, WINGED SEEDS
3mm (1/8") ACROSS

FLAT PODS
6mm (1/4") ACROSS WITH PURPLISH-BROWN MARKINGS

Often as white disks a few centimetres (an inch or two) above the ground, glinting in the sunlight, but may reach 15 cm (6") in height. Close together, may be numerous on roadsides.

During late April–May, note the circular, flat pods, each bounded by a very fine hoop-like strand. It holds 2 thin, papery films marked with small, purplish brown blotches—the enclosed seeds. Unripe pods have a pepper-like flavour attractive to children. A clump of basal leaves, each having 2 lobes near its base. Tiny, 4-petal flowers (March–April). Quickly withering leaves.

TINY, WHITE FLOWERS
3 mm (1/8") LONG

LEAVES
0.6-2.5 cm (1/4-1") LONG
ENTIRE OR LOBED

RANGE BC: Rare on southern Vancouver and Gulf islands. Occasional in southern B.C.

RANGE WA: Common on moist slopes in rocky soils of the sagebrush and bunchgrass habitats. Colville, Fort Simcoe, Goldendale, Columbia Gorge.

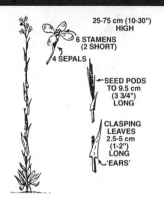

25-75 cm (10-30") HIGH
6 STAMENS (2 SHORT)
4 SEPALS
SEED PODS TO 9.5 cm (3 3/4") LONG
CLASPING LEAVES 2.5-5 cm (1-2") LONG
'EARS'

Mustard Family, *Brassicaceae (Cruciferae)*

DRUMMOND'S ROCKCRESS

Arabis drummondii

Note: Pink rockcresses on p. 263, purple on p. 294.

The many species and varieties of rockcresses are similar; only a few are listed in this book. Flowers assume various pastel shades from area to area. Note the characteristic *Brassicaceae* **cross design of 4 petals**, with 4 sepals and 6 stamens (2 are short). Plants have a single stem. Tiny seeds usually form in 2 rows in **long, narrow, flattened pods**, 3–9.5 cm (1 ¼–3 ¾") in length. Leaves edible, but often bitter. Blooms late spring–summer. Height ranges from 25 to 75 cm (10 to 30").

RANGE BC: Common on dry mountain slopes throughout our area.

RANGE WA: Habitats as for B.C.

TOWER MUSTARD, *A. glabra*: Much coarser than Drummond's rockcress (above), 0.6–1.5 m (2–5') tall, with leaves to 6 cm (2 ½") long and 1.3 cm (½") wide. Basal leaves coarsely toothed. **Slender seed pods, only slightly flattened, tight against the stem.** Pods average 7.5–10 cm (3–4") long. Tiny flowers, white or creamy, form a dense cluster late spring–August. Sporadic throughout, on dry ground at lower elevations and in foothills.

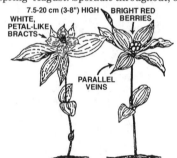

7.5-20 cm (3-8") HIGH
WHITE, PETAL-LIKE BRACTS
BRIGHT RED BERRIES
PARALLEL VEINS

Dogwood Family, *Cornaceae*

BUNCHBERRY [072]

Cornus canadensis

Dwarf Dogwood

Bunchberry, **related to the flowering dogwood tree** (p. 97), has in its typical height of 12.5 cm (5") all the same features except size. See how similar are the somewhat evergreen leaves with their parallel, curved veins. A terminal whorl of 4–7 forms a perfect background for flower or berries. The white 'blossom,' a greenish centre of tiny flowers surrounded by showy white bracts, is about 2.5 cm (1") across.

An abundant forest flower; it carpets many forest dells and glades. Flowers may be out in May and, like the tree, bunchberry sometimes blooms again in late summer. By August, most plants have developed a cluster of **brilliant scarlet red berries**. Insipid and mealy, but not poisonous. Can grow on nitrogen-poor soils, but often on decaying wood and richer sites.

RANGE BC: Throughout, from sea-level shady forest to subalpine terrain.

RANGE WA: Throughout, from sea-level shady forest to subalpine terrain.

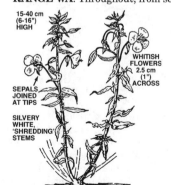

15-40 cm (6-16") HIGH
WHITISH FLOWERS 2.5 cm (1") ACROSS
SEPALS JOINED AT TIPS
SILVERY WHITE, 'SHREDDING' STEMS

Evening-primrose Family, *Onagraceae*

PALE EVENING-PRIMROSE [080]

Oenothera pallida

Pallid Evening-primrose

Note: For an evening-primrose with yellow blooms, see p. 215.

Gnarled, woody and much-branched plant to 40 cm (16") in height. Sandy or **gravelly banks**, or **sand dunes**, with a direct southern exposure to the parching sun is the customary habitat. The 2.5 cm (1") across flowers are a dirty white; buds are pink-tinged. Has **long sepals, generally twisted together at their tips**. Blooms April to early summer.

RANGE BC: East of Cascades in southern B.C.

RANGE WA: East of the Cascades southwards to the Columbia River. Wanapum Road south of Vantage. Columbia Gorge, Snake River.

Evening-primrose Family, *Onagraceae*
DESERT EVENING-PRIMROSE
Oenothera caespitosa

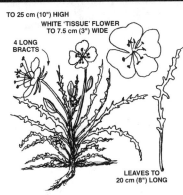

TO 25 cm (10") HIGH

WHITE 'TISSUE' FLOWER TO 7.5 cm (3") WIDE

4 LONG BRACTS

LEAVES TO 20 cm (8") LONG

Evening, May 22, along the Snake River canyon west of Clarkston. Perched on a cliff edge, in the shadows, was an eye-catching cluster of purest white blossoms. Earlier, in the full sun, the flowers would have been closed and the low mat of growth gone unnoticed. How could anything grow in bare ground exposed to direct sunshine day after day?

The crowning glory, the large, 4-petalled 'tissue paper' flowers, are wide and shallowly lobed. **4 long, curved sepals arch down from the base.**

Numerous **long, narrow leaves, notched and toothed in a weird pattern.** A short, stout stem with both seed capsules and slightly higher flower buds ready to lengthen their stems and burst into bloom. All parts hairy, except flowers. Blooms late spring–early summer.

RANGE BC: Not in B.C.

RANGE WA: Dry hills, canyons, desert terrain of eastern Washington.

Bedstraw Family, *Rubiaceae*
NORTHERN BEDSTRAW [077]
Galium boreale

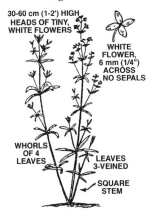

30-60 cm (1-2') HIGH
HEADS OF TINY, WHITE FLOWERS

WHITE FLOWER, 6 mm (1/4") ACROSS
NO SEPALS

WHORLS OF 4 LEAVES

LEAVES 3-VEINED

SQUARE STEM

Bedstraws can be **inconspicuous, garden weeds**, or shady roadside plants both abundant and attractive, with lacy displays of white flowers. Recognize by their **square stems with whorls of 4–8 leaves at the joints**—the number helps in identification. Often, stems and fruit have short, hooked spines that cling to clothing with 'Velcro' stickiness.

Northern bedstraw has **4 very narrow leaves per whorl, each with 3 veins**. To 60 cm (2') tall, it often branches into several erect stems from its base. Flowers, 3–6 mm (⅛–¼") across, have 4 spreading petals, but **no sepals**. In showy, branching end-clusters, they may be white, pinkish or greenish.

Wide altitudinal range, in shady, moist places. Blooms June–September.

RANGE BC: Throughout, except Queen Charlotte Islands; northwards to Alaska.

RANGE WA: Widespread, throughout.

Bedstraw Family, *Rubiaceae*
SWEET-SCENTED BEDSTRAW
Galium triflorum

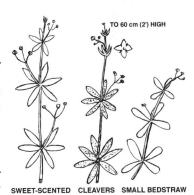

TO 60 cm (2') HIGH

SWEET-SCENTED   CLEAVERS   SMALL BEDSTRAW

The name 'bedstraw' relates back to the time of Christ's birth—it was a plant in the manger.

Sweet-scented bedstraw has **whorls of 5–6 leaves** and a series of **3 tiny flowers per stalk**. To 60 cm (2') long, tends to sprawl or climb weakly. Likes partial shade, wet areas and waste places.

RANGE BC: Lower valleys to mountain slopes.

RANGE WA: Throughout. Habitats as for B.C.

CLEAVERS, *G. aparine*: **6–8 leaves per whorl**. Usually several sweet-scented flower clusters. Damp fields and roadsides; lowlands to mountain slopes. Often in ponderosa pine–Douglas-fir forest. Blooms in early summer.

SMALL BEDSTRAW, *G. trifidum*: A typical bedstraw, **leaves in whorls of 4**. Single, tiny flowers arise from leaf axils. Damp places from lowlands to middle-mountain heights.

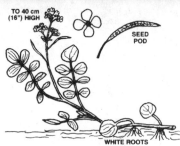

TO 40 cm (16") HIGH

SEED POD

WHITE ROOTS

Mustard Family, *Brassicaceae (Cruciferae)*
COMMON WATERCRESS [083]
Nasturtium officinale
Rorippa nasturtium-aquaticum

Tangy, crisp salad plant brought by Europeans. A convenient food for early miners and settlers. Naturalized over very wide range. In stores and markets. Unlike many plants in or near swamps and ponds, prefers clean, cool water—springs or other running water. On a scorching hot day, a roadside Okanagan irrigation ditch still runs surprisingly cool, its flowing water choked by a green mass of watercress. Small clusters of tiny, white 4-petalled flowers spring–early summer. Individual leaves have 3–9 leaflets. Thin, curved seed pods.

RANGE BC: Cool ditches, springs and streams; lower elevation; southern–central B.C.
RANGE WA: Southwards through Washington to Columbia and Snake Rivers.

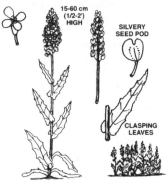

15-60 cm (1/2-2') HIGH

SILVERY SEED POD

CLASPING LEAVES

Mustard Family, *Brassicaceae (Cruciferae)*
FIELD PENNYCRESS
Thlaspi arvense

Different if in flower or in seed. During spring and early summer, the typical young 'mustard' plants form a dense yellow-green growth—a result of the hundreds of seeds dropped the previous fall. Easily recognized **silvery spikes of seed** dominate other vegetation in late summer. An annual, its abundance depends on each year's seed fertility.

Pyramidal flowerhead with a **cluster of tiny, 4-petalled flowers** tops stout stem 15–60 cm (6–24") tall. Large, coarse, alternate leaves clasp stem, like many of its relatives. Subsequent silvery 'penny' seeds; notice notch at top of each.

RANGE BC: An introduced weed widespread throughout our area.
RANGE WA: An introduced weed widespread throughout our area.

ROCK PENNYCRESS, *T. fendleri*: Small, at 10–20 cm (4–8") tall, usually growing densely. Showy, clustered head of very small, 4-petalled flowers. **Basal rosette of tiny 'badminton racket' leaves** about 1.3 cm (½") long. Quite different alternate triangular-to-heart stem leaves clasp stem, climb over halfway. Each alternate, roughly oval, flat **seed pod has tiny projection at end**. Blooms in late May. Moist, rocky places; wide altitudinal range across Washington, not in B.C. Abundant at 1200 m (4000') in Blue Mountains near Anatone.

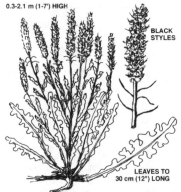

0.3-2.1 m (1-7') HIGH

BLACK STYLES

LEAVES TO 30 cm (12") LONG

Mustard Family, *Brassicaceae (Cruciferae)*
DESERT THELOPODY [115]
Thelypodium laciniatum
Thick-leaf Thelopody

Not too attractive. Invariably grows out of a **near-vertical crack in rock**. On road cuts and canyon walls; its profuse growth belies inhospitable conditions. A great bushy shape of ascending branches. A bleached skeleton often persists until late spring. Usually about 50 cm (20") tall, especially on vertical walls. Stout stem heavily branched from base. Rough rosette of **deeply lobed leaves**. Perhaps a dozen stems per plant, which end in **dense, columnar spikes of very small, dull white or pale purple flowers; projecting styles**. Quite beautiful close up. Blooms May–June.

RANGE BC: Sporadic in south-central B.C.
RANGE WA: On cliffs or rockslides, occasionally sandy soils, of central and eastern Washington. Park and Blue lakes. Vantage, Yakima Canyon. Eastern Columbia Gorge.

Saxifrage Family, *Saxifragaceae*
SAXIFRAGES
Saxifraga spp.

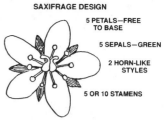

SAXIFRAGE DESIGN

5 PETALS—FREE TO BASE

5 SEPALS—GREEN

2 HORN-LIKE STYLES

5 OR 10 STAMENS

Note: An additional white-flowering saxifrage (*Leptarrhena pyrolifolia*) is on p. 195; there is a yellow *Saxifraga* on p. 218, and a purple one on p. 295.

There are almost 30 species recorded for our area. Trying to correlate both common and Latin names among various references is a maddening puzzle. Names used here follow *The Vascular Plants of British Columbia*, Special Report Series 3 (1991), by Douglas, Straley and Meindinger. A number of saxifrages have so limited a range that they are not included here. Others are widespread. Most have white flowers. In all cases flowers are small, about 1.3 cm (½") across. Nevertheless, the saxifrages form a welcome part of the floral environment. No matter how stark the setting, they appear perfectly designed for it.

The intricate flower design is very attractive. Note that there are **always 5 petals, 5 green sepals and an unusual twin 'horn' style** in the centre. Variations within this design can help quickly identify the species. Leaf shapes are also important.

ALASKA S. [078], *S. ferruginea*: Flower has 2 kinds of petals: 3 broad ones, each with 2 orange-yellow spots, and 2 narrow ones without markings. Note the **rust-coloured sepals**, justifying the *ferruginea* part of the name. Distinctive leaves taper and end in coarse teeth. Moist places on rocks and stream banks, from valley bottom to alpine. From Alaska southwards, on both sides of the Coast Mountains. Also in eastern B.C. and northeastern Washington.

TO 30 cm (12") HIGH

ALASKA S.

TO 30 cm (12") HIGH

TUFTED S.

TUFTED S. [081], *S. cespitosa*: Identify quickly by the **narrow leaves with 3–5 prongs**. Some petals may be notched. Anthers longer than often-purplish sepals. Rocky places, lowland to alpine. Widespread, but most abundant in the Cascades and Olympics.

TO 25 cm (10") HIGH
3 TEETH ON EACH LOBE

MERTENS'S S. [084], *S. mertensiana*: **Large, round leaves** are distinctive. Note the **uniform lobes, each with 3 teeth**. Hairy stems 1–3 times as long as leaf blades. Some white flowers replaced by pink bulblets. Moist, rocky places from lowland to subalpine. Southern and central B.C. Found in Cascades and along coastal strip in Washington.

MERTENS'S S.

CORDATE-LEAVED S., *S. nelsoniana*: Cordate means heart-shaped. Note that **leaves have regular toothing**. The flowers are in a loose, branching head and never have bulblets. There are **10 stamens**. Moist places from middle-mountain to alpine areas. There are a number of varieties of *S. nelsoniana*, with a wide range throughout our area.

TO 15 cm (6") HIGH
10 STAMENS
CORDATE-LEAVED

RUSTY-HAIRED S., *S. rufidula*: As the name suggests, **soft, reddish hair** on the leaves, bracts, flower-stem junctions and sepals is a good clue. The sepals, and sometimes the petals, are purplish tinged. Very similar to western saxifrage (p. 166). Rocky places to alpine heights on Vancouver Island, southwards through Washington to Oregon.

TO 15 cm (6") HIGH
RUSTY-HAIRED

TO 15 cm (6") HIGH
WEDGE-SHAPED PETAL STEM

WEDGE-LEAVED

WEDGE-LEAVED S., *S. adscendens*: Basal rosette of **wedge-shaped, toothed leaves**. Stem leaves have more teeth than do basal leaves. Petals narrow abruptly to short but distinctive stem. Stamen stems shorter than the sepals. Damp subalpine and alpine areas. From the Cascades east to the Rockies, including northern Washington.

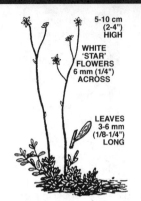

5-10 cm
(2-4")
HIGH

WHITE
'STAR'
FLOWERS
6 mm (1/4")
ACROSS

LEAVES
3-6 mm
(1/8-1/4")
LONG

Saxifrage Family, *Saxifragaceae*
TOLMIE'S SAXIFRAGE [082]
Saxifraga tolmiei

Leaves form a compact mat from which slender, reddish flower stems rise 5–10 cm (2–4"). Round-tipped and 3–6 mm (⅛–¼") long, the **thick, ever-green leaves tend to roll over on the edges**. Generally 1–3 star-like flowers adorn a stem. Each of the 5 narrow, petal-like rods between the real petals supports a pair of stamens. Typically along streams and snowfields at subalpine and alpine heights; quite common in damp, exposed places above timberline.

RANGE BC: Common in the Cascades from northern B.C. southwards. Also on Vancouver Island.

RANGE WA: Common throughout Washington. Olympics.

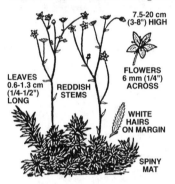

7.5-20 cm
(3-8") HIGH

LEAVES
0.6-1.3 cm
(1/4-1/2")
LONG

REDDISH
STEMS

FLOWERS
6 mm (1/4")
ACROSS

WHITE
HAIRS
ON MARGIN

SPINY
MAT

Saxifrage Family, *Saxifragaceae*
SPOTTED SAXIFRAGE
Saxifraga bronchialis
Prickly Saxifrage, Vesper Saxifrage

Similar to Tolmie's saxifrage, but basal mat has **spiny, sharp leaves**. They are 0.6–1.3 cm (¼–½") long, with hairy edges. The small, white flowers form a galaxy of sparkling stars atop slim stems. The tiny, white-to-cream **petals are artistically specked with distinct maroon and yellow dots**—positive identification.

RANGE BC: Rocky, open places; medium to alpine heights through Cascades; from central B.C. southwards.

RANGE WA: Habitats as for B.C. Cascades, Olympics and Blue Mountains.

THREE-TOOTHED SAXIFRAGE, *S. tricuspidata*: Closely resembles spotted saxifrage, but **petals are marked with orange dots**. Has **tiny, evergreen leaves each tipped with 3 spiny teeth**; they do not have hairy edges. Prefers open, well-drained soils from medium to high elevations. Central B.C. northwards.

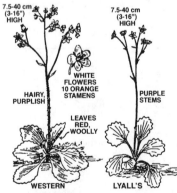

7.5-40 cm
(3-16")
HIGH

7.5-40 cm
(3-16")
HIGH

HAIRY,
PURPLISH

WHITE
FLOWERS
10 ORANGE
STAMENS

PURPLE
STEMS

LEAVES
RED,
WOOLLY

WESTERN

LYALL'S

Saxifrage Family, *Saxifragaceae*
WESTERN SAXIFRAGE [085]
Saxifraga occidentalis

Leaves have very uniform and distinct coarse teeth. Leathery, egg-shaped, serrated leaves, 2.5–7.5 cm (1–3") long and about half as wide, are also good for identification. They have short, broad stems. The branched stem carries a number of **white flowers, beautified by 10 orange stamens**. Older flowers often go cream to greenish. As they wither, it becomes difficult to distinguish them from the emerging reddish brown seeds. Rusty-haired saxifrage (p. 165) is quite similar, but has soft, reddish hair.

RANGE BC: Widespread except for coastal islands and far northeast.

RANGE WA: Widespread except for coastal islands.

RED-STEMMED or **LYALL'S SAXIFRAGE**, *S. lyallii*: Fairly abundant. **Petals have several pale yellow dots. Leaves are a well-defined fan shape, coarsely toothed except at the base** and sometimes covered with black hairs. Often forms a mass of interlocking leaves along damp, subalpine meadows and streamsides. Common across B.C. and north-wards to Alaska. Also plentiful in the northern Cascades and northeastern Washington.

Saxifrage Family, *Saxifragaceae*
GRASSLAND SAXIFRAGE [079]
Saxifraga integrifolia
Northwestern Saxifrage

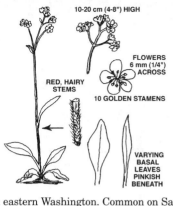

Grassland saxifrage (several varieties) is most common west of the Cascades. **Stem reddish and hairy**. Small, white flowers in packed cluster. **Thick leaves**, many reddish beneath. Leaf shapes and flower colours variable. Shaded plants have leaves to 15 cm (6") long. In mossy, exposed places. Blooms March–April, with shootingstars, satin-flower and spring gold.

RANGE BC: West of and in the Cascades. Common on Vancouver Island and Gulf Islands.

RANGE WA: West of and in the Cascades and in eastern Washington. Common on San Juan Islands. Extends southwards to Columbia Gorge.

BROOK SAXIFRAGE, *S. rivularis*: Small, with **tufted leaves**, it **forms a mat** only 5 cm (2") across. Leaves about 1.3 cm (½") wide, with 3–5 broad lobes—roughly heart-shaped. Slender stems carry 1 or 2 small flowers given a purplish tinge by the calyx. Wet places, subalpine to alpine, from Cascades to Rockies.

Pink Family, *Caryophyllaceae*
THREAD-LEAVED SANDWORT [089]
Arenaria capillaris
Rock Sandwort, Mountain Sandwort

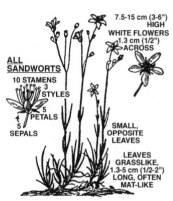

The half-dozen plus wide-ranging sandworts have various names: *Arenaria, Minuartia, Moehringia*. Nearly all have a **tufted or matted base of needle-like leaves**; 2–5 opposite pairs on stem. White flowers have 5 petals, 10 stamens and 3 styles.

Thread-leaved sandwort has purplish **sepals about half the petal length**. Dainty, star-like blooms, about 1.3 cm (½") across, on thin, erect stems to 15 cm (6") tall. Quite common in exposed, dry, rocky places; middle to high elevations.

RANGE BC: Cascades to Rockies.

RANGE WA: Cascades to Rockies. Olympics. Southwards to northern Oregon.

Pink Family, *Caryophyllaceae*
BLUNT-LEAVED SANDWORT
Moehringia lateriflora
Arenaria lateriflora

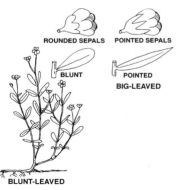

Lacks characteristic clump of spiny basal leaves. Leafy stems to 20 cm (8") tall branch several times. Small, white flowers are single or in clusters of a few. **Petals 2–3 times longer than rounded sepals**.

The **opposite leaves are blunt or round-tipped**, unlike big-leaved sandwort (below).

RANGE BC: Common east of Cascades, valley bottoms to middle-mountain elevations, from northern B.C. southwards.

RANGE WA: Habitats as for B.C.; to California.

BIG-LEAVED SANDWORT, *M. macrophylla*: Very similar in form and flowers to blunt-leaved sandwort, but has **long, sharp-pointed leaves** and **sharp-pointed, petal-length sepals.** In dry situations, meadows to rocky areas, sea level to mountain slopes. Mostly east of Cascades and in the extreme south of B.C., but also along eastern side of Vancouver Island, under Douglas-fir. On both sides of the Cascades in Washington. Blooms March–April.

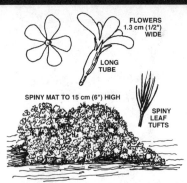

FLOWERS 1.3 cm (1/2") WIDE

LONG TUBE

SPINY MAT TO 15 cm (6") HIGH

SPINY LEAF TUFTS

Phlox Family, *Polemoniaceae*
HOOD'S PHLOX
Phlox hoodii
Grey Phlox

Note: See p. 271 for a white phase of showy phlox.

All phloxes—18 in our area—have symmetrical blossoms with long tubes and 5 spreading petals.

Hood's phlox has **stiff, spine-tipped leaves,** about 6 mm (¼") long, that **tend to be in tufts.** The deep-rooted, **spiny mat,** to 40 cm (16") across, becomes a mass of flowers and buds. The short-lived, attractive flowers are usually white (can tint blue or pink); with a tiny, yellow centre. On hard, dry ground or decorating a rocky exposure. Blooms early in spring, with death-camas and balsamroots.

RANGE BC: Low to middle elevations in sagebrush areas of southeast.

RANGE WA: Dry ground and rocky places in sagebrush zones. Valleys to low mountain heights. Vantage. Mountains that flank Columbia Gorge.

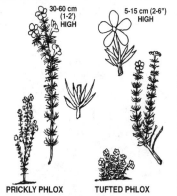

30-60 cm (1-2') HIGH

5-15 cm (2-6") HIGH

PRICKLY PHLOX   TUFTED PHLOX

Phlox Family, *Polemoniaceae*
PRICKLY PHLOX
Leptodactylon pungens

Leptos, Greek for 'thin,' plus *dactylos,* 'finger,' refers to the leaf shape. *Pungens* reflects the aroma. Several of the 18 species of phlox in our area are very similar—in fact, this one was called *Phlox hookeri* for many years. A pink- or red-flowered cousin (see p. 271) is a common neighbour in the dry interior.

Prickly phlox, to 60 cm (2'), is **3 or 4 times as tall as tufted phlox** (below) and is thus easily distinguished. Stout and erect, with short, leafy branches, each carrying a single flower at its tip. However, it **blooms only at night**. (Pollinated by nocturnal moths.) Furthermore, its colour can tinge to pale lavender. **Whorls of thin, spiny leaves** adorn the stem at intervals, earning prickly phlox its name.

RANGE BC: In sagebrush and ponderosa pine areas in southern B.C.

RANGE WA: In sagebrush and ponderosa pine areas throughout Washington.

TUFTED PHLOX, *P. caespitosa*: Like a small prickly phlox, only 5–15 cm (2–6") tall. The usually white flower can be pink to lavender (p. 271), but at least it blooms by day. The **whorls of spiny leaves** are so close together that the large flowers appear to be floating on a spiny mat. Same range as prickly phlox but prefers more shade; also in low mountain forests.

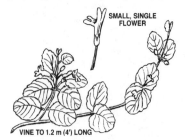

SMALL, SINGLE FLOWER

VINE TO 1.2 m (4') LONG

Mint Family, *Labiatae*
YERBA BUENA [088]
Satureja douglasii

The name 'good herb' was given by early Spanish priests in California. San Francisco's main centre was first known as Yerba Buena after this common little trailing plant. Early settlers made tea with the **aromatic** dried leaves, like the aboriginals did.

In places, yerba buena is the main part of the sparse ground cover, its stems to 1.2 m (4') long, creeping over and through the moss. Size-wise, its twin evergreen leaves fall between those of twinflower and smooth-leaved honeysuckle. Tiny, white 'twin' flowers rise from the paired leaf axils, but note that the **flowers are solitary**, with unusual **2-lobed and 3-lobed lips**.

RANGE BC: Most common west of the Cascades in open coniferous forests, but occasionally on eastern slopes. Southern Vancouver Island, Gulf Islands.

RANGE WA: Habitats as for B.C. San Juan Islands. Southwards into Mexico.

Pink Family, *Caryophyllaceae*
FIELD CHICKWEED [086]

Cerastium arvense

Meadow Chickweed

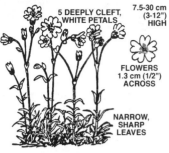

5 DEEPLY CLEFT, WHITE PETALS

7.5-30 cm (3-12") HIGH

FLOWERS 1.3 cm (1/2") ACROSS

NARROW, SHARP LEAVES

The Pacific Northwest has 8 chickweeds; 4 are either garden escapes or quite rare. Field chickweed, a native found worldwide, has flowers to 1.3 cm (½") across. The other 3 natives can be confusing: variable appearance and overlapping ranges. Generally, all are low plants under 30 cm (12") in height. The **5 deeply cleft petals are at least twice as long as the sepals**. Hairy stems carry sharp-pointed, narrow, opposite leaves that may have smaller leaves in their axils. The leaves form a confusing tangle. Blooms April–July.

RANGE BC: Widespread throughout; open and rocky waste areas; low to mid elevations.

RANGE WA: Throughout, in habitats as for B.C.

BERING or **ALPINE CHICKWEED**, *C. beeringianum*: More sprawling; very thin stems, broader leaves. Generally east of Cascades; dry places in subalpine and alpine.

NODDING CHICKWEED, *C. nutans*: No mat of leaves. Flowers nodding. Dampish ground. Alaska southwards through northeastern/central B.C. and Washington to Oregon.

MOUSE-EAR CHICKWEED, *C. fontanum*: Common introduced garden pest. Generally sprawling. Tiny, white flowers, petal-length sepals. Lower elevations.

Pink Family, *Caryophyllaceae*
LONG-STALKED STARWORT [093]

Stellaria longipes

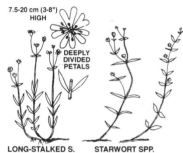

7.5-20 cm (3-8") HIGH

DEEPLY DIVIDED PETALS

LONG-STALKED S.   STARWORT SPP.

Stella or 'star' part of name refers to bright little flowers. 'Wort,' however, from the old English *wyrt*, indicates medicinal value ('sandwort,' 'bladderwort,' 'St. John's-wort'). Starworts have **5 small, deeply notched, white petals**, and usually **5 sepals, 10 stamens, 3 styles**; opposite leaves and many flowers. Many of the 12+ starworts in our area are scarce.

Long, slender flower stalks help give this starwort its name. Mats or clumps. Stemless, narrow, shiny light-green leaves are sharp-pointed. Inspect flower: **petals are sepal length or longer**. Blooms late spring–summer.

RANGE BC: Ranges widely throughout, from low to alpine areas. Favours damp places.

RANGE WA: As for B.C.

CHICKWEED, *S. media*: Common Eurasian garden pest. Trailing, branching stems. **2-part petals shorter than sepals**. Oval, sharp-pointed leaves. Frequent in better soils of southern B.C. and Washington. Likely called 'chickweed' because small birds feed on seeds. You might try it as a succulent salad green in early spring.

OTHER STARWORTS, *Stellaria* spp.: So many have a 'helpless' look to them: slender, twisting stems and tiny flowers. The sketch gives a general idea of what to look for.

Primrose Family, *Primulaceae*
SEA-MILKWORT

Glaux maritima

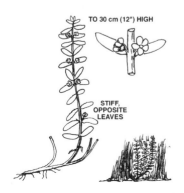

TO 30 cm (12") HIGH

STIFF, OPPOSITE LEAVES

Thick, fleshy stems. **Pretty white or purplish flowers**, near top, from leaf bases. Flowers lack petals, but who cares, when the sepals are so charming? Leaves stiff, short, fleshy; paired on lower stem, alternate above. Often with glasswort. Blooms in early summer. Long ago a brew made from milkwort was given to nursing mothers to increase milk supply.

RANGE BC: Coastal saltwater marshes and tidewater flats, but also in interior saline habitats.

RANGE WA: Habitats as for B.C.

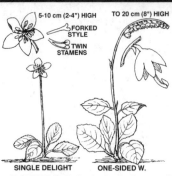

5-10 cm (2-4") HIGH   TO 20 cm (8") HIGH

FORKED STYLE

TWIN STAMENS

SINGLE DELIGHT   ONE-SIDED W.

Wintergreen Family, *Pyrolaceae*
ONE-SIDED WINTERGREEN [105]
Orthilia secunda

Common. Under 20 cm (8") tall. **Toothed, thick, shiny evergreen leaves**, as on most wintergreens.

'One-sided' refers to how the tiny, waxy, 5-petalled **flowers are all on 1 side of the stem.** Initially erect, it becomes archlike under the weight of the row of delicate greenish white 'bell' flowers. Note the decorative **protruding style.** Flowers give way to seed capsules, which dry, reducing the weight, allowing it to straighten somewhat. Summer bloomer.

RANGE BC: Cool, shady forests; lowlands to edge of subalpine. Alaska southwards.

RANGE WA: Habitats as for B.C. Southwards to Mexico.

SINGLE DELIGHT or **ONE-FLOWERED PYROLA**[110], *Moneses uniflora*: Flower may be greenish. Dainty. Once grouped with the pyrolas (similar evergreen leaves) but is only 5–10 cm (2–4") tall. **Nodding saucer-shaped whitish flower; forked beak (style) and twin sets of stamens.** Leaves shiny green and finely toothed. Blooms June–August, in coniferous forests with heavy humus. Wide altitudinal range, sea level to middle-mountain elevations. Both sides of Cascades, Alaska to California. Eastwards to Rockies.

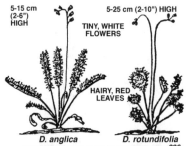

5-15 cm (2-6") HIGH   5-25 cm (2-10") HIGH

TINY, WHITE FLOWERS

HAIRY, RED LEAVES

D. anglica   D. rotundifolia

Sundew Family, *Droseraceae*
SUNDEW
Drosera spp.

Low rosette formed by small leaves, each a stem with a disk at its end. These disks are **bristly with reddish hairs**, each tipped with a **tiny, sticky globe.** They attract and catch insects, which are then absorbed. Inconspicuous white flowers, nodding near the tip of 1 or more thin stems, open only in strong sunlight. Blooms late spring–late summer.

ROUND-LEAVED SUNDEW[096], *D. rotundifolia*: **Small, round leaves.** Sphagnum bogs and lake margins. Widespread throughout our area.

LONG-LEAVED SUNDEW, *D. anglica*: Less common. **Long, narrow leaves** shaped like canoe paddles. Mostly in bogs or wet soils of lakeshores. B.C. southwards to California.

TO 15 cm (6")

Purslane Family, *Portulacaceae*
HEART-LEAVED SPRINGBEAUTY
Claytonia cordifolia
Montia cordifolia

Note: Pink-flowered *Claytonia*s are on p. 270.

All springbeauties bloom early, when their bright flowers and tender dark green leaves command attention. Usually under 25 cm (10") tall. The round, hazelnut-sized corms from which springbeauties grow were dug in number by various aboriginal tribes and were called 'Indian potato.' They were eaten raw or cooked, or stored raw.

The **heart-shaped basal leaves** of *C. cordifolia* distinguish it from all other springbeauties, as do the notched petals. There is an **additional pair of leaves part way up the stem**, as on western springbeauty (below). Flower has 2 sepals, 5 petals, 5 stamens and 1 pistil. This springbeauty grows to 15 cm (6") tall.

RANGE BC: Wet places, meadows, streams and ponds in mid-elevation forests. Most common in southeastern B.C. Rare on Vancouver Island.

RANGE WA: Widespread; habitats as for B.C.

Purslane Family, *Portulacaceae*
WESTERN SPRINGBEAUTY [100]
Claytonia lanceolata

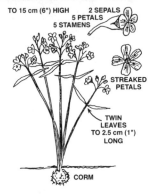

TO 15 cm (6") HIGH   2 SEPALS
5 PETALS
5 STAMENS

STREAKED PETALS

TWIN LEAVES TO 2.5 cm (1") LONG

CORM

Abundant, but flowers **may appear to be beautiful light pink**, because of dark red lines on the petals (see p. 270). Form varies—the one shown here is typical. Each corm may support a number of stems to 15 cm (6") tall.

Like heart-shaped springbeauty, it has 2 sepals, 5 petals and 5 stamens, but the petal tips are only slightly notched and there is a **pair of narrow leaves just below the flower stems**.

RANGE BC: Mostly east of Cascades in southern B.C. From moist ground in sagebrush country to alpine zone. Often seen along melting snowbanks.

RANGE WA: In moist, open forest of Columbia Gorge at medium elevations. Across Washington, from sagebrush valleys to high mountains. Ponderosa pine forests near Cle Elum; blooms in early spring, along with yellow bell and avalanche lilies.

Purslane Family, *Portulacaceae*
MINER'S LETTUCE [098]
Claytonia perfoliata
Montia perfoliata

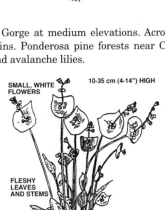

SMALL, WHITE FLOWERS   10-35 cm (4-14") HIGH

FLESHY LEAVES AND STEMS

Miner's lettuce is a name that actually means what it says. Early miners, prospectors and trappers, often at a loss for green vegetables, found it a succulent green. There are about a dozen *Claytonia*s in our area, of which some are rather rare. Miner's lettuce is usually 10–35 cm (4–14") tall but its **basal leaves are very variable in size, colour and shape**. Notice how the **2 stem leaves fuse to form a 'saucer'** that surrounds the stem. Small, 5-petalled white or pinkish flowers grow on thin stems above each of these leaf disks. Among the earliest spring plants to bloom, sometimes showing by late March.

RANGE BC: Widespread throughout southern B.C. in moist, open areas from valley bottom to middle-mountain.

RANGE WA: Wideranging, in habitats as for B.C.

Purslane Family, *Portulacaceae*
SIBERIAN MINER'S LETTUCE
Claytonia sibirica
Montia sibirica

12.5-30 cm (5-12") HIGH

5 NOTCHED, WHITE PETALS WITH RED LINES

You need not go to Siberia to find this plant, although it was first discovered there. It differs from the miner's lettuce above in that the **upper 2 leaves are separate and short stemmed**, instead of joined to form a disk. Lower leaves have long stems.

The flowers, on long thin stems, have **5 petals quite noticeably notched**. Sometimes red lines make them appear pinkish. Blooms April–May, in rich, moist soil of meadows or roadside ditches. Like ordinary miner's lettuce, an excellent salad green.

RANGE BC: Most common west of Cascades; reaches northwards to Prince George, B.C.

RANGE WA: Most common west of the Cascades.

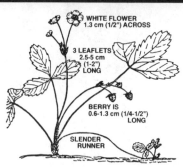

WHITE FLOWER
1.3 cm (1/2") ACROSS

3 LEAFLETS
2.5-5 cm
(1-2")
LONG

BERRY IS
0.6-1.3 cm (1/4-1/2")
LONG

SLENDER
RUNNER

Rose Family, *Rosaceae*
WILD STRAWBERRY [090]
Fragaria virginiana
Blue-leaved Strawberry

Wild strawberry plants are much like the cultivated ones, except smaller, with a smaller, tastier berry. The name 'strawberry' probably derives from the age-old practice of laying straw around cultivated plants to keep the fruit off the ground to prevent rotting. The typical 'strawberry' leaf is quickly recognized. **Everything above ground—leaf, flower, stem or runner—branches directly from a cluster attached to the roots.** Flowers have 5 white petals and a yellow centre; often **among the earliest of spring flowers**.

Each bluish-tinged leaflet of wild strawberry has a very short stem, while the woodland strawberry (below) has stemless leaflets. Also, the fruiting stems are longer than the leaves, while those of the woodland strawberry are shorter. Finally, the terminal tooth of each wild strawberry leaf is narrower and shorter than the adjoining teeth, but on woodland strawberry it is longer than those on either side. If you are old enough and can remember back 60 years or more, you may recall a medicine bottle marked 'Extract of Wild Strawberry.' Produced from the leaves, it was the tried and true remedy for diarrhea—it still has this medicinal use today.

RANGE BC: Common throughout; roadsides and open places to middle-mountain heights.

RANGE WA: Habitats as for B.C.

WOODLAND STRAWBERRY, *F. vesca*: Leaves have yellowish tinge, petals often pinkish. Distinguish from wild strawberry as described above. Common south of Prince George in central B.C. and southwards through Washington; open places to timberline.

COASTAL STRAWBERRY, *F. chiloensis*: Coastal habitat alone is usually enough to identify this one. Often found in seemingly impossible places, such as fine sand baking in the sun. Red runners stretch from plant to plant. **Low, spreading leaves are waxy green; coarsely toothed, prominent veins on silvery undersides**. Common near the shoreline on dunes, rocky bluffs and grassy slopes. Alaska southwards to Chile—even in Hawaii.

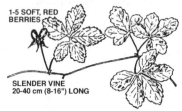

1-5 SOFT, RED
BERRIES

SLENDER VINE
20-40 cm (8-16") LONG

Rose Family, *Rosaceae*
FIVE-LEAVED BRAMBLE [091]
Rubus pedatus
Creeping Raspberry

This tiny, trailing plant, often hiding under false azalea, white rhododendron and huckleberry bushes on higher mountain slopes, could be regarded as one of the more obscure dwarf shrubs in the region. Its appearance is enhanced by its display of symmetrical leaves against a mossy background, bright white flowers and tiny, jewel-like red berries, especially where it is relatively abundant.

The wire-thin, unarmed woody vine may be 20–40 cm (8–16") long. At intervals, a leaf branches off. The leaf, about 2.5 cm (1") across, has just **3 rounded leaflets, 2 of them so deeply lobed that it appears as if it had 5**. The small, white raspberry-like flowers grow singly, with petals lying flat or bent back. They are replaced by **1 to several transparent red berries**, edible and juicy, by September.

RANGE BC: Shady, mossy places in mid-altitude to subalpine coastal forests and interior cedar-hemlock ecosystem. Also north-central B.C. and Rockies. Common along Alaska Highway.

RANGE WA: Shady, mossy places in subalpine elevations of Cascades and Olympic Mountains. Eastwards to Idaho.

Rose Family, *Rosaceae*
SNOW BRAMBLE
Rubus nivalis
Dewberry

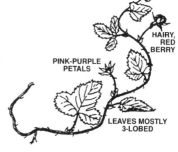

PINK-PURPLE PETALS

HAIRY, RED BERRY

LEAVES MOSTLY 3-LOBED

Short, slender trailer, only a metre or so (a few feet) long. Armed with **numerous small, weak prickles**, as befits a bramble. Evergreen 'blackberry' leaves may have either 3 lobes or 3 leaflets. Flowers usually solitary, with pink or purple petals. The fruit is a finely hairy, **red berry**. Finding snow bramble for the first time is something special.

RANGE BC: Open to shaded moist slopes; subalpine; Vancouver Island, Coast Mountains.

RANGE WA: Most general in moist, exposed places at middle elevations of the Cascades. Olympics. Eastwards to Idaho.

Rose Family, *Rosaceae*
CLOUDBERRY [092]
Rubus chamaemorus

TO 20 cm (8") HIGH

WHITE 'RASPBERRY' FLOWER

YELLOW 'RASPBERRY'

The mountain heights draped in cloud suggested by the common name are in England.

Very attractive, **wiry creeper** worms its way through damp soils and sphagnum moss. At intervals are low clusters of deeply veined, **crinkly leaves; more or less 5-lobed**, something like a maple leaf and coarsely toothed. Stems about 15 cm (6") high bear a single typical **white 'raspberry' bloom**—plants are either male or female. Ripening from light brown, the **raspberry-like yellow fruit** consists of a number of drupelets. Edible, but an acquired taste. Aboriginals across northern Canada valued the fruits as a staple food.

RANGE BC: Decorates sphagnum bogs. Widespread at lower elevations in north half of B.C. Spotty occurrences southwards to Lower Mainland.

RANGE WA: Not known to occur in Washington.

Wintergreen Family, *Pyrolaceae*
INDIAN-PIPE [109]
Monotropa uniflora

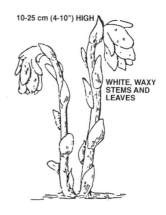

10-25 cm (4-10") HIGH

WHITE, WAXY STEMS AND LEAVES

A person finding Indian-pipe for the first time is sure that his or her senses are being deceived. The waxy white clump of 'pipes' can hardly be likened to any other growing plant. The stout, brittle stems twist over at their tips and droop in a massed cluster of thick, waxy scales. With some imagination, you might deduce that there is a pattern of 5 petals forming a nodding flower. The scale-like leaves are white also and press closely to the stem.

Indian-pipe gets its nourishment by being parasitic through fungal connections to conifers. Look for it in coniferous forests with a mossy floor rich in humus, ranging from low to medium elevations, from late spring to early August. Note that the plant turns black with age.

RANGE BC: Throughout B.C., south of the Peace River but scarce in south-central and central B.C. Common in southern coastal forests. Manning Park.

RANGE WA: Most common in shady coniferous forests west of the Cascades and southwards to California. Columbia Gorge.

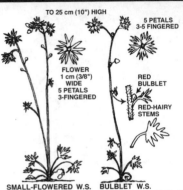

TO 25 cm (10") HIGH

5 PETALS
3-5 FINGERED

FLOWER
1 cm (3/8")
WIDE
5 PETALS
3-FINGERED

RED
BULBLET

RED-HAIRY
STEMS

SMALL-FLOWERED W.S.   BULBLET W.S.

Saxifrage Family, *Saxifragaceae*
SMALL-FLOWERED
WOODLAND STAR [097]
Lithophragma parviflorum
Fringecup, Prairie Star

Note: A pinkish *Lithophragma* appears on p. 270.

Early, dainty flowers in thin-stemmed, white-to-pinkish cluster. **Several flower buds on upper stem. Petal deeply fringed into 3. Leaves from ground**—compare to other woodland stars. Leaves divide into 5 3-lobed parts. Blooms into late June; often with gold stars, shootingstars, yellow bell.

RANGE BC: Open spaces; southeastern Vancouver Island, extreme southern B.C. Wide altitudinal range.

RANGE WA: Wide range, but favours grasslands, sagebrush, and ponderosa pine regions. Vantage, Yakima Canyon, Columbia Gorge.

BULBLET WOODLAND STAR, *L. bulbifera*: Typical woodland star. **Flowers white to pinkish, with 3–5 deep lobes. Reddish stems. Tiny, red bulblets in stem-leaf axils.** Dry, grassy areas in B.C.; wide altitude range. Southeastern Vancouver Island and southern B.C. East of Washington Cascades on dry, open ground, banks, roadsides. Chelan. Columbia Gorge and eastwards throughout ponderosa pine habitat.

SLENDER WOODLAND STAR, *L. tenellum*: To 25 cm (10") tall. **Flowers loosely spaced along stem.** Petals usually 5 lobed. Leaves roundish, deeply lobed, reddish beneath. Limited to ponderosa pine area of southern B.C. (uncommon), but more abundant in Okanogan and Yakima counties. Eastern Washington also.

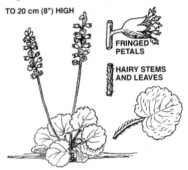

TO 20 cm (8") HIGH

FRINGED
PETALS

HAIRY STEMS
AND LEAVES

Saxifrage Family, *Saxifragaceae*
ELMERA
Elmera racemosa

Neat and little, to 20 cm (8") tall. Creamy white, **bell-shaped flowers alternate along top half of stem. Fringes or fine lobes on 5 tiny petals** (possibly 3–7) make flower fuzzy; 5 stamens. Blooms July–August. Rounded leaves have scalloped edges, wider than long. Short hairs on leaves and stems. Closely resembles mitreworts and fringecups.

RANGE BC: Very rare; on B.C.'s high, rocky slopes. Mt. Cheam.

RANGE WA: Relatively common; high, rocky slopes in Cascades, Olympics. Hart's Pass.

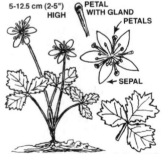

5-12.5 cm (2-5")
HIGH

PETAL
WITH GLAND

PETALS

SEPAL

Buttercup Family, *Ranunculaceae*
THREE-LEAVED GOLDTHREAD
Coptis trifolia

Note: A yellow-flowering goldthread is on p. 220.

Named for its wiry roots. 5–12.5 cm (2–5") tall. Would be lost in moist, mossy places without **small, white, solitary flowers. 'Petals' are really bracts**; between them are **thin petals with tips swollen into nectar glands**. Fuzzy centre of numerous stamens. 3 toothed leaflets per long-stemmed leaf. Blooms late April–May. B.C. only.

Washington has 2 similar goldthreads with **unusual white, strand-like petals and sepals: western goldthread, *C. occidentalis*,** (moist woods, foothills, mountains of west-central Washington) **and coastal goldthread, *C. laciniata*,** (coastal B.C. and Washington).

RANGE BC: Wet woods and bogs, lowlands to middle-mountain forests. Vancouver Island and south coastal B.C. to Alaska. Fairly common west of Cascades.

RANGE WA: *C. trifolia* is not in Washington.

Violet Family, *Violaceae*
CANADA VIOLET [104]
Viola canadensis

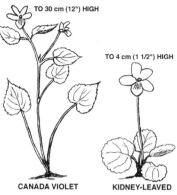

TO 30 cm (12") HIGH

TO 4 cm (1 1/2") HIGH

Usually the violets we see are yellow (p. 219) or blue (p. 315; purple ones on p. 294), so white ones may come as a surprise. The 3 white ones you might find are described here. Their comparative sizes should help you tell them apart.

Canada violet, the largest, has **erect, leafy stems** to 30 cm (12") tall with **long-stemmed, heart-shaped leaves** sharply pointed at their tips. Smaller leaves usually adorn the upper stems. **Dark veins on the lower 3 white petals** of each flower are guides to the yellow petal bases and the spur that

CANADA VIOLET   **KIDNEY-LEAVED**

holds the nectar. The back of the petals has a light purplish tinge. In some areas, the flowers are a very light blue—enough to puzzle you. Blooms May–June.

RANGE BC: Moist forests with rich soils and wetlands, from valley bottoms to middle-mountain elevations throughout our area, except for northwestern B.C.

RANGE WA: In habitats as for B.C.

KIDNEY-LEAVED or **WHITE VIOLET,** *V. renifolia*: This is a low plant, only 2.5–4 cm (1–1 ½") tall, with a single flower to a stem. The **flower has well-defined purple lines on its lower petal**. The leaves are broad or kidney-shaped, with irregular margins. Common in B.C., except for the coastal strip. Possibly in northeastern Washington.

SMALL WHITE VIOLET, *V. pallens*: From 5 to 15 cm (2 to 6") tall. **Very long flower stems and leaf stems rise separately from the ground**. Petals are white, but with the lower 3 having purplish lines. The flower is smaller than Canada violet's, being less than 1 cm (³⁄₈") across. Damp mountain soils and bogs; widespread at middle elevations in our area.

Primrose Family, *Primulaceae*
WHITE SHOOTINGSTAR [102]
Dodecatheon dentatum

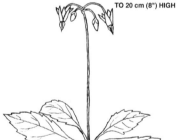

TO 20 cm (8") HIGH

Did I really see a white shootingstar? Probably. Any child seeing a typical (pink) shootingstar (pp. 266–67) for the first time will always remember the dramatic shape, the deep pink petals and the narrow, yellow ring. The combined atmosphere of spring sunshine, sunny, open woods and an abundance of other flowers remains in the memory.

Therefore, a white shootingstar in its shady setting of some cliff damp from spray or along a stream flowing down a mountainside is a surprise. In case

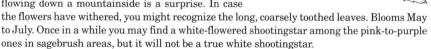

the flowers have withered, you might recognize the long, coarsely toothed leaves. Blooms May to July. Once in a while you may find a white-flowered shootingstar among the pink-to-purple ones in sagebrush areas, but it will not be a true white shootingstar.

Dr. Lyall, a doctor and botanist with the International Boundary Survey in the 1860s, found this plant on the 49th parallel as he crossed the Cascades.

RANGE BC: Shady, damp places on mountainsides on the eastern slope of the Cascades. Southern B.C.

RANGE WA: Shady, damp places on mountainsides on the eastern slope of the Cascades, from B.C. to Oregon. Columbia Gorge.

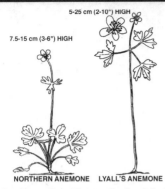

5-25 cm (2-10") HIGH

7.5-15 cm (3-6") HIGH

NORTHERN ANEMONE   LYALL'S ANEMONE

Buttercup Family, *Ranunculaceae*
NORTHERN ANEMONE
Anemone parviflora

Note: Anemones can have 6 'petals' (sepals) or more. Additional white ones on p. 185, yellow one on p. 217, purple ones on p. 300 and blue on p. 313.

Anemones, about 12 in our area (some quite rare), are 'windflowers' as they supposedly open for spring breezes (*anemos* is Greek for 'wind'). Flowers lack petals—the large, shiny sepals can be white, yellow or purple. **Sepals usually blue-tinged on outside. Cottony tuft** seedhead; Western anemone's (p. 185) is most dramatic. Bottom leaves abundant, fern-like: toothed or lobed. **Small leaf cluster part way up stem**.

Dainty northern anemone is 7.5–15 cm (3–6") tall. Clover-like **tri-lobed, toothed basal leaves** form a low clump. Little spray of leaves ⅔ of the way up the stem. Summer bloomer.

RANGE BC: Moist meadows and shady forests; subalpine but also valley bottoms. Common throughout. Rocky Mountains and Cascades.

RANGE WA: Habitats as for B.C. Northern Washington: Rockies and Cascades.

LYALL'S ANEMONE, *A. lyallii*: Slender, less than 30 cm (12") tall, with 3 tri-foliate leaves. **Sepals (usually 5) pure white**. Shady, coastal forests; to subalpine elevations in southwestern B.C.; Olympics, Cascades. Eastern Washington, high in Blue Mountains.

NARCISSUS ANEMONE, *A. narcissiflora*: Abundant north of Prince George, B.C., at higher levels. Also mid- to north-coastal B.C. Extra-large, creamy white flowers in a showy cluster on a central stem. Finely divided 3-lobed leaves. Not in Washington.

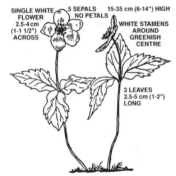

SINGLE WHITE FLOWER 2.5-4 cm (1-1 1/2") ACROSS

5 SEPALS NO PETALS

15-35 cm (6-14") HIGH

WHITE STAMENS AROUND GREENISH CENTRE

3 LEAVES 2.5-5 cm (1-2") LONG

Buttercup Family, *Ranunculaceae*
THREE-LEAVED ANEMONE [095]
Anemone deltoidea
Western White Anemone

Easy to recognize. Large, white 'thimbleberry' blossom. There is a trillium-like **whorl of 3 leaves** just above the half-way point on stem. Lyall's anemone (above) has a similar whorl but leaves are tri-foliate. Sepals variable in size and shape. The **bright green raised centre** is a base for numerous white stamens. Shady places. Depending on elevation, can bloom from April to June.

RANGE BC: Not in B.C.

RANGE WA: Moist woods; lower to middle elevations west of Cascades. Columbia Gorge.

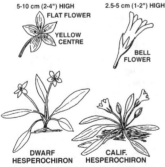

5-10 cm (2-4") HIGH FLAT FLOWER

2.5-5 cm (1-2") HIGH

YELLOW CENTRE

BELL FLOWER

DWARF HESPEROCHIRON   CALIF. HESPEROCHIRON

Waterleaf Family, *Hydrophyllaceae*
DWARF HESPEROCHIRON [101]
Hesperochiron pumilus

Dainty and short. Yellow-centred white flowers, generally 1–5, in a basal clump of narrow, oblong leaves. **Petals lie wide open**, unlike the bowl-shape of California hesperochiron (below), and are sometimes marked with pink lines. **Purple stamens** add beauty. Blooms late spring to early summer.

RANGE BC: Very rare. Near Salmo (southeast).

RANGE WA: Favours moist ground; wide-ranging east of Cascades; valley bottoms to mountainsides. Coulee City, Ellensburg. Also in vicinity of The Dalles in Columbia Gorge.

CALIFORNIA HESPEROCHIRON, *H. californicus*: Like dwarf hesperochiron but **more flowers per plant**. Lavender or purple petals curl to make **bell-shaped flower**; centre not yellow. Wet, alkaline bottomlands and meadows; foothills of eastern Washington.

Wood Sorrel Family, *Oxalidaceae*
OREGON OXALIS [103]
Oxalis oregana

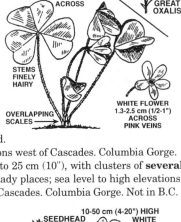

5-18 cm (2-7") HIGH

LEAVES TO 4 cm (1 1/2") ACROSS

NOTCHED PETALS

GREAT OXALIS

STEMS FINELY HAIRY

OVERLAPPING SCALES

WHITE FLOWER 1.3-2.5 cm (1/2-1") ACROSS PINK VEINS

Note: For a yellow-flowering oxalis, see p. 220. Only the 3 most prominent of the half-dozen otherwise small, weedy local oxalises are included here.

When not neatly folded, the **leaves of Oregon oxalis are very clover-like.** All rise from the plant's base. A **single white or pinkish flower,** sometimes with reddish veins, is carried on a stem shorter than the leaves. April and May are the most common blooming months.

RANGE BC: Rare occurrences, coastal. Bamfield.

RANGE WA: Damp, shady woods; lower elevations west of Cascades. Columbia Gorge.

GREAT OXALIS, *O. trilliifolia*: A taller oxalis, to 25 cm (10"), with clusters of **several white flowers per stalk**; notched petals. Damp, shady places; sea level to high elevations. Blooms June to July. Western slope of Washington's Cascades. Columbia Gorge. Not in B.C.

Buttercup Family, *Ranunculaceae*
GLOBEFLOWER [099]
Trollius laxus

10-50 cm (4-20") HIGH

SEEDHEAD BROWN AND GREEN

WHITE FLOWER, 2.5-4 cm (1-1 1/2") ACROSS, WITH GOLDEN CENTRE

Globeflower sometimes pokes its precocious way through the melting snow of alpine meadows. Superficially resembles white marsh-marigold, and do not confuse with the anemones—most have a bluish tinge to outside of 'petals.'

The **showy white flower,** 2.5–4 cm (1–1 ½") across, is composed of 5.to 6 sepals surrounding a beautiful golden centre, a circlet of 7–15 tiny petals. The flower soon discolours and wilts. Plants in Rockies often have pink-tinged sepals.

Under 50 cm (20") in height. The long-stemmed basal leaves are palmately divided into 5 main segments, which are themselves further divided and toothed. Towards the top, the stem leaves get smaller and their stems become shorter. The main stem elongates with age. Note group of leaves just below flower.

RANGE BC: Usually east of Cascades; wet meadows of subalpine and alpine ecosystems.

RANGE WA: Habitat as for B.C., but more widespread across state.

Saxifrage Family, *Saxifragaceae*
FOAMFLOWER
Tiarella spp.

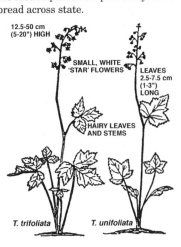

12.5-50 cm (5-20") HIGH

SMALL, WHITE 'STAR' FLOWERS

LEAVES 2.5-7.5 cm (1-3") LONG

HAIRY LEAVES AND STEMS

T. trifoliata   *T. unifoliata*

There is **only 1 leaf on each slender flower stem,** to 50 cm (20") tall, but a group of **long-stemmed leaves grow from the base.** The different leaf shapes easily identify the 3 species in our area. Tiny, white flowers near top of stem dance on short, wire-like stems; 5 petals and 10 stamens. Long blooming season: May–July.

THREE-LEAVED F. [107], *T. trifoliata*: Each **leaf is divided into 3 short-stemmed leaflets.** Generally in moist shady woods, lower elevations to 650 m (2000'). Widely distributed.

ONE-LEAVED F., *T. unifoliata*: Each **leaf has 3 to 5 shallow lobes,** rather like a maple leaf. To elevations of 1050 m (3500'). East of Cascades.

CUT-LEAVED F., *T. laciniata*: **Leaf divided into 3 leaflets, each deeply cut into** several lobes. San Juan, Gulf and Vancouver islands. Puget Sound.

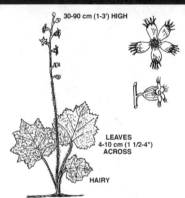

30-90 cm (1-3') HIGH

LEAVES
4-10 cm (1 1/2-4")
ACROSS

HAIRY

Saxifrage Family, *Saxifragaceae*
FRINGECUP [106]
Tellima grandiflora

Note: 'Fringecup' formerly meant the genus *Lithophragma* (woodland stars, pp. 174 & 270).

A distinctive plant, 30–90 cm (1–3') tall, with **finely fringed petals**, and the only *Tellima*. Do not confuse with elmera (p. 174). Flowers, which change colour with age (may be greenish, pinkish or brownish) are scattered along a hairy stem that rises well above large, **heart-shaped, coarsely toothed, long-stemmed leaves**. Moist thickets and damp, shady banks appear ideal for it.

RANGE BC: Generally coastal, from Alaska southwards, and in forested areas of eastern B.C.

RANGE WA: Generally coastal, southwards to California. Also in forested areas of eastern Washington. Columbia Gorge.

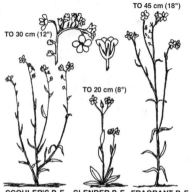

TO 45 cm (18")

TO 30 cm (12")

TO 20 cm (8")

SCOULER'S P. F.   SLENDER P. F.   FRAGRANT P. F.

Forget-me-not Family, *Boraginaceae*
SCOULER'S POPCORNFLOWER [094]
Plagiobothrys scouleri

There are only 3 popcornflowers in B.C., where they are rather rare, but they are more abundant in Washington. All are under 45 cm (18") in height and carry **tiny, 5-petalled white flowers**, the same shape and size as forget-me-not flowers. The 'popcorn' part of the name comes from the several tightly clustered flowers and adjoining coil of buds that give an appearance of kernels of popcorn.

Scouler's popcornflower grows to 30 cm (12") tall, but is **often prostrate**. The unusually arranged **hairy leaves are opposite low on the stem and alternate above**. From the base it branches into thin stems, each having a slightly coiled top with several flowers. Fairly common in moist soils. Blooms late spring–early summer.

RANGE BC: Vancouver Island and south-central B.C.

RANGE WA: Across the state in low, moist habitats.

SLENDER POPCORNFLOWER, *P. tenellus*: Usually under 20 cm (8") tall, this erect little plant scatters itself amidst the fresh spring grass. In southern Washington, it may bloom April through May, along with white plectritis, woodland stars and northwestern saxifrage. Note the **basal clump of small leaves**. The tiny stem leaves are all alternate. While rare on southern Vancouver Island and the Gulf Islands, it is abundant in the Columbia Gorge and eastwards, in dry, open places. Coulee City.

FRAGRANT POPCORNFLOWER, *P. figuratus*: The largest of the 3 popcornflowers in our area—to 45 cm (18") tall. A slender, branching plant with flowers to 1.3 cm (½") across, larger than on other popcornflowers. Notice the yellow eye and the fragrance. Rare on southeastern Vancouver Island and Gulf Islands, but fairly common in lowland Washington, west of Cascades.

Buckbean Family, *Menyanthaceae*
DEER-CABBAGE
Fauria crista-galli

TO 30 cm (12") HIGH

Deer-cabbage and buckbean share soggy habitat and growth patterns. New leaf and flower stems, to a height of 30 cm (12"), arise from growing tips of stout roots. **Star-shaped white flowers**, about 2 cm (¾") across, are in clusters. The **5 or 6 unusual crinkly edged petals** have low, scaly lengthwise ridges. A bad odour attracts pollinating flies.

Broad, kidney-shaped leaves, to 15 cm (6") across, variable stem length, have attractive finely scalloped edges. Blooms during summer.

RANGE BC: Most common in **coastal bogs and seepage areas**; lowlands to subalpine.

RANGE WA: Widespread, in habitats as for B.C., including Olympic Peninsula.

Buckbean Family, *Menyanthaceae*
BUCKBEAN [111]
Menyanthes trifoliata

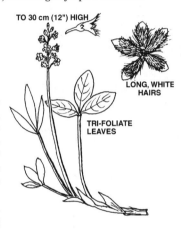

TO 30 cm (12") HIGH

LONG, WHITE HAIRS

TRI-FOLIATE LEAVES

If your feet are getting wet, you are in favourable habitat: bogs and marshes. A small pond or water-way may be completely choked by small buckbean.

Sometimes on muskegs, roots hidden in the mossy floor, but usually with thick roots embedded in the mud of marshes and bogs, stout stems rising to 30 cm (12") above the water. Some have distinctive triple leaflets (hence *trifoliata*), others a loose cluster of rather startling flowers. **Long, white hairs cover the 5 straggly looking petals** on top—no limit to floral design! Blooms spring to early summer.

RANGE BC: Southwards from Alaska. Common throughout B.C.; bogs, marshes and shallow ponds; valley bottoms to subalpine elevations.

RANGE WA: Common throughout; habitats as for B.C.

Buttercup Family, *Ranunculaceae*
WHITE WATER-BUTTERCUP [087]
Ranunculus aquatilis
Water Crowfoot

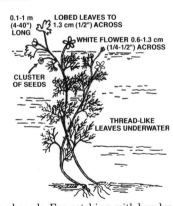

0.1-1 m (4-40") LONG

LOBED LEAVES TO 1.3 cm (1/2") ACROSS

WHITE FLOWER 0.6-1.3 cm (1/4-1/2") ACROSS

CLUSTER OF SEEDS

THREAD-LIKE LEAVES UNDERWATER

Has 2 completely different leaf types: conspicuous, small, deeply lobed, floating ones, 1 to 2 times the size of the flower, and a submerged tangle of thin, multi-branched, short-stemmed, green filaments, limp out of water.

Small, white flowers with bright yellow centres; about 1.3 cm (½") across. **No small scale** at yellow base of each petal, unlike other buttercups.

The complex of stems and leaves, to 1 m (40") long, usually only just protrudes above water.

Masses in ditches and backwaters, sluggish streams and ponds. Eye-catching, with hundreds of blooms; spring–late summer (depends on water temperature). Food of ducks and geese.

RANGE BC: Throughout, lowlands to middle-mountain heights.

RANGE WA: Habitats as for B.C.

STIFF-LEAVED WATER-BUTTERCUP, *R. circinatus*: Like white water-buttercup (above) but with stemless submerged leaves.

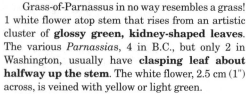

15-30 cm (6-12") HIGH

NORTHERN   NORTHERN

FRINGED   KOTZEBUE'S

SMALL-FLOWERED

Saxifrage Family, *Saxifragaceae*
FRINGED GRASS-OF-PARNASSUS [108]
Parnassia fimbriata

Grass-of-Parnassus in no way resembles a grass! 1 white flower atop stem that rises from an artistic cluster of **glossy green, kidney-shaped leaves**. The various *Parnassias*, 4 in B.C., but only 2 in Washington, usually have **clasping leaf about halfway up the stem**. The white flower, 2.5 cm (1") across, is veined with yellow or light green.

Fringed grass-of-Parnassus has **fringed petal bases** and shiny stamens—a beautiful design! **Petals are about twice the length of the sepals**. Always associated with water: streamsides, wet meadows, bogs. At home from middle-mountain heights to alpine terrain. Plants at higher elevations will still be in bloom after most summer flowers have finished.

RANGE BC: Common throughout.

RANGE WA: Common throughout, in typical habitats.

KOTZEBUE'S or **ALPINE G.**, *P. kotzebuei*: **Sepal-length petals**, usually 3-veined. **No small leaf at mid-stem**. Wet places, from mountain forests to alpine. Most common east of B.C.'s Coast Mountains and north of Prince George. Southwards to just reach Cascades of Okanogan County.

NORTHERN or **MARSH G.**, *P. palustris*: **Unfringed petals, about 2 times as long as sepals**, with 7–13 veins. Stem leaf is below centre. East of Cascades, on wet ground from valley bottom to subalpine, except extreme southern B.C. Not known in Washington.

SMALL-FLOWERED G., *P. parviflora*: Small 'paddle' leaves. **Petals are about 1 ½ times as long as sepals, and are usually 5-veined. Stem leaf is below centre**. Bogs and stream banks in mountain forests of southwestern and south-central B.C. Not known in Washington.

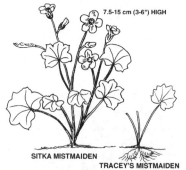

SITKA MISTMAIDEN

TRACEY'S MISTMAIDEN

7.5-15 cm (3-6") HIGH

Waterleaf Family, *Hydrophyllaceae*
SITKA MISTMAIDEN [112]
Romanzoffia sitchensis
Sitka Romanzoffia

A treasure of a plant. Usually 7.5–15 cm (3–6") tall. Usually distinctive enough to easily recognize. Resembles a saxifrage, but the **5 petals are fused at the base to form a tube**. Yellow throated bright white 'buttercup' flowers are 1.3 cm (½") wide.

The plant's appearance—form, unusual leaves and flowers—is enhanced by its beautiful habitat: cool, shady canyons or mossy cliffs with spraying water. Ferns and some of the saxifrages that could be companions also come to mind. Blooms during the summer. Wet places from middle-mountain elevations to the subalpine.

RANGE BC: High mountains; Vancouver Island, Queen Charlottes, Cascades and southeastern B.C. Lower elevations in the north.

RANGE WA: Generally on higher mountains, lower elevations in Columbia Gorge.

TRACY'S MISTMAIDEN, *R. tracyi*: Rather rare, and the only other mistmaiden in our area. **Woolly brown tubers at base** of plant. **Flowers are a soft pink**. Coastal areas around southern end of Vancouver Island and from the Olympics southwards to California.

Saxifrage Family, *Saxifragaceae*
COAST BOYKINIA
Boykinia elata

May be mistaken for saxifrages, foamflowers and alumroots. '**Maple' leaves.** Numerous, finely scattered (not umbel) **small, white flowers; 5 well-separated petals, 5 sepals, 5 stamens; 2 styles form a horn-like structure** (see saxifrages, p.165). **Broad leaves**, 2.5–4 cm (1–1 ½") wide; **brown, hairy stems. Fine bristle on small leaf** where flower stem branches from main stem.

RANGE BC: Moist woods, mountain streams; valley bottom to middle-mountain; south coast.

RANGE WA: Habitats as for B.C.

LARGE-FLOWERED BOYKINIA, *B. major*: Usually much larger than coast boykinia, at 0.3–1.1 m (1–3 ½') tall. Similar, large maple-like leaves reach 20 cm (8"). Loose head of white flowers; broad, oval petals, yellow centres. **Large stipules at leaf junctions on upper stem**. Coastal Washington only; streamsides and wet meadows. Not in B.C.

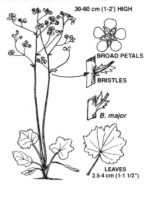

30-60 cm (1-2') HIGH

BROAD PETALS

BRISTLES

B. major

LEAVES
2.5-4 cm (1-1 1/2")

Saxifrage Family, *Saxifragaceae*
SMALL-FLOWERED ALUMROOT [114]
Heuchera micrantha

Named for astringent roots; 5 in our area, flowers white to cream or greenish. Basal group of '**maple' leaves** on long stems. Thin petals, 5 stamens.

Small-flowered alumroot is often along shady roadsides or clinging to rocky walls. Spray of stems, **100s of tiny blooms, 3–5 per side branch** (same for smooth alumroot). **Slender stems reddish, covered with fine, white hairs**. Long-stemmed leaves; rounded lobes, heart-shaped bases. Hairy leaf undersides, stalks. Usually no stem leaves. **Hair-fringed leaf stipules at leaf base**. Rocky habitats.

RANGE BC: Western slope, coastal mountains. Lowland–alpine. Gulf and Vancouver islands.

RANGE WA: Cascades to Columbia Gorge. Northeast and Blue Mountains.

SMOOTH ALUMROOT, *H. glabra*: Similar to *H.micrantha*. Leaf blade as wide or wider than long; several stem leaves. **Sharp-pointed leaf lobes, hairless stem, leaves**. Moist cliffs, canyons, stream banks; coast–alpine. B.C.'s Selkirks, Washington's Wenatchees.

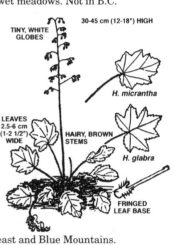

TINY, WHITE GLOBES

30-45 cm (12-18") HIGH

H. micrantha

LEAVES
2.5-6 cm
(1-2 1/2")
WIDE

HAIRY, BROWN
STEMS

H. glabra

FRINGED
LEAF BASE

Saxifrage Family, *Saxifragaceae*
ROUND-LEAVED ALUMROOT [113]
Heuchera cylindrica
Oval-leaf Alumroot

Lightly hairy, **broad-lobed, round leaves**; basal rosette. Stem 30–50 cm (12–20") tall, **tightly clustered small flowers**; white to cream. Distinct petals; yellow stamens deep in throat. May–July. Hardy: **rockslides, talus slopes**, poor soils.

RANGE BC: Taller and most common east of Cascades. Quesnel southwards.

RANGE WA: Most common east of Cascades.

MEADOW ALUMROOT, *H. chlorantha*: Tall, at 40–80 cm (16–32"). Leaf blades slightly wider than long. **Long, brownish hairs on leaves and lower stems. Narrow, smooth leaf stipules**. Tightly packed head of greenish flowers, **pointing upwards**. Moist bluffs, river banks; to middle-mountain forests. Most common west of Cascades. Queen Charlotte Islands; southwards to Oregon.

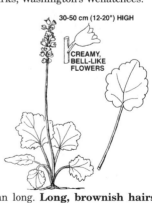

30-50 cm (12-20") HIGH

CREAMY,
BELL-LIKE
FLOWERS

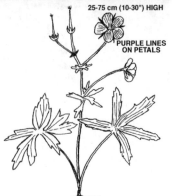

25-75 cm (10-30") HIGH

PURPLE LINES ON PETALS

Geranium Family, *Geraniaceae*
WHITE GERANIUM
Geranium richardsonii

Note: Geraniums with pink flowers are on p. 269.

The **only geranium in our area with white flowers**, pinkish tinged because of purple lines on petals. Flowers almost 2.5 cm (1") across. Long seed pods also add interest. Averages 25–75 cm (10–30") in height. Leaves are quite variable, but dependably divided into 5–7 segments with pointed lobes.

RANGE BC: East of Cascades in moist soils and shady woods. Ranges from valley bottoms to middle-mountain elevations. Central B.C. southwards.

RANGE WA: East of Cascades, habitats as for B.C.

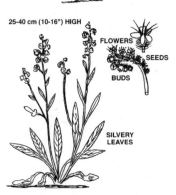

25-40 cm (10-16") HIGH

FLOWERS

SEEDS

BUDS

SILVERY LEAVES

Waterleaf Family, *Hydrophyllaceae*
SILVERLEAF PHACELIA [116]
Phacelia hastata
White-leaved Phacelia

Phacelias are usually easily recognized by their small, typically pastel-coloured flowers (purple on p. 296). **All 5 petals are of uniform size**; usually a fuzz of protruding stamens. They prefer dry terrain: arid valley bottoms to windswept alpine heights.

Silverleaf, 25–40 cm (10–16") tall, has broadly lance-like leaves (possibly with small lobes towards base). True to its name, for **leaves are covered with fine, white hairs that turn them silvery grey**. But its range of flower colours, dull white to lavender, can be confusing. Notice that each **cluster of tiny flowers is in a coil**, a characteristic of phacelias. If you straightened this coil in mid blooming season, you would find buds at the top, flowers next and seed husks below. Blooms in late spring and into summer.

RANGE BC: Dry fields and dry forests to subalpine. South-central/southeastern B.C.

RANGE WA: Habitats as for B.C. Central/eastern Washington; southwards to California.

NARROW-SEPALLED PHACELIA, PHACELIA, *P. leptosepala*: Unlike silverleaf in that it is **often creeping**, and leaves are **dull olive green and hairy**. Erect, unbranched stems rise from a mass of small 'paddle' leaves. May be to 40 cm (16") tall. Flower clusters on short stems developing from most leaf axils. Very **thin sepals with a wide, hairy fringe**.

East of Cascades in southern B.C. and southwards into Washington. A common roadside flower in ponderosa pine and Douglas-fir habitats. Tonasket to Republic on WA 20.

20-40 cm (8-16") HIGH

CLUSTERS OF HAIRY, WHITE FLOWERS, 6 mm (1/4") ACROSS

OLIVE-GREEN LEAVES AND STEM VERY HAIRY

DEEPLY VEINED LEAVES, TO 15 cm (6") LONG

Waterleaf Family, *Hydrophyllaceae*
VARILEAF PHACELIA
Phacelia heterophylla

A large, tousled plant to 40 cm (16") tall. The **deeply veined and hairy leaves**, in a low, dense cluster, vary greatly in shape.

Flowers usually white, but may be light purplish blue; hairy. Blooms middle to late summer, depending on elevation. Look for it in open, arid lowlands and foothills.

RANGE BC: Not in B.C.

RANGE WA: Mostly east of the Cascades, but sporadic on the west side.

Waterleaf Family, *Hydrophyllaceae*
FENDLER'S WATERLEAF
Hydrophyllum fendleri

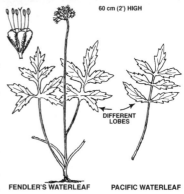

60 cm (2') HIGH

DIFFERENT LOBES

FENDLER'S WATERLEAF   PACIFIC WATERLEAF

Abundant in **subalpine meadows**, but undramatic. Clumps to 90 cm (3') across. Thick mass of large, deeply divided leaves, obscures **dull clusters of small flowers**; long, projecting stamens.Usually in better, moist soils. Middle to subalpine elevations.
RANGE BC: Fairly rare in southern Cascades.
RANGE WA: Very abundant in northern Cascades. Pasayten Wilderness, Cascades to Columbia Gorge. Also Olympics and Blue Mountains.
PACIFIC WATERLEAF, *H. tenuipes*: Closely resembles Fendler's but **white-to-greenish (sometimes purple) flowers at ends of long stalks; different leaf**. Shady, moist places at lower elevations; coastal forests; Vancouver Island southwards to Oregon.

Cucumber Family, *Cucurbitaceae*
BIGROOT [118]
Marah oreganus
Manroot, Wild Cucumber

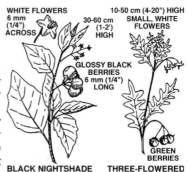

VINE TO 6 m (20')

WHITE MALE FLOWER 1.3 cm (1/2") WIDE

SMALL, SMOOTH MELON

FEMALE FLOWER

LEAVES 5-7.5 cm (2-3") WIDE

TENDRILS

'All species of *Marah* have an immense, manlike, underground tuber,' says one botanical reference. In early spring, a vigorous green vine pushes through the dead vegetation. Unattractive thick, ribbed stems with tendrils. **Leaves, usually less than 7.5 cm (3") across,** small for so robust a vine. Male flowers 1.3 cm (½") across, on short, erect stems. Female flowers not obvious, singly here and there, on same plant. Fruit is a small, bristly melon 2.5–7.5 cm (1–3") long; related to cucumber. Edibility unknown.
RANGE BC: Rare; on Vancouver Island and Gulf Islands.
RANGE WA: Common in western Washington in moist valleys and on moist mountain slopes. Columbia Gorge. Rare in southeastern regions.

Nightshade Family, *Solanaceae*
BLACK NIGHTSHADE
Solanum americanum
Solanum nigrum

WHITE FLOWERS 6 mm (1/4") ACROSS

30-60 cm (1-2') HIGH

10-50 cm (4-20") HIGH SMALL, WHITE FLOWERS

GLOSSY BLACK BERRIES 6 mm (1/4") LONG

GREEN BERRIES

BLACK NIGHTSHADE   THREE-FLOWERED

Note: A purple-flowered nightshade is on p. 297.
Believed poisonous by many; others find **glossy black berries** quite edible when fully mature. Play it safe, just in case you have the wrong plant!
To 60 cm (2') tall, a bushy, erect shrub. Older leaves, to 10 cm (4 ½") long, 5 cm (2") wide, generally with several irregular, shallow teeth. Small, white flowers, in loose, drooping clusters of 3–8; protruding anthers. Related to tomato. Blooms during summer.
RANGE BC: Southern half of Gulf and Vancouver islands. Dry places east of Cascades.
RANGE WA: Dry places east of Cascades. A widespread weed.
CUT-LEAVED or **THREE-FLOWERED NIGHTSHADE**, S. *triflorum*: Sprawling, weedlike. Deeply lobed leaves to 6 cm (2 ½") long. **Clusters of 2–3 white flowers. Green berries**, considered inedible. Fetid aroma. Waste places; lowlands to middle-mountain elevations. Dry areas of southern B.C. and Washington.
HAIRY NIGHTSHADE, S. *sarrachoides*: Similar to cut-leaved nightshade, but **leaf edges merely toothed or wavy, not lobed**. Not aromatic. South American weed. Infrequent in southern B.C.; widespread across Washington. San Juan Islands.

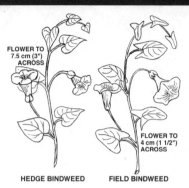

FLOWER TO
7.5 cm (3")
ACROSS

FLOWER TO
4 cm (1 1/2")
ACROSS

HEDGE BINDWEED   FIELD BINDWEED

Morning-glory Family, *Convolvulaceae*
FIELD BINDWEED
Convolvulus arvensis
Orchard Morning-glory

Note: A pink-flowered *Convolvulus* is on p. 267.

There are 6 'morning-glories' in our area. The 2 here, both from Europe, have white flowers, sometimes pinkish. Beautiful? Or a darned pest?

Field bindweed trails and winds, often forming a dense blanket of leaves over fences and low vegetation. Flowers, 1.3–4 cm (½–1 ½") across, can branch 1–2 from a leaf axil. Identify by **arrowhead leaves**, 2–6 cm (¾–2 ½") long.

RANGE BC: Field borders, roadsides and waste places, from valley bottoms to mountainsides. Common throughout; central B.C. southwards.

RANGE WA: Habitats as for B.C. Common in Whitman County.

HEDGE BINDWEED or **WILD MORNING-GLORY,** *C. sepium*: Leaves differ from field bindweed's. Large, white flowers to 7.5 cm (3") across. Likes wetter ground: along ponds, streams and wet meadows; valley bottoms–mountain slopes. Common in southwestern B.C.; extends into Puget Sound and southwards. Rare east of Cascades.

6 PETALS

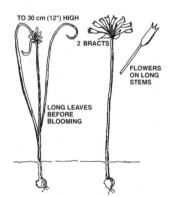

TO 30 cm (12") HIGH

2 BRACTS

FLOWERS
ON LONG
STEMS

LONG LEAVES
BEFORE
BLOOMING

Lily Family, *Liliaceae*
ROCK ONION
Allium macrum

Note: Pink-flowered onions are on pp. 275–77.

Low, less than 30 cm (12") tall. Round, reddish stem; flowerhead branches above 2 large, sharp-pointed bracts. Flower stems within head are 1–2 times as long as flowers. **Sepals and petals taper to long, sharp points** and have a greenish midrib. Stamens protrude to tips of petals. Before blooming, rock onion has 2 deeply grooved leaves, both longer than flower stem. As with several other onions, leaves quickly wither and die, leaving scarcely a trace by May–June blooming. Occurrences seem to be spotty; quite abundant when conditions are right. On very poor, dry, thin rocky soils.

RANGE BC: Not in B.C.

RANGE WA: South-central Washington. Southwards from Kittitas and Douglas counties. About 30 km (20 mi.) west of Yakima, with brittle prickly pear cactus. South of Toppenish, with strict buckwheat and heart-leaved buckwheat.

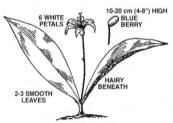

6 WHITE
PETALS

10-20 cm (4-8") HIGH
BLUE
BERRY

HAIRY
BENEATH

2-3 SMOOTH
LEAVES

Lily Family, *Liliaceae*
QUEEN'S CUP [117]
Clintonia uniflora
Blue-bead Lily

A widespread forest flower bound to come to the attention of every mountain hiker. **2–3 shiny green leaves**, 10–20 cm (4–8") in length, mark it immediately. From May to July, a slender stem about 15 cm (6") tall carries **1 pure white flower**, about 2.5 cm (1") across, with a centre of golden stamens. Most unusual, and unpalatable, **oval berry with a deep china-blue colour** not often seen in nature. Shade-tolerant, on nitrogen-poor soils. Most abundant at middle-mountain elevations; it may carpet the mossy forest floor.

RANGE BC: Across B.C. and northwards to the Peace River area.

RANGE WA: Throughout Washington and Oregon, in habitats as for B.C.

Lily Family, *Liliaceae*
WHITE FAWN LILY [119]
Erythronium oregonum
Easter Lily

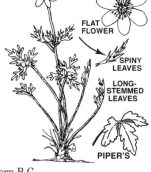

Carpets many forest glades. **Mottled, glossy leaves**. Deep bulb requires 5–7 years before flowering. Seeds dropped close to parent flowers account for great masses. Regal blooms in April.

RANGE BC: Very common on Vancouver Island, from near Comox southwards.

RANGE WA: Very common Puget Sound to Willamette Valley.

WHITE GLACIER or **AVALANCHE LILY**, *E. montanum*: **Pale green leaves**. Rare, subalpine and alpine, of southern Vancouver Island and adjacent mainland southwards. Olympic Mountains. Abundant Mt. Rainier to Mt. Hood.

PINK FAWN or **PINK EASTER LILY** [286], *E. revolutum*: **Beautiful pink**. Resembles white fawn lily; blooms at same time. Moist, shady areas. In B.C., locally common only on southern Vancouver Island. Occasional in coastal Washington, west side of Olympics to Oregon.

Buttercup Family, *Ranunculaceae*
DRUMMOND'S ANEMONE
Anemone drummondii
Alpine Anemone

Note: 'Petal' number varies. 5-sepal anemones, p. 176; yellow, p. 217; purple, p. 300; blue, p. 313.

Anemones have a basal group of toothed or lobed hairy leaves with 1 or 2 leaf whorls on slender stem. Drummond's **6 or more petal-like sepals often bluish**. Numerous protruding stamens. Ball seedhead covered with silky hairs. **Generally under 20 cm (8") tall**, may be dwarfed to 5 cm (2"). Typically half size of cut-leaved (below). Long-stemmed basal leaves divide into leaflets that divide again.

RANGE BC: Subalpine and alpine terrain of southern B.C.

RANGE WA: Subalpine and alpine ecosystems of Olympics and Cascades.

CUT-LEAVED or **GLOBE ANEMONE**, *A. multifida*: **To 40 cm (16") tall**. Similar to Drummond's anemone. 5–9 often blue-tinged sepals. Flowers (June–July) white to greenish, or pink to purple. Young seedhead globe shaped. Common; drier soils to subalpine heights.

PIPER'S or **WESTERN WOOD ANEMONE**, *A. piperi*: Slightly larger than Drummond's anemone. White flower often tinged with purple. **Whorl of leaves high on stem**. Leaf has 3 coarsely toothed leaflets. Very rare in southeastern B.C.; more common in moist mountain forests of eastern Washington.

Buttercup Family, *Ranunculaceae*
WESTERN ANEMONE [120]
Anemone occidentalis
Western Pasqueflower, Tow-headed Baby

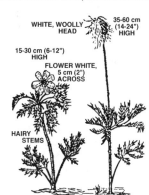

Pops up and blooms immediately after snow has left ground, before visitors get to high country. Number of 'petals' varies—6 is average. Thick stems hairy. Leaves finely divided. A typical anemone flower with blue-tinged white sepals adorns the top of the stem. Flowers 5 cm (2") across. **Fluffy, up-ended 'dust mop' seeds**. Very common in subalpine.

RANGE BC: Southern B.C. Cascades.

RANGE WA: Throughout Cascades. Olympics.

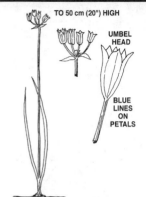

TO 50 cm (20") HIGH

UMBEL HEAD

BLUE LINES ON PETALS

Lily Family, *Liliaceae*

HOWELL'S TRITELEIA [121]

Triteleia howellii

Bicoloured Cluster Lily

Blue *Triteleias* are on p. 318, closely related purple *Brodiaeas* on p. 299.

Flowerhead well above other spring blooms on erect stem to 50 cm (20") tall. Unmissably abundant April–May. (Fool's onion (below) blooms in mid-June.) Favours dryish meadows and gentle slopes; can dot vast areas of hard, dry, stony soil.

Head of up to a dozen buds and flowers on short stems. White flowers with a fair amount of pale blue in them, and a **darker blue line down each petal.** Usual **2 grass-like leaves still green at flowering**, almost as long as flower stem. Native peoples dug up and boiled the marble-sized bulbs.

RANGE BC: Sparse on Vancouver and Gulf islands.

RANGE WA: Through San Juans, Puget Sound, Willamette Valley and southwards. Abundant in midsection of Columbia Gorge. Wenatchee Valley southwards and westwards. High plateaus in Klickitat County.

TO 60 cm (2') HIGH

Lily Family, *Liliaceae*

FOOL'S ONION

Triteleia hyacynthina

Brodiaea hyacynthina

Harvest Lily, White Hyacinth, White Triteleia

Grass-like leaves as long as flower stem disappear at flowering or shortly thereafter. Slender flower stem to 60 cm (2') tall—branches like ribs of a tiny umbrella (umbel); **showy head of tightly packed white flowers** in late spring. Green line down each petal. In bloom it does look like wild onions. Blooms after Howell's triteleia (above).

RANGE BC: Moist meadows; southern Vancouver Island, Gulf Islands, Lower Mainland.

RANGE WA: West of Cascades. San Juan Islands and southwards to California. Moist ground in Columbia Gorge.

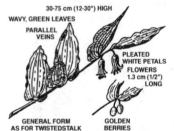

30-75 cm (12-30") HIGH

WAVY, GREEN LEAVES

PARALLEL VEINS

PLEATED WHITE PETALS FLOWERS 1.3 cm (1/2") LONG

GENERAL FORM AS FOR TWISTEDSTALK

GOLDEN BERRIES

Lily Family, *Liliaceae*

HOOKER'S FAIRYBELLS [125]

Disporum hookeri

Oregon Fairybells

Cluster of 1–3 bell-shaped, white flowers hanging from tip of each of several branches—can be almost hidden by glossy-veined leaves. Stems and **leaf margins hairy.** Tips of wide leaves very sharp, aiding rainwater runoff. Flowers May–June. Smooth, egg-shaped berry, **yellow to orange** with age; pulpy centre with 4–6 large seeds. Looks edible, but is not tasty.

RANGE BC: Common at lower elevations; southern half of B.C.

RANGE WA: Lower elevations, southwards to northwestern Oregon. Columbia Gorge.

ROUGH-FRUITED FAIRYBELLS, *D. trachycarpum*: Flowers larger and more greenish than Hooker's fairybells. **Berry ripens a rich red**; rounded, velvety fuzz. Stamens longer than petals. East of Cascades, from B.C.'s Cariboo parklands southwards to northern Oregon.

SMITH'S FAIRYBELLS, *D. smithii*: Similar to Hooker's fairybells except **leaf margins and stems not hairy.** Stamens do not protrude beyond petals. Flowers almost tubular, to 3 cm (1 ¼") long; cream to yellow with green base. Rare on southwestern Vancouver Island; more common in western Washington and coastal Oregon.

Lily Family, *Liliaceae*
CLASPING TWISTEDSTALK [122]
Streptopus amplexifolius

30-90 cm (1-3') HIGH

Twistedstalk is named for a curious **sharp twist in hair-like flower/berry stems** characteristic of this species. **Stem-clasping leaves** and branching stems also distinguish it. Rows of dark green leaves are a common and attractive sight. Dangling white-or-cream flowers, 6 mm (¼") long. Blooms May–June. **Oblong berries bright yellow to red**; not considered edible. Shady, damp woods.

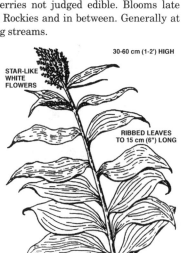

CLASPING LEAVES 5-10 cm (2-4") LONG

TWISTED STALK

CREAMY FLOWERS 6 mm (1/4") ACROSS

RANGE BC: Wideranging; Alaska southwards.
RANGE WA: Wideranging, to California.
ROSY TWISTEDSTALK, *S. roseus*: Look closely—**leaf edges have very short, fine hairs**. **Leaves do not clasp the erect stem**. Seldom branches, to 20 cm (8") tall. Usually a number of single bell-shaped **flowers hang on untwisted stalks** from upper leaf axils. **White-to-rose flowers, spotted with purple**. Small red berries not judged edible. Blooms late spring–summer. Widely distributed in Cascades and Rockies and in between. Generally at middle-mountain elevations in damp forests and along streams.

Lily Family, *Liliaceae*
FALSE SOLOMON'S-SEAL [124]
Smilacina racemosa

30-60 cm (1-2') HIGH

STAR-LIKE WHITE FLOWERS

Note: 2 other false Solomon's-seals: pp. 193 & 200.

Tiny flowers appear 'clustered' but do have 6 petals; plant's general shape is similar to others on page.

May attract attention by reason of its **long, arching stems**, **large, glossy leaves**, or its **plumes of creamy flowers**. Each stout, un-branched stem carries 2 rows of broad, alternate, stem-clasping leaves 6–12.5 cm (2 ½–5") long. **Parallel leaf veins** impart exotic touch. Small, **sweet-scented flowers** packed into pyramidal cluster bloom May–June. Later in season, clusters of mottled greenish red berries slowly turn an unusual red—edible but rather tasteless, with 1 or 2 seeds.

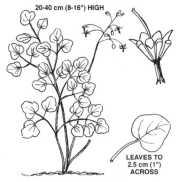

RIBBED LEAVES TO 15 cm (6") LONG

RANGE BC: Damp, shady habitats south of Ft. St. John.
RANGE WA: Damp, shady habitats through Washington to Columbia Gorge.

Barberry Family, *Berberidaceae*
INSIDE-OUT FLOWER [126]
Vancouveria hexandra

20-40 cm (8-16") HIGH

Vancouveria honours Captain George Vancouver, who explored the West Coast in the late 1700s. 3 small, ornate forest flowers, 1 in our area, in south-western Washington, bear his name.

'Inside-out' is very descriptive for this 20–40 cm (8–16") tall annual. The rather complicated flower, reminiscent of shootingstars, appears as if a strong wind has blown it inside-out, leaving its mix of **petals and sepals sharply turned back**. Outer row of 6 wide bracts (or sepals) and inner row of 6 narrow, slightly hooded petals. In ideal situations, can form a loose, green carpet. Attractive glossy green basal leaves. Leaf blade to 2.5 cm (1") across. Blooms late May to early summer.

LEAVES TO 2.5 cm (1") ACROSS

RANGE BC: Not in B.C.
RANGE WA: In dense, moist, coastal woods from Tacoma southwards. Columbia Gorge.

MANY WHITE FLOWERS IN DENSE END-CLUSTERS

TO 2.1 m (7') HIGH

ALTERNATIVE FLOWERHEAD

DULL WHITE FLOWER 1.3-2 cm (1/2-3/4") ACROSS 6 STAMENS

YELLOW-GREEN LEAVES TO 30 cm (12") LONG, RIBBED

STOUT, WOOLLY, UNBRANCHED STEM

Lily Family, *Liliaceae*
WHITE FALSE HELLEBORE [127]
Veratrum californicum
California Corn Lily, White Veratrum

Note: Indian hellebore (green flowers) on p. 336.

'What in Heaven's name is that?' one might ask. False hellebores are the largest mountain-meadow plants, **to 2.1 m (7') tall**, **giants among the wildflowers**. Usually forms a clump of several stems, often extensive patches. Unmistakable **big, prominently ribbed, boat-shaped leaves**, to 30 cm (12") long. Magnificent plumed flowerhead July–August. Small, single flowers are gems of symmetry; **Y-shaped green gland near base of each petal-like part**. Thick roots, shoots considered poisonous. Generally in wet meadows.

RANGE BC: Not in B.C.

RANGE WA: Above 1350 m (4500'); at lower elevations in Columbia Gorge.

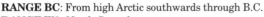

8-MANY PETALS

7.5-20 cm (3-8") HIGH

SERRATE LEAVES 1.3 cm (1/2") LONG

Rose Family, *Rosaceae*
WHITE MOUNTAIN-AVENS [123]
Dryas octopetala
White Dryas

Note: 3 *Dryas* spp. in area; 1 is yellow (see p. 227).

Tough, woody little plant of cold, windswept places. Often forms **evergreen mats** that slowly expand by sending forth new shoots. Identify by **thick, coarsely toothed leaves** 1.3 cm (½") long.

White flower with about 8 petals, 1 per stem, contrasts with drab leaves. Fluffy seedheads permit remote seeding. **Common in alpine;** rocky places, dry tundras. Sometimes on lower gravel bars.

RANGE BC: From high Arctic southwards through B.C.

RANGE WA: North Cascades.

ENTIRE-LEAVED WHITE MOUNTAIN-AVENS, *D. integrifolia*: As 'entire-leaved' suggests, **leaves lack teeth**—except for possibly a few near base of each leaf. Otherwise much like the mountain-avens above. Mountainsides to alpine terrain; southwards from Alaska into central B.C. and eastwards to Rocky Mountains, missing Washington.

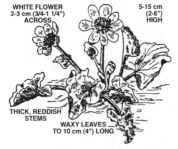

WHITE FLOWER 2-3 cm (3/4-1 1/4") ACROSS

5-15 cm (2-6") HIGH

THICK, REDDISH STEMS

WAXY LEAVES TO 10 cm (4") LONG

Buttercup Family, *Ranunculaceae*
WHITE MARSH-MARIGOLD [130]
Caltha leptosepala

Frigid ice water trickling about its roots does not discourage this succulent subalpine plant with **fleshy leaves and stout, reddish flower stems**. Companions may include glacier lilies, western anemone, buttercups and globeflowers. Abundant leaves rounded or kidney-shaped, often folded and twisted, with fine serrations. Showy white flowers, 1–2 per stem, 2–3 cm (¾–1 ¼") across, **average 8 petals; tinged blue on underside**. Numerous stamens make flower centre greenish yellow.

RANGE BC: Common in high, wet places throughout.

RANGE WA: Common in high, wet places throughout.

TWO-FLOWERED WHITE MARSH-MARIGOLD, *C. leptosepala* var. *biflora*: Biflora suggests **2 flowers, or flower and bud, on each stem**. Leaves as on white marsh-marigold. Cascade passes at about 1200 m (4000'); most common in high, wet places west of summit.

Waterlily Family, *Nymphaeaceae*
FRAGRANT WATERLILY
Nymphaea odorata

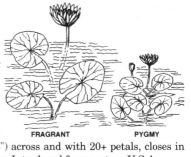

Note: Large, showy, yellow waterlilies on p. 227.
*Nymphaea*s have a wide range of leaf-sizes. Of the 4 in our area, only pygmy waterlily (below) is native. Frequent introductions of non-natives into ponds and lakes close to habitations.

Fragrant waterlily is indeed fragrant—but can you prove it without tipping the canoe? Best in the morning—large white-to-pinkish flower, to 20 cm (8") across and with 20+ petals, closes in afternoon. Typical 'waterlily' leaves 25 cm (10") across. Introduced from eastern U.S.A.

FRAGRANT **PYGMY**

RANGE BC: Sporadic in ponds and lakes in southern B.C.

RANGE WA: Sporadic in ponds and lakes.

PYGMY WATERLILY, *N. tetragona*: Smaller, non-fragrant version of fragrant waterlily; leaves only 4–7.5 cm (1 ½–3") wide, flowers 2.5–5 cm (1–2") across. Rare; ponds and lake edges from central coast/interior of B.C. northwards to Alaska. Not in Washington.

EUROPEAN WHITE WATERLILY, *N. alba*: Very similar to fragrant water-lily but with no odour and fewer petals. Rare; on southeastern Vancouver Island.

TUBULAR FLOWERS

Pink Family, *Caryophyllaceae*
CATCHFLIES AND CAMPIONS
Silene spp.

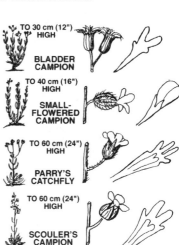

TO 30 cm (12")
HIGH

BLADDER CAMPION

TO 40 cm (16")
HIGH

SMALL-FLOWERED CAMPION

TO 60 cm (24")
HIGH

PARRY'S CATCHFLY

TO 60 cm (24")
HIGH

SCOULER'S CAMPION

Note: Pink-flowering campion on p. 272.

Campions, widespread, about a dozen rather weedy plants, all have **a prominently veined bulb formed by 5-lobed calyx beneath flower. 5 petals, either lobed, divided or fringed; each has a scale at base**, giving the appearance of an inner row of shorter petals. **10 stamens.** Flowers white or pink. Opposite leaves. 'Catchfly' is a *Silene* with sticky stem hairs, which trap small insects.

NIGHT-FLOWERING CATCHFLY, *S. noctiflora*: **Hairy throughout. Deeply 2-lobed white petals.** 2+ flowers per stem. Southern B.C.; common in Washington; dry valleys–forested hillsides.

FORKED CATCHFLY, *S. dichotoma*: Similar to night-flowering catchfly (above) but **flowers lack stems.** A weed pest in southeastern B.C.; limited in eastern Washington.

DOUGLAS'S CAMPION, *S. douglasii*: To 40 cm (16") tall, smooth. **1–3 flowers per stem.** Leaves linear, tufted and matted at base. Common in southern B.C. to high elevations; extends throughout eastern Washington. Also in Olympics.

BLADDER CAMPION [131], *S. vulgaris, S. cucubalus*: European. Very **large, translucent calyx. Petals with 2 protruding lobes.** Blooms during summer. Widely distributed; waste places across southern B.C. and Washington.

SMALL-FLOWERED CAMPION, *S. gallica*: **Only 1 flower per branching stem**, white or pinkish. **2 prongs at petal base.** Bulb, stems and leaves very hairy. West of Cascades. Common on southeastern Vancouver Island, Gulf/San Juan Islands and southwards.

PARRY'S CATCHFLY [128], *S. parryi*: Possibly abundant. Eye-catching. Slender. White flowers age pink. **5 petals, 4 lobes. Paired stem leaves.** Blooms mid–late summer. Lowland to subalpine. South-central/southeastern B.C. Cascades and Olympics in Washington.

SCOULER'S CAMPION or **CATCHFLY**, *S. scouleri*: Erect. Flowers white, may be tinged green or pink. **Petal divided into 4 parts, middle 2 largest.** Spatula-like basal leaves grey and velvety, to 15 cm (6") long. Some plants sticky. Blooms in May. *S. scouleri* ssp. *scouleri* in interior; south-central B.C. southwards. *S. scouleri* ssp. *grandis* is coastal; west of Cascades, on southern Vancouver Island and Gulf/San Juan Islands and southwards.

FLW. WHITE TO PINK, 2.5-7.5 cm (1-3") ACROSS

40-90 cm (16-36") HIGH

FLOWER 0.6-1.3 cm (1/4-1/2") LONG

ROUNDISH END-LEAFLET

ANGLED STEM

3-5 LEAFLETS—MOST ARE TOOTHED

Valerian Family, *Valerianaceae*
SITKA VALERIAN [129]
Valeriana sitchensis
Mountain Valerian

Note: Scouler's valerian (pink), p. 280. Gray's lovage (similar, same habitat), p. 205. Of 8 *Valerians* in area, 1 is a garden escape and these 2 are common:

Mass of small, typically white (may be pinkish) tubular flowers; protruding stamens. **Sweet smell** attracts tiny insects; dirty sock smell to roots—the drug valerian. Leaf shapes identify species.

Sitka valerian is welcome—as you climb to timberline, it foretells promising open glades ahead. Grows singly or mixes in profusion with lupines. A **common tall, high-altitude white flower**, blooms with asters and paintbrush. 40 to 90 cm (16 to 36") tall. Usually 2–5 pairs of large, opposite leaves; **3–5 coarsely toothed leaflets, terminal one largest** and broadest.

RANGE BC: Common throughout higher mountains of B.C.

RANGE WA: High mountains of Washington, mountain forest to alpine terrain.

30-40 cm (12-16")

Valerian Family, *Valerianaceae*
MARSH VALERIAN
Valeriana dioica

30–40 cm (12–16") tall, much smaller than valerian above. Flowerhead of small, white or pinkish 'tube' flowers; faint pleasant smell. Spray of **long-stemmed spoon-shaped bottom leaves**. Just above is a pair of deeply cut leaves, with another, similar pair further up, a leaf placement typical of valerians. Leaf design separates from Sitka (3–7 leaflets) and Scouler's (smooth-edged leaflets).

RANGE BC: Throughout. Common east of Cascades; **wet places**, high forests to subalpine heights.

RANGE WA: Northeastern Washington; habitats as for B.C.

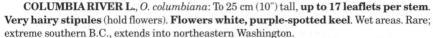

'PEA' FLOWERS

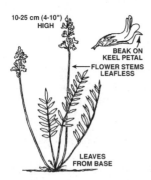

10-25 cm (4-10") HIGH

BEAK ON KEEL PETAL

FLOWER STEMS LEAFLESS

LEAVES FROM BASE

Pea Family, *Fabaceae (Leguminosae)*
LOCOWEEDS
Oxytropis spp.

Note: Locoweeds with purple flowers, p. 305.

About 20 locoweeds in our area, all east of Cascades. Many floral variations. Closely resemble milkvetches, *Astragalus* spp. (pp. 191 & 281); once lumped together, now separate. **Locoweed flowers have small beak on keel, unbranched leaf stems, leaves in loose basal clump**. Some locoweeds poisonous (lack of muscular control, violent behaviour). Field locoweed, *O. campestris*, has been divided into 3 species, flowers white–yellowish:

COLUMBIA RIVER L., *O. columbiana*: To 25 cm (10") tall, **up to 17 leaflets per stem**. **Very hairy stipules** (hold flowers). **Flowers white, purple-spotted keel**. Wet areas. Rare; extreme southern B.C., extends into northeastern Washington.

CUSICK'S L., *O. cusickii*: To 15 cm (6") tall, **more than 17 leaflets per stem, smooth stipules**. Rocky places in subalpine/alpine. Rare. South-central B.C. southwards, east of Cascades, through Washington to northeastern Oregon.

MOUNTAIN L., *O. monticola*: To 25 cm (10") tall, **to 17 leaflets, slightly hairy stipules**. Dry meadows, open forests; low–high elevations. Southwestern/south-central B.C. southwards, generally east of Cascades, through Washington. Olympics.

Pea Family, *Fabaceae (Leguminosae)*
CREAMY PEAVINE
Lathyrus ochroleucus
Woodland Pea, Yellow Pea

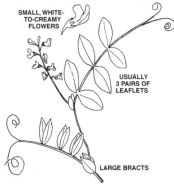

SMALL, WHITE-TO-CREAMY FLOWERS

USUALLY 3 PAIRS OF LEAFLETS

LARGE BRACTS

This peavine can range in length from 30 to 90 cm (1 to 3') as it climbs on supporting shrubbery. Look closely—a **peavine has a style with the upper side hairy from tip to base** ('toothbrush'), while a **vetch has hair only at the style tip** ('bottle-brush'). When you are certain it is a peavine, the colour of the flower cluster should be very conclusive, for this is the only one with **flowers white to cream to yellow but without additional colours**. The small flowers, to 1.3 cm (½") long, are in terminal clusters of 6–15. Leaflets are in 3–4 pairs and are 2.5–5 cm (1–2") long. Note the **large, broad stipules at the base of each stem**. The fruit is a smooth pod.

Moist, open woods and brushy clearings at low to medium elevations suit it best. It should be in bloom from late spring to early summer.

RANGE BC: Widespread. Southwards from the Peace River area, continuing east of the Cascades to southern B.C.

RANGE WA: Moist, open woods east of the Cascades.

COMMON PACIFIC PEA, *L. vestitus*: Generally similar in form to creamy peavine (above), but has **creamy flowers marked with pink or purple**. Usually has **5 pairs of leaflets** instead of 3–4. Open woods in western Washington.

Pea Family, *Fabaceae (Leguminosae)*
LONGLEAF MILK-VETCH [132]
Astragalus reventus

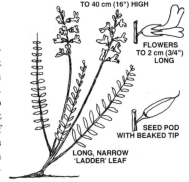

TO 40 cm (16") HIGH

FLOWERS TO 2 cm (3/4") LONG

SEED POD WITH BEAKED TIP

LONG, NARROW 'LADDER' LEAF

Your first reaction on seeing these spikes of whitish flowers on a roadside bank or hillside may be that you are seeing a white-flowering lupine. Perhaps you are (check p. 192). Scramble up for a closer look. Longleaf milk-vetch's **long fronds**, on average to 25 cm (10"), each a **ladder of tiny leaflets about 1.3 cm (½") long**, are far removed from the whorls of leaflets found on lupines. Fortunately, their design is distinctive, a great joy when you are confronted with the 50 or more confusingly similar species of *Astragalus* found in B.C. and Washington.

The 'pea' flowers, which can vary from white to cream or purplish, are about 2 cm (¾") long.

RANGE BC: Not in B.C.

RANGE WA: Ponderosa pine, sagebrush and scablands (the flood-scoured area between Moses Lake and Spokane) of eastern Washington. From Ellensberg southwards. Open, gravelly areas in eastern section of Columbia Gorge. Goldendale to Toppenish, Yakima Canyon.

ALPINE MILK-VETCH, *A. alpinus*: Long stems to 25 cm (10") tall support a **thick cluster of whitish flowers with red-purple markings. Leaflets number 8–11 pairs. Seeds are in hairy, black pods**. Widely ranging in elevation, but generally at subalpine and alpine heights. East of the Coast Mountains and Cascades, from northern B.C. to northern Washington. Rocky Mountains and Wallowa Mountains in Oregon.

TIMBER MILK-VETCH, *A. miser*: Very similar to sketch of longleaf milk-vetch above. Spikes of pink flowers. Open forests. Common in B.C. east of Cascades. Southwards from Prince George to north-central Washington.

10-30 cm (4-12") HIGH

Figwort Family, *Scrophulariaceae*
COW-WHEAT
Melampyrum lineare

'Cow-wheat' is more dramatic than the obscure little plant itself—a related species was eaten by cattle in Europe. The Latin name, 'black wheat,' refers to black, kernel-like seeds.

Slender, single-stemmed annual 10–30 cm (4–12") tall, easily lost among other vegetation. Tubular, 2-lipped, **pea-like flowers, white** but sometimes pinkish, **patch of yellow in throat**, held by green bracts—should make for quick identification. Blooms July–August. Leaves, to 5 cm (2") long, opposite, narrow and sharp-pointed.

RANGE BC: Vancouver Island eastwards. Shady, moist places to middle-mountain elevations. Northwards to Dawson Creek.

RANGE WA: Northeastern Washington.

TO 90 cm (3') HIGH

FINELY HAIRY LEAVES

Pea Family, *Fabaceae (Leguminosae)*
VELVET LUPINE [133]
Lupinus leucophyllus
Woolly-leaved Lupine

Note: For blue lupines and general information, see p.324–27. Yellow, p. 234; 1 is a shrub (p. 149).

Rare. The **only true white-flowered lupine likely**, except south of Snake River or in open uplands of Asotin County (where white-flowering variety of silky lupine (p. 326) is common). Pale pastel variations of other lupines may be mistaken for velvet lupine: e.g., in the southern Okanagan, sulphur lupine (p. 234). As well, from a distance longleaf milk-vetch (see p. 191) can fool you; it also blooms April–late spring.

Some lupines mass profusely over sidehills and meadowlands, but this one often individuals or in small patch. To 90 cm (3') tall. **Soft green** comes from microscopic hairs. **Leaflets and stem really feel velvety. 7 leaflets**. Small flowers clustered in long, showy spikes.

RANGE BC: Very rare! Has been found at Lumby and Lone Pine Creek.

RANGE WA: Eastern Washington, in drier areas: sagebrush and into ponderosa pine habitat. Coulee Dam and southwards here and there to Pasco. Ephrata.

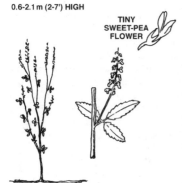

0.6-2.1 m (2-7') HIGH

TINY SWEET-PEA FLOWER

Pea Family, *Fabaceae (Leguminosae)*
WHITE SWEET-CLOVER [135]
Melilotus alba

Note: If flowers yellow rather than white, see yellow sweet-clover (see p. 232).

What is the most common summer roadside flower in our area? Perhaps white sweet-clover, a European import. **Tiny flowers, crowded into spike-like clusters** near top of stiff stems, barely noticed. Perhaps **sweet smell** or noise of bees gets more attention. Along most roads east of Cascades, sweet-clover forms a straggly border for long stretches.

Examine a flower to see its 'sweet pea' structure. 'Clover' reflects tiny 3-leaflet leaves. Grew to over 2.1 m (7') tall in Okanagan fruit orchards where it was planted for composting and nitrogen-fixing. Beekeepers are happy to have this plant near by (by the way, *mel* is Latin for 'honey').

RANGE BC: Sweet-clovers wideranging east of Cascades. Often along road borders.

RANGE WA: Habitats as for B.C.

Mustard Family, *Brassicaceae (Cruciferae)*
SHEPHERD'S PURSE
Capsella bursa-pastoris

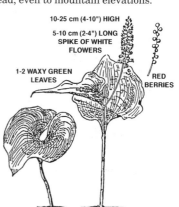

TO 30 cm (12") HIGH

TINY FLOWERS

SMALL SEEDS

CLASPING STEM LEAVES

LEAVES ABOUT 2.5 cm (1") LONG

You might want to ignore this plant as one too small or too weedy. However, the very simplicity and frequent abundance of this spring flower warrant investigation. If the plant is young, then it may be less than 30 cm (12") tall, with a **small, white head of tightly clustered flowers** so small that you can not make out their details. Later, it may reach up to 50 cm (20") in height. Recognition of the **rosette of many odd-shaped basal leaves** is enough to identify it as shepherd's purse.

The real clincher is to find an older plant with most or all of its flowers replaced by distinctive seed pods. At less than 6 mm (¼") across, **pods are very small, but definitely triangular**—to be precise, the shape of a shepherd's purse. Now you know! This is a European import now widely spread and a nuisance as a garden weed.

There are several hundred minor variations of this plant, including leaf and seed shape, so allow some leeway from the sketch of this specimen from the environs of Victoria, B.C.

RANGE BC: Widespread; in open areas around habitations, roadsides, rocky places.

RANGE WA: Often regarded as a weed. Widespread, even to mountain elevations.

Lily Family, *Liliaceae*
FALSE LILY-OF-THE-VALLEY [134]
Maianthemum dilatatum
Two-leaved False Solomon's-Seal

10-25 cm (4-10") HIGH

5-10 cm (2-4") LONG SPIKE OF WHITE FLOWERS

1-2 WAXY GREEN LEAVES

RED BERRIES

Note: Other false Solomon's-seals on pp. 187 & 200.

The **twisting, veiny leaves** are sufficient aid in recognizing this small plant, which generally masses together in shady places. **Each short stem holds 1 or 2 waxy leaves**. During May, a 5–10 cm (2–4") long spike of small, white flowers rises above the leaves. Though this display is short-lived, it is embellished by the **knobby anthers** protruding from each delicately scented flower. **At first mottled with brown, the berries later change to ruby beads**. They were eaten by coastal native peoples. This plant favours shaded forests and moist stream banks, to an elevation of 1000 m (3500').

RANGE BC: Coastal B.C. from Alaska southwards.

RANGE WA: Coastal Washington. Sporadic in eastern Washington. Columbia Gorge.

Lily Family, *Liliaceae*
WILD LILY-OF-THE-VALLEY
Maianthemum canadense

7.5-20 cm (3-8") HIGH

A dainty little plant perhaps **just several centimetres (inches) high** that resembles the garden version. Identified by **1–3 (usually 2) heart-shaped leaves on stem**. Leaves are lightly hairy beneath. Tiny, 4-petalled fragrant flowers are in a loose cylindrical cluster and spaced along the stem. They are replaced by **clusters of small, speckled, pale red berries**.

RANGE BC: Wide range; mossy forests; increasingly abundant across central and northern B.C.

RANGE WA: Not known in Washington.

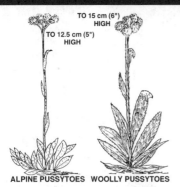

TO 15 cm (6") HIGH

TO 12.5 cm (5") HIGH

ALPINE PUSSYTOES   WOOLLY PUSSYTOES

Sunflower Family, *Asteraceae (Compositae)*
ALPINE PUSSYTOES
Antennaria alpina

Note: One pussytoes is pinkish (see p. 284).

A dozen pussytoes in our area, most widely distributed. Fanciful name refers to soft, roundish flowerheads. **Fine hair or wool—whitish leaves, stem**. In poor soil and open spaces. Most have 'everlasting' flowers. Not colourful, pollinated by flies.

Alpine and woolly pussytoes only in exposed situations on high mountains. Alpine pussytoes, less than 12.5 cm (5") tall, has a single stem with a few wispy leaves, and a tufted mat of thin, grey leaves. Flowers a papery tuft, may be yellowish. A supreme joy of the high country.

RANGE BC: Subalpine and alpine areas throughout.

RANGE WA: Subalpine and alpine areas throughout.

WOOLLY PUSSYTOES [121], *A. lanata*: About 15 cm (6") tall, **tightly packed flowerhead of half a dozen tufts**. **Leafy**: basal leaves long and upright, a few stem leaves. Thick stems and leaves; whitish, woolly. Alpine, like alpine pussytoes; often near snow.

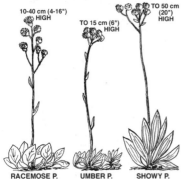

10-40 cm (4-16") HIGH

TO 15 cm (6") HIGH

TO 50 cm (20") HIGH

RACEMOSE P.   UMBER P.   SHOWY P.

Sunflower Family, *Asteraceae (Compositae)*
RACEMOSE PUSSYTOES [137]
Antennaria racemosa

'Racemose' reflects **equally spaced short-stemmed flowers on top third of plant**, rather than usual 1 tuft of soft flowers of most pussytoes; mat-forming basal leaves roundish, not narrow.

RANGE BC: Wide range; lowland–alpine.

RANGE WA: Common in Cascades and east.

UMBER PUSSYTOES, *A. umbrinella*: Closely resembles alpine pussytoes, but **brown-tinged flowers**. Single stem, a few small stem leaves; clump of small, thin basal leaves. Widespread throughout; dry areas; moderate and high elevations, valley bottom to alpine.

SHOWY PUSSYTOES, *A. pulcherrima*: Longish stem, to 50 cm (20"), tightly packed flower cluster. **Thin, lance-like leaves; 3 distinct veins.** Lowlands to high mountains; eastern side of Cascades; Alaska southwards to near B.C.-Washington border.

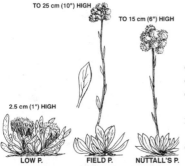

TO 25 cm (10") HIGH

TO 15 cm (6") HIGH

2.5 cm (1") HIGH

LOW P.   FIELD P.   NUTTALL'S P.

Sunflower Family, *Asteraceae (Compositae)*
LOW PUSSYTOES
Antennaria dimorpha

'Tiny'—barely 2.5 cm (1") tall—immediate recognition. 1 woolly, white flower per short stem. Perennial. **Mat of thin, sharp-pointed, tiny leaves** obscure flower stems. Early spring bloomer.

RANGE BC: Dry ground; valley bottom to low mountain slopes. South-central B.C. southwards.

RANGE WA: Throughout eastern Washington; dry ground; valley bottom to low mountain slopes.

FIELD PUSSYTOES, *A. neglecta*: Graceful, to 25 cm (10") tall. **Greenish white bloom**. Several small, narrow stem leaves. **Basal clump of 'spatula' or 'paddle' leaves. Leaves white-woolly beneath**, smooth and green above. Widespread; open, dry forests; low to medium elevations.

NUTTALL'S PUSSYTOES, *A. parviflora*: Under 15 cm (6") tall; **very large, showy flowerheads**. Like field pussytoes, basal mat of small 'spatula' leaves to 4 cm (1 ½") long. Dry ground east of Cascades; valley bottoms–middle-mountain; southwards from southern B.C.. Infrequent on southern Vancouver Island, Gulf/San Juan islands, adjacent mainland.

Rose Family, *Rosaceae*
PARTRIDGEFOOT [138]
Luetkea pectinata
Meadow Spirea

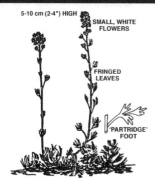

5-10 cm (2-4") HIGH
SMALL, WHITE FLOWERS
FRINGED LEAVES
'PARTRIDGE' FOOT

A small evergreen plant, abundant in open, damp or dry places in subalpine and alpine regions. Fresh green clumps, only to 10 cm (4") tall, composed of a large number of individuals. **Leaves so divided as to appear fringed**; examine a small leaflet to see a fanciful resemblance to a bird's foot or footprint. Slender stems bear **small, white flowers in dense terminal clusters.** Blooms most of the summer. Partridgefoot often masses to form a green carpet marked with a profusion of flowers.

Luetkea honours Count F.P. Lutkia, a 19th-century Russian sea captain and explorer.

RANGE BC: General in subalpine and alpine ecosystems. Southwards from Alaska through the Cascades. North-central B.C. and eastwards to the Rockies.

RANGE WA: General in subalpine and alpine ecosystems. Southwards through the Cascades to California. North-central B.C. and eastwards to the Rockies.

Saxifrage Family, *Saxifragaceae*
LEATHERLEAF SAXIFRAGE
Leptarrhena pyrolifolia

15-30 cm (6-12") HIGH
CLUSTER OF SMALL, WHITE FLOWERS
PURPLISH RED SEEDS AND STEM
THICK, LEATHERY LEAVES TO 7.5 cm (3") LONG
WET GROUND

Note: The rest of the white-flowering saxifrages (*Saxifraga* spp.) are on pp. 165–67; there is a yellow one on p. 218 and a purple one on p. 295.

Most prominent just as snow leaves soggy meadows at timberline and above—its **leathery, glossy green leaves** make a vivid mat pattern against dead grasses of the past year. It also finds suitable habitat along the banks of small creeks.

Deeply veined leaves, dark green top surface, whitish green underside, 2.5–7.5 cm (1–3") long; sheath-like base clasps short, thick main stem. 1 or 2 small, thickish leaves per flower stem. **Dense clusters of small, white flowers at tips of stalks** 15–30 cm (6–12") tall in early summer. As frosts grip the meadows, it adds a final decorative touch: **beautiful purplish red stems and seed capsules** in a tight cluster.

RANGE BC: Wideranging; throughout subalpine; damp meadows, along stream banks.

RANGE WA: Habitats as for B.C.

Sedge Family, *Cyperaceae*
COTTON-GRASS [139]
Eriophorum chamissonis
Chamisso's Cotton-Grass

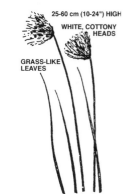

25-60 cm (10-24") HIGH
WHITE, COTTONY HEADS
GRASS-LIKE LEAVES

At least 8 species of cotton-grass in our area; peculiar **white cottony head** and **swamp habitat** are common to all. Often wet meadows are covered so thickly with this sedge they look like fields of cotton, or even like the surface of a pond as seen from a distance. Spreading roots account for concentration.

The springy green stems, from 25 to 60 cm (10 to 24") tall, carry a puffy ball of cotton that obscures the small flowers hidden in its depths. **Leaves short and grass-like**. Wide altitudinal range, from sea level to alpine meadows, where it is part of the main summer floral display.

RANGE BC: Sedgy, wet meadows and shorelines throughout.

RANGE WA: Sedgy, wet meadows and shorelines throughout. Southwards to Oregon.

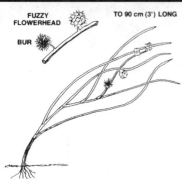

FUZZY FLOWERHEAD

TO 90 cm (3') LONG

BUR

Bur-reed Family, *Sparganiaceae*
NARROW-LEAVED BUR-REED [140]
Sparganium angustifolium

Sticky burs, not in a hot, dry climate, but in our temperate one, floating in water. Where do the burs go, what do they stick to and then what? However, 'there's a reason for everything,' as 'Botany John' Davidson told us long ago at the University of B.C.

An unremarkable plant. Thin, reed-like leaves float with currents. Overall, may reach 90 cm (3') in length. Most noticeable is a line of fuzzy, **round, white-to-greenish balls attached to a short, floating stem**. Leaves floating or submerged; erect when supported by neighbouring vegetation.

RANGE BC: Shallow ponds and sloughs at lower elevations from Alaska southwards.

RANGE WA: Habitats as for B.C., southwards to California.

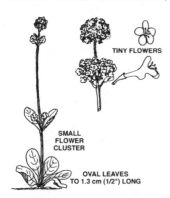

TINY FLOWERS

SMALL FLOWER CLUSTER

OVAL LEAVES TO 1.3 cm (1/2") LONG

Valerian Family, *Valerianaceae*
WHITE PLECTRITIS
Plectritis macrocera

With earliest of spring flowers. Other small, white-flowering plants: slender popcornflower, northwestern saxifrage, woodland stars and vernal whitlow-grass. **Almost a spitting image of sea blush** (pink, p. 284). Favours moist, open areas.

Erect, single stem 5–15 cm (2–6") tall. Flowerhead may look 2–tiered: tightly clustered **flowers form separate rings**. 5 tiny petals; several longer, dark stamens. Flower stems from upper leaf axils. Small, inconspicuous basal leaves with stems; several opposite pairs of clasping stem leaves.

RANGE BC: Spring; moist places among sagebrush, ponderosa pine. Southeastern Vancouver Island. Not common.

RANGE WA: Moist areas, from Strait of Georgia–Puget Sound region to California. Common around seepage slopes in open Garry oak and ponderosa pine forests in Columbia Gorge.

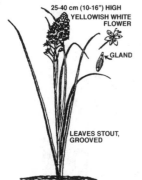

25-40 cm (10-16") HIGH
YELLOWISH WHITE FLOWER

GLAND

LEAVES STOUT, GROOVED

Lily Family, *Liliaceae*
MEADOW DEATH-CAMAS [141]
Zygadenus venenosus

Note: Green-flowered death-camas on p. 336.

Meadow death-camas resembles camas (same habitat; see p. 318): **several long, grass-like leaves** and main-stem flower spike. Native peoples were careful—death-camas is well-named.

Long, thin leaf has deep groove; keel on underside. Creamy flowers less than 1.3 cm (½") across; **6 'petals'—3 are sepals**; tightly packed cluster, unlike panicled death-camas (below). Onion-shaped, dark-coated bulb very poisonous. Leaves often fatal to grazers. Blooms April–June; often dominant over large areas of meadow.

RANGE BC: Southern coast, Gulf Islands, southern half of eastern coastal Vancouver Island. Also east of Cascades, central B.C. southwards.

RANGE WA: San Juan Islands, coastal Washington. Also east of Cascades. Southwards to California. Common in eastern part of Columbia Gorge.

PANICLED DEATH-CAMAS, *Z. paniculatus*: 1 stem to 50 cm (20") tall; **showy white head; small flowers in several upper-stem clusters**. Half-dozen stout, channelled leaves almost to flowerhead. Eastern Washington; sagebrush and ponderosa pine. Columbia Gorge.

Lily Family, *Liliaceae*
STICKY FALSE ASPHODEL
Tofieldia glutinosa
Western Tofieldia

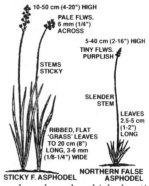

10-50 cm (4-20") HIGH
PALE FLWS.
6 mm (1/4")
ACROSS

5-40 cm (2-16") HIGH
TINY FLWS.
PURPLISH

STEMS
STICKY

SLENDER
STEM

LEAVES
2.5-5 cm
(1-2")
LONG

RIBBED, FLAT
'GRASS' LEAVES
TO 20 cm (8")
LONG, 3-6 mm
(1/8-1/4") WIDE

STICKY F. ASPHODEL
NORTHERN FALSE ASPHODEL

3 false asphodels in B.C. From June to August, watch **lush growth along streams**, on **edges of swamps** and in **damp meadows** for this one's **small, greenish white flowerheads**. Ribbed, grass-like leaves often hidden. Unbranched, **rather dirty looking, sticky flower stem**. Small, white-to-yellow flowers about 6 mm (¼") across in loose cluster; **prominent purplish stamens**. Resembles meadow death-camas but blooms afterwards.

RANGE BC: Widespread, most of B.C. Wet places, meadows, bogs; low–high elevations.

RANGE WA: Throughout most of Washington; habitats as for B.C.

COMMON FALSE ASPHODEL, *T. pusilla*: Non-sticky stem, no stem leaves. Flowers same as sticky false asphodel (above). Not that common: occurs sporadically, north of Quesnel, B.C., on damp mountainsides to alpine heights. Not in Washington.

NORTHERN FALSE ASPHODEL, *T. coccinea*: **Small, greenish purple flowers**; quite rare. Northern Rockies and Fort Nelson area. Not in Washington.

Carrot Family, *Apiaceae (Umbelliferae)*
AMERICAN GLEHNIA [142]
Glehnia littoralis
Beach Carrot, Beach Silver-top

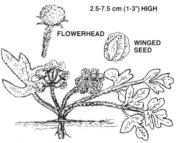

2.5-7.5 cm (1-3") HIGH

FLOWERHEAD

WINGED SEED

Does 'silver-top' refer to the whitish flowerhead, the clusters of scaly seeds or the white undersides of the leaves? Take your choice.

Like many coastal plants found on or near **sand dunes**, glehnia has stout, deep taproot. Broad-based leaves usually clasp stem below sand, thus strengthened against strong winds and severe exposure. **Leaves thick, leathery, crinkled; protective waxy covering on top**; dense, **white-woolly beneath**. Thick, woolly stalk only 2.5–7.5 cm (1–3") tall; umbel of white flowers in early summer; cluster of many-winged seeds.

RANGE BC: Coastal areas with beaches and dunes from Alaska southwards. Scarce occurrences on the Queen Charlottes and Vancouver Island. Pacific Rim National Park.

RANGE WA: Coastal areas with beaches and dunes southwards to California.

Barberry Family, *Berberidaceae*
VANILLA-LEAF [143]
Achlys triphylla
May Leaves, Sweet-after-death

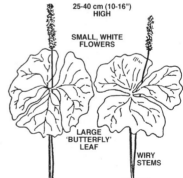

25-40 cm (10-16")
HIGH

SMALL, WHITE
FLOWERS

LARGE
'BUTTERFLY'
LEAF

WIRY
STEMS

Masses of large, 3–winged 'butterfly' leaves here and there at lower elevations throughout shady coniferous forests. Thin, wiry stem about 30 cm (12") tall holds single leaf artistically divided into 3 wavy-edged segments. Leaves start to take shape by mid-April. In May a thin flower stem unerringly pokes through a narrow leaf cleft to support a **spike of small, white flowers. Neither sepals nor petals**—many long stamens provide colour and form.

'Vanilla leaf' and 'sweet-after-death' come from faint vanilla odour of dried leaves. Bundle of dried leaves hung in a room is said to repel flies.

RANGE BC: Low–middle elevations; Vancouver Island, Gulf Islands, coastal mainland.

RANGE WA: Cascades, eastern slope and from western slope to coast and continuing southwards to California. Columbia Gorge.

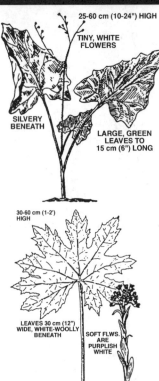

25-60 cm (10-24") HIGH

TINY, WHITE FLOWERS

SILVERY BENEATH

LARGE, GREEN LEAVES TO 15 cm (6") LONG

Sunflower Family, *Asteraceae (Compositae)*
PATHFINDER [144]
Adenocaulon bicolor
Silver Green

Walking through a patch of pathfinder leaves a marked path as weedy, **dark green leaves** expose **conspicuous silvery undersides**—its most noticeable feature, hence 'silver green' and *bicolor*. Silvery sheen created by **mat of fine, white hairs**. Small, white flowers on thin stems not very prominent; sticky seeds catch on clothing. Rich humus and soils.

RANGE BC: Vancouver Island, coastal forests and wetter portions of interior

RANGE WA: Habitats as for B.C.

30-60 cm (1-2') HIGH

LEAVES 30 cm (12") WIDE, WHITE-WOOLLY BENEATH

SOFT FLWS. ARE PURPLISH WHITE

Sunflower Family, *Asteraceae (Compositae)*
PALMATE COLTSFOOT [148]
Petasites palmatus

Among earliest of spring flowers, March onwards. Common. **First cluster of white-to-pink-to-purple flowers often ahead of leaves**. **Sweet-scented flowers** in soft, loose head at top of thick stem about 50 cm (20") tall. Stems lengthen considerably as seeds mature. **Large, fluffy seedheads**. 'Palmate' implies veins spread like fingers on a hand. 7 to 9 toothed lobes per leaf, often split again.

Shady, damp roadsides or edges of swamps, streams, seepage areas. Low–moderate elevations.

RANGE BC: Generally throughout.

RANGE WA: Generally throughout, in suitable habitats.

ARROW-LEAVED COLTSFOOT, *P. sagittatus*: To 50 cm (20") tall; **white flowers**. Broadly triangular, **toothed 'arrowhead' leaves**. Wet places, bogs and marshes; from Alaska southwards, but east of Cascades through B.C. and Washington.

SWEET or **ALPINE COLTSFOOT**, *P. frigidus*: Like arrow-leaved coltsfoot, but leaves have 5–8 broad teeth or deeply lobed into 3–5 coarsely toothed segments. Subalpine and alpine wet areas in northern B.C. Not in Washington.

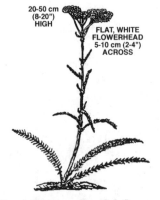

20-50 cm (8-20") HIGH

FLAT, WHITE FLOWERHEAD 5-10 cm (2-4") ACROSS

Sunflower Family, *Asteraceae (Compositae)*
YARROW
Achillea millefolium

Flower or weed? Covers many waste places but has beauty too, and medicinal value. Latin name commemorates Achilles. Aboriginals used it for repelling mosquitoes and for any ailment imaginable. **Often the only white flower**. Adaptable; enlivens sagebrush slopes, canyon sides, forest openings. Pearly everlasting (p. 199) follows. **Leaves finely divided into fringes**, like pipe-cleaners, arching from stout, unbranched stem; **pungent odour if crushed**. Slightly rounded **flowerhead** 5–10 cm (2–4") across; **numerous small, scaly 'flowers'**: white ray flowers, yellow disk flowers; both yield seed. Occasionally pinkish. Blooms from late April in south-central Washington, from May elsewhere; into September at high elevations.

RANGE BC: Throughout, on dry and poor soils, to subalpine heights.

RANGE WA: Throughout Washington and northern Oregon; habitats as for B.C.

SIBERIAN YARROW, *A. sibirica*: **Leaves sharply toothed, not dissected**. Northeastern B.C.; prefers low elevations; stream edges and meadows.

Sunflower Family, *Asteraceae (Compositae)*
PEARLY EVERLASTING [151]
Anaphalis margaritacea

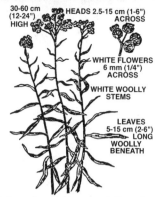

30-60 cm (12-24") HIGH — HEADS 2.5-15 cm (1-6") ACROSS

WHITE FLOWERS 6 mm (1/4") ACROSS

WHITE WOOLLY STEMS

LEAVES 5-15 cm (2-6") LONG WOOLLY BENEATH

Vigorous roadside plant. **Flat-topped mass of white flowers**. Waits until more colourful spring flowers have faded: blooms July–late summer. **Stems and undersides of long, thin leaves covered with white wool** that is usually hidden by the bushy mass of dark green leaves.

Tightly packed flowerhead, perhaps 15 cm (6") across; each pearly white ball has a yellowish or brownish centre flanked by a large number of beautifully arranged, delicately shaded parchment-like scales. Flowers picked in full bloom do not wilt, hence 'everlasting.' Forms main floral decoration for long stretches of roadside, especially at higher elevations, but usually goes unrecognized and little appreciated.

RANGE BC: Throughout, to subalpine. Particularly along roads and near settlements.
RANGE WA: Habitats as for B.C.

Sunflower Family, *Asteraceae (Compositae)*
HOARY FALSE YARROW
Chaenactis douglasii
Dusty Maiden

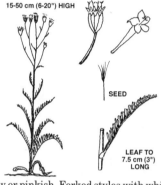

15-50 cm (6-20") HIGH

SEED

LEAF TO 7.5 cm (3") LONG

Not much of a link to yarrow. Rather stiff-looking plant, usually a single stem 15–50 cm (6–20") tall. **White hairs give dusty look. Woolly, olive-green leaves finely dissected**, highly variable in shape, reliably arch away from stem. Fringed edges tilt inwards and upwards. Lower leaves could be 7.5 cm (3") long; higher ones progressively smaller.

Flowers and buds in loose cluster: a few to 24. Greenish, hard-ridged receptacle holds tightly packed head of tiny, tubular flowers; dirty white to yellow or pinkish. Forked styles with white tips protrude about 3 mm (⅛"). Flowerheads, over 1.3 cm (½") wide, have **faint perfume**— probably accounts for names 'dusty maiden,' 'morning bride' and 'bird's bouquet.'

David Douglas found this plant in 1841, at Great Falls along Columbia River, now site of huge dam, at The Dalles, Oregon. It is still there. Quite common in dry, open and rocky places over a wide area; sandy and gravelly soil. Usually grows singly.

RANGE BC: Sagebrush areas east of Cascades; also subalpine and alpine ecosystems.
RANGE WA: East of Cascades; usually lower elevations. Columbia Gorge.

Rose Family, *Rosaceae*
SITKA BURNET
Sanguisorba canadensis
Canada Burnet

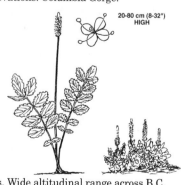

20-80 cm (8-32") HIGH

5 burnets in our area, 2 are common. Sitka burnet produces a **loose cluster of small, shaggy, white 'candles'** on stems 20–80 cm (8–32") tall arising from a mass of leaves. These cylinders, to 6 cm (2 ½") long, consist of **densely packed tiny flowers**. White overall, can be tinged yellow, green or pink. **Main leaves basal: 9–17 coarsely toothed, egg-shaped leaflets.** Swampy areas, stream edges.

RANGE BC: Wet subalpine and alpine meadows. Wide altitudinal range across B.C.
RANGE WA: Not in Washington.

GREAT BURNET, *S. officinalis*: Like Sitka. 9–11 coarsely toothed leaflets. **Tiny purplish blooms.** Under 80 cm (32") tall. Swampy ground; coastal strip, Alaska southwards.

BLACK BERRIES

30-60 cm (1-2') HIGH

SMALL, WHITE STAR-LIKE FLOWERS

APPROX. 9 LEAVES, ABOUT 7.5 cm (3") LONG

Lily Family, *Liliaceae*
STAR-FLOWERED FALSE SOLOMON'S-SEAL [147]
Smilacina stellata

Note: Other false Solomon's-seals on pp. 187 & 193.

Because of similarity of leaf or name, this plant is likely to be **confused with twistedstalk, fairy-bells and regular false Solomon's-seal**. It is fairly uniform in size, being a single stem 30–60 cm (1–2') tall. About 7.5 cm (3") long, the thin, **alternate leaves are clasping and form 2 upward-pointing rows**. The several white, star-like **flowers are at the stem tips**. These tiny sprites make their appearance from April until June. The resulting **green berry develops darkish stripes** and slowly changes to bright red. Found on nitrogen-rich soils and along watercourses.

RANGE BC: Moist woods at middle to higher elevations throughout B.C.

RANGE WA: Moist woods at middle to higher elevations and southwards to California. Columbia Gorge.

THREE-LEAVED FALSE SOLOMON'S-SEAL, *S. trifolia*: Though usually with 3 leaves, this plant may have 2–4. From 7.5 to 20 cm (3 to 8") tall, with star-like white flowers. Most people will not see it for it grows only in wet areas at low elevations in northern B.C.

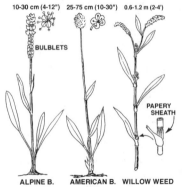

10-30 cm (4-12")   25-75 cm (10-30")   0.6-1.2 m (2-4')

BULBLETS

PAPERY SHEATH

ALPINE B.   AMERICAN B.   WILLOW WEED

Buckwheat Family, *Polygonaceae*
WILLOW WEED
Polygonum lapathifolium
Willow Smartweed

Note: Pink-flowered *Polygonum*s on pp.265 & 283.

The **slender leaves look like those of some willows** and the erect stem, from 0.6–1.2 m (2–4') tall, suggests a willow whip. Perhaps the curving plume of tiny white or pinkish flowers growing from the upper leaf axils is not attractive enough to warrant wildflower status, so it becomes a weed instead. Actually, it is a European introduction.

Note the papery bract surrounding the stem just above each leaf attachment.

RANGE BC: Wet meadows, shorelines and swamps from lowlands to middle-mountain elevations. Most common in southwestern B.C. but infrequent in southern B.C.

RANGE WA: Infrequent in western Washington, in habitats as for B.C.

AMERICAN BISTORT [145], *P. bistortoides*: The **short, cylindrical clusters of tiny white or pinkish flowers** are eye-catching, especially when they dot a high meadow. The slender, erect, jointed stems are 25–75 cm (10–30") tall, generally enough to lift the flowers above the grass and place them on view with other subalpine flora. Basal leaves are long-stemmed. The few further upwards are smaller and stemless. Look for it in damp and dry meadows from high mountain forests to subalpine heights. Mostly sporadic in occurrence in south-central and southeastern B.C., but more widespread in mountains of Washington. Olympics.

ALPINE BISTORT, *P. viviparum*: Usually white or pink flowers on stems 10–30 cm (4–12") tall. **Flowers, sometimes distinctly pinkish, have many long stamens**. After blooming, lower flowers are replaced by bulblets. Moist or thin soils, from high forests to alpine terrain east of the Cascades, from northern Washington into the far north.

Buttercup Family, *Ranunculaceae*
BANEBERRY [146]
Actaea rubra

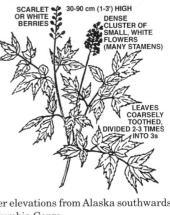

SCARLET
OR WHITE
BERRIES

30-90 cm (1-3') HIGH

DENSE
CLUSTER OF
SMALL, WHITE
FLOWERS
(MANY STAMENS)

LEAVES
COARSELY
TOOTHED,
DIVIDED 2-3 TIMES
INTO 3s

Note: False bugbane (below) is similar.

Baneberry, to 90 cm (3') tall, appears leafy, but actually it has **only 1 long-stemmed basal leaf with 1 or 2 smaller ones on the stem.** Crinkly leaves are divided into 3 sharply toothed segments. **Dense clusters of small, white flowers** appear as fuzzy, white balls because of their many protruding stamens. Blooms April to June. By August, **handsome scarlet berries** (occasionally white), smooth and firm, flaunt their charms. Strong evidence that baneberries are poisonous.

RANGE BC: Widespread. Moist shady places; lower elevations from Alaska southwards.
RANGE WA: Widespread. Habitats as for B.C. Columbia Gorge.

Buttercup Family, *Ranunculaceae*
FALSE BUGBANE [150]
Trautvetteria caroliniensis

25-75 cm
(10-30") HIGH
BUNCHES OF
WHITE TASSELS
6 mm (1/4") LONG

FLOWER
BUDS →

MAPLE-LIKE
LEAF TO
20 cm (8")
WIDE

Prefers shady, moist places that give rise to luxuriant vegetation. Therefore usually hidden except for its head of **small, bristly, white flowers without petals.** 'Bristles' are **tufts of long stamens,** a good distinguishing feature. Blooms during late spring; small cluster of tiny, hooked seeds.

The **single leaves are maple-like,** typically 5–15 cm (2–6") across and generally wider than long, with 5–9 lobes. Stem leaves are few and short-stemmed. Very wet, nitrogen-rich soils. Baneberry, which is similar, is described above.

RANGE BC: Southern coastal B.C. and coastal forest zone from low to middle elevations.
RANGE WA: Coastal, with wide mountain distribution.

Lily Family, *Liliaceae*
BEAR-GRASS [155]
Xerophyllum tenax
Squaw-Grass

TO 1.5 m (5') HIGH

The woody roots of this distinctive plant are reportedly dug up by bear in spring, giving it its common name. Another name, 'squaw-grass' refers to the aboriginal use of the wiry leaves for weaving, which was considered women's work, 'squaw' being a term (no longer in favour) that meant a native woman. David Douglas, the pioneer botanist, wrote, 'Their baskets were formed of cedar bark and bear grass so closely interwoven with the fingers that they are water-tight without the use of gum or resin.'

Memorable **tall tuft of wiry, evergreen leaves** and stout stems to 1.5 m (5') long decorated with **huge, white flower balls**—looks like an escape from a southern desert rather than a plant of high mountainsides. Hundreds of tiny, white flowers in each terminal ball; **pleasant fragrance.** Often forms large, showy roadside or meadow patches.

RANGE BC: Rather rare; open forests–rocky slopes in southeast. Mountains near Salmo and Creston.
RANGE WA: Mostly eastern Washington, where it is abundant at higher elevations. Mt. Spokane, Mt. Rainier. Also in Olympics at subalpine elevations.

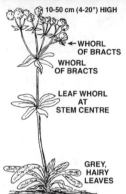

10-50 cm (4-20") HIGH

WHORL OF BRACTS

WHORL OF BRACTS

LEAF WHORL AT STEM CENTRE

GREY, HAIRY LEAVES

Buckwheat Family, *Polygonaceae*
CREAMY BUCKWHEAT [154]
Eriogonum heracleoides
Parsnip-flowered Buckwheat

Note: **May have yellow flowers** (see p. 241).

The most widespread buckwheat in our area. First described in 1834 as growing somewhere near source of Missouri River. Perhaps a 'Hercules' among buckwheats back there, but too much competition in the west. Only 10–50 cm (4–20") tall. Heavy, woody base; large number of flowering stems. Flowerheads vary from white to cream or yellow, some even being a soft lime colour or having a pinkish tinge.

Umbel stems, to 10 cm (4") long, branch from a whorl of leaf-like bracts. A second umbel with shorter stems is quite evident below flowerheads; also has leafy bracts. Key to identification, whorl of leaves below mid-stem, may be missing in Chelan County and environs. Leaf shape regionally variable; remains distinctive. Some plants have basal leaves to 2.5 cm (1") long; others have tuft of 5–8 leaves of varying lengths. All leaves greyish, rather mealy on undersurface. Blooms during May in lower valleys but peaks mid-June to mid-July at higher elevations. Often on rocky terrain and only buckwheat in bloom over large areas

RANGE BC: East of Cascades, very wide range; sagebrush and ponderosa pine habitats.
RANGE WA: Habitats as for B.C. Steptoe Butte.

FLAT-HEADED BUCKWHEAT, *E. heracleoides* var. *minus:* Variety with **long, tapering basal leaves** to 15 cm (6"), **no leaf whorl on the stem** and a flat, white-pinkish flowerhead. Blooms in June. Chelan, Kittitas and Douglas counties. Not in B.C.

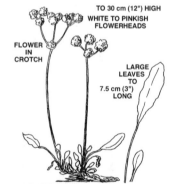

TO 30 cm (12") HIGH
WHITE TO PINKISH FLOWERHEADS

FLOWER IN CROTCH

LARGE LEAVES TO 7.5 cm (3") LONG

Buckwheat Family, *Polygonaceae*
STRICT BUCKWHEAT [153]
Eriogonum strictum

Variable colour and leaves—could be colour-coded as white, cream, yellow or the more abundant pink (see p. 286). Most common combination of **tiny white-and-rose flowers** is eye-catching. Often so abundant that it definitely commands attention.

Like most dryland plants, rises from deep, stout root. May reach 30 cm (12 ") in height. **Basal leaves dull green** because of coating of fine, white hairs. Distinctive flowerhead good for identification. Notice **flower cluster cradled where 3–5 stems branch**.

RANGE BC: Not in B.C.
RANGE WA: Yakima to Cascades. Grant and Douglas counties. Yakima Canyon.

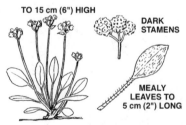

TO 15 cm (6") HIGH

DARK STAMENS

MEALY LEAVES TO 5 cm (2") LONG

Buckwheat Family, *Polygonaceae*
ALPINE BUCKWHEAT
Eriogonum pyrolaefolium

Failing to find a common name for this buckwheat, the author has chosen 'alpine.' In early August, look on dry, rocky sidehills in Cascades, at about 2100 m (7000'). Here, along with Lyall's haplopappus, golden fleabane, alpine anemone and a high-elevation polemonium, firmly anchored here and there in the rocks, is a small buckwheat. **Cluster of whitish flowerheads** liberally **speckled** with dark purple stamens. Cluster of basal leaves, to 5 cm (2") in length; stems a little longer than blades. All stems quite hairy. Mealy texture; lighter colour beneath.

RANGE BC: Not recorded for B.C., though on or near B.C.–Washington boundary.
RANGE WA: Near timberline, higher mountains of Cascades. Hart's Pass.

Buckwheat Family, *Polygonaceae*
TALL BUCKWHEAT
Eriogonum elatum

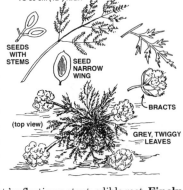

TO 1.2 m (4') HIGH

UMBEL
DESIGN

TINY WHITE
OR PINK
FLOWERS

DARK GREEN
LEAVES TO
40 cm (16")
LONG

Note: **flowers may be pink to pale purple**.

Rather straggly, 0.6–1.2 m (2–4') tall. Use imagination to relate it to usual small, compact buckwheat form. Ungainliness eliminates confusion with anything else. **Basal clump of large, green leaves** to 40 cm (16") long; **like balsamroot or mule's-ears**.

Stout, green stem that branches, again and again; umbel design clearer towards top. Typical small cluster of **tiny, white-to-pinkish flowers**. Use a magnifying glass to see the 6 petals and 6 black dots (stamens).

RANGE BC: Not in B.C.

RANGE WA: Most abundant on eastern slope of Cascades, Okanogan County southwards. Sagebrush through ponderosa pine habitat to mountain ridges. Cle Elum.

Carrot Family, *Apiaceae (Umbelliferae)*
DESERT-PARSLEY
Lomatium spp.

TO 30 cm (12") HIGH

SEEDS
WITH
STEMS

SEED
NARROW
WING

BRACTS

(top view)

GREY, TWIGGY
LEAVES

Perhaps 40 species and varieties of *Lomatium* in B.C. and Washington. Blooms white, yellow (pp. 238–39) or purple (p. 307). Like many plants, prized by aboriginals for large, woody taproots; could be eaten raw but more usually pounded into flour, boiled, steamed or roasted. A trade item. Stored a year or more when dried. Seedheads quite prominent—were gathered as a spice for drinks and food. Common and botanical names vary greatly so 'pay your penny and take your choice' if you check various references.

LARGE-FRUITED DESERT-PARSLEY [152], *L. macrocarpum*: Called 'Indian carrot' and 'biscuitroot,' reflecting a stout, edible root. **Finely dissected, carrot-like leaves off-green to grey or even bluish. Large, dirty white (may be yellow or purplish) flowerheads** (umbels) of individual clusters. **Flowerheads often flat on ground**. Blooms mid-April to mid-May. Seeds exceptionally large, 1.3–2.5 cm (½–1") long; flattened, with a wing narrower than seed outline. Common in south-central B.C. and eastern Washington; open rocky places and foothills. Often with sagebrush.

GEYER'S DESERT-PARSLEY, *L. geyeri*: To 40 cm (16") in height. **Leaves branch into narrow segments. Seeds have short stems**. Extends northwards and westwards into B.C. from main range in drylands of Washington.

Carrot Family, *Apiaceae (Umbelliferae)*
SALT AND PEPPER
Lomatium gormanii
Gorman's Desert-parsley

TO 15 cm (6") HIGH

PURPLE
ANTHERS
ON WHITE PETALS

Easily identified by size, leaf structure and **white flowers (salt) with a sprinkling of purple anthers (pepper)**. Small: only to 15 cm (6") high. Sparse, very thin leaflets do not attract attention like more-common fern-like, bushier species do. **Flower umbels and stems of unequal lengths**. Features shown in drawing should make identification easy. **May be first of local spring flowers**.

RANGE BC: Not in B.C.

RANGE WA: East of Cascades at lower elevations; usually associated with sagebrush. Lincoln and Kittitas counties. Yakima Canyon.

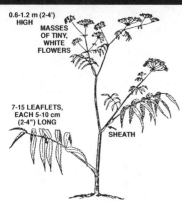

0.6-1.2 m (2-4')
HIGH
MASSES
OF TINY,
WHITE
FLOWERS

7-15 LEAFLETS,
EACH 5-10 cm
(2-4") LONG

SHEATH

Carrot Family, *Apiaceae (Umbelliferae)*
HEMLOCK WATER-PARSNIP [156]
Sium suave

A **common plant of marshes and ponds** most noticeable in July and August because of its large size and its **masses of flattish, white flowerheads.** The 'umbrella' flower stems carry very small flowers less than 3 mm (⅛") across. The **7–15 leaflets are long and finely toothed.** They **do not have prominent veins that point to either tooth tips or bases,** as do a number of the following plants—a clue in identification. The **main leaf stems heavily sheath** the main stem. Sometimes there are a few submerged lower leaves; they differ from the other leaves in being finely fringed. **Seeds are round, with raised ribs.** The roots were a prized food for aboriginal peoples east of the Cascades and also saw some use in coastal areas.

RANGE BC: Widely distributed in wet areas at lower elevations.
RANGE WA: Habitats as for B.C.

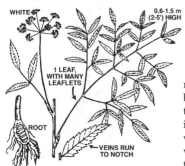

WHITE

1 LEAF,
WITH MANY
LEAFLETS

ROOT

VEINS RUN
TO NOTCH

0.6-1.5 m
(2-5') HIGH

Carrot Family, *Apiaceae (Umbelliferae)*
DOUGLAS'S WATER-HEMLOCK
Cicuta douglasii
Oregon Water-hemlock

Similar in general form to hemlock water-parsnip (above), but with leaves branching into side branchlets that may branch again to short stems that hold 3 leaflets. The **coarsely toothed, tapering leaflets** may be to 10 cm (4") long. Note that water-hemlocks have **leaflet veins that run towards the bottoms of the teeth notches** rather than forward to the points, as in Pacific water-parsley (p. 206) and the *Angelica*s (p. 205). The **flowerhead is a rounded ball of small, white flower clusters. Seeds are globe shaped, with many raised ribs**. When cut open, most of the thick, shallow roots will show horizontal chambers.

Possibly the most poisonous plant in North America. A piece of root the size of a walnut will quickly kill a cow. All parts of plant are poisonous, but the root holds most danger.

RANGE BC: Widespread throughout, but not in the Queen Charlottes or far north; wet meadows, swamps and lake margins.

RANGE WA: Habitats as for B.C.

BULBOUS WATER-HEMLOCK, *C. bulbifera*: A tall plant, to 1 m (40"), that takes its name from **small bulbs in the axils of the upper branching stems**. Lower leaves are thin and much branched. Probably also poisonous. Wet places east of the Cascades; most common north of Prince George, B.C.

POISON-HEMLOCK

POISON-HEMLOCK [157], *Conium maculatum*: This is the plant that was used by the ancient Greeks to put to death their condemned prisoners. Aboriginal peoples sometimes used it in a mixture to poison their arrows. Slowly extending its range along ditches and in waste places, it is more adaptable to drier ground than the water-hemlocks (above). Note that the **stout stem is marked with purplish brown blotches** and that the **leaflets are fringed**, giving a fern-like appearance. **Very small seeds, egg-shaped, with many prominent raised ribs**. Mostly west of the Cascades, from southwestern B.C. southwards through Washington.

Carrot Family, *Apiaceae (Umbelliferae)*
WILD CARROT [158]
Daucus carota
Queen Anne's Lace

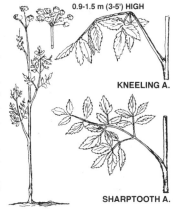

0.3-1.2 m (1-4')
HIGH

WHITE HEADS
5-7.5 cm (2-3")
ACROSS

FORKED
BRACTS

WHITE
FLOWER
3 mm (1/8")
ACROSS

CARROT-LIKE
LEAVES

RIBBED, HAIRY
STEMS

Coarse yet dainty; abundant along roadsides, in waste places; **long blooming time**. To 1.2 m (4') tall. Number of **stout, ribbed leaves; fern-like and finely divided like those of domestic carrot**. According to Dr. Nancy J. Turner, they were 'used and commonly cultivated by the southern Interior peoples.' Variable leaves and height.

Massed white flowerheads 5–7.5 cm (2–3") across; **usually flat-topped, cupped when old; central flower often pink–purple**. 4 petals. Blooms July–September. Queen Anne of England wore its tiny flowers in preference to lace.

RANGE BC: Most visible along roadsides in southwestern B.C.; also east of Cascades.

RANGE WA: Most visible along roadsides, throughout Washington and Oregon.

AMERICAN WILD CARROT, *D. pusillus*: **Smaller** than wild carrot, more-finely dissected leaves; **flowerheads form shallow bowl; 5–12 flowers on each small umbel stem**. Blooms in late summer; west of Cascades, southern Vancouver Island southwards.

GRAY'S LOVAGE, *Ligustichum grayi*: Like wild carrot, **umbel head of small, white flowers, finely divided leaves**. Stout-stemmed, 25–60 cm (10–24") tall. **Conspicuous** in subalpine; grows with lupines, daisies, bistort and magenta paintbrush. **Mistaken for Sitka valerian** (p.190), but leaves fern-like and **flower unscented**. Common; middle to subalpine elevations in Cascades of Washington. Mt. Rainier, Blue Mountains, Columbia Gorge.

Carrot Family, *Apiaceae (Umbelliferae)*
KNEELING ANGELICA
Angelica genuflexa

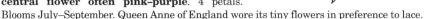

0.9-1.5 m (3-5') HIGH

KNEELING A.

SHARPTOOTH A.

'Kneeling angel' has **peculiar kink in stem that tilts leaf spurs into a bent-over position**.

Angelicas are generally sturdy, with **single, hollow stems**; 0.9–1.5 m (3–5') tall. **Typical white or pinkish umbel flowerhead like that of cow-parsnip or water hemlock**. Lower stem of each leaf has prominent sheath that grasps main stem. **Leaf veins usually run to points of teeth**—distinguishes *Angelica*s from rather similar water-hemlocks (p. 204) and Pacific water-parsley (p. 206). **Seeds almost round, but flattened, with several ribs between the 2 wide wings**.

RANGE BC: Common throughout in moist areas, except Queen Charlottes.

RANGE WA: Common; west of Cascades only.

SHARPTOOTH ANGELICA, *A. arguta*: **Large, compound leaf, 2–3 times divided**; greatly enlarged at base. **Stems not bent. Seeds oblong, flattened; 3 narrow, central ribs per side, between 2 wide wings**. Moist places, valley bottoms–middle-mountain elevations; most common in southern B.C., generally east of Cascades in Washington.

CANBY'S ANGELICA, *A. canbyi*: **Stout; hollow stems; to 1.2 m (4') tall. Often with cow-parsnip**; similar **showy umbel heads**. Leaves with 3 segments; coarsely toothed or deeply lobed leaflets; enlarged stem bases. **Seeds almost round, flattened, with 3–4 ribs through central sections on both sides**. Note **range**: streamsides, moist places of central Washington, westwards into subalpine on eastern slope of Cascades. Wenatchee Mountains,

SEACOAST or **SEA-WATCH ANGELICA**, *A. lucida*: Along coast, beaches and bluffs, and eastwards to Cascades. **Leaves more oval than *genuflexa*'s. No stem kink. Seed narrow, barrel shaped, flat one side; thin-edged longitudinal ribs on rounded side**.

DAWSON'S ANGELICA, *A. dawsonii*: Single yellow flowerhead. Rare; extreme southeastern B.C. and along continental divide, missing Washington.

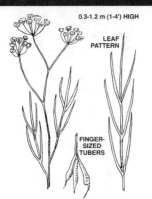

0.3-1.2 m (1-4') HIGH
LEAF PATTERN
FINGER-SIZED TUBERS

Carrot Family, *Apiaceae (Umbelliferae)*
YAMPAH
Perideridia gairdneri
Gairdner's Yampah

Dependable food source for aboriginal peoples and early explorers: Lewis and Clark found 'roots are very palatable either fresh, roasted or dried.' However, yampah is scarce enough to protect. **1–3 tubers the size of a man's finger. Flowers white or pinkish**, several–many in typically umbel heads. Several stem leaves; long, thin; sometimes divided.

RANGE BC: Most common on southeastern Vancouver Island, Gulf Islands. Some in southeast.

RANGE WA: San Juan Islands and adjacent mainland. Also in eastern Washington.

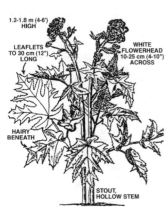

1.2-1.8 m (4-6') HIGH
LEAFLETS TO 30 cm (12") LONG
WHITE FLOWERHEAD 10-25 cm (4-10") ACROSS
HAIRY BENEATH
STOUT, HOLLOW STEM

Carrot Family, *Apiaceae (Umbelliferae)*
COW-PARSNIP [160]
Heracleum lanatum

Often 1.8 m (6') tall at low elevations. **May be our largest perennial**; rivals white false hellebore (p. 188). Often with similar Pacific water-parsley.

Distinctive **very large leaves** arch from main stem through large sheath to hold **3 leaflets**, to 30 cm (12") long, palmately lobed, coarsely toothed. Single thick, coarse main stem. Main stalks divide into umbrella of short stems; each branches again. Small, white flowers. **Massive, slightly rounded flowerhead**, 10–25 cm (4–10") across. Blooms in May at coast; later in mountains. **Large number of flattened, oval seeds** without distinct ribs (see water-hemlocks, p. 204 and angelicas, p. 205).

Parts are mildly toxic, but is eaten by cattle, and was eaten by aboriginals. Seepage areas, ditches, stream banks, sometimes roadsides. Moist, rich soils; sea level to subalpine.

RANGE BC: Throughout, at low to subalpine elevations.

RANGE WA: Throughout, at low to subalpine elevations. Hart's Pass.

GIANT COW-PARSNIP, *H. mantegazzianum*: Comparatively new to B.C. and Washington, from Asia. Truly a giant, to 3 m (10') tall; dense growth. **Flowerhead immense**. No portion edible—can cause severe skin rashes if handled. Southern Vancouver Island, Gulf Islands and in Puget Sound area.

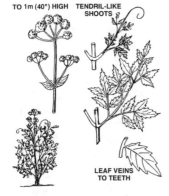

TO 1m (40") HIGH
TENDRIL-LIKE SHOOTS
LEAF VEINS TO TEETH

Carrot Family, *Apiaceae (Umbelliferae)*
PACIFIC WATER-PARSLEY
Oenanthe sarmentosa
American Oenanthe

Leaves and flowers confused with those of cow-parsnip (above), water-hemlocks (p. 204) and angelicas (p. 205). **Soft, weak stems, either lying on the ground or more erect** (entwined with other vegetation). Look-alikes have stout single stems. **Leaf veins run to teeth tips**, like angelicas. Leaf 3 times pinnately divided. Terminal spray of small, white flowers in compact cluster. **Seed barrel-shaped, ridged with wide ribs**. Wet habitats: streams, may be in seasonally wet areas. Low–middle elevations.

RANGE BC: Alaska southwards, in west of Coast Mountains and Cascades. Chilliwack.

RANGE WA: West of Cascades. Western Columbia Gorge, Western Klickitat County.

Sunflower Family, *Asteraceae (Compositae)*

SILVERBACK LUINA [162]
Luina hypoleuca

10-17 DULL YELLOW FLOWERS

15-30 cm (6-12") HIGH

LEAVES SILVERY BENEATH

Note: Truly yellow-blooming *Luinas* on p. 242.

While you might argue as to whether the **flowers are white, creamy-buff or light yellow**, there is no mistaking the abundant 'silverback' leaves. **Dark green on top**, thick and stiff, less than 6 cm (2 ½") long; so **thickly covered with fine, white hairs below as to be silvery**. Stems hairy too.

As befits a rockery plant, it is seldom more than 30 cm (12") tall. However, it is nevertheless conspicuous with its wide-spreading form and its flowerhead consisting of a spray of about a dozen blooms. Rocky places, slides, cliffs and bluffs.

RANGE BC: From Cascades through southwestern B.C. Fairly common on southern Vancouver Island.

RANGE WA: From the Cascades westwards.

Sunflower Family, *Asteraceae (Compositae)*
WHITE HAWKWEED [159]
Hieracium albiflorum

TO 90 cm (3') HIGH

Note: Our only white-flowered hawkweed. Other hawkweeds are yellow or orange (pp. 244–45 & 261).

Rather weedy, unbranched, to 90 cm (3') tall, **loose head of a dozen or more flowers**, each to 1.3 cm (½") across, separated by loose branching. **Ray flowers only**, like dandelion. **Bracts in 1 row**. Blooms around middle of the summer.

Clump of basal leaves survives over winter. Lower leaves slender and very hairy; to 15 cm (4") long. Stem has long, loose hairs at base; is smooth above. **Hawkweeds exude a milky juice if stem**

LOWER LEAVES AND STEM HAIRY

is split, but often flow is very meagre. You might experiment by collecting some of this latex, congealing it in the sun and turning it into a chewing-gum. Sow-thistles bleed also, but have yellow flowers. The peculiar name 'hawkweed' relates to a Greek myth that hawks would tear apart a related plant to obtain its juice; they washed their eyes with it to improve their sight.

RANGE BC: Common; dry woods to middle elevations. Most abundant south of Prince George.

RANGE WA: Cascades and westwards, from low to medium elevations.

Sunflower Family, *Asteraceae (Compositae)*
HOOKER'S THISTLE
Cirsium hookerianum

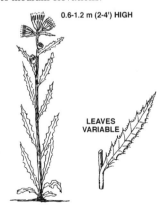

0.6-1.2 m (2-4') HIGH

LEAVES VARIABLE

Recognize as a thistle by its **spine-pointed leaves**. **The only white-flowered thistle** you will encounter; that should be enough for positive identification! Stem and leaves may be covered with a cobwebby hair. Plants usually form large, thick masses because new plants spring up from rootstocks of older plants. White-to-creamy flowers of summer go to seed in late summer, creating large balls whose fluff is widely dispersed by wind. Like most thistles, can become a serious agricultural pest.

RANGE BC: Common; fields and waste ground in southern B.C., east of Cascades and along Rockies.

RANGE WA: Extends into northern Washington.

WHITE TO PINKISH
FLOWERS
2-2.5 cm
(3/4-1")
ACROSS

15-30 cm
(6-12")
HIGH

FINE HAIRS
ON LEAVES
AND STEMS

Sunflower Family, *Asteraceae (Compositae)*

TUFTED FLEABANE

Erigeron caespitosus

White Fleabane

Note: More white *Erigerons* on p. 210; also yellow (p. 250), pink (p. 287), purple (p. 310) and blue (p. 330).

Fleabanes bloom spring–summer; have **equal-length flowerhead bracts and very narrow, more numerous ray 'petals.'Asters similar but bloom late summer–autumn**; with **several unequal rows of bracts**. 'Fleabane' and 'daisy' almost interchangeable. Said to repel fleas.

Tufted fleabane attractive. **Clusters of round, white flowers**, sometimes pinkish; **numerous thin ray flowers**; yellow-green centres. Long, narrow leaves and branching stems dull green; fine hairs prevent water loss. Leaves to 7.5 cm (3") long, under 3 mm (⅛") wide; upper ones wispy. Roadsides, waste places; arid parts to open forests. Blooms May–July.

RANGE BC: Sporadic occurrences; east of Cascades; valley bottoms–alpine heights.

RANGE WA: Habitats as for B.C.

ARCTIC DAISY, *E. humilis*: **White ray flowers, may be purplish; yellow centre**. **Single-flowered short, hairy stems**. To 20 cm (8") tall. High-mountain damp areas, especially coastal; sometimes low-elevation bogs. Common in much of B.C. Not in Washington.

5-15 cm (2-6") HIGH

WHITE OR
PURPLE
FLOWERS
TO 2 cm (3/4")
ACROSS

HAIRY
LEAVES

Sunflower Family, *Asteraceae (Compositae)*

CUT-LEAVED DAISY [165]

Erigeron compositus

Dwarf Mountain Fleabane

Dainty. **Ray flowers white or pink** with yellow centre, less than 2 cm (¾") across. Generally low and many branched; **leaves velvety, branching into thin fingers**. Blooms midsummer onwards. Several varieties; minor differences in leaves and hairiness.

RANGE BC: Across province, dry mountain slopes, medium to high elevations.

RANGE WA: Dry, gravelly soils; sagebrush and bunchgrass ecosystems.

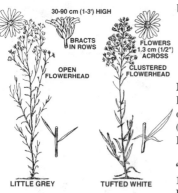

30-90 cm (1-3') HIGH

BRACTS
IN ROWS

OPEN
FLOWERHEAD

FLOWERS
1.3 cm (1/2")
ACROSS

CLUSTERED
FLOWERHEAD

LITTLE GREY   TUFTED WHITE

Sunflower Family, *Asteraceae (Compositae)*

LITTLE GREY ASTER

Aster falcatus

Aster or fleabane? See tufted fleabane (above). **Miniature blooms, late summer–fall**, (unlike earlier-blooming fleabanes) identify this aster (and the one below). Fine, much-branched stems; 30–90 cm (1–3') tall. Leaves continue up stem to flowerheads. Leaves very narrow, about 2.5 cm (1") long.

Pure white flowers, with yellow or brown 'button' centres, about 1.3 cm (½") across. **About 17 ray flowers per head**, but number variable. Flowerhead much looser than for one below.

RANGE BC: Much of B.C. east of Coast Mountains.

RANGE WA: Open dryland areas at lower elevations of eastern Washington.

TUFTED WHITE PRAIRIE ASTER, *A. ericoides* ssp. *pansus*: Like little grey aster; similar flowers, under 1.3 cm (½") across, in denser clusters. **Larger leaves, up to flowerheads**, to 5 cm (2") long and 6 mm (¼") wide. 30–75 cm (1–2 ½') tall. Sometimes a **tousled clump of tiny, white flowers**. Common east of B.C.'s Cascades; open dryland areas; north to Ft. St. John; Cariboo parklands, Chilcotin; also south-central. Possibly eastern Washington.

Sunflower Family, *Asteraceae (Compositae)*
OXEYE DAISY [166]
Leucanthemum vulgare
Chrysanthemum leucanthemum

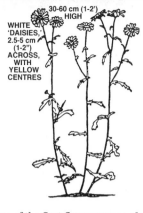

WHITE
'DAISIES,'
2.5-5 cm
(1-2")
ACROSS,
WITH
YELLOW
CENTRES

30-60 cm (1-2')
HIGH

Sometimes entire fields are taken over by oxeye daisy; massed blooms are unforgettable shimmering in sunshine or rippling in a breeze.

Naturalized and widely distributed throughout B.C. and Washington, except more arid regions. Many roadsides have ribbons of these blooms for long stretches. Could be most abundant of showy flowers.

Erect, graceful stems, branch near top. **Symmetrical white daisies**; yellow centre, white ray flowers; to 5 cm (2") across. **Dark green, lobed leaves**. Blooms May–August, a lovely addition to the landscape after spring flowers have gone. 'Daisy,' likely one of the first flower names of childhood, comes from an English flower that closed at night, opened at dawn—the 'day's-eye'.

RANGE BC: From central B.C. southwards, except sagebrush and bunchgrass ecosystems; roadsides and fields to 1200 m (4000') elevation.

RANGE WA: Habitats as for B.C.

OLYMPIC MOUNTAIN ASTER, *Aster paucicapatus*: Close resemblance to oxeye daisy, but not as tall; **smaller flowers average 11 ray flowers**; usually a **dense clump on erect stems**. Leaves numerous all the way up stem. Open slopes and roadsides; most common at higher elevations. Sporadic on central Vancouver Island but common in Olympics.

Sunflower Family, *Asteraceae (Compositae)*
ENGELMANN'S ASTER
Aster engelmannii

0.6-1.5m (2-5')

Can be **very tall for an aster**, up to 1.5 m (5'), a good clue. Attractive flowerheads: yellow centres; **from 11 to 17 white-to-pink ray flowers;** rather tousled look. Flower may be 6 cm (2 ½") across.

Leaves, sharp-pointed, a little larger than on many other asters, at 4–9 cm (1 ½–3 ½") long. Blooms during summer, after fleabanes.

RANGE BC: Favours moderate soils, forested slopes to subalpine, east of Cascades to Rockies.

RANGE WA: Habitats as for B.C. Eastern slope of Cascades.

Sunflower Family, *Asteraceae (Compositae)*
CORN CHAMOMILE [167]
Anthemis arvensis
Field Chamomile

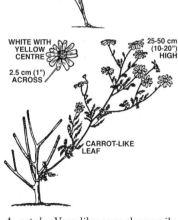

WHITE WITH
YELLOW
CENTRE

2.5 cm (1")
ACROSS

25-50 cm
(10–20")
HIGH

CARROT-LIKE
LEAF

Introduced daisy-like flower; to 50 cm (20") tall. **Flowers not so large and crisp** as oxeye daisy's. More tousled form, more branched, **lacks long, erect stems**. Buds continue to develop, prolonging blooming, April–September, perhaps through mild coastal winter. Carrot-like **finely dissected leaves**.

RANGE BC: Waste places: roadsides and fields; throughout southern coastal B.C.

RANGE WA: Common in coastal regions.

STINKING CHAMOMILE or **DOG FENNEL**, *A. cotula*: Very like corn chamomile, except for **unpleasant smell if you crush leaves**. Finely divided **leaves appear lacy**. Upper stems much branched; wealth of flowers. Blooms all summer. Roadsides, pastures and waste ground east of Cascades. Whitman County roadsides.

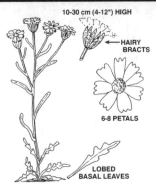

Sunflower Family, *Asteraceae (Compositae)*
WHITE LAYIA [164]
Layia glandulosa
Tidy Tips

A typical small, white daisy; beautiful showy mass of flowers in May. Plants either single-stemmed or branched; 10–30 cm (4–12") in height. **Each petal has 3 distinct lobes**. But how many petals? Most plants have 6, some 8; some have overlapping petals, others seem to be missing some. Small **basal leaves distinctly lobed**. **Stem leaves white-hairy** and thin; some opposite, some alternate, on same plant. In the open, on sandy soil.

RANGE BC: A historical note—and an opportunity for some sharp-sighted person to once again add it to B.C. flora: 'First reported by Macoun (1886) but no material available.'

RANGE WA: Mostly confined to central Washington, but also in eastern Washington, near Pasco. Sparse in eastern section of Columbia Gorge.

Sunflower Family, *Asteraceae (Compositae)*
ENGLISH DAISY [161]
Bellis perennis

It was only a moment's walk to the edge of my lawn to dig up an unwelcome English daisy to draw. Given any freedom, it spreads and spreads until a large area can be as white as snow.

Many flowers pinkish, but with yellow 'button' centre. Some plants hug ground; others raise spoon-shaped basal leaves 5 cm (2") into air. Thrives with suitable soil, shade and sufficient water. Flower stem variable in length, but seldom over 15 cm (6"). Blooms as early as February.

RANGE BC: Persistent intruder in lawns and semi-cultivated grounds. Coastal areas.
RANGE WA: Habitats as for B.C. Southwards through Pacific states.

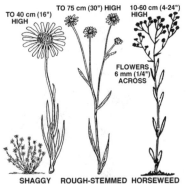

Sunflower Family, *Asteraceae (Compositae)*
SHAGGY FLEABANE [163]
Erigeron pumilus

Note: More white *Erigerons* are on p. 208; yellow (p. 250), pink (p. 287), purple (p. 310) and blue (p. 330).

Over 30 fleabane species, and additional varieties, occur in our area. **Flowers range through white, yellow, pink, purple and blue**. Shadings and age produce confusing variations.

Shaggy fleabane can be pinkish or even blue-tinged. Many branching stems, to 40 cm (16") tall, yield mass of flowerheads. **Leaves thin, hairy**. Blooms late spring to early summer.

RANGE BC: Southern B.C.
RANGE WA: Southwards through sagebrush and bunchgrass ecosystems.

ROUGH-STEMMED FLEABANE or **DAISY FLEABANE**, *E. strigosus*: To 75 cm (30") high. Blooms all summer. **Leaves very narrow**. **Hairy bracts**. Ranges from central B.C. southwards into sagebrush areas of B.C. and Washington but not into foothill country.

HORSEWEED, *Conyza canadensis*: *Erigeron*-like. Common introduced weed. Stiffly erect; many-branched stem, **large number of white 'fleabane' flowers about 6 mm (¼") across**. **Short, white ray flowers** take second place to yellow disk flowers. Numerous, alternate leaves slender, to 6 cm (2 ½") long. Southern B.C. and southwards, in sagebrush, bunchgrass and coastal ecosystems of Washington.

Wintergreen Family, *Pyrolaceae*
WHITE-VEINED WINTERGREEN [170]
Pyrola picta

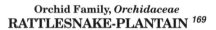

Clearly marked leaf-vein system, white on dark green; underside purplish. Drooping, waxy flowers, white to greenish, singly along stem; **style curved**. Evergreen. Rich humus; shady forests, lowland–middle-mountain heights. Blooms in summer.

RANGE BC: Widespread in southern B.C.

RANGE WA: Widespread; also southwards.

LESSER WINTERGREEN, *P. minor*: Like most pyrolas, has a **single, slender stem**; may reach 15 cm (6") in height. Leaves thin, dark green and rounded; some faintly heart-shaped. Flowers white to pinkish, with 5–20 per stem. **Protruding straight style**, not curved. Moist, mossy forests throughout.

GREEN WINTERGREEN, *P. chlorantha*: **Greenish white flowers**. Leaf blades thick, clear green and broadly oval, slightly shorter than stems. **Flowers few** and well spaced. Style protrudes in upward curve. Summer bloomer. Moist mid-elevation forests, Alaska to California. In B.C., most common east of Cascades.

Orchid Family, *Orchidaceae*
RATTLESNAKE-PLANTAIN [169]
Goodyera oblongifolia

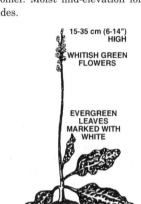

Long, ovalish leaves dark green, laced with criss-cross network of white; prominent white mid-vein; occasionally mid-vein only. Markings supposedly resemble those of a rattlesnake.

Leaves, close to ground, often only feature to be seen—does not bloom every year. **Small greenish white flowers** loosely scattered along spike on thickish stalk. **Flowers tend to favour 1 side of stem**. Blooms July–August. Mossy, dry coniferous forests at low to middle-mountain elevations.

RANGE BC: Widespread throughout.

RANGE WA: Widespread throughout.

DWARF RATTLESNAKE ORCHID, *G. repens*: Half-sized edition of rattlesnake-plantain. **Leaves have white criss-cross network, but no white mid-vein**. White-to-greenish flowers hang on 1 side of stem. Open forests east of Cascades, southern B.C. northwards.

Orchid Family, *Orchidaceae*
MOUNTAIN LADYSLIPPER [172]
Cypripedium montanum

Note: Yellow one p. 257; green-brown, p. 340.

Stout stem. Number of **large, waxy green leaves, parallel veins**. 1–3 pure white 'slippers' **veined with purple**, to 5 cm (2") long. Petals and other parts surrounding slipper are brown; **2 sepals twisted like coppery ribbons. Exquisite perfume**. Blooms mid-May–late June. A floral jewel to be protected—takes 12+ years to grow from seed.

RANGE BC: Moist, open woods, river valleys, mountain slopes. Scattered, into far north.

RANGE WA: Sporadic, east of Cascades, valley bottom to middle-mountain heights.

SPARROW'S-EGG or **SMALL WHITE LADYSLIPPER**, *C. passerinum*: **Pure white slipper, purple tinge inside**. Blooms in July. Southeastern B.C. northwards along Rockies; most abundant in far north of B.C. **Moist–boggy soils**, lower elevations. Not in Washington.

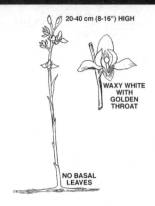

20-40 cm (8-16") HIGH

WAXY WHITE WITH GOLDEN THROAT

NO BASAL LEAVES

Orchid Family, *Orchidaceae*
PHANTOM ORCHID [173]
Cephalanthera austinae
Eburophyton austinae

Really has phantom qualities: hideaway location in mossy forest glade and waxy, ghost-like appearance. **Whole plant waxy white, except for spot of gold in flower's throat. Faint, sweet perfume.** Lovely and rare, so leave undisturbed.

A saprophyte; leaves persist as small scales on stem. Summer bloomer. Deep, mossy woods to dry slopes thickly forested in conifers.

RANGE BC: Very rare; in southwestern B.C.
RANGE WA: Fairly rare; Cascades, Olympics.

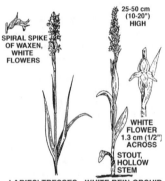

SPIRAL SPIKE OF WAXEN, WHITE FLOWERS

25-50 cm (10-20") HIGH

WHITE FLOWER 1.3 cm (1/2") ACROSS

STOUT, HOLLOW STEM

LADIES' TRESSES   WHITE REIN-ORCHID

Orchid Family, *Orchidaceae*
WHITE REIN-ORCHID [171]
Platanthera dilatata
Habenaria dilatata
White Bog-orchid

Note: Green *Platanthera* spp. on pp. 339–40.

About 12 rein-orchids in our area. *Habena,* 'bridle' or 'rein,' alludes to narrow lip of some. **Most rein-orchids like moist soils,** hence 'bog-orchid.'

White rein-orchid has **pure white flowers, delicate perfume.** Upper sepals and 2 petals curve to form head. **Slender spur longer than lip.** Stem stout, leafy. Bog and swamp habitats.

RANGE BC: Widespread, Alaska southwards.
RANGE WA: Wide distribution, southwards to California.

SEASIDE REIN-ORCHID, *Platanthera greenei* or *Piperia maritima*: Stout-stemmed; 20–40 cm (8–16") tall; **2–3 broad lower leaves, wither at flowering**. Higher leaves small, sharp-pointed, scab-like. **Flowers clustered, definitely fragrant. Spur twice lip-petal length.** Southeastern Vancouver Island, Gulf/San Juan Islands, southwards along coast.

LADIES' TRESSES, *Spiranthes romanzoffiana*: 1.3 cm (½") long, white–cream **flowers spiral around stem in 3 vertical rows; sweet fragrance**. 7.5–40 cm (3–16") tall, half size of white rein-orchid. **Narrow basal leaves.** Marshes, bogs, streamsides; low–medium elevations. Blooms July–August. Widespread, except arid regions. Alaska to California.

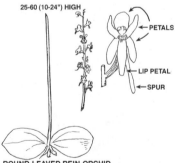

25-60 (10-24") HIGH

PETALS

LIP PETAL

SPUR

ROUND-LEAVED REIN-ORCHID

Orchid Family, *Orchidaceae*
ROUND-LEAVED REIN-ORCHID [168]
Platanthera orbiculata
Habenaria orbiculata

Superficially identical, rein-orchids are usually differentiated by flower structures. Fortunately, this one is easily recognized by **2 large, roundish, fleshy leaves flat on ground** at base of leafless 25–60 cm (10–24") tall stem—or without a stem. In loose pattern, **flowers may have greenish tinge**. Blooms during summer. Variable habitat: swamps, wet meadows and dry, mossy forests.

RANGE BC: Alaska southwards, both sides of Cascades; most common on eastern side.
RANGE WA: Both sides of Cascades; most common on eastern side.

ROUND-LEAVED ORCHIS, *Amerorchis rotundifolia*: Orchis (orchid) with 1 rounded leaf and lovely **pink–white flowers with purple dots. Flower lip is 3-lobed.** Spotty occurrences in B.C.'s lowland wet areas east of Cascades and northwards. Not in Washington.

Mustard Family, *Brassicaceae (Cruciferae)*
FEW-SEEDED WHITLOW-GRASS
Draba oligosperma

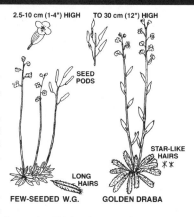

2.5-10 cm (1-4") HIGH   TO 30 cm (12") HIGH

SEED PODS

STAR-LIKE HAIRS ✳✳

LONG HAIRS

FEW-SEEDED W.G.   GOLDEN DRABA

Note: White-flowering drabas are on p. 161.
'Whitlow-grass' and 'draba' are almost interchangeable. Drabas are all quite similar; **a few low, slender stems from a base of very small leaves.** In this case, 4-petalled **flower is pale yellow. Small leaves in mat-like clump; prominent midribs, covered with star-like hairs.** On gravel and in rocky places, dry hillsides to high, open ridges.
RANGE BC: Widespread but relatively scarce.
RANGE WA: Abundance increases southward.
GOLDEN DRABA or **GOLDEN WHITLOW-GRASS**, *D. aurea*: Yellow blooms on slender stems to 30 cm (12") tall. **Narrow, flat seed pods, often with a twist. Hairy leaves** about 2.5 cm (1") long. Common in B.C. east of Cascades; dry places, middle-mountain forests to alpine. Southwards along western slope of Rockies into Montana and Idaho; not in Washington.
ALPINE DRABA, *D. alpina*: Alpine; northern B.C. Usual basal tuft of leaves; 5–12.5 cm (2–5") tall. **Small, bright yellow flowers** in loose clusters.

Mustard Family, *Brassicaceae (Cruciferae)*
COLUMBIA BLADDER POD [174]
Lesquerella douglasii

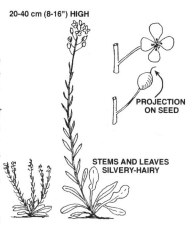

20-40 cm (8-16") HIGH

PROJECTION ON SEED

STEMS AND LEAVES SILVERY-HAIRY

Low: 20–40 cm (8–16") tall. Branches from base into an open form. Each stem crowned with **loose cluster of yellow-orange flowers** about 6 mm (¼") across. Blooms April through May.
Stems and leaves silvery hairy. Leaves of basal rosette small, oval and long-stemmed. Stem leaves much smaller, alternate and very narrow. Alternate flowers yield **oval-to-round seeds, each with a fine projection**. Not particularly eye-catching—loose branching and small flowers—but does give a touch of colour to open roadside banks.
RANGE BC: Sagebrush, ponderosa pine areas.
RANGE WA: Sagebrush and ponderosa pine ecosystems. Republic, Loomis, Columbia Gorge.

Mustard Family, *Brassicaceae (Cruciferae)*
AMERICAN WINTER CRESS
Barbarea orthoceras

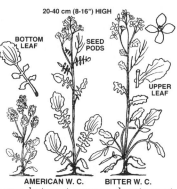

20-40 cm (8-16") HIGH

BOTTOM LEAF   SEED PODS

UPPER LEAF

AMERICAN W. C.   BITTER W. C.

Can be rather obscure or quite showy, depending on number of branching stems and flowerheads. **Small yellow flowers** quickly reveal it as member of mustard family; leaves and long, thin seed pods (siliques) also typical. Lower leaves are easiest identification feature: **large, oval terminal leaflet with several pairs of small, narrow leaflets beneath**. Upper leaves more fringed, clasping the thick stem. Blooms early spring to middle of summer.
RANGE BC: Stream banks, damp meadows and moist forest areas; at lower elevations throughout our area, but most common close to coast.
RANGE WA: Habitats as for B.C.
BITTER WINTER CRESS, *B. vulgaris*: Does *vulgaris* imply a common, no-good weed? Botanists seem to use it thus. Eurasian; very similar to American winter cress, but **higher leaves lobed, not fringed**. Wet places. Common south-central B.C. and Puget Sound region.

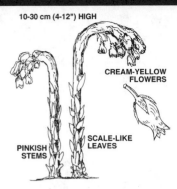

10-30 cm (4-12") HIGH

CREAM-YELLOW FLOWERS

SCALE-LIKE LEAVES

PINKISH STEMS

Heather Family, *Ericaceae*
PINESAP [175]
Hypopitys monotropa

Note: Other saprophytes include candystick and pinedrops (red flowers, p. 289). Indian-pipe, similar form, always white, on p. 173.

Distinctive, quickly recognized. **Flowers cream to dull yellow or even pinkish.** Several stout, tawny–to–pinkish-white stems may be in a cluster; seldom reaches 30 cm (12") in height. **3–10 large flowers hang from curving stem.** 4 petals fairly hairy. Round, brown seed capsules. Pinesap becomes more erect and **turns black with age,** as does Indian-pipe. First described in Europe over 2 centuries ago. Favours coniferous forests from low to medium elevations; high-humus soil.

RANGE BC: From northern B.C. southwards.
RANGE WA: Southwards to California.

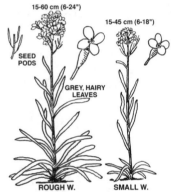

15-60 cm (6-24")

15-45 cm (6-18")

SEED PODS

GREY, HAIRY LEAVES

ROUGH W.   SMALL W.

Mustard Family, *Brassicaceae (Cruciferae)*
ROUGH WALLFLOWER [176]
Erysimum asperum
Western Wallflower

Single stem to 60 cm (2') tall or much branched. **Base is a thick mass of leaves**; top forms large, showy golden flowerhead. **Leaves grey-green because of hairy covering.** Wallflower leaves characteristically long and thin. 'Rough' because **leaves feel scruffy.** 4 petals—typical for mustard family. **Thin, upright seed pods** (all wallflowers), to 5 cm (2") long. **Sandy banks, weedy dunes.**

RANGE BC: Rather sparse; at low to medium elevations east of Cascades.

RANGE WA: East of Cascades; abundant in many dry, sandy areas. Columbia Gorge.

SMALL WALLFLOWER, *E. inconspicuum*: 'Inconspicuous' says Latin name, relating to **pale yellow flowers, small compared to rough wallflower's**; 1.3 cm (½") across; **loose cluster.** Usually branchless—possible few on upper stem; 15–45 cm (6–18") tall. Leaves narrow, to 7.5 cm (3") long. Dry, open areas; lower elevations. Sometimes prefers alkaline habitat. At times in foothill country. Fairly abundant; throughout, east of Cascades.

PALE WALLFLOWER, *E. occidentale*: **Small,** to 40 cm (16") tall. **Large, pale yellow flowers,** to 2.5 cm (1") across. Washington, east of Cascades; sagebrush. Not in B.C.

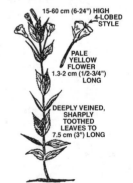

15-60 cm (6-24") HIGH
4-LOBED STYLE

PALE YELLOW FLOWER 1.3-2 cm (1/2-3/4") LONG

DEEPLY VEINED, SHARPLY TOOTHED LEAVES TO 7.5 cm (3") LONG

Evening-primrose Family, *Onagraceae*
YELLOW WILLOWHERB [177]
Epilobium luteum

Luxuriant along mountain streams; habitat and form similar to pink monkey-flower's (p. 279). Stems erect, to 60 cm (2') tall. **Leaves to 7.5 cm (3") long, heavily veined, sharply toothed**; generally in opposite pairs. 'Willowherb' relates shape of leaves to those of weeping willow.

Modest **pale yellow flowers**, erect on long stems near top; 1.3–2 cm (½–¾") long. **4 folded petals suggest closed flower,** supported by 4 long, narrow sepals. Depending on elevation, blooms from early to late summer. Moist meadows and stream edges, middle mountain to high elevations.

RANGE BC: From Alaska southwards through Cascades and Selkirks.
RANGE WA: Olympics. Southwards through Cascades.

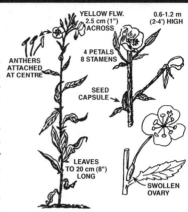

Evening-primrose Family, *Onagraceae*
COMMON EVENING-PRIMROSE
Oenothera biennis

Generally so widely scattered that they are missed unless a person looks closely for them, even though the stout and possibly branched, glandular stem may reach 1.2 m (4') in height. Several confusing garden escapes, so check and note ranges.

About 2.5 cm (1") across, **satin-like flowers are an exquisite soft yellow,** but note that **sepals are green.** Evening-primrose is **at its best evening or early morning,** or on a cloudy day. See how the **flower stem swells at its base,** enclosing the ovaries, eventually to wither away to leave an **upright, 4–sided seed capsule.** Numerous leaves to 20 cm (8") long; lower ones usually have 'wings' near bases. Blooms late spring into summer.

RANGE BC: Mostly west of Cascades.

RANGE WA: Mostly east of Cascades.

YELLOW EVENING-PRIMROSE, *O. villosa*: A coarse, **grey-hairy stem** to 90 cm (3') tall with **delicate yellow blossoms** 1.3–2.5 cm (½–1") across. **Flowers open in evening, close at sunrise.** Swollen section at base of flower stem contains ovaries. Scattered occurrences in southern B.C. east of Cascades. Not in Washington.

Mustard Family, *Brassicaceae (Cruciferae)*
YELLOW BEE PLANT
Cleome lutea

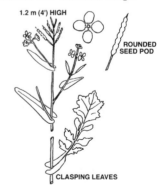

Widely ranging in form—perhaps a single stem or much-branched—from 0.3 to 1.2 m (1 to 4') in height. **Small, bright yellow flowers,** each with **6 protruding stamens that give flowerhead a fuzzy appearance.** Leaves have an umbrella form: **5 leaflets radiate from end of a long stem.** Some leaves higher on plant may have only 3 leaflets. **Flat, curved seed pods.** Blooms from late spring to August.

RANGE BC: Not in B.C.

RANGE WA: Sandy and rocky ground of eastern Washington. Sagebrush and bunchgrass ecosystems. Adjacent to Columbia River, extends into Columbia Gorge.

Mustard Family, *Brassicaceae (Cruciferae)*
RAPE or FIELD MUSTARD [180]
Brassica campestris

Of the half-dozen or more mustards (*Brassicas*) in our area, this is the only one that is not comparatively rare. Just the opposite: **old pastures and abandoned fields turn bright yellow with this rank plant.** To 1.2 m (4') tall. This European import or a close relative, *B. nigra*, decorates the countryside with **solid blocks of yellow where it is under cultivation.** You may see such a tended field of tall, yellow plants that are being grown for their oil. Called 'canola,' this variety of *B. campestris* has seeds that yield a culinary oil chemically superior to rapeseed oil, and with a more-marketable name besides. Turnips, rutabagas, broccoli and cauliflower are all close relatives.

Most people casually recognize mustards by the 4-petalled cross-shaped yellow flowers. However, leaf shape is more specific: notice how **lower leaves are distinctly different from upper ones.** Rape is **among the earliest flowers to bloom**—continues all summer.

RANGE BC: Common; roadsides and waste areas, from central B.C. southwards.

RANGE WA: Common; roadsides and waste areas, southwards from central B.C.

5-15 cm (2-6") HIGH

2 LEAF SHAPES

SAGEBRUSH BUTTERCUP   SMALL YELLOW WATER-BUTTERCUP

Buttercup Family, *Ranunculaceae*
SAGEBRUSH BUTTERCUP [179]
Ranunculus glaberrimus

Of course, everyone recognizes a buttercup—or do they? There are over 25 species in our area. Buttercups have **'varnished' golden petals with a very small scale at the base of each one.** This is the nectary that gives buttercups their name. Leaves vary: some are lobed and some finely dissected. Possible confusion might come from *Potentillas* (cinquefoils), but they have leaves made up of leaflets. Some buttercups actually grow in the water, some hug the ground in the driest of places and still others live on the edge of alpine snowbanks. Wherever they grow, these 'little frogs' (a literal translation of *Ranunculus*) are a cherished part of the landscape.

I have boyhood memories of early spring in the Okanagan Valley, of ground-hugging sagebrush buttercup and its bright petals. A first sign of spring, followed by yellow bell and then by long-leaved phlox—all plants of sagebrush country. To make sure of identification, look for **2 differently shaped leaves: 1 entire and 1 with 3 fingers.** May have more than 5 petals.

RANGE BC: Sagebrush areas mostly, but into the low foothills also.

RANGE WA: Habitats as for B.C.

SMALL YELLOW WATER-BUTTERCUP [178] *R. gmelinii*: Look for yellow buttercup flowers brightening a **loose mat of thin stems floating in a pond** or a shallow backwater of a stream. (White water-buttercups are on p. 179.) The flowers, about **half the size of most buttercups** but otherwise typical, also have a varnished yellow sheen. Leaves vary in shape but are distinctly lobed. Stems are finely hairy. There should be no confusion with this little gem. Ponds, slow-flowing water and creeping onto wet ground. Common across B.C., but rare along the coast. Eastern and western Washington.

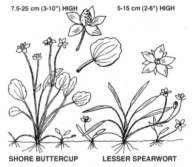

7.5-25 cm (3-10") HIGH   5-15 cm (2-6") HIGH

SHORE BUTTERCUP   LESSER SPEARWORT

Buttercup Family, *Ranunculaceae*
SHORE BUTTERCUP
Ranunculus cymbalaria

Two aspects taken together should serve to quickly identify this widespread buttercup. First, the name reflects a **wet habitat**: shores, but also the edges of muddy ponds and saline meadows. Second, the **flower has a distinctive cone-shaped seed mass** in the centre that holds up to 200 tiny, beaked seeds. Nearby plants are often joined by runners, much like strawberries. The long-stemmed leaves are also distinctive, being oval-shaped with the top half scalloped.

RANGE BC: Favours saline or alkaline shorelines of ponds, marshes and meadows at lower elevations. Common in southern half of B.C.

RANGE WA: Habitats as for B.C. Extends southwards to Grant and Whitman counties.

LESSER SPEARWORT or **SMALL CREEPING BUTTERCUP**, *R. flammula*: Another wet-ground buttercup possibly partly submerged and often in the same saline habitats as shore buttercup. This dainty plant is recognized by its **very thin, string-like leaves**, **long runners** and **very small flowers**. Common at lower elevations throughout our area.

Buttercup Family, *Ranunculaceae*
WESTERN BUTTERCUP
Ranunculus occidentalis

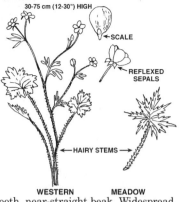

30-75 cm (12-30") HIGH

←SCALE

←REFLEXED SEPALS

←HAIRY STEMS →

WESTERN   MEADOW

Prominent, attention-getting. **High flowers. 5 reflexed sepals fall off** while in bloom. **Stems finely hairy.** Long-stemmed 3-part basal leaves, partly divided again. Blooms March–late spring.

RANGE BC: Common west of Cascades; coastal bluffs, moist meadows; to middle-mountain slopes.

RANGE WA: Habitats as for B.C.

MEADOW BUTTERCUP, *R. acris*: European. **To 90 cm (3') tall, branches freely.** Flowers spread widely apart. Leaves large, finely divided, **like cut-leaf maple; hairy stems** to 20 cm (8") long. **Petals much longer than sepals; not reflexed.** Seeds smooth, near-straight beak. Widespread, except northern B.C.; moist ground; valley bottoms to mountain forests. Roadsides.

STRAIGHT-BEAKED BUTTERCUP, *R. orthorhynchus*: Many, **larger-than-average flowers, paler yellow than most buttercups. Petals narrow, spaces between.** Tall stems branch, branch again. **Seeds with straight beak.** Blooms April–June, sometimes with camas. Favours wet meadows and richer soils. Alaska southwards.

Buttercup Family, *Ranunculaceae*
OTHER BUTTERCUPS
Ranunculus spp.

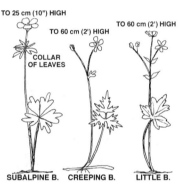

TO 25 cm (10") HIGH

TO 60 cm (2') HIGH

TO 60 cm (2') HIGH

COLLAR OF LEAVES

SUBALPINE B.   CREEPING B.   LITTLE B.

SUBALPINE BUTTERCUP, *R. eschscholtzii*: Bright gold flower bejewels fresh greenery at **snowbank edges.** Several lower leaves fancifully lobed; **narrow stem leaves in high circlet.** Perhaps the most common buttercup above timberline. Blooms early to late summer as snow recedes. Common on higher mountains; Alaska to California.

SNOW BUTTERCUP, *R. nivalis*: **Brown hair dense on sepals.** High mountains; east of Cascades; north of Ft. St. John, B.C. Not in Washington.

CREEPING BUTTERCUP [181], *R. repens*: **Creeping stems** root, form **masses.** Stem, leaves slightly hairy. **Leaves divided into 3 leaflets; 3-lobed or 3-parted.** Bright golden yellow flower. Most common around **damp areas** west of Cascades. Fraser Valley roadsides.

LITTLE or **BONGARD'S BUTTERCUP**, *R. uncinatus*: **Small, pale yellow flowers under 8 mm (⅓") across.** 3-leaflets, toothed lobes. Stems usually hairy. Seeds have small, hooked beak. Throughout, south of Ft. St. John, B.C.; shady, moist low-elevation forests.

YELLOW ANEMONE, *A. richardsonii*: **Only yellow anemone in our area** (white, pp.176 & 185; purple, p. 300; blue, p. 313); confined to damp mountain areas east of Coast Mountains and north of Quesnel, B.C. **Easily mistaken for buttercup**; 15 cm (6") tall or less.

Lily Family, *Liliaceae*
YELLOW BELL [194]
Fritillaria pudica
Yellow Fritillary

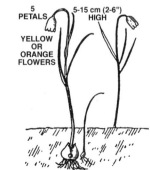

5 PETALS

5-15 cm (2-6") HIGH

YELLOW OR ORANGE FLOWERS

Benches and slopes of grasslands, sagebrush terrain and ponderosa pine country. Blooms April–May. Seldom over 10 cm (4") tall. **1 nodding, fragrant, yellow-to-orange flower**, 1.3 cm (½") long. 2–3 narrow, olive-green leaves start well up on stem. Tiny, white bulb 2.5 cm (1") beneath ground. Like fawn lily, it suffers from thoughtless over-picking.

RANGE BC: Central drylands and southwards.

RANGE WA: Drylands on both sides of the Cascades. Columbia Gorge.

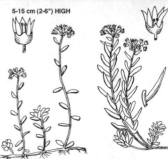

5-15 cm (2-6") HIGH

SPREADING   LANCE-LEAVED

Sedum Family, *Crassulaceae*

SPREADING STONECROP [189]

Sedum divergens

About 7 stonecrops in our area have a loose cluster-head of 5-petalled yellow flowers; 1 is rosy-purple (see p. 305). Most stonecrops under 15 cm (6") tall and favour dry, exposed, rocky areas.

'Spreading' describes how this one **forms mats on rocky surfaces**: horizontal root probes along cracks; succession of stems, some bear flowers, some are sterile. Smooth, bright green foliage; parts may be red if very exposed. Stubby, **thick, oval leaves** mostly alternate; **rosettes at top of sterile stems**. Clustered flowers. **Stamens a little shorter than lance-shaped petals**. Dry, rocky places. Very colourful in July.

RANGE BC: Higher mountains—subalpine to alpine, but can be in lower open areas.

RANGE WA: Habitats as for B.C. Cascades and Olympics.

LANCE-LEAVED STONECROP, *S. lanceolatum*: Same size as spreading stonecrop, similar habitat, overlapping range. Commonly basal rosettes. **All leaves narrow, lance-like; rounded cross-section. Thin, sharp-pointed petals, twice stamen length**. Dry soils, rocky places; medium–alpine elevations. Widespread. Mt. Spokane over 1000 m (3500').

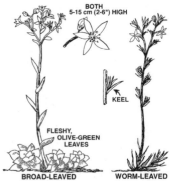

BOTH
5-15 cm (2-6") HIGH

KEEL

FLESHY,
OLIVE-GREEN
LEAVES

BROAD-LEAVED   WORM-LEAVED

Sedum Family, *Crassulaceae*

BROAD-LEAVED STONECROP

Sedum spathulifolium

Pacific Sedum

Irregular head of star-like flowers, bursting buds. Enlivens mossy bluffs with dabs of bright yellow May–June. **Thick, fleshy, spoon-shaped basal leaves sage-green to red**; may form flattish rosettes, quite different from alternate, thinner leaves on stem. Fleshy leaves help tide plant over dry periods. Rocky areas only.

RANGE BC: Common; coastal bluffs, ravines; southern Vancouver Island, Gulf Islands.

RANGE WA: Common; coastal bluffs and canyons on San Juans and west of Cascades throughout Washington. Olympics, Columbia Gorge.

OREGON STONECROP, *S. oreganum*: Leaves egg- to spoon-shaped. Seed capsule segments erect. Range similar to broad-leaved stonecrop's, plus inland mountain slopes.

WORM-LEAVED STONECROP [188], *S. stenopetalum*: Open head of bright yellow flowers. Stems tinged yellow and pink. **Thin, worm-like leaf; noticeable keel on lower side**. Leaves generally rounded on bottom, flat on top. Rocky or gravelly slopes; valley bottoms to subalpine heights. Central B.C. east of Cascades, southwards through eastern Washington. Columbia Gorge, Mt. Spokane.

5-15 cm (2-6") HIGH

STEMLESS
LEAVES
TO 1.3 cm (1/2")
LONG

Saxifrage Family, *Saxifragaceae*

EVERGREEN SAXIFRAGE

Saxifraga aizoides

Yellow Saxifrage

Note: General notes, white-flowered species on pp. 165–67 & 195; a purple one on p. 295.

Low profile. Showy **5-petalled yellow flowers with orange spots. 3-veined petals slightly longer than sepals**. Mat or tuft of stiff, narrow, stemless leaves about 1.3 cm (½") long. Brightens damp moraines, gravel bars, and wet, rocky areas mostly in alpine tundra ecosystem.

RANGE BC: Most common in Selkirks and Rockies.

RANGE WA: Not recorded.

Violet Family, *Violaceae*
YELLOW VIOLETS
Viola spp.

Note: White violets are on p. 175, purple ones on p. 294 and blue ones on p. 315. There are a dozen or so yellow violets. With violets coming in a variety of colours and sizes, perhaps you should make sure you are looking at a violet. Each **small pansy-like flower has 5 showy petals: 2 upper ones, 2 side ones and a larger lower lip petal.** Sometimes the lip and the 2 side petals have darker streaks on them. The flowers produce a seed capsule, but to make sure of reproduction, an unusual arrangement sees another set of tiny, greenish flowers, without petals, borne next to the ground or under it and so going unnoticed. They do not open and thus bear self-fertilized seeds. **Leaves are single and oval or roundish.**

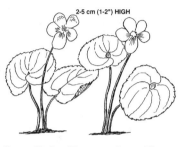

2-5 cm (1-2") HIGH

Violet Family, *Violaceae*
ROUND-LEAVED VIOLET [192]
Viola orbiculata

Possibly the common name says enough. If not, check to make sure that the **flower stems and leaf stems come out of the ground individually,** rather than branching off a main stem. Flowering stems range from 2.5 to 5 cm (1 to 2") in height. The **3 lower petals are marked with brown lines.** The leaves may remain green over winter, giving rise to the alternative name 'evergreen violet,' one more usually applied to *V. sempervirens.* Flowers are out during summer.

    RANGE BC: Moist places, middle-mountain heights to alpine. Central B.C. southwards.
    RANGE WA: Habitats as for B.C. Southwards to Oregon. Cascades and Olympics.
    YELLOW SAGEBRUSH VIOLET, *V. vallicola*: Watch for this plant with sagebrush and ponderosa pine in the dry interior. Only 5–10 cm (2–4") tall, it has several long-stemmed ovate leaves that may hide the flower. The **upper 2 petals usually have brown backs.** Common in south-central B.C. and southwards to Oregon.
    YELLOW MONTANE VIOLET [191], *V. praemorsa*: This violet and yellow sagebrush violet (above) were formerly considered as varieties of Nuttall's violet, *V. nuttallii.* Both have since been 'elevated' to species rank. Neither has the heart-shaped leaves common to most violets. Montane violet has thicker, fleshier leaves, usually conspicuously hairy. West of Cascades, southern Vancouver Island to California.
    STREAM VIOLET, *V. glabella*: The **most common yellow violet.** Starts to bloom on edges of the melting snows. **2 or 3 toothed leaves branch out near top of each stem,** leaving lower third leafless. The **2 top petals are pure yellow; lower 3 have brown or purplish lines.** Wide range in damp soils, from valley bottom to alpine ecosystem.
    TRAILING YELLOW or **EVERGREEN VIOLET**, *V. sempervirens*: This bright yellow violet captures attention in moist woods along the western flank of our coastal mountains. **Leaves and flowers often originate from runners,** in the fashion of a strawberry. The thick, leathery leaves often persist through winter, hence the name 'evergreen.' B.C. and southwards through Oregon.

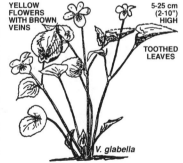

YELLOW FLOWERS WITH BROWN VEINS

5-25 cm (2-10") HIGH

TOOTHED LEAVES

V. glabella

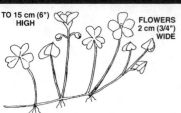

TO 15 cm (6")
HIGH

FLOWERS
2 cm (3/4")
WIDE

Wood Sorrel Family, *Oxalidaceae*
WESTERN YELLOW OXALIS [193]
Oxalis suksdorfii

Provides a beautiful ground cover on shady forest floors beneath the giant conifers of southwestern Washington. A **slender, creeping vine** that roots at the nodes, providing an abundance of plants, especially in moist areas. Too slow to see, the 3 heart-shaped leaflets unfold, turn and close as they adjust to the filtered sunlight.

Single flowers, yellow to pale pinkish, rise on stems to 15 cm (6") tall, barely clearing the leaves. They are about 2 cm (¾") across. Watch for these blooms from spring to summer.

RANGE BC: Not in B.C.

RANGE WA: Lower forest areas of southwestern Washington.

YELLOW OXALIS, *O. corniculatus*: If you are a west-coast gardener, then you may know this European import as Dutch clover, a persistent lawn weed with tiny clover leaves and a nub of yellow flowers.

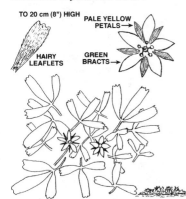

TO 20 cm (8") HIGH   PALE YELLOW
PETALS →

HAIRY
LEAFLETS

GREEN
BRACTS →

Rose Family, *Rosaceae*
SIBBALDIA [183]
Sibbaldia procumbens

Mountain hikers above tree line will likely find **bright green mats** of sibbaldia while tramping rocky ridges, angling across rockslides or relaxing on tundra slopes. **Tough and shrubby,** with yellow flowers—looks much like a cinquefoil. But there are **only 5 or 10 stamens**; cinquefoils have many.

Small, pale yellow flowers, about 1.3 cm (½") across, have a strange design: **5 slim, very short yellow petals, dwarfed by 5 much larger green calyx lobes**; stamens accentuate symmetry. Stems to 20 cm (8") tall. Flowers early summer–August.

For positive identification, look among the tufts of mat-forming leaves to find that each **3-branched leaflet has 3 teeth at its tip**. It may be purplish beneath. Leaves are hairy—protection against the severe weather of subalpine and alpine slopes.

RANGE BC: High mountains from Alaska southwards.

RANGE WA: High mountains southwards to California.

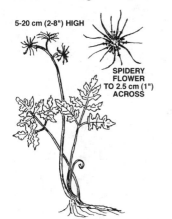

5-20 cm (2-8") HIGH

SPIDERY
FLOWER
TO 2.5 cm (1")
ACROSS

Buttercup Family, *Ranunculaceae*
FERN-LEAVED GOLDTHREAD
Coptis asplenifolia

Note: Three-leaved goldthread, with the same range, has white flowers (p. 174).

The intriguing 'goldthread' part of the name comes from the bright yellow colour of the thin roots when peeled. Who might have discovered that colourful bit of natural history? Only 5–20 cm (2–8") tall, a miniature jewel with **glossy dark green, fern-like leaves with 5-toothed leaflets**. Flower is **pale greenish yellow** and has a spidery appearance. This distinctive floral design comes from the arrangement of **5–8 sepals and 5–7 petals, both grass-thin and long**; sepals are longest. The effect is enhanced by an abundance of stamens and pistils.

There are **2–3 nodding flowers per stalk**. Goldthread blooms late April–May.

RANGE BC: Wet woods and bogs, lowlands to mountain forests. Alaska to Vancouver Island and southern coastal B.C.

RANGE WA: Extends into Washington along the Cascades to Mt. Pilchuck. Olympics.

Rose Family, *Rosaceae*
SILVERWEED [185]
Potentilla anserina

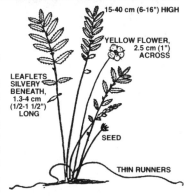

15-40 cm (6-16") HIGH

YELLOW FLOWER, 2.5 cm (1") ACROSS

LEAFLETS SILVERY BENEATH, 1.3-4 cm (1/2-1 1/2") LONG

SEED

THIN RUNNERS

Note: White cinquefoil is here with the yellows; see p. 222. marsh cinquefoil, red-purple, on p. 297.

There are about 30 attractive *Potentilla* species in our area; most are called cinquefoils. But first, is the specimen a *Potentilla*? These plants generally have **small, nonwaxy 'buttercup' flowers. Petals lie flat or are slightly dished.** There are a large number of long stamens. **Many *Potentillas* have notched petals**; buttercups never do. Usually (a weasel word very useful in botanical descriptions), the part holding the 5-petalled flower, the **calyx, appears to have 10 parts: 5 bracts alternating with 5 green sepals that just show their tips.** This flower design is similar to that of large-leaved avens (p. 223).

Several features should make for easy identification of silverweed. The 'silver' part of the name comes from the **silvery appearance of the underside of the leaves.** Many rather buttercup-like **flowers, singly borne on leafless stems** less than 15 cm (6") long. Notice the **system of thin runners,** like strawberry plants. **Roots develop at nodes or joints— another plant gets started.** Often forms a thick, extensive growth. Long, narrow roots beneath an old plant are edible if boiled or roasted. In past times they were called 'Indian sweet potato.' They were an important food source for aboriginal peoples from B.C. to Oregon.

At the coast, frequently in wet places, such as edges of ponds, streams, estuaries and salt marshes; sometimes on sand dunes. Inland specimens (sometimes called *P. pacifica*) also favour wet places, including bogs, mudflats, meadows and alkaline areas.

RANGE BC: From Alaska southwards and throughout B.C. in suitable habitat.

RANGE WA: Generally confined to east of the Cascades.

Rose Family, *Rosaceae*
STICKY CINQUEFOIL [186]
Potentilla glandulosa

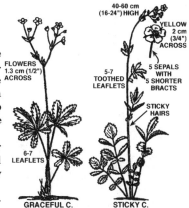

40-60 cm (16-24") HIGH

YELLOW 2 cm (3/4") ACROSS

FLOWERS 1.3 cm (1/2") ACROSS

5-7 TOOTHED LEAFLETS

5 SEPALS WITH 5 SHORTER BRACTS

STICKY HAIRS

6-7 LEAFLETS

GRACEFUL C.   STICKY C.

Has all the typical floral characteristics of the cinquefoils, as noted with silverweed (above). **Noticeably sticky,** perhaps the easiest clue. Also note the strong branching of the upper stems, giving a showy display of the small, **flowers of pale to deep yellow. Petals are about the same length as the mass of stamens** they surround.

Stem leaves are few; usually a small cluster where branching occurs. The more abundant basal leaves are pinnate: **leaf stems carry 4–8 coarsely toothed, opposite leaflets and a terminal one.**

A wideranging plant of meadows and open forests but most abundant in the bunchgrass and ponderosa pine ecosystems.

RANGE BC: From Prince George southwards, east of the Cascades.

RANGE WA: Habitats as for B.C. Wideranging across state, except northwestern part.

GRACEFUL or **NUTTALL'S CINQUEFOIL**, *P. gracilis*: Both 'graceful cinquefoil' and 'five-finger' (an alternative name) help identification. But be careful: there **may be 7 highly dissected leaflets!** Slender stems to 60 cm (2') tall branch near top to carry clusters of attractive flowers, each only 1.3 cm (½") across. Blooms in summer, to considerable elevation in meadows, grasslands and open forests. Common in southern B.C. but not on coast. Several varieties give wide coverage across Washington. Columbia Gorge, Blue Mountains.

BROOK CINQUEFOIL, *P. rivalis*: To 50 cm (20") tall; spreading crown. Distinguished by **hairy bracts, coarsely toothed lower leaves, 5 leaflets—upper ones with 3—**and **small petals about half sepal length.** Damp soils in Okanagan and Kootenay valleys of B.C.; continuing southwards through Washington to California.

TO 70 cm (28") HIGH

Rose Family, *Rosaceae*
SULPHUR CINQUEFOIL
Potentilla recta

Rather unobtrusive, to 70 cm (28") tall, growing here and there with other sparse vegetation on dryland benches. Sometimes so abundant and widespread in fields and pastures that some farmers and ranchers consider it a weed. The **overall dull green leaves and stems** are blessed with a terminal tuft of a dozen or more small 'buttercup' flowers each about 1.3 cm (½") across. **Sepals almost as long as petals.** Perhaps these **distinctive pale sulphur-yellow flowers** will be the main clue in identification. Distinctive finger-like fanning of leaves should also help distinguish this cinquefoil from others. While most leaves reflect the 'cinque' (meaning '5') part of the name in having 5 coarsely toothed leaflets, larger leaves near base of plant may have 7 or 9 leaflets. An early summer bloomer.

RANGE BC: Common east of Cascades.

RANGE WA: Common east of Cascades. Ferry County.

SILVERY CINQUEFOIL [187], *P. argentea*: Quickly comes to attention of a person strolling through open forest: **disturbed leaves shine bright silver from lower surfaces**, shape like **very incised 'maple' leaf**. Flowers similar to other potentillas. Usually grows to 1200 m (4000') elevation. Locally frequent European import. In southern B.C., east of Cascades in dryland areas. In Washington, Stevens and Spokane counties and Douglas-fir–ponderosa-pine forests of Blue Mountains.

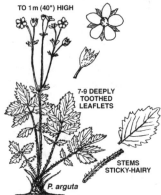

TO 1 m (40") HIGH

7-9 DEEPLY TOOTHED LEAFLETS

STEMS STICKY-HAIRY

P. arguta

WHITE CINQUEFOIL, *P. arguta*: Pale yellow to white—the **only whitish cinquefoil**. Relatively tall, to 10 cm (40"). Lower leaves have 7–9 leaflets; sticky, hairy and deeply toothed. Flowerhead is narrow and flat-topped. Petals are slightly larger than bracts and sepals. Common throughout in coastal meadows and lower mountain slopes.

Rose Family, *Rosaceae*
FAN-LEAVED CINQUEFOIL [184]
Potentilla flabellifolia

This very common plant of alpine meadows might be mistaken for a robust buttercup unless the **notched petals and green sepals** are noticed. The flower, which is adorned by a ring of a **dozen yellow stamens**, may be 2.5 cm (1") across. Usually several flowers branch from the main stem but overall the plant may have a mat form, especially in early spring when it often lies pressed to the ground. **Coarsely toothed and white-woolly beneath, leaves are like a strawberry plant's in having 3 leaflets**. Wedge-shaped at base. Look for bright yellow flowers, especially in moist areas among other spring- and summer-blooming subalpine flowers.

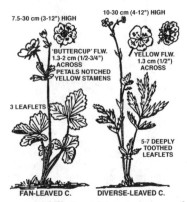

7.5-30 cm (3-12") HIGH

10-30 cm (4-12") HIGH

'BUTTERCUP' FLW. 1.3-2 cm (1/2-3/4") ACROSS PETALS NOTCHED YELLOW STAMENS

YELLOW FLW. 1.3 cm (1/2") ACROSS

3 LEAFLETS

5-7 DEEPLY TOOTHED LEAFLETS

FAN-LEAVED C.   DIVERSE-LEAVED C.

RANGE BC: Very common in the Cascades throughout. Sparse on Vancouver Island.

RANGE WA: Very common in the Cascades throughout. Olympics.

DIVERSE-LEAVED or **MOUNTAIN MEADOW CINQUEFOIL**, *P. diversifolia*: Grows to 30 cm (12") tall and is sparsely hairy. Long-stemmed leaves carry **5–7 deeply toothed leaflets all arising from 1 point**. Common in moist habitats above and below timberline, from Cascades to Rockies and far northwards. High mountains of Washington. Olympics.

Rose Family, *Rosaceae*
PRAIRIE CINQUEFOIL
Potentilla pensylvanica
Pennsylvanian Cinquefoil

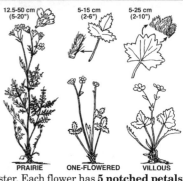

12.5-50 cm (5-20") — 5-15 cm (2-6") — 5-25 cm (2-10")

PRAIRIE — ONE-FLOWERED — VILLOUS

The first collection of this plant was in 1767, its origin simply recorded as 'Canada.' A branching **grey-green plant** with several flowering stems 12.5–50 cm (5–20") tall. **Stems and leaves are covered with a mat of hair** but the leaves are nevertheless green above and whitish beneath. There are 5–9 'carroty' pinnately divided basal leaves and a few that climb the stem to the flowerheads. There are several 'buttercup' flowers in each small terminal cluster. Each flower has **5 notched petals and 20 stamens**.

The dry, rocky habitat is significant; in sagebrush areas and on lower mountain slopes.

RANGE BC: Common in south-central and eastern B.C.; extends into Cariboo parklands ecosystem as far north as the Peace River.

RANGE WA: Dry places at lower elevations in the Cascades.

ONE-FLOWERED CINQUEFOIL, *P. uniflora*: Low, rather **mat-like silvery-green plant 5–15 cm (2–6") tall, with greyish hair throughout** but not woolly as in villous cinquefoil (below). Basal leaves 3-lobed and deeply toothed. 1 or 2 flowers adorn each short, slender stem. Fairly small flowers with petals shallowly notched; 20 stamens. From Alaska southwards, in rocky places in alpine terrain, north Cascades to Wallowa Mountains. Absent from coast.

VILLOUS or **HAIRY CINQUEFOIL** [182], *P. villosa*: A low, tufted plant 5–25 cm (2–10") tall. Stems, **lower surfaces of leaves and calyx are woolly** (villous). **Leaflets thick, crinkled and strongly veined**. 2–5 large flowers, **petals longer than sepals**. Ranges from Alaska southwards on ridges and rock slopes at subalpine and alpine heights. Also frequent on rocky seashore bluffs. Southwards through Cascades of B.C. and Washington. Olympics.

ARCTIC CINQUEFOIL, *P. hyparctica*: A **prostrate** form, to 10 cm (4") tall. Leaves 3-foliate, not divided but toothed and with **long, soft hairs**. Alpine ecosystem of northern B.C. and southwards through the Coast Mountains.

Rose Family, *Rosaceae*
LARGE-LEAVED AVENS [190]
Geum macrophyllum

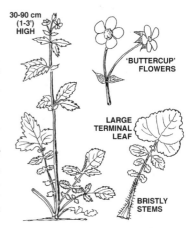

30-90 cm (1-3') HIGH

'BUTTERCUP' FLOWERS

LARGE TERMINAL LEAF

BRISTLY STEMS

Large-leaved avens, a common plant, produces **small, bright yellow, saucer-shaped 'buttercup' flowers** that develop into round burs with hooked prickles. The flowers, either singly or in a few-flowered cluster, grow at the tips of the branches.

While avenses generally have flowers with 5 petals and 5 green sepals separated by 5 small bractlets, so do cinquefoils (pp. 221–23). However, the leaves of this plant are distinctive. The **large, coarse basal leaves**, to 30 cm (12") long, are a prime factor in identification. Each **basal leaf has a ragged scattering of mismatched leaflets on its lower stem and terminates in a large, oval-lobed leaflet**.

Blooms during May on the southern part of the coast but as late as July in the north. Most abundant in moist, shady woods and along stream banks.

RANGE BC: Widespread except for drier areas, e.g., sagebrush and ponderosa pine regions.

RANGE WA: Habitats as for B.C., southwards through the Pacific states.

CLUSTERED YELLOW FLOWERS IN TIP

20-60 cm (8-24") HIGH

YELLOW FLOWER 1 cm (3/8") LONG

FINELY HAIRY LEAVES, 2.5-7.5 cm (1-3") LONG

Forget-me-not Family, *Boraginaceae*
LEMONWEED [197]
Lithospernum ruderale
Western Gromwell, Stoneseed

Typically 30–60 cm (1–2') tall; **rough, hairy appearance**. Generally **bushy** because of very leafy stems. Alternate **leaves bunch together towards top**, become tightly packed. **Small, pale yellow, funnel-like flowers, near-hidden** among many leaves at stem tips; only 6 mm (¼") across. Small seeds teeth-like. From May to July, in open country east of Cascades. Most abundant in dry valleys and foothills, typical range country, but will go higher.

RANGE BC: Central B.C.

RANGE WA: Southwards and eastwards of Cascades, through Washington and Oregon.

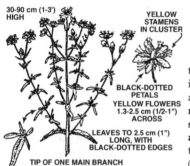

30-90 cm (1-3') HIGH

YELLOW STAMENS IN CLUSTER

BLACK-DOTTED PETALS YELLOW FLOWERS 1.3-2.5 cm (1/2-1") ACROSS

LEAVES TO 2.5 cm (1") LONG, WITH BLACK-DOTTED EDGES

TIP OF ONE MAIN BRANCH

St. John's-wort Family, *Hypericaceae*
COMMON ST. JOHN'S-WORT [195]
Hypericum perforatum

European import, becoming a serious weed. Said to bloom on June 24, St. John the Baptist's Day. Used in Europe for medicinal purposes, but may be moderately poisonous. *Hypericum*s have yellow flowers, **many protruding yellow stamens. Clustered stamens in this case. Faintly black-dotted petal and leaf margins**. Flowers may mass into tousled, top-heavy head. Contrast with tansy butterweed, (blooms later, p. 247). Could be plant most widespread and visible along roadsides July and August. Perennial. Dies back to dark brown stem; ragged head of husks lasts through winter into early summer. Particularly noticeable in Ferry and Spokane counties; along roadside to summit of Mt. Spokane, 1763 m (5878').

RANGE BC: Usually at low elevations across B.C.

RANGE WA: Usually at low elevations; southwards to California.

BOG ST. JOHN'S-WORT, *H. anagalloides*: Very short stems, 2.5–7.5 cm (1–3") long, **forms creeping mats in moist places**. Paired tiny, oval leaves. **Flowers yellow-orange**. Moist meadows and pond edges at lower elevations; B.C. and Washington.

WESTERN ST. JOHN'S-WORT, *H. formosum*: Upright, to 75 cm (2 ½'); paired opposite oval leaves, often with dark purple dots. Leaves small, 2.5 cm (1") long. **Flowers margins fine-toothed**; 2.5 cm (1") across. Moist places; coast–subalpine; southern B.C., Washington.

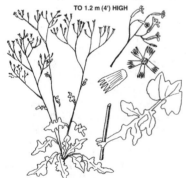

TO 1.2 m (4') HIGH

Sunflower Family, *Asteraceae (Compositae)*
WALL LETTUCE [198]
Lactuca muralis

There are 3 local yellow-flowered *Lactuca*s and 2 blue ones (p. 319). *Lactuca* means lettuce (cultivated kind is *L. sativa*). **Milky juice**, sprays of small flowers, leafy stems, oddly shaped leaves.

Tall, very weedy-looking, with fireworks display of tiny yellow flowers. Possibly 100+ yellow flashes on stems so thin they might be almost invisible. Wall lettuce often on walls in native Europe. Leaf shape beggars description. **Each leaf is unique**. To 20 cm (8") long, upper ones smaller. Flowers seem to sleep at times (partially close). Only 6 mm (¼") across, **actually 5 separate flowers, each with 1 stamen**. Rectangular **petal, 5-tooth fringe**. Frequent along roadsides, trails, in gardens.

RANGE BC: Most common in southwest, to middle elevations

RANGE WA: Habitats as for B.C., through coastal Washington.

Loosestrife Family, *Lythraceae*
TUFTED LOOSESTRIFE [199]
Lysimachia thyrsiflora

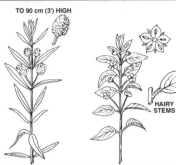

TO 90 cm (3') HIGH

Note: Invading purple loosestrife on p. 299.
At least 7 loosestrifes, all marsh plants, in our area: several garden escapes and 2 fairly rare ones. Stout stems to 90 cm (3') tall. Leaves opposite, lance-shaped, to 7.5 cm (3") long. Often in standing water. 'Tufted' must refer to **yellow balls of tiny flowers; projecting stamens add fuzzy appearance**. Wetlands, lower valleys and mountains.

HAIRY STEMS

TUFTED LOOSESTRIFE   FRINGED LOOSESTRIFE

RANGE BC: Across our area.
RANGE WA: Across our area.
FRINGED LOOSESTRIFE [200], *L. ciliata*: **Hairy leaf stems**, flowers larger, 1.3 cm (½") across. Otherwise like above. Wet areas; south-central B.C., occasionally eastern Washington.

Loasa Family, *Loasaceae*
BLAZING-STAR [201]
Mentzelia laevicaulis

TO 90 cm (3') HIGH

TO 30 cm (12") HIGH

Gorgeous blaze of colour on a ragged plant—always a surprise. Sometimes solitary; number of plants may be scattered on **steep, eroding gravel or rocky slope**. Usually much-branched; may have a dozen very attractive yellow flowers to 10 cm (4") across. **Prefers to hide its glory unless sun is bright**. Narrow, sharp-pointed petals. **About 50 long, slender filaments**. Dull green, heavily toothed, **sandpapery leaves**. Basal leaves to 10 cm (4") long. Blooms late spring and into summer.

BLAZING-STAR   SMALL-FLOWERED

RANGE BC: With sagebrush, ponderosa pine. Mostly east of Cascades; lower elevations.
RANGE WA: Habitats as for B.C.
SMALL-FLOWERED BLAZING-STAR, *M. albicaulis*: Unexpected: **tiny, yellow flowers** less than 1.3 cm (½") across and **shiny white stems** to 30 cm (12"). Range as above.

Figwort Family, *Scrophulariaceae*
GREAT MULLEIN [196]
Verbascum thapsus

0.6-1.5 m (2-5') HIGH

FLOWER SPIKE 2.5 cm (1") THICK

SEED AND SEPALS STICKY AND WOOLLY

SCATTERED FLOWERS

YELLOW FLW. 6 mm (1/4") ACROSS

Noticeable, rapidly spreading dryland plant from Europe. Most abundant of 3 mulleins in our area. Tall, **often to 1.5 m (5'); large, dull green, flannel-like leaves; unpleasant coarse, woolly texture** (fine hairs). Several short branches near top may become twisting, contorted. Large leaf rosette in first year, stem raised in second. Tiny, yellow flowers, tightly packed cluster; progressive ragged bloom July onwards. Common in fields, roadsides and waste places to 900 m (3000') elevation. Most abundant on alkaline soils (good indicator).

BRIGHT GREEN LEAVES

FLANNELLY LEAVES TO 30 cm (1') LONG

Ancient Romans rolled thick, woolly stalks in tallow for use as torches. Similarly used by early miners in west; once known as 'miner's candle.'

GREAT MULLEIN   MOTH MULLEIN

RANGE BC: Widespread, except for subalpine and alpine ecosystems.
RANGE WA: Widespread, except for subalpine and alpine ecosystems.
MOTH MULLEIN, *V. blattaria*: Like above but **leaves, coarsely toothed or lobed; bright green**. Scattered bright yellow flowers. Sporadic; southern B.C., eastern Washington.
WOOLLY MULLEIN, *V. phlomoides*: Like great mullein, but **hairy, greyish leaves** and short flower branches. **Crinkled petal edges**. Rare; Gulf Islands, adjacent mainland. Not recorded in Washington.

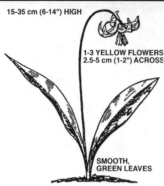

15-35 cm (6-14") HIGH

1-3 YELLOW FLOWERS
2.5-5 cm (1-2") ACROSS

SMOOTH,
GREEN LEAVES

Lily Family, *Liliaceae*
YELLOW GLACIER LILY [203]
Erythronium grandiflorum
Glacier Lily, Snow Lily

Note: Other *Erythronium*s are white or pink (see p. 185).

The **2 glossy green leaves** contrast with **1–3 bright flowers** of purest yellow each with **6 long, recurved petals**. Like western anemone, it **blooms soon after snow has left**. Sometimes large subalpine meadows become sheets of gold, a floral spectacle of great beauty. Hundreds of thin, dead stems from the previous year, each with a long capsule on top, are lost amid following flowers. But even after the flowers wither, a grizzly knows where to dig for bulbs in late summer. I watched such a grizzly in the Trophy Mountains of Wells Gray Park, B.C. Was it the sight of a dried stem or a faint smell that led to a steady digging that could have uprooted a hundred bulbs and left the meadow roughly cultivated behind?

RANGE BC: Generally considered a plant of subalpine heights but sometimes at valley elevations. Northern B.C. southwards. Coastal and Vancouver Island mountains. Chase.

RANGE WA: The opposite of B.C.: common at low elevations, even to the sagebrush ecosystem, and extending to mountain slopes but not usually reaching subalpine heights. Leavenworth, Mt. Spokane.

15-45 cm (6-18") HIGH

Iris Family, *Iridaceae*
GOLDEN-EYED GRASS
Sisyrinchium californicum

The name tells you that this unusual grass-like plant grows in California. Extends northwards along the coast in lakes, bogs and seepage areas, reaching its northerly limit on Vancouver Island.

There is no other plant to confuse with this early summer bloomer. Flowers, several on a stem, have **6 petals, each with half a dozen dark veins**. Flowers are about 2 cm (¾") across and in bloom from early summer to August. You stand about a 50-50 chance of seeing them: **flowers are open only to about midday**. Very narrow, dull green leaves.

RANGE BC: Southern Vancouver Island, Gulf Islands.

RANGE WA: San Juan Islands; southwards near coast through Washington and Oregon.

30-90 cm (1-3') HIGH

Iris Family, *Iridaceae*
YELLOW IRIS [204]
Iris pseudacorus
Yellow-Flag

Note: 2 blue-purple irises are on p. 319.

A flash of bright yellow among the dark green plants bordering ditches and streams should quickly pinpoint this introduced plant. It looks so much **like a garden iris** that it is unmistakable, the **only large 'wild' yellow iris in our area**. The **flowers are enhanced with light streakings of purple**. Occasionally white flowers occur. It blooms from May into June.

RANGE BC: Vancouver Island, lower Fraser Valley, Creston Wildlife Centre.

RANGE WA: Very common in wet bottomlands around Ellensberg. Small ponds in Yakima Canyon.

NOTE: Also see 'Daisy' section for blooms to 5 cm (2") across, or 'Sunflower' for larger ones.

Rose Family, *Rosaceae*
YELLOW MOUNTAIN-AVENS [202]

Dryas drummondii

Yellow Dryas

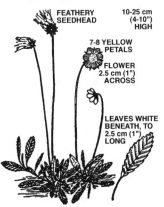

FEATHERY SEEDHEAD

10-25 cm (4-10") HIGH

7-8 YELLOW PETALS

FLOWER 2.5 cm (1") ACROSS

LEAVES WHITE BENEATH, TO 2.5 cm (1") LONG

Note: White-flowered species are on p. 188. **Distinctly crinkled evergreen leaves densely hairy beneath, appear silvery.** Round, **pale yellow flowers** have **7–8 petals**. Seedhead is a fluffy, umber tuft that drifts off, possibly to start another colony. Usually on high rockslides or hard-packed gravel bars of mountain streams.

RANGE BC: Alaska southwards; all high mountains except in southwest.

RANGE WA: Habitats as for B.C.; all high mountains except Olympics.

Sunflower Family, *Asteraceae (Compositae)*
GOLD STAR [212]

Crocidium multicaule

Gold Fields

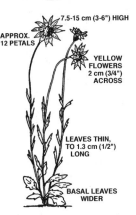

7.5-15 cm (3-6") HIGH

APPROX. 12 PETALS

YELLOW FLOWERS 2 cm (3/4") ACROSS

LEAVES THIN, TO 1.3 cm (1/2") LONG

BASAL LEAVES WIDER

March to April, gold star suffuses many a Washington field and rocky slope with a golden hue. Makes up for lack of stature by extensive patches of flowering stems. **Abundance itself should identify.** Wool on stems when young; often persists in leaf axils. Small, alternate stem leaves have a few carelessly spaced teeth; a few broader leaves grow at stem bases. Flower yellow throughout. **12 or so tapering petals**, slightly pleated; **faint perfume**.

RANGE BC: Spotty; eastern Vancouver Island south of Campbell R.; Gulf Islands. Mt. Finlayson.

RANGE WA: Whidbey and Orcas islands, vicinity Ellensburg, through central Washington, WA 14 (along Columbia River) and Blue Mountains.

Waterlily Family, *Nymphaeaceae*
YELLOW WATERLILY [207]

Nuphar polysepalum

Pond Lily

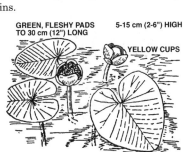

GREEN, FLESHY PADS TO 30 cm (12") LONG

5-15 cm (2-6") HIGH

YELLOW CUPS

Note: A large, yellow waterlily on Vancouver Island or west of Cascades will be this one. Rocky Mountain cow-lily is more eastward, but there is a central band of overlap and hybrids. A Mexican yellow waterlily (*Nymphaea mexicana*) has dark green leaves with brown blotches. Very rare in our area, recorded only on southeastern Vancouver Island.

Large, scaly roots anchored in mud; long-stemmed, large, floating leaves, often seem tossed about, some on edge and perhaps well in air in shade. Showy flower far more complicated than it looks: Yellow cup, to 10 cm (4") across, of waxy sepals plus 4 small, green ones. Large, crown-like, yellow stigma. **Small petals mostly hidden by thick, purple stamens.** Edible seeds were important for coastal peoples. Blooms May onwards, throughout summer.

RANGE BC: Throughout, but generally west of Cascades.

RANGE WA: Throughout, but generally west of Cascades.

ROCKY MOUNTAIN COW-LILY, *N. variegatum*: Like yellow waterlily, but **stamens yellow**. Mostly east of Cascades; quite certain if east of Okanagan River. Large leaves, often flat on water, crimped edges—looks a respectable waterlily. Ponds, marshes; lower elevations.

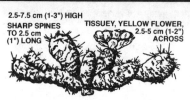

2.5-7.5 cm (1-3") HIGH
SHARP SPINES
TO 2.5 cm
(1") LONG
TISSUEY, YELLOW FLOWER,
2.5-5 cm (1-2")
ACROSS

Cactus Family, *Cactaceae*

BRITTLE PRICKLY PEAR CACTUS [206]

Opuntia fragilis

The spiny clumps of wrinkled, fleshy cactus 'pads' are characteristic of the arid sagebrush and ponderosa pine rangelands, more so in B.C. than in Washington. The beautiful yellow tissuey blossoms with numerous **red-stalked stamens** give this otherwise drab cactus a brief period of glory from June to July. The spiny mat, 30 cm (12") or more across, is often covered with these bright blooms.

The thick, spongy stem carries out the functions normally performed by leaves and is also designed for storing precious water. The **upper stem segments break apart easily** and are thus 'brittle.' Spines to 2.5 cm (1") long and smaller, yellowish bristles arise from bumps known as areoles. If the **areoles are white-woolly**, chances are that you have this plant. (But if they are rusty-woolly or not woolly at all, then check out many-spined prickly pear below.) Spines are slightly barbed, which explains the strong pull needed to get them loose from a shoe. They provide excellent protection from grazing animals.

I am amazed that there are so few cacti in Washington. Vast areas of bunchgrass and rocky sagebrush appear ideal and yet one has to search to find them. In extreme southern B.C., by contrast, comparable terrain with similar vegetation has this cactus in abundance. In the northern part of its range it is often reduced to a few 'nubbins.'

RANGE BC: Similkameen, Okanagan and Nicola valleys, extending northwards to Kamloops and Clinton. Along the Fraser southwards to Lytton. Found on some Gulf Islands; Saturna and Hornby. Spotty occurrences to Campbell River.

RANGE WA: Mostly east of the Cascades, but only random occurrences in the typical drylands of the sagebrush and ponderosa pine community. Yakima Canyon. Also on dry slopes on a few of the San Juan Islands and on hillsides on the eastern side of Puget Sound.

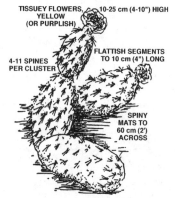

TISSUEY FLOWERS,
YELLOW
(OR PURPLISH)
10-25 cm (4-10") HIGH
4-11 SPINES
PER CLUSTER
FLATTISH SEGMENTS
TO 10 cm (4") LONG
SPINY
MATS TO
60 cm (2')
ACROSS

Cactus Family, *Cactaceae*

MANY-SPINED PRICKLY PEAR CACTUS [205]

Opuntia polyacantha

The largest of the cacti growing in our area and generally similar to brittle prickly pear (above), this ragged mass of long-spined cactus is to the unwary person about as dangerous as a rattlesnake. Needle-sharp spines, which may be 5 cm (2") in length, have a terrific holding quality when stuck in flesh or shoes. Clumps of spines are just less than 1.3 cm (½") apart, rising from small, rounded knobs known as areoles. **Areoles should be rusty-woolly or not woolly** at all, else see brittle prickly pear (above). A stem segment, or 'pad,' may be 10 cm (4") long.

The beautiful tissue-like flower is usually yellow, although **pink and orange blossoms do occur**. Blooming time is May to July—a phenomenon of colour that all too few people see.

RANGE BC: Generally the same as for prickly pear above, but east of Cascades. Isolated occurrences on north bank of Peace River, where it is very robust.

RANGE WA: Sagebrush ecosystem; along the Snake River from Clarkston westwards to the Columbia River. Very abundant along WA 209 following the Snake River southwards. Yakima River Valley.

Sunflower Family, *Asteraceae (Compositae)*
WOODLAND TARWEED [208]
Madia madioides

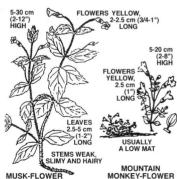

Tarweeds, a half-dozen or more in our area, are 15 cm (6") to 1.2 m (4') tall; leaves generally narrow, often sticky; may be alternate. Tiny yellow 'sunflowers'; **noticeably 3-fingered petals**, number varies. Key identifying feature: bristly circle of bracts between **ray (petal-like) flowers and disk (button centre) flowers**, seen clearly only on dissection.

Woodland tarweed has single erect stem; top branched. Yellow flowers to 2.5 cm (1") across; **8+ fingered ray flowers**. Fairly few, narrow leaves, may have several teeth. Dry, low-elevation open areas.

RANGE BC: West of Cascades. Southern Vancouver Island and Gulf Islands.

RANGE WA: West of Cascades. San Juan Islands and southwards to California.

LITTLE TARWEED, *M. exigua*: Small, erect. Alternate narrow leaves. **Stems glandular, hairy. Small, single, 2-lobed yellow flowers** at tips of branching stems. Sometimes masses. Extreme southern B.C. only. Common in Washington; low-elevation dry, open places.

SHOWY TARWEED, *M. elegans*: To 1.2 m (4') tall, large 'sunflower' bloom. **Petals deeply 3-fingered. Ray flowers with reddish base form circle around central 'button.'** Showy roadside weed in dry, open areas of southwestern Washington.

TUBULAR FLOWERS

Figwort Family, *Scrophulariacea*
YELLOW MONKEY-FLOWER [209]
Mimulus guttatus

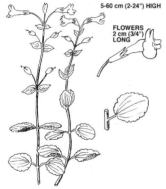

Sea level to alpine, **usually at water's edge**. Clear yellow 'snapdragon' flower about 2 cm (¾") long; **small brown or crimson dots in a bearded throat**. Petals often drop right after picking. Short-stalked, toothed lower leaves almost round. Upper leaves clasp stem. Blooms from May to August.

RANGE BC: Widespread throughout; wet or seepage habitats. Commonly on wet roadside rocks.

RANGE WA: Habitats as for B.C.

CHICKWEED M.F. or **BABY M.F.** [210], *M. alsinoides*: To 20 cm (8") tall; oval leaves. Curious little yellow face, **lower lip has 1–2 purplish dots. Moist, mossy places**, shady cliffs, lower forests. Blooms mostly March–May. Coastal, Campbell River, B.C., southwards: Gulf Islands, Washington, Oregon. Southern Cascades.

Figwort Family, *Scrophulariacea*
MUSK-FLOWER
Mimulus moschatus

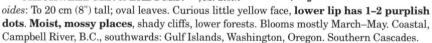

Least attractive monkey-flower. **Weak, drooping stems sticky, hairy, often slimy**. Musky odour, strength varies. 2-lipped yellow flowers: 2-lobed upper, 3-lobed lower. Very narrow, **hairy throats; dark lines or spots**. Shallow-toothed leaves. Wet places; low–mid-elevations. Blooms May–late July.

RANGE BC: Gulf Islands ecosystem and southwards. Also, central B.C. southwards.

RANGE WA: Coastal. Also, central Washington from B.C. southwards through Oregon.

MOUNTAIN or **ALPINE M.F.**, *M. tilingii*: Lovely. **Thrives along cold streams**; often low, bright-gold mats of large flowers. **Subalpine meadows**; wet places; high mountains.

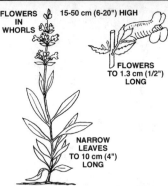

FLOWERS IN WHORLS

15-50 cm (6-20") HIGH

FLOWERS TO 1.3 cm (1/2") LONG

NARROW LEAVES TO 10 cm (4") LONG

Figwort Family, *Scrophulariaceae*
YELLOW PENSTEMON
Penstemon confertus

Note: Most penstemons are purple (see p. 302) while others are blue (p. 322). Some are shrubs (p. 133).

Whorls of tubular flowers and opposite leaves identify as a penstemon—this is the **only cream–pale yellow penstemon** (may appear whitish, but scorched penstemon, *P. deustus*, the only true white one, favours dry areas in central Washington). May have 1 stem or several. Flowers tightly clustered. Leaves, especially lower ones, may have stems and possibly teeth. Usually in moist ponderosa pine regions or on surrounding slopes. Blooms June–early July.

RANGE BC: Across southern B.C. east of Cascades.
RANGE WA: In habitats as for B.C.

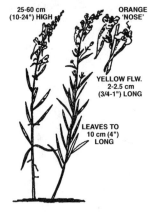

25-60 cm (10-24") HIGH

ORANGE 'NOSE'

YELLOW FLW. 2-2.5 cm (3/4-1") LONG

LEAVES TO 10 cm (4") LONG

Figwort Family, *Scrophulariaceae*
BUTTER-AND-EGGS
Linaria vulgaris
Common Toadflax

Unmistakable. Intricate **yellow flower with orange 'nose'** closely matches yellow of butter and deeper egg yolk. Blooms in rich profusion at top of erect stalk. Flowers have graceful **yellow spur and 2 lips**: upper is 2-lobed, pouch-shaped lower is 3-lobed. 'Egg' colour comes from **orange swelling that almost closes throat**. Leaves narrow, to 10 cm (4") long; considered alternate, may appear opposite. Naturalized European weed.

RANGE BC: Widespread, in exposed places.
RANGE WA: Widespread, in exposed places.

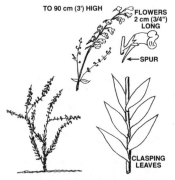

TO 90 cm (3') HIGH

FLOWERS 2 cm (3/4") LONG

SPUR

CLASPING LEAVES

Figwort Family, *Scrophulariaceae*
DALMATIAN TOADFLAX [211]
Linaria genistifolia

May be most conspicuous plant east of Cascades in early summer. Ungainly, in some ways beautiful. Mediterranean escape, spreads rapaciously—potential enemy of more-fragile natives. In many areas whole hillsides are now yellow July–August. **Bright yellow flowers, touch of orange in throat**. Very noticeable on gravel banks and dry waste areas. **Row of flowers with tell-tale spurs** is attractive, but akimbo branching and drying flowers detract.

Stem has many **symmetrically arranged egg-shaped, clasping leaves**; about 4 cm (1 ½") long at base; higher ones shorter. **Leaves, stems olive green**. May be confused with butter-and-eggs (above), but leaf design much different. 'Toad' part of name may reflect squeezed blossom opening wide like toad's mouth.

RANGE BC: Generally east of Cascades; dry plains and hillsides. Most common south-central B.C. but also northwards to Dawson Creek. Hillsides in Princeton area. Oliver.

RANGE WA: Generally east of Cascades; on dry plains and hillsides. Southwards from B.C. through central Washington to Oregon.

BLUE TOADFLAX, *L. canadensis*: Native. Erect, single stem to 50 cm (20") tall. **Terminal cluster of sticky, light blue flowers; curved spurs**. Number of stems with tiny leaves form distinctive basal rosette about 12.5 cm (5") across, flat on ground. In moist, sandy areas. Southwestern B.C., southwards through Washington.

Figwort Family, *Scrophulariaceae*
YELLOW RATTLE [213]

Rhinanthus minor
Rhinanthus crista-galli
Rattlebox

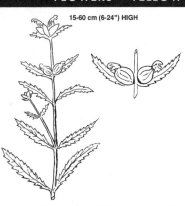

15-60 cm (6-24") HIGH

A double joy in the yellow rattle: it is the only *Rhinanthus* in our area and it is easily identified with the help of its name, for the **ripe seeds rattle in the greatly swollen pod.** The flowers have lips that form an exact entry tube for a particular species of foraging bee, thus ensuring efficient pollination. Leaves are opposite and toothed. An early summer bloomer. Prefers fields and meadows with some moisture.

RANGE BC: From Alaska southwards near the coast. Also in central B.C. and southwards in suitable habitats; especially common along road edges.

RANGE WA: Coastal Washington and Oregon. Also southwards through central Washington in suitable habitats.

Balsam Family, *Balsaminaceae*
COMMON TOUCH-ME-NOT [214]

Impatiens noli-tangere
Jewelweed

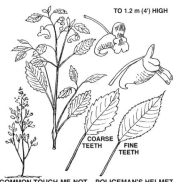

TO 1.2 m (4') HIGH

COARSE TEETH   FINE TEETH

COMMON TOUCH-ME-NOT   POLICEMAN'S HELMET

With a half-dozen 'touch-me-nots' in our area, a person with a perverse streak would dearly want to see why not. Lacking any additional information, you might guess that the pretty flowers, each hanging by a hair-thin stem, would drop off at a touch or at least be too delicate to pick. The *Impatiens* part of the name might be a useful clue for, as you might guess, it means 'impatient,' referring to the way a **ripe seed capsule could explode at a touch.**

Another name, jewelweed, brings a memory of seeing these truly beautiful yellow-to-orange jewels, a remarkable design of nature, along a shady roadside border. The plant might be up to 1.2 m (4') tall, with **stout stems that are swollen at every major branching point.** Leaves are alternate, long-stemmed and coarsely toothed. Several flowers and buds can be on display, hung on thin stems and partly obscured by leaves. The flower is a **complicated design of 3 sepals, of which 1 forms a curled spur**—for the record, there are **4 petals.** Flowers are about 2.5 cm (1") long. The **swollen tubular section is yellow,** the **top and bottom petals are orange.** There is some colour variation in certain areas.

RANGE BC: Never very abundant, but widely dispersed; moist lowlands.

RANGE WA: Habitats as for B.C. Westport.

POLICEMAN'S HELMET, *I. glandulifera*: An introduction from Asia. Sometimes found growing very rankly in shady ditches. Its large, showy, 'trumpet' **flowers are a mix of white, pink and purple and have a sharply twisted spur** or tail. Note that the leaves are finely toothed. Southwestern B.C., mostly in the Fraser Valley, and southwards along the coast of Washington. Chilliwack, B.C. Occasionally inland; Castlegar, B.C.

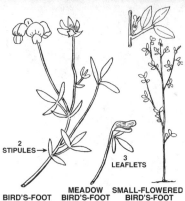

2 STIPULES →

3 LEAFLETS

MEADOW   SMALL-FLOWERED
BIRD'S-FOOT   BIRD'S-FOOT   BIRD'S-FOOT

Pea Family, *Fabaceae (Leguminosae)*
BIRD'S-FOOT TREFOIL [218]
Lotus corniculatus

Related to peavines, vetches, clovers and lupines. All members of this family are distinctive with attractive pea flowers. Most *Lotuses* are blue, pink or purple, but these 3 yellow ones are nicely distinguished.

Bird's-foot trefoil has **thin, wiry stems** that tend to bend or twist or even creep along ground, as do most *Lotuses* and vetches. Positive identification as bird's-foot by **2 large, leaf-like stipules at base of each stem**. This *Lotus* has **3 clover-like leaflets**; an exception, most have more. From Eurasia.

RANGE BC: Rapidly spreading. Favours moist soil. Wide range. Roadsides. Fraser Valley, Vancouver Island.

RANGE WA: Favours moist soil. Spreading rapidly in western Washington.

MEADOW BIRD'S-FOOT TREFOIL, *L. denticulatus*: Wiry, rather erect plant with short branch spurs carrying 3–4 tiny leaflets. Simple, stemless **flowers cream coloured, with some purple**; this plant is grouped here with like species. Pods to 1.5 cm (½") long. Common in B.C. on light, drier soils, from valleys to middle-mountain elevations. On both sides of Cascades in Washington.

SMALL-FLOWERED BIRD'S-FOOT TREFOIL, *L. micranthus*: Another common name is 'miniature bird's-foot trefoil.' Look in late spring and summer: **less than 30 cm (12") tall; tiny, single, light yellow flowers**, often with pinkish tint. **1–3 leaf-like bracts just below each flower**. Sandy soils and rocky bluffs preferred. Ranges from moist to sandy soils at lower elevations. Common on southwestern Vancouver Island and in western Washington.

MOUNTAIN FALSE LUPINE, *Thermopsis montanum*: Called 'mountain golden bean' in *Vascular Plants of B.C.* Erect single stem to 60 cm (2') tall, with **showy head of large 'lupine' flowers** each about 2.5 cm (1") long. Alternate leaf stems; **3 large, oval leaflets. Pair of wing-like bracts at leaf–stem junction**. Erect, somewhat hairy seed pods, 2–5 beans. Valley meadows; to mid-elevations in ponderosa pine ecosystem. South-central B.C., Rock Creek. Sporadic southwards through Washington. At 900 m (3000') in Blue Mountains.

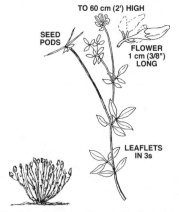

TO 60 cm (2') HIGH

SEED PODS

FLOWER 1 cm (3/8") LONG

LEAFLETS IN 3s

Pea Family, *Fabaceae (Leguminosae)*
YELLOW CLOVER
Trifolium aureum
Hop-Clover

Note: Similar looking bird's-foot trefoil is above. Other *Trifoliums* are on p. 282.

Eye-catching—deserves special mention. In early summer, you may wonder at a low, green shrubby clump topped by bright yellow flowers. Surprisingly, each 'flower' is in fact a **cone of 6–7 tiny pea-like flowers**, each about 1 cm (³⁄₈") long. This arrangement probably rings a bell, but a quick leaflet count may find 5 instead of the 3 typical of clovers. But there are indeed 3—the other 2 are actually leaflet-like stipules growing from where leaf attaches. In contrast, most bird's-foot trefoils do not have these stipules.

Probably some flowerheads will have gone to seed, leaving several thin rods. Sometimes a considerable spread of plants grows from falling seeds. An alien likely to become a pest.

RANGE BC: Roadsides and open places in southern B.C. Noticeable on Hope–Princeton Highway around 1200 m (4000') elevation.

RANGE WA: Not seen in Washington.

YELLOW SWEET-CLOVER *Melilotus officinalis*: Often with white sweet-clover (p.192), form and perfume similar. Wideranging, often along dry roadsides east of Cascades.

Pea Family, *Fabaceae (Leguminosae)*
SMALL HOP-CLOVER
Trifolium dubium
Shamrock

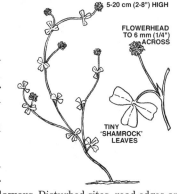

5-20 cm (2-8") HIGH

FLOWERHEAD
TO 6 mm (1/4")
ACROSS

TINY
'SHAMROCK'
LEAVES

Weedy little roadside plant. Blooms from early spring and abundant, so can not escape attention. **Small 'shamrock' leaves, tiny, yellow flower 'balls'** each with up to 25 flowers; minute, but technically 'pea' flowers.

RANGE BC: Adaptable alien well distributed near coast, along roadsides and in dry fields.

RANGE WA: Habitats as for B.C.

LOW HOP-CLOVER, *T. campestre*: Slightly larger copy of small hop-clover, but **flower cluster about 1 cm (³/₈") across, contains over 30 tiny flowers**. Disturbed sites, road edges and fields. Southwestern B.C. and lower Fraser Valley; south through Washington to California.

Bladderwort Family, *Lentibulariaceae*
GREATER BLADDERWORT [217]
Utricularia vulgaris

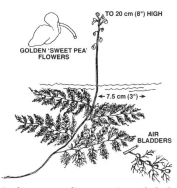

TO 20 cm (8") HIGH

GOLDEN 'SWEET PEA'
FLOWERS

7.5 cm (3")

AIR
BLADDERS

'Wort' is a corruption of an old English word for a plant with special properties. There are 3 bladderworts of note in our area. All are **aquatic and free-floating**; part of main stem may lie on pond bottom. Alternate 'leaves' about 7.5 cm (3") long consist of tiny branches with hair-like fringes. Tiny air sacs or bladders are scattered amongst fringes. A remarkable design: they float the plant in a desirable habitat, and each one is a clever trap. When a small aquatic creature touches a hair on the door, it opens. A partial vacuum then sucks the victim inside. Mosquito larvae can disappear in such fashion—a plant that merits special blessing.

Seemingly delicate growth with a stem to 20 cm (8") above water. Adorned with **half-dozen or more golden 'sweet pea' flowers** but with broad petals for lip and hood. May be in bloom anytime during summer. Lake edges, ponds and backwaters; at lower elevations.

RANGE BC: Central B.C. southwards.

RANGE WA: Central Washington, from B.C. southwards

Bleeding Heart Family, *Fumariaceae*
GOLDEN CORYDALIS [215]
Corydalis aurea

30 cm (12") HIGH

FLOWERS
TO 4 cm (1 1/2")
LONG

Close relative to bleeding hearts, so imagine similar fern-like foliage, pushed down, **sprawling, much-branched**, with **several plumes of golden blooms** late spring–summer. Sprawling habit does not allow flowers to rise more than 30 cm (12") from ground. A dozen or more flowers per stem, rather untidy looking; **1 upper petal arches up to form hood**. Seed capsules have shiny black seeds.

RANGE BC: Generally east of Cascades, in low forests throughout.

RANGE WA: Habitats as for B.C.

OTHER CORYDALISES, *Corydalis* spp.: 3 other corydalises to consider, all with intricate branching leaf structure, consistent flower shape. Blue corydalis, *C. pauciflora*, is a high mountain plant with scattered occurrences in northern B.C. Scouler's corydalis, *C. scouleri*, pure pink flowers, stands erect, to 1.2 m (4') tall; shady, moist areas; southern Vancouver Island, western Washington. Pink corydalis, *C. sempervirens* (pink flowers with touch of yellow) is in dry meadows and open forests; common in B.C., but not in Washington.

TO 60 cm (2') HIGH

Pea Family, *Fabaceae (Leguminosae)*
YELLOW HEDYSARUM
Hedysarum sulphurescens
Yellow Loments

The thin stems, vetch-like leaves and pea-like flowers suggest plants such as vetches and locoweeds. However, the particular combination of **leaves divided into pairs of narrow leaflets** and the **cascade of drooping yellow 'sweet pea' flowers** should identify this plant. After flowering, the **seeds form in a chain of flattened pods**, with a constriction between each of the 2–4 seeds. This design separates the hedysarums from the vetches, locoweeds etc., which have pods much like garden peas.

RANGE BC: Dry valleys and open forests at middle elevations east of the Cascades. Most common in southeastern B.C.

RANGE WA: Dry valleys and open forests at middle elevations on the eastern slope of the Cascades, throughout Okanogan County.

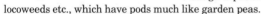

TO 75 (30") HIGH

Pea Family, *Fabaceae (Leguminosae)*
SULPHUR LUPINE
Lupinus sulphureus

Note: For information on lupines and the common blue-flowered species, see pp. 324–27. There is a white lupine on p. 192, and a shrub on p. 149.

STEMS AND LEAVES FINELY HAIRY

This is the **main yellow-flowered lupine**—one without identification problems. Sulphur lupine is a sturdy plant to 75 cm (30") tall. Its **stems have noticeable silky hairs**. **Flowers are pale yellow**, less than 1.3 cm (½") long and irregularly arranged on the flower stems. **Leaves are silvery-grey, finely hairy**, and with very narrow leaflets. Favours drylands from valley bottoms to middle-mountain heights. Blooms in spring or early summer.

RANGE BC: South-central B.C.
RANGE WA: East of the Cascades, Palouse, extending to California.

CLUSTERED FLOWERS

Four O'clock Family, *Nyctaginaceae*
YELLOW SAND-VERBENA [219]
Abronia latifolia

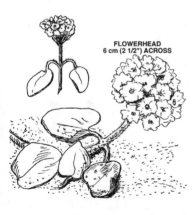

FLOWERHEAD 6 cm (2 1/2") ACROSS

One can not easily confuse this plant with anything else. Look for it on **coastal sand dunes**. Just look! Do not harm it, for it is quickly growing scarce. The **small, bright, fragrant flowers pack into a yellow ball** 6 cm (2 ½") across. Stems rise from a common base.

The plant may spread its **rubbery stems** over a metre or so (several feet) with its thick roots going deep into the sand. The thick, fleshy, opposite leaves and **stems are so sticky that they hold grains of sand**. The plant, often partly buried and with heavy leaves, is well protected from strong winds.

RANGE BC: Coastal dunes.
RANGE WA: Coastal dunes southwards to California.

Figwort Family, *Scrophulariaceae*
YELLOW OWL-CLOVER
Orthocarpus luteus
Yellow Orthocarpus

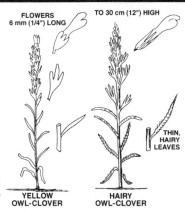

FLOWERS
6 mm (1/4") LONG

TO 30 cm (12") HIGH

THIN,
HAIRY
LEAVES

YELLOW
OWL-CLOVER

HAIRY
OWL-CLOVER

Note: Pink-purple owl-clovers are on p. 283.

A fanciful name—no resemblance to owls or clover! There are many owl-clovers; a wide range of colours. General overall appearance, especially pink and purple ones, reminiscent of paintbrushes, *Castilleja* spp. (below, and pp. 291–92). Generally shorter plants, many only to 20 cm (8") tall. Recognize as owl-clover by **tiny, tubular flowers** with **4-segment calyx that holds flower tube. Flower itself divided into upper lip that narrows to beak and lower one that is inflated and lobed. Both lips about same length** whereas in paintbrushes upper lip is much longer.

Yellow owl-clover is slim-stemmed, to 30 cm (12") tall. Leaves alternate and wispy; become bract-like higher on stem. **Tiny flowers, pale yellow to white**, only 6 mm (¼") long, peek out over large, **greenish bracts with 4 divisions**. Blooms July–August. In open places; favours low meadows and adjoining mountain slopes.

RANGE BC: Northern B.C. southwards, east of Cascades.

RANGE WA: East of Cascades through Washington.

HAIRY OWL-CLOVER, *O. hispidus*: Same general form as yellow owl-clover, **flowers pale yellow or white**. However, leaves very thin, long, covered with hair. **Upper leaves divide into several fingers**. Ranges in moist meadows and seepage areas, at lower elevations, from southern Vancouver Island across central B.C. and through Washington on both sides of Cascades.

BEARDED OWL-CLOVER, *O. barbatus*: A yellowish green owl-clover with obscure yellow flowers and large greenish yellow bracts. To 25 cm (10") high. Leaves cleft into 3–5 parts and covered with hairs. Limited range in central Washington, in sagebrush areas. South Okanogan through Grand Coulee and into Clark County. Not in B.C.

Figwort Family, *Scrophulariaceae*
GOLDEN PAINTBRUSH
Castilleja levisecta

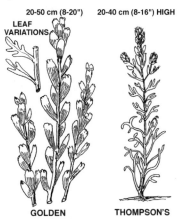

20-50 cm (8-20")   20-40 cm (8-16") HIGH
LEAF
VARIATIONS

GOLDEN

THOMPSON'S

One usually thinks of paintbrushes as being bright red and growing in high mountain meadows. This one does not conform: it is yellow and grows on dry sites—meadows and benches—at low elevations. What you see are bright yellow tips of sheathing bracts that obscure the real flower. Note **distinct alternate leaves, 3–6 lobed or with toothed tips**, a characteristic that distinguishes it from Thompson's (below; almost same range).

RANGE BC: Lower elevations west of Cascades.

RANGE WA: As for B.C.

THOMPSON'S PAINTBRUSH [220], *C. thompsonii*: Often associated with sagebrush. Conforms to harsh conditions by having tough, woody base, which supports 1 or several stiff stems, often branching. Yellow flowers do show, despite **rough, hairy bracts** that protect them. Alternately spaced leaves, a rather unattractive tangle, narrow and somewhat hairy on lower part of stem and divide into fingers higher up. While range is spotty through sagebrush areas of southern B.C., more abundant in same general habitat in Washington.

SULPHUR PAINTBRUSH, *C. sulphurea*: Clumps, to 50 cm (20") tall, with colourful **pale yellow bracts** enclosing flowers. In general, distinct paintbrush form. Profusion of **narrow, alternate leaves—not forked**, unlike golden paintbrush. Common in southeastern B.C. in high mountain meadows. Rocky Mountain Trench. Not recorded in Washington.

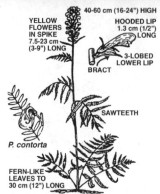

YELLOW FLOWERS IN SPIKE 7.5-23 cm (3-9") LONG

40-60 cm (16-24") HIGH

HOODED LIP 1.3 cm (1/2") LONG

3-LOBED LOWER LIP

BRACT

SAWTEETH

P. contorta

FERN-LIKE LEAVES TO 30 cm (12") LONG

Figwort Family, *Scrophulariaceae*
BRACTED LOUSEWORT [221]
Pedicularis bracteosa
Wood Betony

Unusual spike of **flowers: hook-beaked oddities of pale yellow, may be pink-tinged**. Quaint little faces, one peeping overtop the other is typical of louseworts; twisted beak, **prominent bract** at base of flower. 'Fern-look' leaves, mostly on flowering stems, finely divided into **rows of saw-edged leaflets**. If flowers faded, note unbranched stems; leafy throughout. High, moist mountains.

RANGE BC: Throughout.

RANGE WA: Throughout.

COIL-BEAKED or **CONTORTED LOUSEWORT**, *P. contorta*: Soft yellow flowers 1.3 cm (½") long; **remarkably twisted beak**. 20–40 cm (8–16") tall. **Leaves cut to midrib between leaflets**. Dry places; high mountains, southeastern B.C. and across Washington, from meadows and open forest to subalpine. Mt. Rainier.

LABRADOR LOUSEWORT, *P. laboradorica*: **Yellow flowers, may be reddish; loose cluster. Long, straight upper petal hooded, 2 thin teeth near tip**. Stem leaves deeply cut, coarsely toothed. North of Quesnel, B.C., in shady forest or meadow. Up to subalpine.

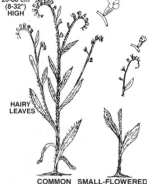

20-80 cm (8-32") HIGH

HAIRY LEAVES

COMMON SMALL-FLOWERED

Forget-me-not Family, *Boraginaceae*
COMMON FIDDLENECK
Amsinckia intermedia
Note: An orange-flowered fiddleneck is on p. 259.

Excellent common name fits curving flowerhead. Several similar fiddlenecks all have small, yellow-to-orange flowers. Common fiddleneck usually erect; may branch. Tiny, **5-lobed yellow flowers ride outside curve of 'fiddleneck'**; may have orange touch. Stony nutlet seeds. Leaves alternate, slender; to 15 cm (6") long. Leaves, stems have **bristly hair**. Grow close together—weedy, yellow masses.

RANGE BC: Mostly east of Cascades, but also on southern Vancouver Island.

RANGE WA: Southwards through Pacific states to Mexico. Mostly east of Cascades.

SMALL-FLOWERED FIDDLENECK, *A. menziesii*: Small copy of common fiddleneck. **Tiny, pale yellow flowers**. Southern B.C., east of Cascades; both sides in Washington.

15-60 cm (6-24") HIGH

YELLOW CAPE

GREENISH FLOWERS

LEAVES TO 1.4 m (4 1/2') LONG

Arum Family, *Araceae*
SKUNK CABBAGE [225]
Lysichiton americanum
American Skunk Cabbage

Most noticeable March–April, before nearby shrubs come into leaf; **perhaps first plant to herald spring**. Almost every **swampy or mucky place** dotted with soft yellow sheaths, each one actually a large bract; **sickish sweet smell**. Thick, fleshy spadix (club) inside sheath with large number of small greenish flowers; becomes knobby. **Leaves probably largest of any native plant**: some to 1.4 m (4 ½') long, 60 cm (2') wide. Aboriginals used large, thick leaves to line steam-cooking pits. Black bear may eat all or part of plant.

RANGE BC: Generally in wet ground under or near cedar trees. Most common west of Cascades but also plentiful in wet areas inland.

RANGE WA: Habitats as for B.C.

Carrot Family, *Apiaceae (Umbelliferae)*
PACIFIC SANICLE [216]
Sanicula crassicaulis

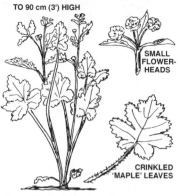

TO 90 cm (3') HIGH

SMALL FLOWER-HEADS

CRINKLED 'MAPLE' LEAVES

Thick, coarse stems, abundant leaves and rather uninteresting flowerhead. However, this sanicle keeps turning up in shady places and is difficult to ignore. One of those peculiar plants that **may have flowers of alternate colour: maroon purple.** To 90 cm (3') tall; **long-stemmed, hairy 'maple' leaves.** Bloom is an uneven cluster of **tightly packed flowerheads.** Flowers have 5 petals almost too small to see. Blooms in late spring, with camas, Menzies's larkspur, monkey-flowers and sea blush. While some plants bloom as early as May, others prolong the blooming period. Open forests and shady margins of thickets, at low elevations.

RANGE BC: Frequent on Vancouver and Gulf islands and adjacent mainland. Victoria, Hornby Island.

RANGE WA: San Juan Islands, western Washington and Columbia Gorge.

SIERRA SANICLE, *S. graveolens*: Tighter flowerhead and leaves with 3 irregularly lobed leaflets. Wide range at lower elevations throughout B.C. and Washington.

Sunflower Family, *Asteraceae (Compositae)*
CANADA GOLDENROD [222]
Solidago canadensis
Meadow Goldenrod

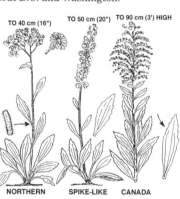

TO 40 cm (16") TO 50 cm (20") TO 90 cm (3') HIGH

NORTHERN SPIKE-LIKE CANADA

Rejoice, rejoice! These 3 goldenrods—along many roadsides—can be distinguished by distinctive forms. A closer look will find more clearly distinguishing features.

The 100 or so species of goldenrod in North America include some with exceptionally beautiful flowers. Enjoy them, for the widespread belief that pollen from goldenrods causes hay fever has been proven wrong.

Canada goldenrod is quickly noticed, **often 90 cm (3') tall** and **bearing masses of golden blooms,** during July and August. Wide, **diamond-shaped mass of small flowers** was once a source of permanent yellow dye, extracted through boiling. Stems covered by partly toothed leaves. Each leaf clearly has 3 veins, a positive identification feature.

RANGE BC: Across our area in better soils from lowlands to lower mountain slopes.

RANGE WA: Habitats as for B.C.

SPIKE-LIKE or **DWARF GOLDENROD,** *S. spathulata*: Stem 25–50 cm (10–20") tall, often with bend near base. Marked by **narrow spike of rather large, yellow flowers.** Note shape of partly toothed, hairless basal leaves. Handle a plant to confirm that it is **strongly resinous and aromatic.** Common throughout, from low to high elevations, except for most arid areas.

ALPINE GOLDENROD, *S. spathulata* var. *nana*: Attractive. **Far shorter than other goldenrods,** at less than 12.5 cm (5"), with **short, compact flowerheads.** Lower leaves not hairy, which helps differentiate it from northern goldenrod (below). Subalpine and alpine.

NORTHERN GOLDENROD, *S. multiradiata*: To 40 cm (16") tall, but in some terrain less than half that. **Tiny, yellow flowers** at top of **reddish stems,** often in close cluster but sometimes in more open spray. Leaves alternate and clasping. **Lower leaves finely fringed along long stems,** in contrast to smooth stems of alpine goldenrod. Lower leaves may be noticeably toothed while upper (smaller) ones may be entire. Blooms July to September. Fairly common throughout at higher elevations but a tundra plant in northern B.C. Subalpine in Olympics.

60-90 cm (2-3') HIGH
FLAT, YELLOW FLOWERS
6 mm (1/4")
ACROSS

Sunflower Family, *Asteraceae (Compositae)*
COMMON TANSY [224]
Tanacetum vulgare

Introduced. Very prominent when large, **flat-topped cluster of yellow flowers** bursts into full bloom. Each tiny 'flower' is only 6 mm (¼") across, but bunched grouping shows very effectively above **thick mass of dark, carroty leaves**. Flowers late August to mid-September. Prominent roadside flower; turns up unexpectedly along most highways. Height and large, flat flowerheads make identification easy. Emits a strong odour when crushed.

RANGE BC: Common in southern B.C.

RANGE WA: Common throughout Washington.

WESTERN DUNE TANSY, *T. bipinnatum*: Habitat—**sand dunes along coast** (B.C. to California) should be sufficient to identify. Also, **minutely dissected leaflets, each with 2–5 pointed lobes**. In contrast, common tansy has at least twice that number. Flowers of dune tansy form broad, flat heads.

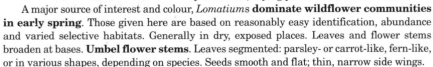

UMBEL FLOWERS

7.5-30 cm (3-12") HIGH

CARROT-LIKE
LEAVES

SMALL,
YELLOW
FLOWERS

Carrot Family, *Apiaceae (Umbelliferae)*
SPRING GOLD [223]
Lomatium utriculatum
Fine-leaved Desert-parsley

There are 30–40 species of desert-parsley in the Pacific Northwest, confused further by different specimens of same species having flowers in any 1 of 2 or 3 different colours, typically **white, yellow or purple** (see p. 203). While highly dissected leaves are not very useful for distinguishing species, skilled botanists may depend on size and shape of seeds. But by seed time, drying plants attract little interest.

A major source of interest and colour, *Lomatium*s **dominate wildflower communities in early spring**. Those given here are based on reasonably easy identification, abundance and varied selective habitats. Generally in dry, exposed places. Leaves and flower stems broaden at bases. **Umbel flower stems**. Leaves segmented: parsley- or carrot-like, fern-like, or in various shapes, depending on species. Seeds smooth and flat; thin, narrow side wings.

'Spring gold' is very applicable—**among earliest to bloom**. Clusters of bright yellow flowers against a bouquet of dark green, fern-like leaves make it very special. **Often a low tuft of carroty leaves**, sometimes almost prostrate. Finds secure footing in mossy and rocky situations it prefers. Plants may grow to 30 cm (12") as seedhead develops. Companions might be seablush, Menzies's larkspur, camas, muskflower and buttercup.

'Fine-leaved' part of name can be appreciated only by a close look at a leaf and noting it is a **filigree of green**. How can a leaflet have segments as fine as thread?

RANGE BC: Southwestern B.C., west of Cascades.

RANGE WA: West of Cascades southwards to California.

COUS, *L. cous*: Sometimes called biscuitroot. On dry, open slopes, often associated with sagebrush. An important aboriginal food that turns up in many historical writings. Although a plant of southeastern Washington, its roots were traded widely.

To quote Nancy Turner from her book *Food Plants of B.C. Indians*, '[Cous roots] were commonly used by the Southern Okanogan, Flathead, Kalispel, Nez Perce, Kootenay and other peoples of the northwestern United States. The roots were dug up with bitter-root in April and May, during or just after flowering, and eaten raw, boiled or steam-cooked in pits. They could be dried on mats in the sun—and could be stored up to three years.'

Under 30 cm (12") tall. Flowers yellow to purple. Observe a lower leaf to see how each branches divides again, creating several narrow **fingers with rounded tips**.

Carrot Family, *Apiaceae (Umbelliferae)*
FERN-LEAVED DESERT-PARSLEY [227]
Lomatium dissectum

A very vigorous *Lomatium*. Rounded outline of fern-like leaves, mostly slightly under 60 cm (2') tall; very decorative on rocky ridges and slopes. Yellow or purple blooms (p. 307), April–May, on flowering stems that poke up through leaves. By early summer, mass of **foliage is obvious olive green**. Thick, aromatic roots were once roasted by aboriginal peoples.

RANGE BC: Common in south-central and southeastern B.C., in dry, rocky places.

RANGE WA: Frequent roadside plant east of Cascades; rocky areas and exposed soils. Vantage, Yakima Canyon, Columbia Gorge.

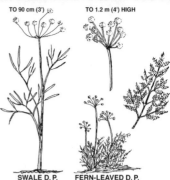

TO 90 cm (3')   TO 1.2 m (4') HIGH

SWALE D. P.   FERN-LEAVED D. P.

Carrot Family, *Apiaceae (Umbelliferae)*
SWALE DESERT-PARSLEY
Lomatium ambiguum

Very tall, to 90 cm (3') at maturity, from thick, woody taproot—possibly used as food by aboriginals, like most desert-parsleys. Smooth, erect flowering stems; umbels of **yellow flowers, may be purple tinged** in **loose spray of 8–18 per flowerhead**. Leaves branch from sheath on stem, divide into ungainly sprays of **thin, paddle-like leaves**. Might be confused with narrow-leaved desert-parsley (below)—note definite difference in leaf pattern.

RANGE BC: Widespread east of Cascades, on dry mountainsides of southern B.C.

RANGE WA: Widespread east of Cascades, on dry mountainsides.

Carrot Family, *Apiaceae (Umbelliferae)*
NARROW-LEAVED DESERT-PARSLEY [228]
Lomatium triternatum
Nineleaf Biscuitroot

Happy thought! Leaf shape and division provide definite identification: often with **long, thin 9-segment leaves**. To 60 cm (2') tall, but most less than half that. Can **cover large areas**; yellow flowerheads provide dominant landscape colour early–late spring. Very wide distribution. In sagebrush zone, blooms with sagebrush buttercup, shootingstars, yellow bell and woodland stars, Howell's triteleia and barestem desert-parsley.

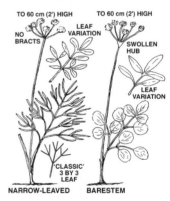

TO 60 cm (2') HIGH   TO 60 cm (2') HIGH

NO BRACTS   LEAF VARIATION   SWOLLEN HUB

LEAF VARIATION

'CLASSIC' 3 BY 3 LEAF

NARROW-LEAVED   BARESTEM

RANGE BC: Scarce occurrences on southern Vancouver Island. Southern B.C. in sagebrush and ponderosa pine ecosystems.

RANGE WA: Sagebrush, ponderosa pine zones throughout. Prolific Columbia River sidehills, eastern Columbia Gorge. Abundant blooms April–May, Toppenish–Bickelton and south.

BARESTEM DESERT-PARSLEY or **INDIAN CONSUMPTION PLANT** [226], *L. nudicaule*: **Considerable variation in leaf shape, but broad**, unlike relatives. Leaf shape and **swollen hub (with age) just below umbel branching**, make recognition easy. Clustered flowers, variable stem length, mid-April onwards. Spicy, aromatic seeds used by aboriginals: fumigant, house deodorant, mosquito repellent. Locally common; southeastern Vancouver Island, Gulf and San Juan islands. Most abundant east of Cascades; sagebrush and ponderosa pine habitats, also higher, open mountain slopes. Klickitat County. Columbia Gorge.

GRAY'S DESERT-PARSLEY, *L. grayi*: Similar to other fern-leaved, yellow-flowered *Lomatiums*. Usually under 40 cm (16") tall. Many past years' growth accumulates as **heavy, dry matting at plant base. Leaves, many times dissected**, have unusual thickness; **distinctive odour when crushed**. Perhaps a dozen flower stems. Blooms in April. Open, rocky areas; low–middle elevations east of Cascades. Steptoe Butte, Komiak Butte and north.

10-25 cm (4-10") HIGH

'SULPHUR' FLOWERS

LEAVES TO 2.5 cm (1") LONG

SULPHUR B.

Buckwheat Family, *Polygonaceae*
SULPHUR BUCKWHEAT [230]
Eriogonum umbellatum
Sulphur Eriogonum

Most abundant at high elevations, where winds beat at the compact mass of leaves and stout flower stems. **Unusual sulphur-yellow flowers**. Sometimes a **rosy tinge** suffuses yellow. Do not confuse with heart-leaved buckwheat (below). Colour differences of several varieties could create more confusion.

Leaves like tiny, broad paddles about 2.5 cm (1") long, and **whorl of smaller leaves at umbel juncture** good for identification—use in case of colour variations.

RANGE BC: Cascades; arid valleys to alpine areas

RANGE WA: Cascades; arid valleys to alpine, drier mountains near Columbia Gorge.

SUBALPINE BUCKWHEAT, *E. umbellatum* var. *subalpinum*: Closely related to sulphur buckwheat. Distinguished mainly by whitish-to-bright-yellow flowers. Same range.

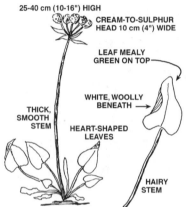

25-40 cm (10-16") HIGH

CREAM-TO-SULPHUR HEAD 10 cm (4") WIDE

LEAF MEALY GREEN ON TOP

WHITE, WOOLLY BENEATH

THICK, SMOOTH STEM

HEART-SHAPED LEAVES

HAIRY STEM

Buckwheat Family, *Polygonaceae*
HEART-LEAVED BUCKWHEAT [231]
Eriogonum compositum

Heart-shaped or triangular leaf, mealy green above, white-woolly beneath, is key. Stems to twice blade length. Stout flower stems may be 25–40 cm (10–16") tall; branch into half-dozen or more umbel stems 2.5–5 cm (1–2") long that hold **tight cluster of tiny flowers; protruding anthers. Flowers cream to pale yellow to sulphur; definite perfume.** Whorl of half-dozen tiny leaves at main umbel branching point. Clear-cut style—all features in plain view.

Blooms May to July, on rocky ground and canyon walls. Covers long stretches of landscape and may be dominant yellow or sulphur component.

RANGE BC: Not in B.C.

RANGE WA: Thin, rocky soils of sagebrush and bunchgrass ecosystems. Yakima Canyon, Toppenish, Goldendale, Columbia Gorge.

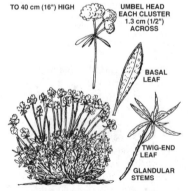

TO 40 cm (16") HIGH

UMBEL HEAD EACH CLUSTER 1.3 cm (1/2") ACROSS

BASAL LEAF

TWIG-END LEAF

GLANDULAR STEMS

Buckwheat Family, *Polygonaceae*
ROUND-HEADED BUCKWHEAT
Eriogonum sphaerocephalum

Showy lemon-yellow flower clusters replace Carey's sunflower, line-leaved fleabane and dwarf sunflower. Tough and twiggy; **as much shrub as flower**. From a distance, a broad dome (hence 'round-headed') about 40 cm (16") tall. Bottom half of dome is dull green mass; large number of stems rise from it about 15 cm (6"), each has cluster of several compact heads. Truly eye-catching! Blooms mid-May in Kittitas County, mid-June south of Coulee City.

Great variation in small basal leaves. **Stems, leaves finely hairy; leaves whitish beneath.**

Possible narrow leaf part-way up stem. **Whorl of leaves just below flowerhead.**

Over 150 years ago, David Douglas was first botanist to collect it, along Columbia River.

RANGE BC: Not in B.C.

RANGE WA: Rocky soils with sagebrush and borders of ponderosa pine. South Chelan County. Columbia River drylands south of Vantage.

Buckwheat Family, *Polygonaceae*
CREAMY BUCKWHEAT
Eriogonum heracleoides
Parsnip-flowered Buckwheat

Often with yellow flowers, this widespread buckwheat is colour-coded as white (see p. 202) since it could appear to be this colour over parts of its range.

CUSHION BUCKWHEAT or **CUSHION ERIO-GONUM** [229], *E. ovalifolium*: Very low; compact mass of **small, oval 'paddle' leaves to 1.3 cm (½") long. Leaves and stem white with woolly hairs**. Soft, yellow flowers, sometimes touched with red. Locally frequent; alpine; south-central B.C. and southwards. Cascades, Olympics and Rocky Mountains.

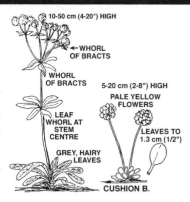

10-50 cm (4-20") HIGH
← WHORL OF BRACTS
WHORL OF BRACTS
LEAF WHORL AT STEM CENTRE
GREY, HAIRY LEAVES
5-20 cm (2-8") HIGH
PALE YELLOW FLOWERS
LEAVES TO 1.3 cm (1/2")
CUSHION B.

'DANDELION' FLOWERS

Sunflower Family, *Asteraceae (Compositae)*
HAIRY CAT'S-EAR
Hypochaeris radicata

European. **May be most common 'dandelion' flower of fields, lawns and roadsides** in southwestern B.C. and coastal Washington; lawn pest on poorer soils. Like dandelion only in form and colour of flower and seeds. **Ray-flower flowerhead, open regardless of weather**; thin, wiry, green stem, often branched, unlike thick, tubular, milky ones of dandelions. To 40 cm (16") tall in shady areas.

Scalloped, hairy leaves spread fan-like near ground in fanciful rosette. Extended blooming season: early summer to late September; often beyond.

RANGE BC: Common in southwestern B.C.
RANGE WA: Common west of the Cascades.

SMOOTH CAT'S-EAR, *H. glabra*: Smaller than hairy cat's ear: to 30 cm (12") tall. **Flowers open only in full sunlight**; less than 2 cm (¾") across. Leaves not hairy. Vancouver Island, southwards along coast to California.

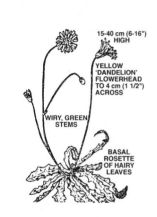

15-40 cm (6-16") HIGH
YELLOW 'DANDELION' FLOWERHEAD TO 4 cm (1 1/2") ACROSS
WIRY, GREEN STEMS
BASAL ROSETTE OF HAIRY LEAVES

Sunflower Family, *Asteraceae (Compositae)*
SHORT-BEAKED AGOSERIS [232]
Agoseris glauca
Pale Agoseris, Smooth Agoseris

The 6 agoserises in our area have yellow flowers, except orange agoseris (p. 261). Fortunately, leaf shapes vary. Do not confuse with butterweeds (flower clusters on short, branching stems).

Short-beaked has showy **pale yellow flower, purplish tinge beneath**; 5 cm (2") diameter. **Smooth, unlobed leaf**. 1 stem. Blooms April–May.

RANGE BC: Dry valleys to high elevations.

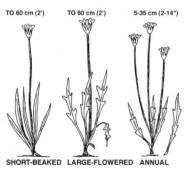

TO 60 cm (2') TO 60 cm (2') 5-35 cm (2-14")
SHORT-BEAKED LARGE-FLOWERED ANNUAL

RANGE WA: Habitats as for B.C. Olympics, Dry Falls, Grant County.

LARGE-FLOWERED AGOSERIS [237], *A. grandiflora*: 1 stout stem; dwarfed in subalpine. **1 showy yellow flower per stem; may age pinkish**; about 4 cm (1 ½") across. Stems and bracts woolly on young plant. Leaves 4–25 cm (1 ½–10") long.

Sporadic occurrences in drylands of extreme southern B.C. More abundant southwards through Washington. Subalpine in Olympics and Cascades.

ANNUAL AGOSERIS, *A. heterophylla*: Annual, unlike perennials above. **Several stems. Partly closed yellow flowers**. Open areas; southern B.C. southwards to California.

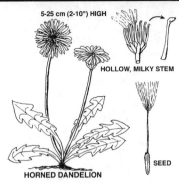

5-25 cm (2-10") HIGH

HOLLOW, MILKY STEM

SEED

HORNED DANDELION

Sunflower Family, *Asteraceae (Compositae)*

HORNED DANDELION

Taraxacum ceratophorum

Why put a dandelion in a wildflower book? Is it not perhaps the best-known 'weed' in the U.S.A. and Canada? But 1 of the 5 species is a native, most commonly found in high mountains.

That thick taproot that resists efforts to pull it up has been used over the centuries for medicinal purposes and of course as a coffee substitute. Young leaves have been a food item for an equally long period and many profess that the flowers make a fine wine. A valuable food for a number of birds and animals. 'Dandelion' derives from French *dent-de-lion* (lion's tooth), but what is its significance?

Horned dandelion gets its unusual name from a minute feature: long **bracts have 'horn' or twist at top**. Otherwise, if you find a single plant on a mountain top with **leaves flat on ground** and **short-stemmed flower**, you can be quite certain.

RANGE BC: Scattered occurrences; high mountain meadows. Sometimes to valley elevations in northern B.C.

RANGE WA: Habitats as for B.C.

COMMON DANDELION [238], *T. officinale*: Most abundant dandelion. Spring blooms of bright yellow make as spectacular a display as California poppy in south and arrow-leaved balsamroot in dry interior. Distinguish from red-seeded dandelion (below) by seeds being olive to brown at maturity. Widespread.

RED-SEEDED DANDELION, *T. laevigatum*: **Name says it all**. Otherwise identical to common dandelion. Most common in southern B.C. but also eastern and western Washington.

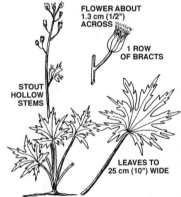

FLOWER ABOUT 1.3 cm (1/2") ACROSS

1 ROW OF BRACTS

STOUT HOLLOW STEMS

LEAVES TO 25 cm (10") WIDE

Sunflower Family, *Asteraceae (Compositae)*

SILVERCROWN [236]

Cacaliopsis nardosmia

Luina nardosmia

Cutleaf Luina

Note: There is a white-blooming *Luina* on p. 207.

'Silvercrown' must relate to seedheads, but 'cutleaf' is also descriptive. Conspicuous large, dark green leaves, silvery beneath, break from forest floor mid-April to mid-May—you may wonder what will follow. A tall plant, to 1.2 m (4'), with **cluster of large, dissected 'maple' leaves** at base; cluster of **bright yellow, bristly flowers** at top. Very often a mass of plants grows together, an interlocking pattern of leaves with stems to 20 cm (8") long. Leaves appear sensitive to weather: on a hot day, edges turn over on top, rather than drooping.

Flowerhead, about 1.3 cm (½") across, is a **uniform mass of disk flowers—no ray flowers (petals)**—like many closely related other members of *Senecio* tribe. Another *Senecio* feature: **1 row of thin, sharp bracts, a stout cup,** support flowerhead. Silvercrown flowers from 600 m (2000') elevation to edge of subalpine, where both height and leaf size decrease. Flowers give way to **fluffy, white seedhead**. Stalk collapses and big leaves quickly yellow and wither—a rather unsightly display often appearing by late June at lower elevations.

RANGE BC: Very rare. Manning Park. Make your own discovery.

RANGE WA: Eastern slope of Cascades. Yakima County, Hart's Pass, vicinity of Winthrop. South side Columbia River. The Dalles, Oregon.

SLENDER LUINA, *L. stricta*: Quite different from silvercrown; **long-stemmed basal leaves** widen into **narrow, tongue-shaped leaves** to 30 cm (12") long. Higher stem leaves progressively smaller. Erect stem, to 90 cm (3') tall, topped by **long spike packed with upward-pointing yellow flowers**. Moist subalpine areas; common on Mt. Rainier on open slopes below Paradise Visitor Center, southwards in similar habitats, but sporadic.

Sunflower Family, *Asteraceae (Compositae)*
APARGIDIUM
Microseris borealis
Bog Microseris

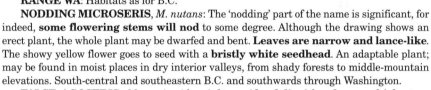

BOTH 10-40 cm (4-16") HIGH

LANCE-LIKE LEAVES

APARGIDIUM   NODDING MICROSERIS

Apargidium has a name neither memorable nor with any associations of value. Perhaps it is best to associate it with damp places, **sphagnum bogs** and **wet meadows**. In these places its basal tuft of **dark green, grass-like leaves** and bright yellow flowers will quickly attract attention. Notice that there is **only 1 flower on each of the leafless stems**, which are from 10 to 40 cm (4 to 16") tall and often have a noticeable bend near the base.

RANGE BC: Fairly common along coast, in suitable habitat of sphagnum bogs and wet meadows.

RANGE WA: Habitats as for B.C.

NODDING MICROSERIS, *M. nutans*: The 'nodding' part of the name is significant, for indeed, **some flowering stems will nod** to some degree. Although the drawing shows an erect plant, the whole plant may be dwarfed and bent. **Leaves are narrow and lance-like**. The showy yellow flower goes to seed with a **bristly white seedhead**. An adaptable plant; may be found in moist places in dry interior valleys, from shady forests to middle-mountain elevations. South-central and southeastern B.C. and southwards through Washington.

FALSE AGOSERIS, *M. troximoides*: A **large 'dandelion' head on a thick stem** 15–50 cm (6–20") tall, rising from a **mat of narrow, serrated and crinkled leaves** to 25 cm (10") long. Blooms in April–May on dry hillsides, along with other plants of the bunchgrass ecosystem. Extreme south-central B.C., but more abundant in Washington. Brewster, Chelan. Eastern section of Columbia Gorge.

Sunflower Family, *Asteraceae (Compositae)*
YELLOW SALSIFY [233]
Tragopogon dubius
Oyster Plant

30-60 cm (1-2') HIGH
SEEDHEAD 7.5 cm (3") ACROSS
YOUNG SEEDHEAD
YELLOW FLOWER

Note: *T. porrifolius* has purple flowers (see p. 300).

This weedy plant, widely cultivated throughout Europe and North America, is generally called salsify. The thick root is edible in early spring or late fall and the young stem and leaves may be eaten as a cooked vegetable.

A stout **flower stem that thickens near its top** carries a bright yellow **dandelion-like flower about 5 cm (2") across**. The seedhead that follows is similar in shape and form to that of the dandelion but may be 7.5 cm (3") in diameter. **Several long, alternate grass-like leaves clasp the stem** near the base. Flowers will appear around May 1 in southern Washington, several weeks later in B.C.

RANGE BC: Vancouver Island and eastwards across southern B.C., in dry waste places at lower elevations. A common roadside plant in central B.C.

RANGE WA: Widespread across the state at lower elevations.

MEADOW SALSIFY [234], *T. pratensis*: Similar to yellow salsify (above) but **does not have a thickened stem** below the more-orange flower. It is much branched and grows to 1.2 m (4') tall. The flowers of our salsifies close early in the afternoon, leading to the common name Jack-go-to-bed-at-noon! Although regarded as quite rare in B.C., it grows abundantly both north and south of Dog Creek in the Cariboo. Likely represented across Washington.

Sunflower Family, *Asteraceae (Compositae)*
HAWKWEEDS
Hieracium spp.

Hawkweeds can easily be confused with hawksbeards, *Crepis* spp. (p. 245), hairy cat's ear, *Hypocharis* spp. (p. 241) and sow thistles, *Sonchus* spp. (p. 246). All have **yellow 'dandelion' flowers**, have **stems that exude a milky juice when broken** and produce **fluffy seedheads**. Illustrations and photos should solve most problems. A hawkweed is a perennial with a short, shallow root and thin seeds whose **top bristles (pappus) are brownish**. A hawksbeard, conversely, is usually an annual with a stout tap root and pappus (fluff or hairs attached to the top of the seed) typically white.

Note: There is a white hawkweed (p. 207) and also a dramatic orange one (p. 261).

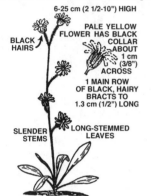

6-25 cm (2 1/2-10") HIGH

PALE YELLOW FLOWER HAS BLACK COLLAR
BLACK HAIRS
ABOUT 1 cm (3/8") ACROSS
1 MAIN ROW OF BLACK, HAIRY BRACTS TO 1.3 cm (1/2") LONG
SLENDER STEMS
LONG-STEMMED LEAVES

Sunflower Family, *Asteraceae (Compositae)*
SLENDER HAWKWEED
Hieracium gracile

A quick identification among the hawkweeds can be made from the size. Easily overlooked were it not for its abundance; a **small, slender hawkweed, only 6–25 cm (2 ½–10′) tall**. The small flowers, about 6 mm (¼") across, may be solitary at stem ends or in loose clusters. Basal rosette formed by **light green, spoon-shaped leaves**; wide stems.

RANGE BC: Locally abundant in our area on thin, exposed soils of the subalpine ecosystem.

RANGE WA: Habitats as for B.C.

NARROW-LEAVED HAWKWEED, *H. umbellatum*: In sharp contrast to slender hawkweed (above), this plant may be **to 1.2 m (4′) tall**. Usually a single stem that loses its basal leaves. Middle leaves to 10 cm (4") long, alternately spaced, with a few irregular teeth. Fairly common across southern B.C. in dry, open areas at lower elevations. Extending into Washington through to the Puget Sound area.

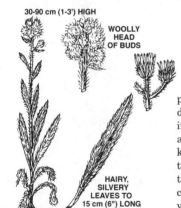

30-90 cm (1-3') HIGH

WOOLLY HEAD OF BUDS

HAIRY, SILVERY LEAVES TO 15 cm (6") LONG

Sunflower Family, *Asteraceae (Compositae)*
WESTERN HAWKWEED [240]
Hieracium scouleri var. *albertinum*
Hieracium albertinum
Hairy Hawkweed

How do you decide whether to include a certain plant in a book? An example: While I was driving down Steptoe Butte, that strange volcano-like peak in the Palouse country, the sun tipped over a ridge and back lit a steep hillside into a fantasy of sparkling silver. Night rains had saturated the vegetation, especially one plant so **thick with white hair** that it looked like a kitten's fur. Granted that while conditions like this would be a rarity, this plant provided an unforgettable floral spectacle. And besides, it is the hairiest plant I have ever seen and **almost pure silver except for the yellow flowers**. Therefore it has to be included. The hairiness may be enough for identification, but hawkweeds are variable and other specimens may not have the dense hair I encountered on Steptoe Butte on the 4th of July.

Before flowers appear, head is a soft, lumpy, white ball so hairy that you can barely see that it contains individual developing flowerheads. Flowers, only about 1 cm (⅜") across, are in a loose cluster like most other hawkweeds. Each **petal (ray flower) has 5 minute teeth**. Stocky; 30–90 cm (1–3') tall. Leaves fairly dense at bottom but also continuing up stem.

RANGE BC: Common east of the Cascades, in open, dry foothills and adjacent mountain slopes. Southern B.C. Idaho Peak.

RANGE WA: Habitats as for B.C. South to the Palouse, where it is most abundant.

Sunflower Family, *Asteraceae (Compositae)*
HOUND'S TONGUE HAWKWEED
Hieracium cynoglossoides

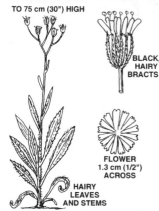

TO 75 cm (30") HIGH

BLACK, HAIRY BRACTS

FLOWER
1.3 cm (1/2")
ACROSS

HAIRY LEAVES
AND STEMS

On joining a nature walk in a western park, I several times heard the guide mutter under her breath, 'Another G.D.C.!' in response to a query about some plant. Her frustrated tone of voice prompted me to ask her what she meant. 'Another God-Damned Compositae,' was the brief answer, one that keeps repeating endlessly as the amateur struggles with confusing plants (see the general description of the hawkweeds, facing page).

This rather common plant reaches about 75 cm (30") in height and looks much like Scouler's hawkweed (below). The **smallish flowers in a loose cluster**, a helpful identification feature, are 1.3 cm (½") across and supported by **long, thin, bracts covered with black hairs**. The name 'hound's tongue' could come from the shape of the lower leaves, some being to 20 cm (8") long and sharply pointed, but is a hound's tongue sharply pointed? Stem and leaves have varying degrees of fine hairiness.

RANGE BC: Southern B.C. east of the Cascades. Also sporadic in the Cascades.

RANGE WA: Valleys and foothills of eastern Washington. Also sporadic in the Cascades.

SCOULER'S HAWKWEED, *H. scouleri* var. *scouleri*: Such a close copy of hound's tongue (above) that about all one can say is that the leaves are wider and they may not be as hairy. Common to medium elevations in drylands of southern B.C. east of the Cascades. In Washington the range widens to include the Cascades and Puget Sound area.

Sunflower Family, *Asteraceae (Compositae)*
SMOOTH HAWKSBEARD [235]
Crepis capillaris

25-60 cm (10-24")   20-60 cm (8-24") HIGH

10-30 cm (4-12")

LEAF VARIATIONS

BRACTS 2-RANKED

Note: See 'Hawkweeds' (facing page) for information on similar appearing plants.

What is it that turns off the interest in identifying some types of plants? The hawksbeards must be such a collection, for they do not rate mention in most wildflower books. Probably it is because they have a weedy 'dandelion' look, their flowers are small and they are most abundant on wastelands. (Or perhaps the authors gave up on trying to adequately describe each one.)

There are 11 in our area but only 5 or 6 have a wide range; 3 typical ones are described here.

SMOOTH HAWKSBEARD   SLENDER HAWKSBEARD   ELEGANT HAWKSBEARD

Hawksbeards are told apart from some other species of *Asteraceae* by the **leafy and branched flower stems, thin milky juice, bracts in 1 or 2 rows, yellow flowers** and the fact that the head contains **ray flowers only** (no centre of disk flowers).

Smooth hawksbeard has **toothed leaves** and **bright yellow flowers**.

RANGE BC: Southeastern Vancouver Island, Gulf Islands and adjacent mainland.

RANGE WA: Mostly west of the Cascades, in fields and waste places.

SLENDER HAWKSBEARD, *C. atrabarba*: Several common leaf variations are shown. The most distinctive feature is the **long, stringy leaf lobes**. Numerous bright yellow flowers. Common east of the Cascades in southern B.C. and continuing through Washington.

ELEGANT HAWKSBEARD, *C. elegans*: Note leaf shape and also the **2-ranked bracts**. Favours moist areas, stream banks and pond margins, but also found on bluffs and low mountain slopes. Widespread in B.C. Extends into eastern Washington.

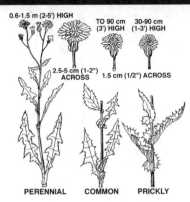

PERENNIAL   COMMON   PRICKLY

Sunflower Family, *Asteraceae (Compositae)*
PRICKLY SOW-THISTLE [244]
Sonchus asper

Easily separate sow-thistles from true thistles, *Cirsiums*, by **milky sap from broken stem or torn leaf**; also, no *Cirsium* has yellow flowers. (See also 'Hawkweeds,' p. 244) Sow-thistles have **thin, yellow ray flowers—no disk flowers. Seeds topped by silky white hairs**—a fluffy ball about 1.3 cm (½") across. All 3 here are European weeds. Name reflects a mistaken belief that pigs ate them.

Distinguish prickly sow-thistle from common by prickly leaf edges—perennial has larger flower.

RANGE BC: Roadsides and waste places.

RANGE WA: As for B.C. Throughout much of North America.

COMMON or ANNUAL SOW-THISTLE, *S. oleraceus*: Much-branched, untidy-looking. Lower, weakly prickly leaves clasp smooth stem. Small, **yellow flowers, 1.3 cm (½") across**, white seed ball. Common on southern Vancouver Island. Widespread in Washington.

PERENNIAL SOW-THISTLE, *S. arvensis*: Note **height and large flowerheads**. Fine, prickly clasping leaves to 25 cm (10") long. **Most common yellow roadside flower in southeastern B.C.**; also coastal uplands, sloughs. Cranbrook, Rockies. Cariboo parklands ecosystem. Fields and waste places in Washington.

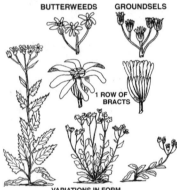

BUTTERWEEDS   GROUNDSELS

1 ROW OF BRACTS

VARIATIONS IN FORM

Sunflower Family, *Asteraceae (Compositae)*
BUTTERWEEDS, GROUNDSELS AND RAGWORTS
Senecio spp.

The naming of *Senecio*s is an area for confusion. Some books call them all 'butterweed,' some 'groundsel,' while others use both and 'ragwort' too, without apparent reason. A proposed solution: Most *Senecio*s flowers are yellow to orange—butter-like. Add their generally weedy look to give 'butterweed.'

But the **flowers are in 2 distinct designs**. Here, **butterweeds are those with rather untidy heads of ray flowers**, usually well spaced, easily counted. **Groundsels, though, have tightly clustered flowerheads**—like a dandelion made columnar (see illustration). Therefore, ray-flowered *Senecio*s, butterweeds, are in the 'Daisy' section; the others, groundsels, are in the 'Dandelion' section. 'Ragwort' is not used, except for species for which it is in common use.

*Senecio*s all have the **flowerhead in a receptacle—row of sharp-pointed bracts**; so do hawkweeds and arnicas. But arnicas have opposite leaves and usually a larger flowerhead. **Generally, *Senecio*s have alternate stem leaves** that change in size and shape up stem.

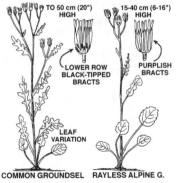

LOWER ROW BLACK-TIPPED BRACTS

PURPLISH BRACTS

LEAF VARIATION

COMMON GROUNDSEL   RAYLESS ALPINE G.

COMMON GROUNDSEL, *S. vulgaris*: Low—to 50 cm (20")—and **no ray flowers**. Often variable, **odd-shaped leaves** to 10 cm (4") long. **Leaves continue up stem to flowerheads. Shorter row of black-tipped bracts around flower base**—the best identification (but other *Senecio*s also have black bracts). Coastally, may bloom year-round. Very common along roadsides and gardens in southwestern B.C. and western Washington.

RAYLESS ALPINE GROUNDSEL, *S. pauciflorus*: High country. **Rayless flowerheads, bracts purple tinged**. Note 2 kinds of leaves. Wet places in subalpine and alpine; higher mountains.

Sunflower Family, *Asteraceae (Compositae)*
CANADIAN BUTTERWEED [241]
Senecio pauperculus

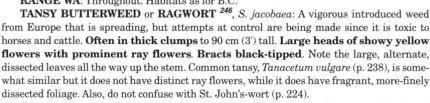

10-50 cm (4-20") 40-90 cm (16-36") 15-45 cm (6-18")

CANADIAN BUTTERWEED TANSY BUTTERWEED WOOLLY BUTTERWEED

Note: For an introduction to butterweeds, groundsels and ragworts, see facing page.

Canadian butterweed, 10–50 cm (4–20") tall, superficially looks like other butterweeds, but it ranges from **valley bottoms to middle-mountain elevations**. Most others are found in the subalpine and alpine ecosystems. Notice the shape of the long-stemmed basal leaves. Most are lightly toothed. The few leaves on the stem differ greatly in shape.

RANGE BC: Common. Favours moist ground, swamps, shady forests.

RANGE WA: Throughout. Habitats as for B.C.

TANSY BUTTERWEED or **RAGWORT** [246], *S. jacobaea*: A vigorous introduced weed from Europe that is spreading, but attempts at control are being made since it is toxic to horses and cattle. **Often in thick clumps** to 90 cm (3') tall. **Large heads of showy yellow flowers with prominent ray flowers. Bracts black-tipped**. Note the large, alternate, dissected leaves all the way up the stem. Common tansy, *Tanacetum vulgare* (p. 238), is somewhat similar but it does not have distinct ray flowers, while it does have fragrant, more-finely dissected foliage. Also, do not confuse with St. John's-wort (p. 224).

Common in waste places on southern Vancouver Island and adjacent mainland and southwards along the coast through Washington. The most common roadside flower on the Olympic Peninsula during August.

WOOLLY BUTTERWEED [239], *S. canus*: The name relates to the **woolly covering of fine, white hairs on stems and leaves**. That one feature should be enough to identify this plant of 15–45 cm (6–18") in height. Flat cluster of **flowers with 5–8 ray flowers** each. Blooms June and July. Ranges in the drier mountain zones of southern B.C. east of the Cascades and across Washington.

STREAMBANK BUTTERWEED or **WESTERN GOLDEN RAGWORT**, *S. pseudaureus*: Another confusing butterweed with little to tell it apart from others except for the shape of its leaves and habitat. Note the difference between **long-stemmed lower leaves and dissected stem leaves**. This plant, to 40 cm (16") tall, has a rather tight cluster of a half-dozen or more bright yellow flowers. **Ray flowers are short** and the **centre of the flower tends to be brownish**. It favours moist areas from **lowland stream banks up to lower mountain slopes**. From northern B.C. southwards, generally east of the Cascades but in Washington diverging to include both slopes of the Cascades.

TO 40 cm (16") HIGH
ORANGE-YELLOW FLOWERS 1.3-2 cm (1/2-3/4") ACROSS
MAIN BRACTS IN 1 ROW THICKENED ON BACK
SOFT, BRISTLY SEED
LEAVES 5-15 cm (2-6") LONG
30-90 cm (1-3') HIGH

STREAMBANK   ARROW-LEAVED

ARROW-LEAVED BUTTERWEED or **GIANT RAGWORT** [243], *S. triangularis*: A colourful and common flower in the **subalpine floral display, it is a butterweed to know**. Grows 60–90 cm (2–3') tall; leafy to the top and crowned with **flat-topped heads of small, yellow flowers with ray flowers. The triangular, coarsely toothed leaves** are a key feature, but do check American sawwort (below). A plant of damp ground from medium to high mountain elevations. Blooms in late summer. From north-central B.C. through the mountain systems and continuing through Washington.

AMERICAN SAWWORT, *Saussurea americana*: Noted here because the leaves and habitat are so similar to those of arrow-leaved butterweed (above). However, once it blooms, the **flowers are purple**. Favours **subalpine mountain meadows**. Northern coastal B.C., southern Vancouver Island, Olympics and Washington Cascades.

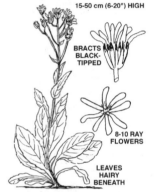

WESTERN B.   **TALL B.**

30-60 cm (1-2') HIGH

YELLOW FLOWERS 1.3 cm (1/2") ACROSS

YELLOW FLOWERS 1.3 cm (1/2") ACROSS

8+ PETALS

BRACTS BLACK-TIPPED

BRACTS IN 1 SERIES

SAWTOOTH LEAVES TO 10 cm (4") LONG

BOTTOM LEAVES WIDE

LONG, HAIRY STEMS

Sunflower Family, *Asteraceae (Compositae)*
WESTERN BUTTERWEED
Senecio integerrimus

Note: Introduction to *Senecio*s on p. 246.

Distinguish from groundsels by **5–12 rather disorderly ray flowers.** Single-stemmed, usually about 40 cm (16"). Long-stemmed leaves near base mostly round-tipped; may be 12.5 cm (5") long; smaller and narrower up stem. Jumble of butter-coloured flowers; loosely or tightly clustered. **Bracts in 1 main row; triangular black tips.** April–June. Widespread with sage and ponderosa pine. To subalpine.

RANGE BC: Common in southern B.C. east of Cascades. Idaho Peak, at 1500 m (5000').

RANGE WA: From bunchgrass to subalpine ecosystems. Blue Mountains.

TALL BUTTERWEED, *S. serra*: Common name gives best clue: **can reach 1.8 m (6') in height. Leaves sharply saw-toothed,** hence Latin name. **Numerous small, yellow flowers, ray flowers only**; dense, domed cluster July–August. Dry, open areas; lowlands–mountain slopes; east of Cascades; very rare in B.C. but quite common along Washington Cascades.

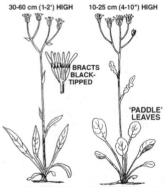

15-50 cm (6-20") HIGH

BRACTS BLACK-TIPPED

8-10 RAY FLOWERS

LEAVES HAIRY BENEATH

Sunflower Family, *Asteraceae (Compositae)*
ELMER'S BUTTERWEED [242]
Senecio elmeri

Not attractive but takes on special significance in late spring, on steep, rocky slopes of subalpine mountainsides after earliest blooms have gone, with fan-leaved cinquefoil for company.

Stout stem; **a butterweed because of ray flowers.** *Senecio*'s characteristic **single row of thin bracts** to support flower—not unique. No thin, milky juice when you break a stem, as with hawkweeds or hawksbeards. Still confused? Note sidehill habitat—not a wet meadow—and height: **15–50 cm (6–20").** Also, **among first plants to bloom. Tight cluster of up to 10 flowers,** 1.3–2 cm (½–¾") across; **8–10 ray flowers. Thickish, coarse basal leaves**; some toothing, wide stems. **Leaves hairy beneath,** to 15 cm (6") long. Plant stem has bend at base.

RANGE BC: Subalpine slopes in southern Cascades.

RANGE WA: Rocky places; high in Cascades, Wenatchee Mountains, Hart's Pass.

BLACK-TIPPED B.   **ALPINE MEADOW B.**

30-60 cm (1-2') HIGH

10-25 cm (4-10") HIGH

BRACTS BLACK-TIPPED

'PADDLE' LEAVES

Sunflower Family, *Asteraceae (Compositae)*
BLACK-TIPPED BUTTERWEED
Senecio lugens

A butterweed because of **ray flowers.** 'Black tipped' refers to **distinctive markings on row of flower bracts** that are characteristic of *Senecio*s. Note **hairy, irregularly toothed basal leaves,** 5–20 cm (2–8") long; narrow, tapering gradually out of a long stem. Several other summer-blooming butterweeds and groundsels are also subalpine.

RANGE BC: Mostly high elevations, east of Cascades.

RANGE WA: Seems rare, but in Olympics.

ALPINE MEADOW BUTTERWEED, *S. streptanthifolius*: High elevations. Can be under 25 cm (10") tall in subalpine; taller, leafier in mountain forests. **Broad, toothed 'paddle' leaves at base.** Variable number of flowerheads. Moist places; middle-mountain forests–subalpine. Widespread but not in coastal strip.

Sunflower Family, *Asteraceae (Compositae)*
LYALL'S GOLDENWEED
Haplopappus lyalli

5-15 cm (2-6") HIGH

FLOWER 2.5 cm (1") ACROSS

STEMS AND LEAVES STICKY AND HAIRY

LEAVES 3-VEINED

Note: Goldenweed, though it might be confused with golden fleabane (see p. 250), has much larger leaves on its stem; note too that arnicas (p. 251) usually have opposite leaves.

One could object to implying that this cheery little plant is a weed. Despite its small size, it is one of the floral joys of the high country. The hairy-glandular **leaves continue well up on its stem**.

RANGE BC: A variety of soils, from meadows to rocky places, in the subalpine and alpine ecosystems of southern B.C.

RANGE WA: Habitats as for B.C. Olympics.

Sunflower Family, *Asteraceae (Compositae)*
NARROW-LEAVED HAPLOPAPPUS
Haplopappus stenophyllus

7.5-15 cm (3-6") HIGH

YELLOW FLOWER 2 cm (3/4") ACROSS 6-8 RAGGED PETALS

1-2 STEM LEAVES

LEAVES ABOUT 6 mm (1/4") LONG OLIVE GREEN

WOODY STEMS AND ROOTS

A pretty **flower less than 2.5 cm (1") across** that looks like a miniature edition of a sunflower with its **6–8 large, yellow petals and a spongy yellow centre with protruding yellow tubes**. As it is only from 7.5 to 15 cm (3 to 6") tall, its mat-like growth of **narrow, pointed leaves, dark olive-green, finely hairy** and about 6 mm (¼") long, appears quite in keeping. Several leaves often adorn the upper flower stem. The base of the plant is woody and branches from a long, stringy root. This haplopappus blooms in May on dry, rocky ground and has as common companions thyme-leaved eriogonum, daggerpod and sagebrush violet.

RANGE BC: Not in B.C.

RANGE WA: Sagebrush and bunchgrass ecosystems bordering Cascades. Wenatchee. Ellensburg, Yakima.

Sunflower Family, *Asteraceae (Compositae)*
BROWN-EYED SUSAN [245]
Gaillardia aristata

35-50 cm (14-20") HIGH

3-TIPPED PETALS

FLOWERS YELLOW, WITH RED-BROWN CENTRE

OLIVE-GREEN LEAVES

When most of the early flowers of the drylands have disappeared and the sidehills and fields are taking on the dehydrated look of summer, brown-eyed Susan appears to charm the eye. To most people it is a **bright yellow 'sunflower' with a reddish brown centre**. The drab olive-green, alternate leaves are clasping. The flower reposes without detraction at the top of a long, erect stem, of which there may be 1 or a number arising from a common base. The bloom is from 4 to 5 cm (1 ½ to 2") across and the **yellow ray flowers are deeply notched at their tips**. Dry, open fields and sidehills from June to August.

RANGE BC: East of Cascades in drylands; nowhere is it plentiful.

RANGE WA: Range and abundance expands in Washington and is usually associated with the bunchgrass and ponderosa pine ecosystems. Sidehills above the Snake River.

BLACK-EYED SUSAN, *Rudbeckia hirta*: As the name implies, this 'sunflower' has a **central button that is so deep brown or purple as to appear black**. Generally same height and form as brown-eyed (above), but **orange ray flowers (petals) are not notched**. While infrequent in southern B.C.—along roadsides, on old farms and in meadows—it is wideranging at lower elevations through Washington and southwards to California.

YELLOW FLOWERS
2.5-4 cm
(1-1 1/2")
ACROSS

10-40 cm (4-16")
HIGH

WOOLLY,
OLIVE-GREEN
LEAVES
3-7
LOBED

Sunflower Family, *Asteraceae (Compositae)*
WOOLLY SUNFLOWER
Eriophyllum lanatum

Note: The **3 widespread, probably confusing flowers** here make up most of spring–summer gold in landscape. **Notice differences in leaves!**

Bright little sunflower, to 40 cm (16") tall, supported on **slender, hairy stems above a twisted mat of olive-green leaves. Leaves very woolly beneath**, giving flower its common name; can be much sparser on interior plants. Wide, irregularly lobed lower leaves. **Stem leaves narrow; 3–7 fingers**. Stems branch—hence showy mass of bloom.

Flower 2.5–4 cm (1–1 ½") across; **8–11 broad, golden-yellow petals, beady yellow centre**. Notice that **flower is on long stem above uppermost leaves**. Sunny locations; grassy banks, rocky places; lowlands–subalpine. Depending on elevation, blooms spring to late summer.

RANGE BC: Gulf and Vancouver islands. Rare; east of Cascades.

RANGE WA: San Juan Islands and both flanks of Cascades. Common at middle to alpine heights in Olympics. Columbia Gorge.

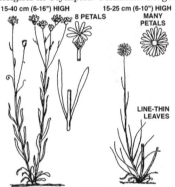

15-40 cm (6-16") HIGH

8 PETALS

15-25 cm (6-10") HIGH
MANY
PETALS

LINE-THIN
LEAVES

OREGON SUNSHINE   LINE-LEAVED FLEABANE

OREGON SUNSHINE [247], *E. lanatum* var. *integrifolium*: 15–40 cm (6–16") tall; blooms have **about 8 ray flowers. Forms clumps** that produce **showy masses of flowers**. The numerous **stems carry narrow 'paddle' leaves to 2 cm (¾") long well up each stem**. While it does reach southern B.C., it is very abundant east of the Cascades in Washington, where it is the **dominant yellow decorating rocky slopes, coulees, bluffs and canyons**. Overlaps blooming period of species above and below; best from late April into June. Common east of Cascades in southern B.C. Keremeos. East of Cascades in Washington's sagebrush areas.

LINE-LEAVED FLEABANE, *Erigeron linearis*: **Large number of thin ray flowers, about 24, compared to approximately 8 wide ones of Oregon sunshine**. This large number of petals is a common design feature of *Erigerons*. **Leaves are line-thin**, as name implies. Usually **15–25 cm (6–10") tall**; blooms late April-May. Common east of Cascades in sagebrush areas in southern B.C. Keremeos. Same habitat in Washington. Grand Coulee region, Yakima County, Steptoe Butte.

5-15 cm (2-6") HIGH

YELLOW FLOWER
1.3-2.5 cm (1/2-1")
ACROSS

HAIRY, PINKISH
BRACTS

STEMS AND
LEAVES
FINELY
HAIRY

'MEALY'
LEAVES 1.3-6 cm
(1/2-2 1/2") LONG

Sunflower Family, *Asteraceae (Compositae)*
GOLDEN FLEABANE [248]
Erigeron aureus

Note: *Erigerons* also come in white (pp. 208 & 210), pink (p. 287), purple (p. 310) and blue (p. 330). Also see Lyall's goldenweed (p. 249).

This cheery midget 'daisy' greets high-mountain hikers. **Only a few centimetres (inches) tall**, it adorns rocky ridges and gravelly pockets. Finely hairy long-stemmed oval leaves; basal except for 1–3 small ones on stem. **Flowerheads single, to 2.5 cm (1") across**, with many ray flowers and a spongy yellow centre. Note **pinkish to purple bracts beneath flowerhead**. May be seen from snow melt until late summer, often growing with silky phacelia, Lyall's goldenweed and alpine anemone

RANGE BC: Subalpine and alpine terrain across B.C. to the Rockies.

RANGE WA: Cascades in northern Washington. Hart's Pass.

Sunflower Family, *Asteraceae (Compositae)*
ALPINE ARNICA
Arnica angustifolia

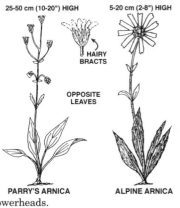

25-50 cm (10-20") HIGH   5-20 cm (2-8") HIGH

HAIRY BRACTS

OPPOSITE LEAVES

PARRY'S ARNICA   ALPINE ARNICA

There are at least a dozen *Arnica*s in our area; 5 are described in this book. Whether on robust plants in lowland forests or on dwarfed plants in alpine terrain, these prominent yellow flowers—except for orange arnica (below)—do more than their share of complementing the colours of their companion wildflowers. Like the closely related *Senecio*s, **Arnicas have a single row of flower bracts—always covered with scattered hairs**. Other features that distinguish *Arnica*s are **opposite leaves** and **flowers more than 2.5 cm (1") across**; *Senecio*s, however, have alternate leaves and small flowerheads.

Alpine arnica is **less than 30 cm (12") tall**, with **1–3 flowers on a stiff stem**. The main leaves form in 2–3 pairs and are densely hairy, as is the stem.

RANGE BC: Mountain slopes throughout, but most abundant at high elevations. Manning Park.

RANGE WA: Not in Washington, although B.C.'s Manning Park is on its northern border.

PARRY'S ARNICA, *A. parryi*: This plant, 25–50 cm (10–20") tall, has the usual stout single stem of the *Arnica*s. But its **flowerheads are less widely opened and lack the usual ray flowers**, so they are not as showy as most. **Heads that are still in bud tend to nod**. The long-stemmed basal leaves are quite distinctive in shape. Lower mountain slopes to subalpine. Widespread in B.C. and Washington.

RYDBERG'S ARNICA, *A. rydbergii*: Rather similar to alpine arnica (above), but **only to 20 cm (8") tall**. Often branched, with 1 to several blooms, each with **about 8 ray flowers**. Basal leaves with short stems, plus **3–4 pairs of sessile stem leaves. Leaves have 3–5 veins**. High country—subalpine and alpine regions. General in our area.

Sunflower Family, *Asteraceae (Compositae)*
MOUNTAIN ARNICA
Arnica latifolia
Broad-leaf Arnica

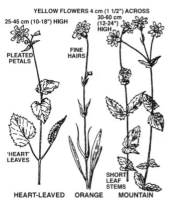

YELLOW FLOWERS 4 cm (1 1/2") ACROSS
25-45 cm (10-18") HIGH   30-60 cm (12-24") HIGH

PLEATED PETALS   FINE HAIRS

'HEART LEAVES

SHORT LEAF STEMS

HEART-LEAVED   ORANGE   MOUNTAIN

The straight stem, to 60 cm (2') tall, is unbranched and carries flowers from 2.5 to 5 cm (1 to 2") across. The **showy blooms have a light brown centre and 8–12 ray flowers with prominent teeth at their squared-off tips**. The large, coarsely toothed leaves are oval, though rarely heart-shaped. They branch opposite from one another, the lower ones having short stems and the upper ones being stemless. This is the **main yellow bloom of the multicoloured subalpine meadows**, to which it adds a special touch. However, it does range down to near sea level in the Columbia Gorge. Note: The medicinal arnica used for cuts and bruises is from a European species.

RANGE BC: Widespread throughout, from middle elevations up to subalpine heights, where it reaches its greatest abundance.

RANGE WA: Habitats as for B.C.

HEART-LEAVED ARNICA [249], *A. cordifolia*: **Lower elevations** but still in moist mountain terrain. Only 25–45 cm (10–18") tall but has slightly larger flower than mountain arnica; **often 3 blooms per stem. Several long-stemmed, large, heart-shaped leaves, roughly toothed**, on lower stem. Higher stem leaves have short stems or none. From Alaska southwards, widely distributed throughout; foothills and mountainsides.

ORANGE ARNICA, *A. fulgens*: **Sticky, hairy**; differs from mountain arnica by having **orange-tinged flowers** and **long, 'paddle' leaves; 3–5 veins**. Lower elevations. Blooms late spring–August. Northern B.C. southwards throughout; in almost every mountain park.

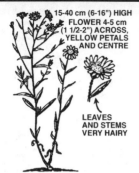

15-40 cm (6-16") HIGH
FLOWER 4-5 cm
(1 1/2-2") ACROSS,
YELLOW PETALS
AND CENTRE

LEAVES
AND STEMS
VERY HAIRY

Sunflower Family, *Asteraceae (Compositae)*
GOLDEN-ASTER
Heterotheca villosa
Chrysopsis villosa

Here is a plant with **flowers much like an aster's**, except it has **yellow rays** rather than white-purple or violet ones. A bushy clump, it also has a peculiar **overall greyish tinge to the stems and leaves**, which results from fine, matted, spreading hairs, protection from intense exposure and drought. The fibrous roots penetrate deeply into the soil. Note that each stem is usually branched near the top and that the **flowers are entirely yellow, with 10–16 ray flowers**. Leaves are numerous and may be with short stems or without. No part of the plant is sticky, so **do not confuse it with curly cup gumweed**, *Grindelia squarrosa* (below). Watch for this plant in dryland regions, from river bottoms and sidehills to well up on mountainsides. The blooming period is July onwards.

RANGE BC: Common in southern B.C., east of the Cascades.

RANGE WA: Generally east of the Cascades in open lands above the range of sagebrush. Eastern section of Columbia Gorge.

OREGON GOLDEN-ASTER, *H. oregona*: Resembles the golden-aster above in terms of leaves and general form, but its **bloom has disk flowers only**. It also lacks the greyish hairs of the one above. Another clue comes from its liking to be along streams, on sand bars and in gravelly places. Not in B.C. Most abundant east of the Cascades, in southern Washington.

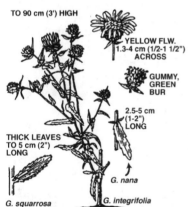

TO 90 cm (3') HIGH

YELLOW FLW.
1.3-4 cm (1/2-1 1/2")
ACROSS

GUMMY,
GREEN
BUR

2.5-5 cm
(1-2")
LONG

THICK LEAVES
TO 5 cm (2")
LONG

G. nana

G. squarrosa   *G. integrifolia*

Sunflower Family, *Asteraceae (Compositae)*
PUGET SOUND GUMWEED [250]
Grindelia integrifolia

This plant seems to appear someplace along the way if you walk any open area, especially one that is close to the sea. It is **sprawling to bushy** in form and can often reach 90 cm (3') in height. The appeal of the bright yellow flowers is lessened by the **gummy, green cups** that hold them, but touch one to confirm that it really is sticky—the name 'gumweed' is well chosen. This bur has **5 or 6 rows of gummy bracts that curl partly outwards at their tips**.

The broad-stemmed **lower leaves are dotted with resin**. Up to 40 cm (16") long, they may have teeth or be entire. Stem leaves are alternate and often clasping at their bases. There is a prolonged blooming season from May to October.

RANGE BC: Southern coastal areas of B.C.

RANGE WA: Puget Sound region and northwards.

CURLY CUP GUMWEED, *G. squarrosa*: The 'curly cup' part of the name relates to the fact that below the petals the **bracts are fully curled back**, rather than partly, as for Puget Sound gumweed (above). Otherwise it is much the same in form, but the range is quite different, being east of the Cascades. A common roadside plant in central B.C., with a prolonged summer blooming season. Southwards east of the Cascades in waste places and along roadsides at lower elevations.

COLUMBIA GUMWEED, *G. columbiana*: Small, yellow flowers about 1.3 cm (½") across and **no ray flowers**. **Very gummy. Dry sand or gravel areas**, east of the Cascades. Columbia Gorge. Not in B.C.

SMALL-FLOWERED GUMWEED, *G. nana*: Contradicting its name, this plant has a **large flowerhead to 4 cm (1 ½") across**. Leaves to 15 cm (6") long and slightly toothed. Open places, banks, dry meadows at lower elevations east of the Cascades. Not in B.C.

Sunflower Family, *Asteraceae (Compositae)*
NODDING BEGGARTICKS [251]
Bidens cernua
Bur Marigold

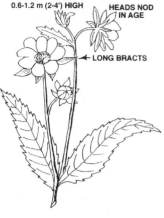

0.6-1.2 m (2-4') HIGH
HEADS NOD IN AGE
LONG BRACTS

A bright 'sunflower' plant **limited to wet or boggy soil, marshes and lake edges. Some aging flowerheads will nod** or droop. Notice that there are **6–8 showy ray flowers** (petals). Leaves are long and lance-like with toothed margins. It is a late summer bloomer. The flattened seeds with their hooked barbs will grab steadfastly onto your clothing should you pass too close—hence the name!

RANGE BC: Common east of the Cascades and well-known in the Rockies. Spotty on southern Vancouver Island and adjacent mainland.

RANGE WA: Here and there, coastal.

VANCOUVER ISLAND BEGGARTICKS, *B. amplissima*: **Leaves are deeply divided into 3 parts and are toothed**—very different from those of nodding beggarticks (above). The **ray flowers are very short**, barely exceeding the central floral parts. Most common in wet sites on southeastern Vancouver Island and adjacent mainland, where it is a comparatively rare plant. Ranges southwards to Oregon, in and near aquatic habitats.

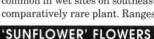
'SUNFLOWER' FLOWERS

Sunflower Family, *Asteraceae (Compositae)*
MULE'S-EARS [255]
Wyethia amplexicaulis

TO 75 cm (30") HIGH
GLOSSY GREEN LEAVES TO 40 cm (16") LONG

The genus *Wyethia* is named after Nathaniel Wyeth, an adventurous explorer who crossed the continent to Oregon in 1834. This showy yellow flower can easily be confused with the balsamroots (following pages), with which it mixes in a number of areas, but it **blooms as balsamroots are finishing**. A plant that appears similar to the balsamroots, with **vigorous growth** and **flowers in bloom in late May**, will very likely be mule's-ears. At high elevations it may bloom into July.

The **glossy green leaves, to 40 cm (16") long, stand erect** and are shaped like a mule's-ears. As **leaves climb the stem, almost to the flowerhead**, they decrease in size. Notice the very thick, yellow midrib on both sides of the leaf. A flower stem carries **1 large, handsome 'sunflower' at its top and several smaller ones lower down**.

RANGE BC: Not in B.C.

RANGE WA: A wide, spotty range in the ponderosa pine ecosystem. Leavenworth, road to Hart's Pass, hills east of Coulee City, Colfax, Blue Mountains.

NARROWLEAF MULE'S-EARS, *W. angustifolia*: Sometimes called narrow-leaved sunflower, it differs from mule's-ears in having **narrow, dullish leaves** and **usually just 1 large flower to a stem**. Not in B.C. Generally west of the Cascades in Washington, but confined to a small area north of and along the Columbia Gorge. Extends southwards through Oregon to California.

☐ DELTOID BALSAMROOT, *B. deltoidea*. Rare!
☐ CAREY'S BALSAMROOT, *B. careyana*
■ ARROW-LEAVED BALSAMROOT, *B. sagitatta*

Sunflower Family, *Asteraceae (Compositae)*
BALSAMROOTS
Balsamorhiza spp.
Sunflowers

Note: True sunflowers, *Helianthus* genus (see pp. 255–56), are scarce in our area; usually distinguished from other 'sunflower' plants by seed shape.

Balsamroots' dramatic golden blazonry brightens benchlands and mountain slopes from April into May. But, especially in Washington, identification can be puzzling: shapes and colours of leaves vary and much hybridization occurs in some areas; moreover, mule's-ears (p. 253) might be misidentified. After looking at at least 1000 plants here and there, a possible solution dawned: a map to show main species in Washington, based on personal observation.

In B.C., only arrow-leaved balsamroot occurs east of Cascades. West of Cascades, deltoid balsamroot is very rare; on southeastern coast of Vancouver Island.

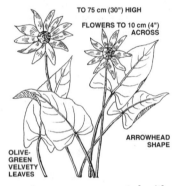

TO 75 cm (30") HIGH
FLOWERS TO 10 cm (4") ACROSS
ARROWHEAD SHAPE
OLIVE-GREEN VELVETY LEAVES

Sunflower Family, *Asteraceae (Compositae)*
ARROW-LEAVED BALSAMROOT [252]
Balsamorhiza sagittata
Spring Sunflower

A dozen or so large flowers, to 10 cm (4") across. Clump of **large, velvety, olive-green leaves. Solitary flowers on leafless stems. Pale green, woolly look to both sides of leaf, heart-shaped base.** Soon after blooming, leaves curl into twists of parchment, to remain for many weeks as drab evidence of former glory.

Aboriginals ate the rich, oily seeds either raw or mixed with deer fat and boiled. Deep-growing roots were eaten raw or were roasted, with much care and ceremony, in large pits.

RANGE BC: Low–middle elevations throughout ponderosa pine zone. Osoyoos to Kamloops, Williams Lake and Soda Creek. Magnificent on mountains north of Chase. Grand Forks, Cranbrook and northwards to Columbia Lake.

RANGE WA: Southwards from south-central B.C. through sagebrush and ponderosa pine habitats. A possible change near Swauk Pass, south of Leavenworth, to Carey's balsamroot. Boundary continues eastwards to near Coulee Dam and southwards to Waitsburg, near Walla Walla.

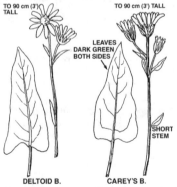

TO 90 cm (3') TALL
TO 90 cm (3') TALL
LEAVES DARK GREEN BOTH SIDES
SHORT STEM
DELTOID B.
CAREY'S B.

DELTOID BALSAMROOT or **NORTHWEST SUNFLOWER**, *B. deltoidea*: **Triangular leaf shape. Possible second, smaller bloom below main one; perhaps 1 or 2 very small leaves on stem.** Showy flowerhead. Only balsamroot west of Cascades (rare). In B.C., very rare: southeastern coast of Vancouver Island only. In Washington, see map for approximate range. Extends eastwards in Columbia Gorge to M73.

CAREY'S BALSAMROOT, *B. careyana*: **Leaves shiny dark green on both sides!** Colour remains into late summer. Do not confuse with mule's-ears (p. 253). Often **several showy blossoms** per stem. **Appears to bloom longer than other balsamroots. May be most widespread balsamroot in Washington;** not in B.C. Largely replaces arrow-leaved balsamroot in southern Washington; **generally above sagebrush zone on sidehills and in open forests.** Also see Hooker's balsamroot on next page.

Sunflower Family, *Asteraceae (Compositae)*
HOOKER'S BALSAMROOT [253]
Balsamorhiza hookeri

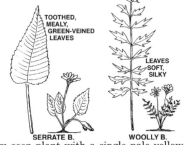

50cm (20") HIGH
FLOWERS 5 cm (2") ACROSS

TWISTED, SERRATE LEAVES

While balsamroots on p. 254 are relatively clear cut in form and range, others are a great puzzle, and Hooker's, most of all, has a surprising variation in design.

About 50 cm (20") tall, each stem carries a single flower about 5 cm (2") across with a deep golden 'button' centre. Bracts are leafy and stick out at all angles. Leaves to 50 cm (20") long are deep green, deeply cut to the midrib and twisted. Blooms April–May.

RANGE BC: Not in B.C.

RANGE WA: Coulees above Vantage, southern third of Yakima Canyon. Klickitat County.

HOOKER'S DWARF BALSAMROOT: A new name used for easy reference. This plant is regarded as **a dwarf form of the above**. Blooming April–May. To 30 cm (12") tall with stems tending to droop. Flowers to 7.5 cm (3") across. Leaves to 10 cm (4") long, a mass of twisted leaflets; may hug the ground in a rosette pattern. Small plants sprouting nearby in the stony soil shared by rigid sagebrush. Not in B.C. but in abundant bloom at the turn-off to Sun Lakes, the Vantage Highway and open slopes above Cle Elum–Ellensberg #10 Road.

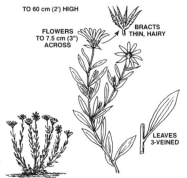

TO 40 cm (16") HIGH
TOOTHED, MEALY, GREEN-VEINED LEAVES
SERRATE B.

TO 60 cm (2') HIGH
LEAVES SOFT, SILKY
WOOLLY B.

SERRATE BALSAMROOT, *B. serrata*: The degree of serration is variable: **broad, mealy green, heavily veined leaves may be wavy-edged only or quite deeply cut**. To 40 cm (16") tall, **1 flower per stem**. Rather rare, but still listed in some books as occurring on rocky knolls of southern Washington, east of Cascades. Not in B.C.

WOOLLY BALSAMROOT, *B. incana*: A seldom-seen plant with a single pale yellow flower per stem. Woolly hairs make stems olive-green. Leaves cut to midrib. Moist sidehills of southeastern Washington.

ROSY BALSAMROOT, *B. rosea*: Under 30 cm (12") tall; **small flower ages pinkish. Narrow leaves coarsely toothed**. Limited occurrences in Yakima and Walla Walla counties. Consider it a rarity! Not in B.C.

Sunflower Family, *Asteraceae (Compositae)*
CUSICK'S SUNFLOWER
Helianthus cusickii

TO 60 cm (2') HIGH
FLOWERS TO 7.5 cm (3") ACROSS
BRACTS THIN, HAIRY
LEAVES 3-VEINED

Note: Close relatives are Rocky Mountain helianthella (p. 257) and Lyall's goldenweed (p. 249).

A large number of plants have showy 'sunflowers'; the trick is to find a few features to identify the 2 true sunflowers. Common sunflower (p. 256), is a giant, with large, wide leaves. **Cusick's is only to 60 cm (2') tall**, with branching stems, but may be spreading in form. Flowers, which so brightly adorn a hillside, have **8–15 sharp-pointed, pale yellow petals that point upwards**—a shallow bowl effect. Petals finely pleated. Head is supported by thin, densely hairy, sharp-pointed bracts.

Leaves are a good clue too: either alternate or opposite, depending on specimen, without stems and to 7.5 cm (3") long. **Leaves have slight 3-vein pattern**. Blooms in May, with Hooker's onion and Douglas buckwheat.

RANGE BC: Not in B.C.

RANGE WA: Dry, open plains and foothills in south-central Washington. Vantage to Ellensburg and southwards. Sunnyside, Steptoe Butte, Yakima.

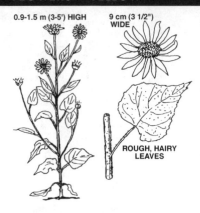

0.9-1.5 m (3-5') HIGH

9 cm (3 1/2") WIDE

ROUGH, HAIRY LEAVES

Sunflower Family, *Asteraceae (Compositae)*
COMMON SUNFLOWER [257]
Helianthus annus
Kansas Sunflower

Note: Other 'sunflower' blooms are on pp. 250 & 253–55.

A 'sunflower' at first glance? But what is a sunflower? The few true examples we have (*Helianthus* spp.) are so rare that only this one and Cusick's (p. 255) have been included. Other plants, such as brown- and black-eyed Susans, *Gaillardia aristata* and *Rudbeckia hirta* (p. 249), and Rocky Mountain helianthella, *Helianthella uniflora* (facing page), look like our general idea of sunflowers. The real sunflowers increase in variety and abundance as you go southwards to California.

The state flower of Kansas does not make much of an impression in B.C. but it is very common in south-central Washington. When you do see one, the size of the flower is impressive and can only be described as a sunflower. Plants I have seen average **0.9–1.5 m (3–5')** in **height** and carry a **flower typically to 9 cm (3 ½") across.** The bright yellow petals surround a rich brown centre.

This coarse plant is much branched and flowers are single at the end of stout stems. Stems and **large, triangular leaves are sandpapery** and hairy. Leaves are rather disorganized, the lower being opposite but changing to alternate higher on the stem. Since it is an annual, and thus dependent on the previous year's seeds for its survival, the number of plants in an area can vary. This sunflower is widespread in the U.S.A. and it is a native in the western states. Many of the domesticated sunflowers, including those grown specifically for their edible seeds, are cultivars of *H. annus*.

RANGE BC: Sporadic; Fraser Canyon, southern Okanagan, south of Dog Creek in the Chilcotin.

RANGE WA: Widespread in arid regions, especially in the south-central region. Eastern area of Columbia Gorge. Lyon's Ferry.

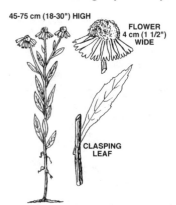

45-75 cm (18-30") HIGH

FLOWER 4 cm (1 1/2") WIDE

CLASPING LEAF

Sunflower Family, *Asteraceae (Compositae)*
MOUNTAIN SNEEZEWEED [256]
Helenium autumnale
Western Sneezeweed

A good sniff of dried, pulverized sneezeweed will supposedly bring on enough sneezes to clear your nose and lungs. Aesthetically, the botanical name is associated with Helen of Troy. It should be easily identified by the **globe-like orange-yellow head with a number of bright yellow 'lazy' petals drooping from it**. Notice that the dozen or so **ray flowers are triple-notched**. Leaves are many and large, with clasping bases that extend down the stem. Some may have a wavy-tooth design. It is a late summer bloomer and often grows in a mass with stems 45–75 cm (18–30") tall. In damp ground from low to middle elevations.

RANGE BC: Comparatively rare; southern B.C.

RANGE WA: Comparatively rare. Bottomlands of the Columbia Gorge.

Sunflower Family, *Asteraceae (Compositae)*
ROCKY MOUNTAIN HELIANTHELLA [258]
Helianthella uniflora
One-Flowered Little Sunflower

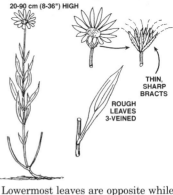

20-90 cm (8-36") HIGH

THIN, SHARP BRACTS

ROUGH LEAVES 3-VEINED

The only *Helianthella* in our area and a close relative to the true sunflowers, *Helianthus* spp. (pp. 255–56), but how do we distinguish it from other sunflower-looking plants? Begin by noticing the **erect form** and **single flowerhead** and the **lack of very hairy bracts**—all features that separate *Helianthella* from Cusick's sunflower.

In general, it can grow 20–90 cm (8–36") tall and usually has several unbranched stems from one base. Lowermost leaves are opposite while those above are alternate. They are 7.5–15 cm (3–6") long, with **3 veins that are especially distinctive on lower leaves**. Allow for some variation in leaf shape from region to region, but all **leaves are sandpapery** to the touch.

The bright yellow flowerheads, usually 1 (*uniflora*) to a stem, are from 4 to 6 cm (1 ½ to 2 ½") across and average **about 13 petals**. Note that the petals are sharp-pointed. In B.C., where it is often associated with aspen, it blooms in July and August. Also found with ponderosa pine and Douglas-fir. In Washington it blooms in early June and into July in higher forests.

RANGE BC: Locally frequent in south-central B.C., from valley bottoms to mountainsides.

RANGE WA: Common roadside plant along eastern flank of the Cascades. Leavenworth, Yakima Canyon. Also eastern Washington to 1200 m (4000'). Blue Mountains.

'ORCHID' FLOWERS

Orchid Family, *Orchidaceae*
YELLOW LADYSLIPPER [259]
Cypripedium calceolus
Moccasin Flower

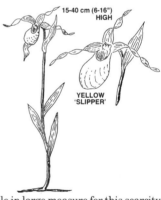

15-40 cm (6-16") HIGH

YELLOW 'SLIPPER'

Note: For 2 ladyslippers with white blooms, see p. 211; with a green-brown one, p. 340.

The **pouch-like yellow flower of satiny texture** can very well be likened to a dainty moccasin. **Claret-purple streaks and markings on the flower's lining** enhance its beauty. A bizarre touch is added by the **long, twisted sepals** with tints of green and brown. Flowers have a **fragrant perfume**. The several leaves are rich green and form sheaths on the stem. The moccasin flower is comparatively scarce now. No doubt collectors are responsible in large measure for this scarcity. It takes over a dozen years for this orchid to raise a flower, so please leave it undisturbed.

RANGE BC: Wet places at lower elevations. Generally east of the Cascades. It is interesting to note that nearly 80 years ago they were recorded growing at Lytton, Okanagan Valley and Golden and no doubt much more abundant than at present. I remember them being along Shingle Creek, west of Penticton, a long time ago!

RANGE WA: East of the Cascades; damp, shady places, stream banks and marshy areas.

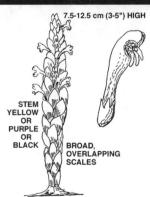

7.5–12.5 cm (3–5") HIGH

STEM
YELLOW
OR
PURPLE
OR
BLACK

BROAD,
OVERLAPPING
SCALES

Broomrape Family, *Orobanchaceae*
GROUNDCONE [254, 337]
Boschniakia hookeri
Pogue

A parasitic oddity easily overlooked in the green richness of salal growth. Look for groundcone on the forest floor in late spring, beneath its **only host, salal** (see p. 150). The thick, scaly stem is 7.5–12.5 cm (3–5") tall and can be **yellow, black or red to purple**. Scales are broad and overlapping. 'Pogue' is a name that several coastal native tribes used for this plant, which they considered edible.

Tiny, tube-shaped flowers peek over the scales. A hood petal partly hides the slender filaments. The lower petal has 2 or 3 lobes.

RANGE BC: With coastal salal.
RANGE WA: With coastal salal.

NORTHERN GROUNDCONE, *B. rossica*: Larger than regular groundcone (above), northern groundcone bears a remarkable **resemblance to a long, thin conifer cone** standing on end. It is more brown in colour than yellow. It is parasitic on shrubby alders and various other shrubs, rather than on salal. Its range—**extreme northern B.C.**, north of the Queen Charlottes—is important in its identification.

Poppy Family, *Papaveraceae*
CALIFORNIA POPPY [260]
Eschscholzia californica

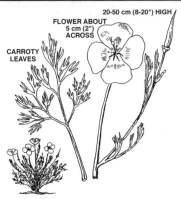

20-50 cm (8-20") HIGH

FLOWER ABOUT 5 cm (2") ACROSS

CARROTY LEAVES

California's state flower. Does very well in gardens or as an escape on southern Vancouver Island and some of Gulf Islands. Seemed exotic when I first saw it. What a show of **golden and orange blooms** amid **fern-like foliage**! Once tried to dig a few—found a thick, tapering root that never stopped growing down. Blooms for many months, sometimes from April. **Bowl-shaped flowers open wide when sun is shining**. A prized part of my garden for years now; after blooming, trim back for fresh growth and flowers in late summer (and in fall for early spring). **Long seed pods split** to release seeds that germinate readily in dry, rocky places.

RANGE BC: Most abundant on southeastern Vancouver Island and Gulf Islands; sporadic on Lower Mainland.

RANGE WA: Local occurrences but widely distributed, especially along roadsides. Colourful display near mid-section of Columbia Gorge, April–May.

5 PETALS

Mallow Family, *Malvaceae*
MUNRO'S GLOBEMALLOW [261]
Sphaeralcea munroana
White-leaved Globemallow, Desert Mallow
Note: For pink globemallows, see p. 273.

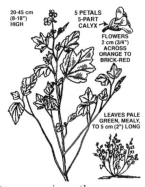

20-45 cm (8-18") HIGH

5 PETALS 5-PART CALYX

FLOWERS 2 cm (3/4") ACROSS ORANGE TO BRICK-RED

LEAVES PALE GREEN, MEALY, TO 5 cm (2") LONG

Arid places, often with sagebrush. Bloom of **unusual orange or brick-red**. Note very pale green leaves and stem, and rather loose form of several branching, fibrous stems. **Roughly 'maple' leaves; mealy texture and lobing** (3–5) characteristic of mallows; to 5 cm (2") in length, stems almost as long. Flowers and buds form showy clusters in leaf axils. 5 colourful petals held in 5-parted calyx; numerous stamens. Blooms in May in extreme southeast, late summer in north.

RANGE BC: South-central; sagebrush area.

RANGE WA: Sagebrush habitat throughout. Coulee County, Yakima, Wenatchee.

Forget-me-not Family, *Boraginaceae*
RIGID FIDDLENECK [262]
Amsinckia retrorsa

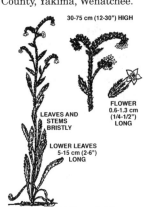

30-75 cm (12-30") HIGH

FLOWER 0.6-1.3 cm (1/4-1/2") LONG

LEAVES AND STEMS BRISTLY

LOWER LEAVES 5-15 cm (2-6") LONG

Note: 'Amsinckia' and 'tarweed' are other names for these half-dozen species **largely in dry terrain east of Cascades**, though fiddlenecks should not be confused with unrelated tarweeds of genus *Madia* (p. 229). Fiddlenecks with yellow flowers are on p.236.

Recognize *Amsinckia*s by **bristly leaves and stems** and **long flowering head twisted in a curl like a violin neck**. Funnel-like **orange-to-yellow flowers**. 5 stamens usually hidden in throat, below petals. Bristly husks along outer edge of curved stem carry horned seeds or nutlets shaped like broad arrowheads. Rigid fiddleneck blooms early spring–late summer; new flowers keep forming.

RANGE BC: Not in B.C.

RANGE WA: Open ground or fields; preference for some moisture. Mostly in sagebrush and bunchgrass ecosystems. Columbia Gorge.

45-90 cm (1 1/2-3')
HIGH

Phlox Family, *Polemoniaceae*
LARGE-FLOWERED COLLOMIA
Collomia grandiflora

Note: 2 other native collomias (rather obscure) are more accurately described as pink (see p. 279).

    This flower is difficult to colour-code—it is often described as 'salmon or pink.' Some specimens may look to be **pale yellow or with an orange shading**, which seems to be the most accepted description of their colour.

    Large-flowered collomia is **45–90 cm (18–36")
tall**, is unbranched and has narrow leaves to 5 cm (2") long alternating on the stem. It deserves its common and botanical names, for it has a **large, showy head of tubular flowers** atop its stem. Each is about 2.5 cm (1") long. Although individual flowers bloom and fade, the large number of flowers in a head allows a plant to hold its colour throughout early summer. It favours well-drained ground from sagebrush areas to middle-mountain forests.

RANGE BC: Dry places on Vancouver Island, but most common east of the Cascades in southern B.C.

RANGE WA: Most common east of the Cascades. Entiat River.

6 PETALS

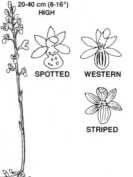

20-40 cm (8-16")
HIGH

SPOTTED   WESTERN

STRIPED

Orchid Family, *Orchidaceae*
SPOTTED CORALROOT [263]
Corallorhiza maculata

Note: Some references treat spotted coralroot and western coralroot (below) as subspecies of *C. maculata*. These coralroots are **considered as orange based on the overall colour of the flower**, not on the primary colour of the petals.

    Coralroot's name comes from its mass of **gnarled, knobby roots**, which resembles a piece of coral. The plant is saprophytic (lives on dead organic matter in the soil) and therefore can do without leaves, though they do persist as small scales.

    The coralroots—4 in our area—**belong to the orchid family**; **beautiful small flowers** bring that association quickly to mind. Stout stems, 20–40 cm (8–16") tall, are adorned on the top third by **spaced single flowers, each with 6 petals**. Coralroots bloom during May and June and are found in rich forest soils at lower elevations. They are quickly distinguished from each other by colouring or shape of flower. **Spotted coralroot has a white tongue marked with crimson spots**.

RANGE BC: Across the province in suitable habitats south of Ft. St. John.

RANGE WA: Throughout eastern and western Washington.

WESTERN CORALROOT, *C. mertensiana*: Resembles spotted coralroot (above), with the same general **orange-to-reddish tinge to stems and flowers**. Note that **only the lip petal is marked with red lines**, which will eliminate confusion with striped coralroot (below). Prefers moist coniferous woods. Blooms from mid- to late summer. From Prince George, B.C., southwards throughout the province and in eastern and western Washington. More abundant at low elevations, west of the Cascades.

STRIPED CORALROOT, *C. striata*: Same general form as other coralroots, but both the **lip and the other petals are striped with red**. Range is as for western coralroot (above), in rich, moist forest soils.

YELLOW CORALROOT, *C. trifida*: **Smallish yellow 'orchid' flowers**. Although it is widespread across B.C. and Washington in moist, shady forests, finding one is a rare experience. More common to north.

Lily Family, *Liliaceae*
TIGER LILY
Lilium columbianum
Columbia Lily

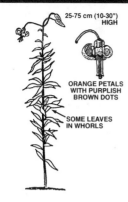

25-75 cm (10-30")
HIGH

ORANGE PETALS
WITH PURPLISH
BROWN DOTS

SOME LEAVES
IN WHORLS

Very suggestive of cultivated tiger lily. Usually 2–3 nodding blooms, sometimes 6–9. Orange flower spotted with purplish brown; long anthers stand out. Low to high altitudes— blooming ranges June–August. Most of the **narrow leaves are in whorls up and down stem**. Large, deep-rooted, scaly, white bulb. When conditions are right, abundant and remarkably decorative. Native peoples dug bulbs from spring to fall—generally steamed or boiled; also dried for winter use.

RANGE BC: Sea level to 1200 m (4000'); damp soils. Widespread throughout.

RANGE WA: Habitats as for B.C.

'DANDELION' FLOWERS

Sunflower Family, *Asteraceae (Compositae)*
ORANGE AGOSERIS [265]
Agoseris aurantiaca

10-30 cm (4-12") HIGH

BRACTS
'SHINGLED'

Note: Half-dozen agoseris in our area. Those with yellow flowers are on p. 241.

The first time you or a friend exclaims, 'That's a **peculiar dandelion, it's orange!**' you have found orange agoseris. There is nothing quite the same colour. **Bloom consists only of ray flowers; no 'button' centre.** Head may turn brown or pink. Height, like dandelion is 10–30 cm (4–12"). Fluffy, white seedhead. Notice **thin leaves, more upright than toothed leaves of dandelions' basal rosette**. Bracts create shingled effect; dandelions have 1 main row with outer, shorter row bent back.

Do not confuse with apargidium (p. 243)—yellow, toothed ray flowers. Like dandelions, **agoserises have milky juice**. Aboriginals had preferred species of agoserises from which they made a chewing gum. Orange agoseris usually on dry ground, medium–high elevations.

RANGE BC: Widely distributed; open forest, meadows; middle to high elevations.

RANGE WA: Habitats as for B.C.

Sunflower Family, *Asteraceae (Compositae)*
ORANGE HAWKWEED [264]
Hieracium aurantiacum
Devil's Paintbrush

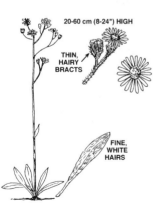

20-60 cm (8-24") HIGH

THIN,
HAIRY
BRACTS

FINE,
WHITE
HAIRS

Note: White-flowered hawkweeds are on p. 207 while the more common yellow ones are on pp. 244–45).

Distinctive **heads of vivid orange-red flowers—streaked yellow centres**—are sure identification. A European weed pest now spreading in southern B.C and various regions of Washington. Vivid flowers, to 2.5 cm (1") across; more eye-catching because **plants usually densely mass** above spreading roots. Some fine hair on stems and leaves. After flowering, **bristles (pappus) at top of seeds are brownish on hawkweeds, but white on hawksbeards** (p. 245).

RANGE BC: Roadsides, fields, waste places; central–southern; both sides of Cascades.

RANGE WA: Habitats as for B.C.; mostly east of Cascades. Eastern slope Snoqualmie Pass. Republic.

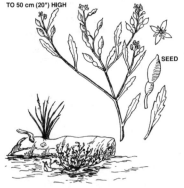

7.5-25 cm (3-10") HIGH

DIVIDED LEAF

ENTIRE LEAF

OAKS TOOTHWORT   SLENDER TOOTHWORT

Mustard Family, *Brassicaceae (Cruciferae)*
OAKS TOOTHWORT [266]
Cardamine pulcherrima var. *pulcherrima*
Large-flowered Bittercress

Recognize 2 varieties of toothwort by different basal leaves. Fortunately, **unusual leaf design on upper part of stem** is enough to identify. **Usually 3 leaves, 3–5-fingered; may have 2 leaves, 3-fingered**. Flowers 1.3 cm (½") across, in loose cluster at top. **4 pink petals veined with rose lines**: an insect's guide to the nectar. Warning: **flower may be white, with lines on petals not noticeable**, especially in shady areas. Blooms March to June.

'Oaks' refers to favoured habitat under Garry oaks. In Gulf Islands zone, blooms with shootingstars, satin-flower, white fawn lily; in Columbia Gorge: snow queen, ballhead waterleaf, Columbia kittentails, shootingstars, woodland stars, avalanche lilies, slender popcornflower and white plectritises.

RANGE BC: Southern Vancouver Island, Gulf Islands and adjacent mainland.

RANGE WA: West of Cascades, San Juan Islands, Olympic Peninsula, Mt. Rainier. Shady woods of Klickitat and Yakima counties. Columbia Gorge.

SLENDER TOOTHWORT, *C. pulcherrima* var. *tenella*: Very similar to oaks toothwort except **basal leaves tend to be lobed rather than deeply divided.** Range overlaps.

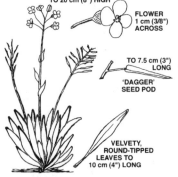

TO 20 cm (8") HIGH

FLOWER 1 cm (3/8") ACROSS

TO 7.5 cm (3") LONG

'DAGGER' SEED POD

VELVETY, ROUND-TIPPED LEAVES TO 10 cm (4") LONG

Mustard Family, *Brassicaceae (Cruciferae)*
DAGGER POD
Phoenicaulis cheiranthoides

Small flowers, **typically pink to reddish purple**, branch out from stem. Massed head of colourful flowers is very conspicuous, although **seldom reaches 20 cm (8") in height. Thin, dagger-like seed pods**—hence common name. Usually 3 flowering stems poke up from **olive-green basal tuft—may have 30 or more thick, velvety leaves**.

Look for these early spring flowers in the vicinity: gold stars, woodland stars, ballhead waterleaf and fern-leaved desert-parsley.

RANGE BC: Not in B.C.

RANGE WA: Rocky soils at low elevations in sagebrush and ponderosa pine habitats. Eastern section of Columbia Gorge. Ellensburg.

TO 50 cm (20") HIGH

SEED

Mustard Family, *Brassicaceae (Cruciferae)*
AMERICAN SEAROCKET [265]
Cakile edentula

Among logs at high tide line. Untidy looking; though sprawling, 15–50 cm (6–20") tall. Some branches close to sand, others upright. **Thick, fleshy leaves; irregularly lobed or wavy-edged**.

Little stem-top clumps or strings of **small pink–purple flowers**—rather modest for a plant flamboyantly named 'searocket.' You might recognize 4 small petals as being typical of mustards. Flowers July–August; distinctive **yellow seed pods, 1.3–2.5 cm (½–1") long**. Sometimes leaves drop, leaving seed pods as only clue to what grew here.

RANGE BC: Usually close to high tide line. From northern B.C. southwards.

RANGE WA: May be most abundant beach plant on Washington coast. Exceptionally so at Long Beach, Washington. Southwards to California.

Mustard Family, *Brassicaceae (Cruciferae)*
HOLBOELL'S ROCKCRESS
Arabis holboellii

TO 90 cm (3') HIGH   25-75 cm (10-30") HIGH

Note: White rockcresses on p. 162. For a common purple rockcress, see p. 294.

So **many of the rockcresses have pinkish, white or pale purple flowers** that colour by itself is poor identification. Rather, it is the unusually long seed pods that provide the best clue. Holboell's has **pods that are only slightly curved but hang down at a steep angle.** Also note the **arrowhead leaves that clasp the stem.** This rockcress may reach 90 cm (3') in height. A tip for experts: use a hand lens to see star-shaped hairs on the basal leaves.

HOELBOELL'S   SPREADING-POD

RANGE BC: From the eastern slope of the Cascades eastwards, in dry rocky and gravelly places. Mainly in ponderosa pine and sagebrush areas but ranging to subalpine heights.

RANGE WA: Habitats as for B.C.

SPREADING-POD ROCKCRESS, *A. divaricarpa*: **Pods are long and straight and spread upwards and outwards,** a good identification mark for this rockcress and for Lyall's (see p. 294). However, at 25–75 cm (10–30"), this plant is twice the height of Lyall's, which averages 10–25 cm (4–10"). Note too that lower stems are hairy. The **pink flowers are quite small and form a little cluster at the top of the stem.** Range as for Holboell's.

Evening-primrose Family, *Onagraceae*
PINK FAIRIES [267]
Clarkia pulchella
Clarkia, Deerhorn Clarkia

15-40 cm (6-16") HIGH

BRIGHT PINK FLOWER 4-5 cm (1 1/2-2") ACROSS

SEPALS JOINED

DULL GREEN LEAVES 2.5-5 cm (1-2") LONG

FAREWELL-TO-SPRING

Are pink fairies not the oddest looking blooms you might find? While the bright pink-purplish flowers look as if torn to ribbons, they actually consist of 4 deeply lobed petals. Also unusual is that the sepals join at their tips. Dull green leaves 2.5–5 cm (1–2") long almost go unnoticed.

Clarkia gets its name from William Clark of the famed Lewis and Clark Expedition to the West Coast in 1808.

Favours dry, exposed places, rocky slopes and dried-out areas with thin or sandy soils. When conditions are favourable it grows abundantly but tends to be obscured by grasses and other native plants. It blooms from mid-June to mid-August.

RANGE BC: East of the Cascades. Grand Forks, Castlegar, Christina Lake and Columbia Valley.

RANGE WA: Most abundant in the southeast, but in north-central Washington too. From 900–1500 m (3000–5000') in the Blue Mountains. Also along eastern slope of the Cascades and southwards to southern Oregon. Columbia Gorge.

FAREWELL-TO-SPRING, *C. amoena*: A good name for this plant because it blooms just when spring is over. The colour varies greatly but is **generally pink, with blotches of deeper red on each petal.** The flower size, to 6 cm (2 ½") across, is variable too. Strong, thin stems to 50 cm (20") high hold alternate narrow, lance-like leaves. In B.C. it makes a rare occurrence west of the Cascades on dry, rocky low-elevation slopes and extends from southern Vancouver Island through western Washington to the Columbia Gorge.

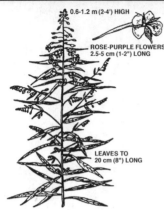

0.6-1.2 m (2-4') HIGH

ROSE-PURPLE FLOWERS
2.5-5 cm (1-2") LONG

LEAVES TO
20 cm (8") LONG

Evening-primrose Family, *Onagraceae*
FIREWEED [268]
Epilobium angustifolium
Rosebay Willowherb

Note: One *Epilobium* is yellow (p .214). White-flowered *E. lactiflorum* is under 30 cm (12") tall; flowers under 1.3 cm (½") across, petals notched.

Certainly fireweed and alpine fireweed (broad-leaved willowherb, below) are 2 native plants that capture attention early. They **can turn entire mountainsides into blazes of rose-purple**, especially if a forest fire (or logging) has gone through a year or so before. And for the high-mountain hiker crossing a wet gravel bar, there is a spreading carpet of broad-leaved (alpine) fireweed so uniquely bright mauve and magenta that it stops a person in his or her tracks. While these 2 fireweeds are the ones most people will see, there are over 2 dozen species of *Epilobium* in our area, out of which only a half-dozen that might interest the amateur botanist.

This fireweed **averages 1.2 m (4') in height**, with a leafy, unbranched stem. Leaves alternate, to 20 cm (8") long; paler beneath, with distinct veining. **Large, pink-purple flowers bloom successively on ascending tips**. Later, seed pods split open and disgorge many thousands of **fluffy, white seeds**. Seeds can be wind-borne for many kilometres (miles), accounting for remote burned-over areas suddenly becoming alive with fireweed. As well, a scattering of dwarfed plants may be found on remote high, rocky subalpine slopes. Spreading roots produce dense growth. Fireweed honey rates high with beekeepers. Blooms mid-June to late summer.

A plant as abundant as fireweed had its uses for aboriginals: the pith of young shoots was eaten raw and often the whole stem was boiled. Roots, 'bark' and even the seed fluff had value.

RANGE BC: Common throughout on old burn and waste areas. From lowlands to subalpine zones. In small patches along highways. Not in the more arid regions.

RANGE WA: Habitats as for B.C.

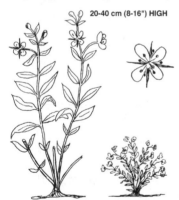

20-40 cm (8-16") HIGH

Evening-primrose Family, *Onagraceae*
BROAD-LEAVED WILLOWHERB
Epilobium latifolium
Alpine Fireweed

One could consider this willowherb to have **flowers of pink, red, purple, magenta, etc.** It is the **showy plant of subalpine and alpine areas, where it favours wet gravel bars and stream banks**. At some times you see a single plant, perhaps rather dwarfed and sprawling, at others a great expanse of colour. The **large, reddish purple flowers** are enhanced by the **soft blue-green of the thick, smooth leaves**. Usually alternate, they can be opposite now and then. To make sense out of the name 'willowherb,' note that the leaves supposedly resemble those of a weeping willow.

RANGE BC: Hikers in most of the subalpine and alpine areas of our region will come across this beautiful plant. In northern B.C. it drops to lowland moraines and river banks.

RANGE WA: Habitats as for B.C.

SMALL WILLOWHERB, *E. glaberrimum*: Opposite leaves and **flowers with 4 deeply notched pinkish petals** might be enough for you to relate it to other willowherbs, but unusual colours of leaves and stems could throw you off. A **powdery bloom makes leaves a distinct greyish green; flower stems are as pink as the several small flowers**. From 15 to 50 cm (6 to 20") high, this plant usually grows in wet places, in meadow drainages or along stream edges, from high forests to the subalpine ecosystem. Common in southwestern B.C. and in Olympics; southwards to California.

Buckwheat Family, *Polygonaceae*
BEACH KNOTWEED
Polygonum paronychia

5-25 cm (2-10") HIGH

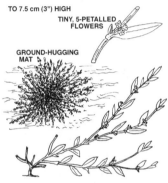

This shrubby little perennial has tough, woody stems and **narrow, curled-under leaves with prominent midribs beneath**. The clusters of **small pinkish-to-white flowers** at the stem ends are not impressive, but this plant does arouse interest because of its ability to live on **exposed beaches** and **wind-blown sand dunes**. The flower has 5 petals, each with a dark centre line, 8 stamens and a 3-cleft style. It blooms July to September.

TINY, PINK FLOWERS 3 mm (1/8") ACROSS

LEAVES 0.6-2.5 cm (1/4-1") LONG, ROLLED OVER

DARK BROWN, SHRUBBY STEMS

'Knotweed' refers to the many-jointed stems.

RANGE BC: Sandy beaches and dunes along the coast, southern Vancouver Island and adjacent mainland.

RANGE WA: Sandy beaches and dunes along the coast to California.

Buckwheat Family, *Polygonaceae*
YARD KNOTWEED
Polygonum aviculare

TO 7.5 cm (3") HIGH

TINY, 5-PETALLED FLOWERS

GROUND-HUGGING MAT

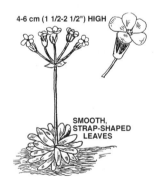

An 'unnoticeable' plant, although you probably walk on it quite often. Even if you casually noticed it, the quick impression is that it is so tough that you would not harm it. Usually it is a **dark green mat** about half a metre (1 or 2 feet) across and so tight to the ground that you step on it without a thought.

Unless you pick a piece and give it close scrutiny, you would never notice the **tiny pink, white or purple petals** mixed in confusion. Flowers are pinched into a leaf axil and have 5 well-defined petals. Leaves match tiny flowers, being only about 3 mm (⅛") long. Blooms July to September.

The plant has a single deep root from which the many ground-hugging stems spread out to form a mat. Commonly a part of dry trails, exposed yards and parking lots.

RANGE BC: Widespread in southern B.C.

RANGE WA: Widespread. Southwards to Mexico.

Primrose Family, *Primulaceae*
SMOOTH DOUGLASIA [271]
Douglasia laevigata
Cliff Douglasia

4-6 cm (1 1/2-2 1/2") HIGH

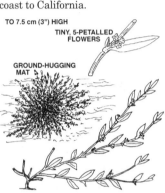

Few people will have heard of this tiny flower, much less seen one. Its name commemorates David Douglas, a pioneer botanist. Smooth douglasia is included because if seen, it will certainly capture the attention. It is relatively rare, but its **pink-rose flower** makes it especially noticeable. Some stonecrops have similar habitat and leaves, but they bear yellow flowers of different design.

SMOOTH, STRAP-SHAPED LEAVES

If you find smooth douglasia, it will be a small, dainty plant only 4–6 cm (1 ½–2 ½") high. **Leaves are basal and in a rosette**; adjoining plants usually intermingle to form a thick clump. The narrow leaves are from 0.6 to 2 cm (¼ to ¾") long and with edges smooth or few-toothed. The short stem has a cluster of **5-petalled flowers, usually 2–5 in number**, each only 1 cm (⅜") across. After blooming, there is still an attractive pinkish fringe to the flowerhead. **Blooms in July, ahead of the main subalpine floral display**.

RANGE BC: A typical moist-rockery plant that favours coastal slopes from lowland to alpine. Queen Charlotte and Vancouver islands. Manning Park. Cascades.

RANGE WA: Habitats as for B.C. Cascades southwards to Columbia Gorge. Olympics.

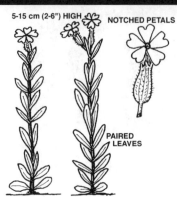

5-15 cm (2-6") HIGH

NOTCHED PETALS

PAIRED LEAVES

Phlox Family, *Polemoniaceae*

PINK MICROSTERIS
Microsteris gracilis
Midget Phlox, Slender Phlox

Simplicity could be the keynote of this trim little plant only 5–15 cm (2–6") high. Sometimes branching. Beautiful, eye-catching **bright pink flowers, yellow centres, under 6 mm (¼") across**. Clues to watch for are **notched petal tips and long, slender, hairy tubes** from which flowers project. **Leaves neatly packaged in pairs**. Open, grassy meadows and edges of moist areas.

RANGE BC: Lower elevations of southern B.C. East and west of Cascades.

RANGE WA: Habitats as for B.C.

Primrose Family, *Primulaceae*

SHOOTINGSTARS
Dodecatheon spp.
Peacock

Note: A white shootingstar is on p. 175.

Such attractive flowers! Of 8 species found in B.C. and Washington, northern shootingstar, *D. frigidum*, is found only in northern B.C. and poet's shootingstar (below) has a limited range in Washington. Although trying to distinguish some of the 8 species requires magnification and considerable imagination, those listed here are recognizable mainly by leaf shape. As distinctive as they are, shootingstars vary greatly within a species, so unless you have a good botanical key, probe, magnifying glass and fresh specimens, expect some confusion.

Some shootingstars have 1 or 2 small, pink or purple flowers; others display a half-dozen or more gaudy blooms. Some are found blooming standing in ice water. Fortunately, flower shape remains true to form: Gaudy pink-to-purple petals stream out like a headdress. Dark stamens and style cling together to form a spear-like point. Yellow petal bases are a final decorative touch, but there is a purpose: such an unusual flower shape requires a special technique for pollination: a bumblebee grasps this yellow band while hanging upside down. It then gives a quick buzz of its wings, which shakes pollen out of the flower's anthers (dark purple tubes) and onto its abdomen. The bee now scratches this pollen backwards into sacs on its legs. But some is left on the abdomen—when the bumblebee visits the next shootingstar, the thin stigma protruding from the tube is placed perfectly to receive the pollen.

Shootingstars generally prefer moist areas and may cover a meadow with thousands of blooms. However, they are even in arid sagebrush areas, concentrated where moisture persists a little longer, they bloom and go to seed rapidly. They are a very special tribute to spring.

RANGE BC: Generally throughout at lower elevations; a few species extend to alpine.

RANGE WA: Habitats as for B.C.

DESERT SHOOTINGSTAR [273], *D. congugens*: Very similar to few-flowered (next page) but in **sagebrush and ponderosa pine ecosystems**. Usually 5–20 cm (2–8") high and showy. Can bloom in early April with first arrow-leaved balsamroots. Flower has yellow ring; yellow part way down the dark tube. **Leaves narrow, light green and with distinct stem**. Widespread on drier habitats through Okanagan Valley of B.C. and into Washington's Okanogan.

POET'S SHOOTINGSTAR, *D. poeticum*: Exceptionally rich flower colour: **bright pink-purple petals and showy yellow ring**. Leaves might be confused with those of Jeffrey's and few-flowered shootingstars but **stem and leaves are finely glandular and hairy**. Possible suggestion of teeth on some leaves. Note **limited range**: east of Cascades, but only on moist slopes close to Columbia River. Both sides of river have park areas in which it may be found, e.g., Tom McCall Nature Reserve. Not in B.C. Blooms late March to early spring.

Primrose Family, *Primulaceae*
HENDERSON'S SHOOTINGSTAR [269]
Dodecatheon hendersonii
Broad-leaved Shootingstar

The **only shootingstar with oval leaves and short stem**. The most common shootingstar west of the Cascades. Blooms in early spring, with spring gold, trillium, miner's lettuce, skunk cabbage, field chickweed and gold star. May bloom for 2 months. Open areas, rocky knolls, forest glades.

RANGE BC: Common in Gulf Islands ecosystem.

RANGE WA: Common southwards into Oregon.

Primrose Family, *Primulaceae*
JEFFREY'S SHOOTINGSTAR
Dodecatheon jeffreyi
Tall Mountain Shootingstar

JEFFREY'S FEW-FLOWERED

Note that the **leaves slowly narrow to a wide stem**. So do those of poet's shootingstar (above), but the leaves of Jeffrey's are glandular and finely hairy. Check ranges too. Its height, which can reach 50 cm (20"), is above average compared to most other shootingstars. **Leaves, to 15 cm (6") long, are dull green. Flower with a white ring, not a yellow one**. A preference for damp mountain meadows and stream banks. Blooms April to June, depending on elevation—it ranges to high mountains.

RANGE BC: Most abundant west of the Cascades, from Alaska southwards. Also in southeastern B.C.

RANGE WA: Most abundant west of the Cascades. Also in eastern Washington.

Primrose Family, *Primulaceae*
FEW-FLOWERED SHOOTINGSTAR
Dodecatheon pulchellum
Dodecatheon pauciflorum

Leaf narrows gradually to a long, thin stem. While this plant may be 50 cm (20") tall at lower elevations, it can dwarf to 10 cm (4") high when it stands in ice water at the edge of a subalpine snowbank. Generally leaves are light green and from 5 to 10 cm (2 to 4") long. **Flower with 2 yellow rings**. Moist meadows, stream banks and shady coastal bluffs, but adaptable to high elevations.

RANGE BC: Widespread.

RANGE WA: Widespread.

Morning-glory Family, *Convolvulaceae*
BEACH MORNING-GLORY [270]
Convolvulus soldanella
Beach Bindweed

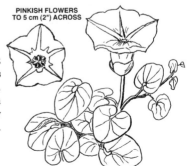

PINKISH FLOWERS
TO 5 cm (2") ACROSS

A native of **coastal dunes and beaches**, but relatively scarce. The **showy pinkish flowers open on sunny days** but may close on dull days. Clumps are usually fairly compact, at 60–90 cm (2–3') across. The thick leaves are alternate, fleshy and sand-hugging, like so many other dune plants. May be in bloom from May to September.

RANGE BC: Sporadic, along the coast.

RANGE WA: Sporadic, along the coast.

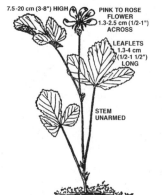

5-10 cm (2-4") HIGH

PINK 'TWIN' FLOWERS
6 mm (1/4") LONG

FINE → NOTCHES

GLOSSY EVERGREEN LEAVES 1.3 cm (1/2") LONG

Honeysuckle Family, *Caprifoliaceae*
TWINFLOWER
Linnaea borealis

So dainty and gay with flowers during spring and summer. that it is **usually thought of as a flower**, although it is a **shrublike woody vine**.

You will not go far into shady, cool woods without seeing **thin, spreading vines**, perhaps 0.9–1.2 m (3–4') long, crawling over the forest floor. **Little, evergreen leaves, less than 1.3 cm (½") long and almost round**; shiny dark green above, much paler on underside. **Several well-separated nicks or teeth along upper half of leaves** will clear up any identification problem. Occasionally a dense mat with hundreds of flowers.

In June and July, numerous slender flower stems shoot up from the vine. Each carries a **pink 'twin' flowers** so **delicately scented** that their fragrance will not soon be forgotten. Linnaeus, famous Swedish botanist and founder of the current system of botanical classification, chose twinflower as his favourite plant.

RANGE BC: Cool, moist woods throughout, from sea level to timberline.

RANGE WA: As for B.C.

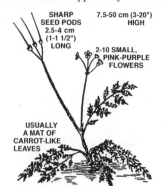

7.5-20 cm (3-8") HIGH

PINK TO ROSE FLOWER
1.3-2.5 cm (1/2-1") ACROSS

LEAFLETS
1.3-4 cm
(1/2-1 1/2")
LONG

STEM UNARMED

Rose Family, *Rosaceae*
NAGOONBERRY [274]
Rubus arcticus
Arctic Raspberry

Crinkled, 3-parted leaves and **pink-to-rose blossom with narrow petals** are typically 'raspberry.' However, this is an unarmed dwarf representative of the family that seldom reaches 15 cm (6") in height. **Technically a shrub**, but difficult to recognize as such. **Bright red berry of fair flavour**. In May and June, dainty pink blooms bring welcome dabs of warm colour to the fresh green of spring growth. Most commonly on borders of **mossy, boggy places**. 'Nagoonberry' is a mystery name.

RANGE BC: Wetter portions east of Cascades to Rockies and northwards to Alaska. From low to high elevations. Wells Gray Park. Lake Louise, Alberta.

RANGE WA: Apparently does not reach Washington.

SHARP SEED PODS
2.5-4 cm
(1-1 1/2")
LONG

7.5-50 cm (3-20")
HIGH

2-10 SMALL,
PINK-PURPLE
FLOWERS

USUALLY
A MAT OF
CARROT-LIKE
LEAVES

Geranium Family, *Geraniaceae*
COMMON STORK'S-BILL [276]
Erodium cicutarium
Filaree, Heron's-bill

From Europe, easily passed off as a little weed; appears widely scattered on dry, open ground that could support little else. It grows as a mat of finely cut leaves with flower stems 2.5–5 cm (1–2") high. However, may reach 30 cm (12") or more in height and be a mass of dark green fern-like leaves. **2–10 small, bright, pink-to-purplish flowers per stem. Has 5 stamens, matching the 5 petals, while geraniums have 10. Long, sharp seed pods, 2.5–4 cm (1–1 ½") long**, are the most distinctive feature. They stick up in groups, reminiscent of a bird's nest full of hungry, long-billed young. Flowers and seed pods may be found at the same time on a plant. The flowers, once recognized, will be seen in an amazing number of places. Usually blooms March–May but flowers might be found all summer.

RANGE BC: Common throughout on drier ground, along road edges and in waste places.

RANGE WA: Common on drier ground, road edges, waste places throughout.

Geranium Family, *Geraniaceae*
GERANIUMS
Geranium spp.

Note: A white-flowered geranium is on p. 182.

There are many geraniums in our area; except for the white-flowered one, they are all pinkish to purple in colour. A broad identifying feature is petal length: **0.6–1.3 cm (¼–½") petals for small-flowered geraniums and petals twice as long, 1.3–2.5 cm (½–1"), for the others.** The different species have minute floral differences that are beyond the scope of this book. Many have 'buttercup' leaves—quickly visible clues.

BICKNELL'S G., *G. bicknellii*: **Small, pink-purple flowers** and **seeds in pairs**; **bristle-tipped sepals. Deeply 5-parted leaves.** Dry areas, from valleys to mountains, east of the Cascades. Also southeastern Vancouver Island, Gulf and San Juan islands.

CAROLINA G., *G. carolinianum*: **Small, pink flowers** with **slightly notched petals. Style tips are yellowish green. Leaves round-tipped.** Dry, rocky places; valley bottoms to mountains. South-central B.C. and southwards into Washington. Vancouver Island, Gulf and San Juan islands.

DOVEFOOT G. [277], *G. molle*: A low plant with very tiny flowers, **deeply notched petals, sepals not bristle-tipped** and **seed-pod stems with a downward twist.** 'Dovefoot' supposedly relates to the shape of the leaves. Does anyone have a dove's foot? Southwestern B.C. but with much wider distribution southwards.

NORTHERN G., *g. erianthum*: **Large flowers, sometimes bluish**, among **large, dissected leaves.** From middle-mountain to subalpine in northern B.C. Coastal Mountains. Not in Washington.

ROBERT G. or **HERB-ROBERT** [282], *G. robertianum*: Note the **bristles on sepals, entire petals, flowers and seeds in 2s**, dissected leaf. Stems often reddish in colour. **Usually blooming with buttercups.** Dry places and roadsides, from valley bottom to mountain. Common in southwestern B.C. and southwards.

SMALL-FLOWERED G., *G. pusillum*: A European weed with **very small flowers. Flowers and seed pods in pairs.** Waste places, most common west of the Cascades.

Geranium Family, *Geraniaceae*
STICKY GERANIUM
Geranium viscosissimum
Cranesbill

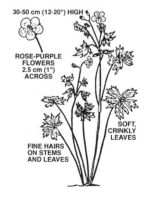

30-50 cm (12-20") HIGH

ROSE-PURPLE FLOWERS 2.5 cm (1") ACROSS

SOFT, CRINKLY LEAVES

FINE HAIRS ON STEMS AND LEAVES

Sticky geranium's coarse appearance comes from a branching of **stout stems and lobed leaves, all covered with sticky hairs.** The most attractive feature is the **rose-purple flowers**, from 2.5 to 4 cm (1 to 1 ½") across. **Petals finely veined.** The blooming period is from late May to July.

RANGE BC: East of Cascades, in open forests and aspen glades, from Cariboo southwards to Washington. Scarce on southern Vancouver Island.

RANGE WA: East of Cascades in open forests to 1200 m (4000'). Blue Mountains.

CUT-LEAVED GERANIUM, *G. dissectum*: A rather sprawling plant, often 60 cm (2') high; an introduced weed. **Petals are pink to purplish. Style tips are purple.** The flowers are small—only 6 mm (¼") across. The **leaves are deeply dissected—cut to the midrib**—and could serve as identification. Notice the hairy stems. Like many weeds, this geranium grows on roadsides and in waste places. Common west of Cascades, in B.C. and Washington.

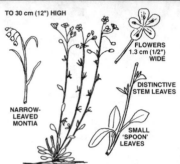

TO 30 cm (12") HIGH

FLOWERS 1.3 cm (1/2") WIDE

DISTINCTIVE STEM LEAVES

NARROW-LEAVED MONTIA

SMALL 'SPOON' LEAVES

Purslane Family, *Portulacaceae*
SMALL-LEAVED MONTIA [278]
Montia parvifolia
Streambank Springbeauty

A simple little wildflower—with great potential for confusion. **Generally pink flower, with darker pink veins,** but you might judge it **white, with no veins, on Gulf and San Juan islands.** Shape, however, is constant. **Petals are slightly notched. Little, alternate, worm-like clusters of leaves on stem** are a clincher. Furthermore, note **basal cluster of small 'spoon' leaves.** Spindly stems to 30 cm (12") high in shade. In drier areas, such as under Douglas-fir, may be only a few centimetres (an inch) above the ground, numerous thin stems lying flat; a final twist presents flowers to the light of day. Then again, a shady rock face rising high in a forest may display a veritable shower of flowers. **Many new plants sprouting from runners of older ones** result in abundant bloom.

    RANGE BC: Common in moist or shady places, at low to medium elevations.

    RANGE WA: Moist habitats, wet rocks and stream sides; in lower forest areas.

    NARROW-LEAVED MONTIA, *M. linearis*: **Tiny white 'bell' flowers—under 6 mm (¼") wide—may be missed. Row of flowers hangs from 1 side of stem.** Long, narrow leaves. Fairly common in lowlands; favours sandy areas. Southern B.C., through Washington.

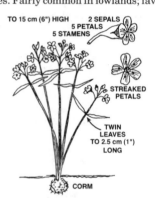

TO 15 cm (6") HIGH

2 SEPALS
5 PETALS
5 STAMENS

STREAKED PETALS

TWIN LEAVES TO 2.5 cm (1") LONG

CORM

Purslane Family, *Portulacaceae*
WESTERN SPRINGBEAUTY
Claytonia lanceolata

Could have white flowers (see p. 171 for details) or more rarely, orange ones. Quite widespread with **beautiful light pink petals, darker lines.** Form varies; one shown is typical. **2 sepals, 5 stamens and 5 petals; tips very slightly notched. Pair of narrow leaves below flower stem.** Moist ground in sagebrush country to alpine zone.

    RANGE BC: East of Cascades, in south.

    RANGE WA: Moist, open forest of Columbia Gorge at medium elevations. Ponderosa pine forests. **Blooms in early spring.**

    ALASKA SPRINGBEAUTY, *C. sarmentosa*: Usually a definite pink; petals more deeply streaked. Mountains of northwestern B.C. Not in Washington.

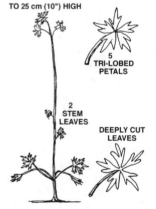

TO 25 cm (10") HIGH

5 TRI-LOBED PETALS

2 STEM LEAVES

DEEPLY CUT LEAVES

Saxifrage Family, *Saxifragaceae*
SMOOTH WOODLAND STAR [275]
Lithophragma glabra
Smooth Prairie Star

Note: 3 other early *Lithophragma*s are keyed as 'white,' with notes about companions (see p. 174).

    Flower colours can vary, so check other woodland stars for comparison. *Lithophragma*s all have deeply lobed flowers—recognize this one by **deeply cut lowest leaves branching from the stem near the ground,** and by the fact that there are **only 2 short-stemmed stem leaves.** To 25 cm (10") tall or even double that. During a cool, wet spring, may bloom for 2 months. **Petals age to white.**

    RANGE BC: Dry, grassy slopes; lowlands to subalpine heights. Southeastern Vancouver Island and Gulf Islands. South-central B.C.

    RANGE WA: East of Cascades, Chelan and Kittitas counties. Columbia Gorge.

Phlox Family, *Polemoniaceae*
SPREADING PHLOX [280]
Phlox diffusa
Carpet Pink

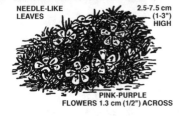

NEEDLE-LIKE LEAVES

2.5-7.5 cm (1-3") HIGH

PINK-PURPLE FLOWERS 1.3 cm (1/2") ACROSS

Although this beautiful mat of a plant is easy to identify, note that moss campion (next page) has several similar features. However, spreading phlox **does not have notched petals**. If you catch spreading phlox in full bloom, the sight can amaze you. The many dozens of pink flowers grow so closely together they almost hide the **ground-hugging mat of small, spiny leaves, which are fused in pairs at their bases.**

Flower colours can range from a seldom-seen white through shades of pink and lavender. The **5-petalled flowers open flat and are about 1.3 cm (½") across.** When in full bloom they have a **noticeable perfume**. The nectar at the base of each 'trumpet' attracts a variety of insects. Since the altitudinal range is quite variable, you may find plants in bloom at different times throughout early summer.

Phlox is Greek for 'flame,' which nicely describes the brilliant colours of this species.

RANGE BC: Dry ground from middle-mountain forests to alpine heights. Across southern B.C. Scarce in the dry interior regions.

RANGE WA: Across the state in dry, open areas; high forests to alpine terrain. Cascades.

TUFTED PHLOX, *P. caespitosa*: **Sometimes white** (see p. 168). Very similar to spreading phlox (above) but **loosely erect (tufted)**. Botanical differences are slight, but the difference in ranges does the trick. Found **scattered under ponderosa pine** or here and there with antelope bush, death camas and hound's tongue, but will spread into low mountain forests. Southwards from eastern B.C. to central and southeastern Washington.

Phlox Family, *Polemoniaceae*
LONG-LEAVED PHLOX [281]
Phlox longifolia

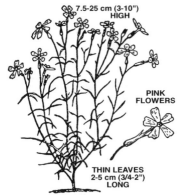

7.5-25 cm (3-10") HIGH

PINK FLOWERS

THIN LEAVES 2-5 cm (3/4-2") LONG

This abundant phlox is a **welcome spring wildflower of the dry interior**. It is at home on the lower mountain slopes of ponderosa pine and sagebrush areas, taking over where buttercups, yellow bells, and balsamroot have bloomed earlier or are still in bloom. The plant forms a rounded clump 25 cm (10") high and is covered with a **mass of small pink or lavender flowers**. Usually, that is; to confuse matters, the **flowers may be pure white**, especially in Washington. Most phloxes have a **long flower tube; the green sepals are almost the same length** for long-leaved phlox. Flowers are 1.3 cm (½") across. Leaves are long and thin and up to 5 cm (2") long. Blooms from late April into June. Afterwards, this phlox becomes quite inconspicuous amongst the range plants.

RANGE BC: South-central B.C. Lower elevations of the Okanagan Valley.

RANGE WA: Dry rocky terrain east of the Cascades from lowlands to mountain forests.

SHOWY PHLOX, *P. speciosa*: Plants to 40 cm (16") high, with **narrow, opposite leaves to 7.5 cm (3") long**. Easily distinguished by its **notched petals**. The bright pink flowers are typically sized for phloxes: about 1.3 cm (½") across. They are in bloom from April into June and have a **faint perfume**. From ponderosa pine and sagebrush areas of southern B.C. southwards through similar habitats into Oregon. Yakima Canyon, Steptoe Butte, eastern half of Columbia Gorge.

A large showy phlox can illustrate the confusion that results from trying to fit particular specimens into a definite slot. North of Ellensburg on U.S. 97, a considerable length of sidehill is bright with large clumps of pure white flowers from mid-April to mid-May. The flowers are almost 2.5 cm (1") across, petals are notched and the sepals—disregarding the typical pattern—are as long as the flower tube instead of half the length. And yet, according to good advice, it has to be showy phlox.

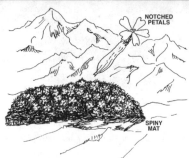

Pink Family, *Caryophyllaceae*
MOSS CAMPION [279]
Silene acaulis

Note: White *Silene*s are on p. 189.

Among the wealth of flowers that adorn high places, this one is sure to be remembered. Sometimes only a **domed or flat little mat** that your hand could cover—or perhaps bigger, but still small enough to span with a wide-brimmed hat. However, **bright green leaves**—tiny enough to mistake for moss—complement massed **bright pink flowers** to create a miniature floral masterpiece. Its beauty is further enhanced by its dramatic setting—often wedged between rock slabs.

Flower colour varies from light pink to deep rose; sometimes a lavender hue or even (rarely) white. Flowers are less than 1.3 cm (½") across, single and appear to sit tight against **short, thin, stiff, spiny leaves**. Moss campion is a treasure of the **alpine tundra**.

RANGE BC: From Alaska southwards. In most of the high mountain parks.

RANGE WA: Southwards to Oregon. A distinctive and appealing plant, in most of the high mountain parks except the Olympics. Mt. Spokane at 1740 m (5800'), near the summit.

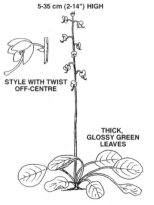

Heather Family, *Ericaeae*
PINK WINTERGREEN [283]
Pyrola asarifolia
Pink Pyrola, Bog Wintergreen

Note: White- to cream-coloured *Pyrola*s on p.211.

The **only pink *Pyrola***—overshadows other *Pyrola*s in both size and lovely colour. Like most *Pyrola*s, has a distinct **basal spray of thick, leathery, glossy green leaves** that retain their colour over winter, hence 'wintergreen.' **Drooping flowers arranged singly along and around stem**; faint perfume. **Thick, waxy petals are greenish white to red. Long, protruding beak or style with downward twist**. Pink wintergreen spreads its attractiveness across moist, wooded sites from lowlands to near subalpine heights. Blooms July–August.

RANGE BC: Common throughout.

RANGE WA: Common throughout.

Heather Family, *Ericaeae*
PRINCE'S PINE [284]
Chimaphila umbellata
Pipsissewa

Often grows in association with *Pyrola*s in **cool evergreen forests**. 12.5–25 cm (5–10") high, with attractive, **sharply toothed, leathery evergreen leaves that cluster around the stem in loose whorls. 3–9 flowers, bunched near the top of the stem**, bloom from June to August. Waxy petals, ranging from pink to white, give an artificial look.

RANGE BC: Throughout; in cool, damp, evergreen forests. Manning Park. Common northwards as far as Prince George.

RANGE WA: Throughout; in cool, damp, evergreen forests.

MENZIES'S PIPSISSEWA [285], *C. menziesii*: **Smaller than prince's pine**, to 15 cm (6") high. Fewer leaves, **1–3 flowers that may be pale enough to be called white**. Petals usually reflexed. Coniferous forests from B.C. southwards to California.

Sunflower Family, *Asteraceae (Compositae)*
RUSH-LIKE SKELETON-PLANT
Lygodesmia juncea
Rushpink

An oddball type of plant, as you might infer from its 2 common names. Its height, less than 40 cm (16"), could allow it to be lost among the grasses but for its **small, pink flowers. Single stems branch, a slender flower at each tip.** 5-petalled flowers, usually a half-dozen or so, form a loose, flat-topped head. **Seedheads form a tight, bristly ball.**

Finely ridged, slim stems have inconspicuous scale-like leaves, so the appearance is both skeletal and rush-like. **Broken stems exude a milky juice.** The only plant of its genus, its design makes it a real curiosity. Most commonly found with **sagebrush and cactus,** it blooms during summer.

RANGE BC: Southern;. especially with sagebrush or cactus. Lillooet, Oliver, Invermere.

RANGE WA: Open, arid places fringing sagebrush habitat.

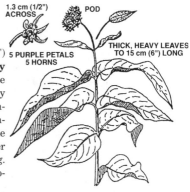

15-30 cm (6-12") HIGH

PINK
5 PETALS

RUSH-LIKE
STEMS

Milkweed Family, *Asclepiadaceae*
SHOWY MILKWEED
Asclepias speciosa

Coarse-looking, stout clusters 0.6–1.2 m (2–4') high, along roadsides and in fields. **Thick, fleshy leaves to 15 cm (6") long.** The smallest nick in the finely hairy leaves or stem releases a thick, milky fluid—protection against sharp-clawed climbing insects. However, you might notice several colourful insects nibbling quite happily on these parts. One might be the larvae of the monarch butterfly, another a kind of beetle and yet one more, the milkweed bug. Somehow they have a method for overcoming the potential disaster of becoming embedded in latex.

1.3 cm (1/2")
ACROSS

POD

5 PURPLE PETALS
5 HORNS

THICK, HEAVY LEAVES
TO 15 cm (6") LONG

Light-pink-to-purplish flowers bunch into a knobby head a few centimeters (inches) across. Flower is complicated: **5 purple petals,** with **5 curved horns protruding from the central stamen tube.** Usually blooms July–August. In fall, **long seed pods release thousands of silky parachutes.** Aboriginals ate the tender, young shoots like asparagus.

RANGE BC: Driest parts. Roadsides. Oliver, Ashcroft and eastern Manning Park.

RANGE WA: Common roadside plant in association with ponderosa pine and sagebrush.

Mallow Family, *Malvaceae*
STREAMBANK GLOBEMALLOW
Iliamna rivularis
Mountain Hollyhock

Perhaps you have marvelled at the **large, pink 'hollyhock' flowers.** Certainly beautiful enough to warrant a close look. Usual form is a **thick, robust perennial bush about 1.5 m (5') high,** topped with spikes of blossoms in pink to lavender. Blossom is a full 2.5 cm (1") across. **5 petals surround an unusual style** that is almost a miniature flower in itself, with **many stigma tips or side branches.** Blooms mid-June through July. Complete identification by noting **large 'maple' leaves to 12.5 cm (5") wide.**

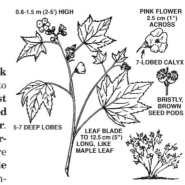

0.6-1.5 m (2-5') HIGH

PINK FLOWER
2.5 cm (1")
ACROSS

7-LOBED CALYX

BRISTLY,
BROWN
SEED PODS

5-7 DEEP LOBES

LEAF BLADE
TO 12.5 cm (5")
LONG, LIKE
MAPLE LEAF

Usually found on good soils near streams or meadows, but also on high mountain slopes.

RANGE BC: Sporadic occurrences in southern B.C. east of Cascades.

RANGE WA: East of Cascades through Washington to Oregon. Pend Oreille River valley.

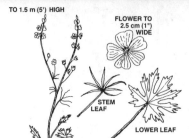

TO 1.5 m (5') HIGH

FLOWER TO 2.5 cm (1") WIDE

STEM LEAF

LOWER LEAF

Mallow Family, *Malvaceae*
OREGON SIDALCEA
Sidalcea oregana
Oregon Checker-Mallow, Marsh Hollyhock

Perhaps you will get just a glimpse of a few light pink flowers among sparse vegetation on a dry hillside, without much of a clue as to what they are attached to. Looking closer, you find that there are **1 or 2 spindly stems to 1.5 m (5') tall, each with a spike of clustered attractive flowers**. They have a 'mallow' look to them and could remind you of the flowers of streambank globe-mallow (pevious page). Flowers are about 2.5 cm (1") across and are light pink with darker veins; petals are lightly lobed.

Generally **round in outline, but with big-toothed, deep lobes, the basal leaves are large**, to 7.5 cm (5") across. Leaves higher on the stem are smaller and finely fingered.

This is a plant of drylands such as the sagebrush and ponderosa pine habitats found east of the Cascades. However, it is adaptable, as shown by its occurrence in damp areas of the Columbia Gorge.

RANGE BC: Sparse in south-central B.C.

RANGE WA: Relatively abundant through eastern Washington. Draws leading to the Snake River.

HENDERSON'S CHECKER-MALLOW, *S. hendersonii*: Although comparatively rare throughout its limited range, this **showy deep pink 'hollyhock'** deserves mention. Its **flowers are about 2.5 cm (1") across**. The **5 broad petals are shallowly lobed at the tip**. A large number of yellow stamens add to the flower's beauty. The short-stemmed flowers are scattered along the top part of the stem. A tall plant, at 0.6–1.2 m (2–4'), it has **hollow stems** whereas those of Oregon sidalcea (above) are solid. **Mostly in coastal tidal marshes**, flats and meadows at low elevations; southern Vancouver Island, Gulf Islands and adjacent mainland southwards to Oregon.

15-50 cm (6-20") HIGH PINK-PURPLE FLOWERS WITH CREAM CENTRE

WHITE, FEATHERY SEEDHEAD

Rose Family, *Rosaceae*
OLD MAN'S WHISKERS [272]
Geum triflorum
Long-Plumed Avens

This plant is a real puzzle to colour-code! The overall impression of **flower colour can range from pink to purple**. The **globe of 5 almost-unnoticeable yellow petals is enclosed by 5 arching sepals and other, shorter ones**.

Sometimes not a particularly plentiful wildflower, but at other times very abundant. Old man's whiskers is distinctive enough to command attention and can not be confused with anything else. A plant of open slopes at low and middle elevations, where it may be found with **companions such as ponderosa pine, lupines and sticky geranium**. The bright green leaves grow in a tuft from a thick stem and are deeply and finely lobed. The stems usually bend at the top, allowing most flowers to nod.

The **purplish stems bear 3 (*triflorum*) peculiar flowers, which appear pink because of the conspicuous long sepals that flank the paler petals**. After blooming in late May or June, the flower turns into a **white-plumed seedhead** very like that of the western anemone (p. 176) or pasqueflower (p. 300).

RANGE BC: East of the Cascades, ranging in open woodlands from the Peace River southwards through the Cariboo; also on the edge of ponderosa pine into mixed forest.

RANGE WA: Eastern and western Washington; favours the ponderosa pine ecosystem but ranges from moist sagebrush habitats to high mountains. Blue Mountains.

Iris Family, *Iridaceae*
SATIN-FLOWER [287]
Sisyrinchium douglasii
Grass Widow

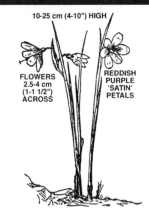

10-25 cm (4-10") HIGH

FLOWERS
2.5-4 cm
(1-1 1/2")
ACROSS

REDDISH
PURPLE
'SATIN'
PETALS

Note: A close relative of blue-eyed (p. 298) and golden-eyed (p. 226) grasses, which bloom much later.

The southern part of Vancouver Island, the Gulf and San Juan islands and the Puget Sound area produce an abundance of spring wildflowers. Camases, stonecrops, saxifrages, fawn lilies, violets, trilliums and shootingstars are some common ones; no doubt satin-flower should be included. Almost every rocky bluff with moss on it is decorated in April by clumps of beautiful satin-flower. From a **spray of long, thin leaves** that look like stout grass blades, **1 or 2 stems** arise, bearing **several reddish purple flowers** 2.5–4 cm (1–1 ½") across. Unlike most other flowers, they last just a few days, during a very short **March–April blooming** season.

Botanist David Douglas found this plant in 1826, near the Columbia River's historically famous native fishing area of Celilo Falls (now flooded); satin-flower can still be found nearby.

RANGE BC: West of Cascades: southern Vancouver Island, Gulf Islands.

RANGE WA: West of Cascades, southwards from San Juan Islands to Oregon. Columbia Gorge and extending northwards along eastern base of Cascades to northern Kittitas County. U.S. 97 south of Toppenish. Klickitat County; high ridges to over 1500 m (5000') elevation.

Iris Family, *Iridaceae*
GRASS WIDOW [288]
Sisyrinchium inflatum

15-40 cm (6-16")
HIGH

BLUE OR
PURPLE-TO-
REDDISH
FLW. 2.5-4 cm
(1-1 1/2")
ACROSS

Closely resembles satin flower, but has **sharp-pointed petals and grass-like leaves shorter than the main stem**. Grass widow is a prolific bloomer. Its large, starry blooms beautify moist slopes. One sees a wide variety of colours: **flowers may range from pale purple to blue or pink but are also often white**, which adds a touch of fantasy.

RANGE BC: Quite rare; east of Cascades. Chase, Castlegar.

RANGE WA: Sagebrush ecosystem, through the bunchgrass ecosystem and into the ponderosa pine forests, where it is most common. Wilbur, Spokane, lower edges of Blue Mountains.

6 PETALS (UMBEL)

Lily Family, *Liliaceae*
DOUGLAS'S ONION
Allium douglasii

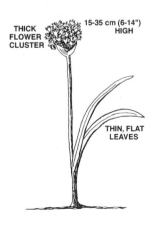

THICK
FLOWER
CLUSTER

15-35 cm (6-14")
HIGH

THIN, FLAT
LEAVES

Note: There is a white-flowered onion on p. 184.

A beautiful variation on typical wild onions, with massed flowerhead, but with usual form. Tends to form a **mass of pink. Usually a noticeable onion smell**. Squarish stems. The **2 grooved leaves persist unwithered**, unlike with many other wild onions. **Tiny, star-shaped flowers, so numerous they hide the umbel structure**.

RANGE BC: Not in B.C.

RANGE WA: Along border of Columbia Basin and southwards. Kittitas County, Columbia Gorge and Blue Mountains.

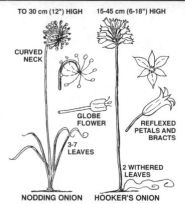

TO 30 cm (12") HIGH   15-45 cm (6-18") HIGH

CURVED NECK

GLOBE FLOWER

3-7 LEAVES

NODDING ONION

REFLEXED PETALS AND BRACTS

2 WITHERED LEAVES

HOOKER'S ONION

Lily Family, *Liliaceae*
NODDING ONION [289]
Allium cernuum

Nodding onion, which usually stands about 30 cm (12") high, is easily recognized by its **nodding flowerhead of a dozen or so pink blooms**. The flower has **6 stamens that protrude far out of the pink cup on hair-thin stems**. Each petal and sepal has one thin vein down its centre line. The few 'grass' leaves rise from clustered bulbs something similar to a green onion—the **onion smell** is there, too. **Possibly the most common, most widespread** of the almost 30 wild onions in our area. It prefers dry soils and open spaces. Generally blooms will be found May 15–July 15; sometimes in late June along with harvest brodiaea in rocky coastal areas. Native peoples found the bulbs edible after steaming them in covered pits.

RANGE BC: Central and coastal B.C.

RANGE WA: Central and coastal Washington and into Oregon. Common at middle-mountain to subalpine heights along the Deer Park road in the Olympics.

Lily Family, *Liliaceae*
HOOKER'S ONION [290]
Allium acuminatum

Easily distinguished from nodding onion (above) if you notice that the **flowers are mostly erect**. Also note that both **sepals and petals are fine-pointed and curl outwards**. At flowering, the **2 leaves have withered**; it takes a search to find the remnants. Most plants range from 15 to 45 cm (6 to 18") in height and carry a **beautiful head of rose-purple flowers**. The **stems match the flowers' colour**. In places this onion is so abundant it **gives a reddish blush to a hillside** and an onion smell to the air. May is the best month to see blooming.

RANGE BC: Favours very dry ground on Vancouver Island and the Gulf Islands.

RANGE WA: Northwest Washington and east of Cascades, southwards to California.

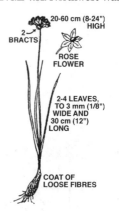

20-60 cm (8-24") HIGH

2 BRACTS

ROSE FLOWER

2-4 LEAVES, TO 3 mm (1/8") WIDE AND 30 cm (12") LONG

COAT OF LOOSE FIBRES

Lily Family, *Liliaceae*
GEYER'S ONION
Allium geyeri

The odour might tell you that you have found onions, but telling onions apart can be very difficult. Many are separated by the cell pattern of the dried husk that covers the onion bulb. Recognize Geyer's by the **2–4 narrow leaves that remain green during the blooming** period and by the **cluster of rose-coloured flowers, often mixed with pink-tinged bulblets**. The 2 broad bracts under the flowerhead are usually joined on one edge. Notice that the **stamens are usually shorter than the tiny, triangular petals**.

RANGE BC: Variable, from moist ground to exposed rocky places. Sporadic on Vancouver Island. East of the Cascades in southern B.C.

RANGE WA: Variable from moist ground to exposed rocky places. East of the Cascades. Common in eastern Washington north of the Snake River and westwards to the Cascades. Kittitas and Yakima counties.

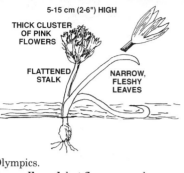

5-15 cm (2-6") HIGH

THICK CLUSTER
OF PINK
FLOWERS

FLATTENED
STALK

NARROW,
FLESHY
LEAVES

Lily Family, *Liliaceae*
SCALLOPED ONION [294]
Allium crenulatum
Notched Onion, Olympic Onion

2 thick, flat leaves protrude from short, squarish stem; twisted appearance. Distinctive low, clustered umbel heads. Small, pinkish flowers. Dry, rocky habitat. Bulbs deeply buried; leaves succulent. Late spring bloomer.

RANGE BC: Rare. Gravelly, rocky places; higher mountains of Gulf and Vancouver islands.

RANGE WA: Cascades. Wenatchee Mountains. Olympics.

ROBINSON'S WILD ONION, *A. robinsonii*: **Like scalloped,** but flowers may be more white. **Limited range**: sand/gravel along Columbia River below Vantage, Washington. Not in B.C.

7 PETALS–MANY

10-20 cm (4-8") HIGH

PALE PINK
FLOWERS
WITH 7 PETALS

4-7 LEAVES
IN WHORL

Primrose Family, *Primulaceae*
BROAD-LEAVED STARFLOWER [292]
Trientalis latifolia
Western Starflower

A forest sprite both neat and delicate, **only 10–20 cm (4–8") tall. Simple, slender stem. Symmetrical whorl of 4–7 thin, oval, glossy green leaves** each 2.5–7.5 cm (1–3") long, above which rise **1–4 thin, graceful stems, each with 1 flower** of charming simplicity. Flowers may appear to float in the air. **Usually pinkish—sometimes white. Pointed petals; typically 7,** but possibly 5–8. In shady, moist forests from May to July.

RANGE BC: Common on Vancouver Island and adjoining mainland.

RANGE WA: Coastal. Also in low forests of eastern Washington.

NORTHERN or **UPLAND STARFLOWER**, *T. arctica*: **Much smaller than broad-leaved**—to 15 cm (6") tall—but same-sized flower, 1.3 cm (½") across, **waxy white**. On **sphagnum moss or boggy soils**. Blooms June–July. Note **several small leaves on stem**. Cascades southwards to northern Oregon. Locally in eastern Washington.

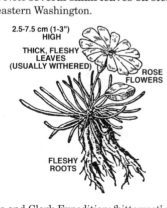

2.5-7.5 cm (1-3")
HIGH

THICK, FLESHY
LEAVES
(USUALLY WITHERED)

ROSE
FLOWERS

FLESHY
ROOTS

Purslane Family, *Portulacaceae*
BITTERROOT [293]
Lewisia rediviva
Rock-rose

Note: More *Lewisias* on next page. Several local ones are quite rare and are omitted.

Montana's state flower. **Abundant spring bloom** in **most-exposed parts** of **arid regions**. Magically appears in May, disappears in same fashion. **As flowers appear, tufts of fleshy leaves dry up**. Thick, fleshy root once used in large quantities by aboriginal peoples. Boiled, they swell and become jelly-like. An early explorer noted a sack of roots would buy a good horse. *Lewisia* commemorates Lewis and Clark Expedition; 'bitterroot' reflects their opinion of the taste. Very thin soils in dry, rocky places of sagebrush areas.

RANGE BC: Sporadic; east of Cascades. Okanagan Valley. Cranbrook.

RANGE WA: Sporadic; east of Cascades. Southwards from Okanogan Valley through Oregon. Grand Coulee region. Mile 74 (km 119) on WA 14 in Columbia Gorge.

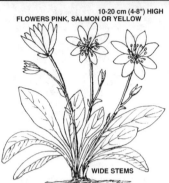

10-20 cm (4-8") HIGH
FLOWERS PINK, SALMON OR YELLOW

WIDE STEMS

Purslane Family, *Portulacaceae*
TWEEDY'S LEWISIA
Lewisia tweedyi

Did you marvel at the beauty of more common bitterroot (p. 277)? Flowers of this close relative can be **pale pink, salmon or pale yellow**; about 4 cm (1 ½") across; **7–9 petals, usually 8**. Usually **several buds or flowers per stem**; unbranched. Clump of thick basal leaves seem to come direct from ground; 10–20 cm (4–8") long, wide, tapering stem. Watch for this beauty in late spring–early summer.

    RANGE BC: Just barely in B.C.; Manning Park.

    RANGE WA: Mostly east of Cascades summit. Rocky slopes, crevices; mostly with sagebrush and ponderosa pine. Wenatchee Mountains. Southern Chelan County and northern Kittitas County.

    THREE-LEAVED LEWISIA, *L. triphylla*: Low—to 10 cm (4") high. **Loses basal leaves by flowering, but retains group of 3 leaves high on stem**, hence name. **Small flowers may be pink or white; 5–9 petals**; about 1.3 cm (½") across. Moist gravel areas, middle-mountain to alpine. Rare in B.C.—on Vancouver Island and in south-central B.C. In Washington, east of Cascades, from sagebrush and ponderosa pine to mountain snowbanks.

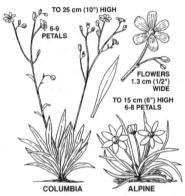

TO 25 cm (10") HIGH
6-9 PETALS
FLOWERS 1.3 cm (1/2") WIDE
TO 15 cm (6") HIGH
6-8 PETALS
COLUMBIA   ALPINE

Purslane Family, *Portulacaceae*
COLUMBIA LEWISIA [291]
Lewisia columbiana

Varies from pink to almost white; 6–9 petals always have veins of darker rose. Prefers harsh surroundings. Fleshy roots force into cracks and crevices to anchor plant. Thick tuft of narrow, fleshy leaves. Leaves remain light green until snowfall.

    Can grow in considerable numbers in **preferred habitat—exposed, thin, rocky soil**—but not always found. May bloom June to August.

    RANGE BC: Both sides of Cascades. High mountain slopes to alpine terrain.

    RANGE WA: Habitats as for B.C. Hart's Pass.

    ALPINE LEWISIA, *L. pygmaea*: Flowers, **6–8 petals**, nestled among leaves, each on short, unbranched stem; about 2 cm (¾") across; **almost pure white through pink to lavender**. Thick roots anchor plant tight to ground. Distinctive **narrow leaves to 15 cm (6") long**. Solitary dwarf near/above timberline; high mountains; southern B.C. and Washington.

TUBULAR FLOWERS

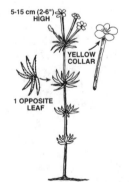

5-15 cm (2-6") HIGH
YELLOW COLLAR
1 OPPOSITE LEAF

Phlox Family, *Polemoniaceae*
BICOLOURED LINANTHUS [295]
Linanthus bicolor
Babystars

Very small. Unusual design—**quickly recognized either by leaves or flowers**. 1 erect stem, or branches at base. **Several 'whorls' of leaves on stem are sets of opposite leaves, each with 3–7 narrow, sharp fingers bearing stiff hairs**. Unusual flower: thin tube over 2.5 cm (1") long supports flat disk of 5 petals. Yellow underside of flower complements overall pink, supporting Latin name. Most common in sunny, open places.

RANGE BC: Southern Vancouver Island and Gulf Islands.

RANGE WA: San Juan Islands. Continues southwards through Washington.

Mint Family, *Lamiaceae*
PURPLE DEAD-NETTLE [299]
Lamium purpureum
Red Dead-nettle, Red Henbit

Small, erect alien, prominent in waste places but also on rock bluffs and in open forests; tenacity and appeal earn recognition. **Often masses—crown of purplish leaves, an unusual colour.** Relationship to mints is shown by **square stems, crinkly leaves** and **tiny, tubular flowers in loose whorls.**

A few small leaves near ground level and a pair about one third up stem. **Top part of stem bears heart-shaped leaves in several whorls—very attractive pinkish to reddish purple flowers**

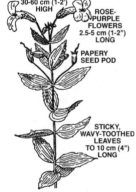

peek out. Best seen with magnification; **2 lips, each with a black dot, and 4 purple dots (stamens) framed by hood petal. Early spring bloomer**, along with shootingstars and *Erythroniums*. 'Dead-nettle' because, unlike the *Urticas* (p. 339), does not sting when touched.

RANGE BC: Most common on waste ground and roadsides; adaptable to other situations. Southern Vancouver Island, Gulf Islands.

RANGE WA: Widespread throughout in habitats as for B.C. San Juan Islands.

COMMON DEAD-NETTLE, *L. amplexicaule*: Like purple dead-nettle, same height and habitat, but **flowers distinctly red-purple, more visible.** Compare leaf shapes; rounded upper ones clasp stem. Frequent in southwestern B.C., sporadic in Washington.

Figwort Family, *Scrophulariaceae*
PINK MONKEY-FLOWER [297]
Mimulus lewisii
Lewis's Monkey-flower, Red Monkey-flower

Often displays a large, **spectacular sheet of colour.** Common in alpine terrain **along wet banks or stream edges.** Stems cluster together; number of **opposite, pink-to-rose or purplish red 'snapdragon' flowers 2.5–5 cm (1–2") long; 5 petals, 2 turn up, 3 down.** Throat has touch of yellow, may be hairy. Succession of new buds keeps it a profusion of bloom July–August. Large, opposite leaves, clasping stems. Parallel ribs add to beauty.

RANGE BC: Wet places; middle to high mountains throughout.

RANGE WA: Habitats as for B.C.; southwards through Oregon.

Phlox Family, *Polemoniaceae*
COLLOMIAS
Collomia spp.

Inconspicuous, with **tubular pink flowers.** Large-flowered collomia, most common and showiest of the genus (p. 260).

NARROW-LEAVED COLLOMIA, *C. linearis*: Small, slender, to 40 cm (16") tall; **stem leaves narrow, without teeth.** Dense clusters of **small flowers usually pink**, can shade to blue or white. Summer bloomer from lowlands to mountain slopes, chiefly east of Cascades. Southern B.C. to California.

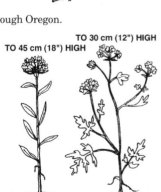

VARI-LEAVED COLLOMIA, *C. heterophylla*: **Leaf shapes vary; most lobed or toothed.** Much-branched. Clusters of **tiny, 'trumpet' flowers of rose to white. Shady, damp areas.** Locally frequent on southern Vancouver Island and southwards west of Cascades in similar habitats. Western part of Columbia Gorge.

TO 60 cm (2') HIGH

Valerian Family, *Valerianaceae*
SCOULER'S VALERIAN
Valeriana scouleri

Note: Similar valerians, but with white flowers, are on p. 190.

Distinctive **globe cluster of pinkish flowers about 2.5 cm (1") across**. Leaves may be lost in neighbouring vegetation but are noticeable because of **3–5 leaflets—terminal one is by far the longest. Small, single, rounded basal leaves**. Blooms April and May; in wet places, along roadsides.

RANGE BC: Southern Vancouver Island and adjacent mainland.

RANGE WA: West of Cascades, southwards to coastal California. Western part of Columbia Gorge.

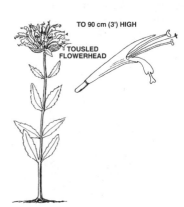

TO 90 cm (3') HIGH

TOUSLED FLOWERHEAD

Mint Family, *Lamiaceae*
WILD BERGAMOT [296]
Monarda fistulosa
Horsemint

Typically mintlike, with **slim, square, erect stems** and opposite leaves to 7.5 cm (3") long. Very attractive, **roundish head of rose-purple flowers surrounded by large, leaf-like bracts**, at stem top. A most unusual design: overall like **tousled mass of short, twisted ribbons**—perhaps a large purple aster gone wild. **Each 'ribbon' is a single flower**; starts off as **long, narrow tube that distends to produce 1 drooping, notched petal**. Rest of tube continues as long sheath for projecting 2 stamens and pistil. No set pattern for flowers—each appears to twist and turn at will.

Summertime plant. Although relatively common, every sighting will be something special. Favours moist soils but will occur on dry hillsides and sagebrush areas.

RANGE BC: Relatively common east of Cascades.

RANGE WA: Not in Washington.

0.6–2.1 m (2-7') HIGH

FLOWERS WHITE-PINK-RED-PURPLE 5 cm (2") LONG

BASAL LEAVES 20-40 cm (8-16") LONG

Figwort Family, *Scrophulariaceae*
FOXGLOVE [298]
Digitalis purpurea

One evening, while travelling, I parked my small camper at a road-widening below a steep, logged-over hillside. I imagined what a desolate sight I would likely see next morning. Instead, there was a once-in-a-lifetime floral spectacle: that mountainside was a sheet of magnificent colour, in hundreds of shades **from white to deepest purple**! Climate and soil were perfect for foxglove.

Single plants or small groups can thrive from valley bottoms to high on remote mountainsides.

Stout, erect stem 0.6–2.1 m (2–7') tall. Long-stemmed, finely toothed basal leaves; progressively shorter-stemmed, smaller stem leaves. Top third of plant is a **cascading column of lovely flowers**; tubular, wide lobed, about 5 cm (2") long. **Throat marked with darker dots**. Long-flowering spike; sequence of blooms early June–August. Source of **heart stimulant digitalis**. *Digitus* is Latin for 'finger,' so botanical name translates as 'purple finger.' Develop your own theory. Roadsides and open areas.

RANGE BC: Most common west of Cascades, but inland here and there.

RANGE WA: Habitats as for B.C., southwards to California.

5-15 cm (2-6") HIGH

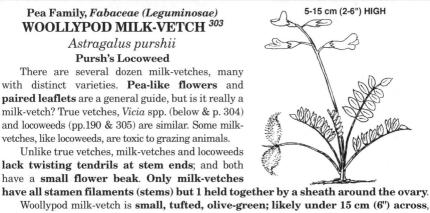

Pea Family, *Fabaceae (Leguminosae)*
WOOLLYPOD MILK-VETCH [303]
Astragalus purshii
Pursh's Locoweed

There are several dozen milk-vetches, many with distinct varieties. **Pea-like flowers** and **paired leaflets** are a general guide, but is it really a milk-vetch? True vetches, *Vicia* spp. (below & p. 304) and locoweeds (pp.190 & 305) are similar. Some milk-vetches, like locoweeds, are toxic to grazing animals.

Unlike true vetches, milk-vetches and locoweeds **lack twisting tendrils at stem ends**; and both have a **small flower beak**. Only milk-vetches have all stamen filaments (stems) but 1 held together by a sheath around the ovary.

Woollypod milk-vetch is **small, tufted, olive-green; likely under 15 cm (6") across**, close to the ground. **Flowers white to yellow, or a shade of pink, rose or purple. 7–19 leaflets**. Distinctive, and among earliest blooms. Eye-catching **fuzzy white seed pods** in May.

RANGE BC: South-central B.C. southwards.

RANGE WA: Grasslands and sagebrush areas, to central California. Vantage area.

Pea Family, *Fabaceae (Leguminosae)*
BROAD-LEAVED PEAVINE [304]
Lathyrus latifolius
Everlasting Pea

Note: Creamy peavine, p. 191; purple, p. 303.

LARGE FLOWER CLUSTER PINK-PURPLE

Boisterous, eye-catching display, and easily identified. Lush. **Climbs over fences and bushes.** Colourful **clusters of pink-purple flowers**, on stems above leaves. **Unusual wide wings border stem. Leaves end in tendrils.** Leaves no wider than on most peavines—why is it 'broad-leaved,' not 'wide-stemmed'? Blooms early June throughout summer. Along coastal bluffs and scattered along roadsides.

WIDE WINGS ON STEMS

RANGE BC: Common around Victoria; along Island Highway; Gulf Islands; Lower Mainland.

RANGE WA: Most abundant west of Cascades. Isolated occurrences on both sides of Columbia Gorge and near settlements throughout much of state.

NARROW-LEAVED EVERLASTING PEAVINE, *L. sylvestris*: An escape very similar to broad-leaved, **also with wide wings on stems. Much smaller, reddish flowers.** Leaflets narrow. Sporadic appearances, especially around habitations and towns.

Pea Family, *Fabaceae (Leguminosae)*
GIANT VETCH [302]
Vicia gigantea

Coarse, strongly zig-zagging; long, heavily ridged, hollow stems overgrow beach debris. **Tight, faded pink-brown flower clusters**, may be bright pink-purple; sometimes seem to be dying. Blooms late spring–summer. **Stem has 18–26 leaflets**, opposite or alternate. Bright green leaflets have very short stems; to 5 cm (2") long. Each **stem ends in several strong tendrils**. Small, **sharp-pointed leaf stipules** (where leaf stem joins main stem). Smooth pods to 4 cm (1 ½") long. **Most common on beach logs; extends up coastal hillsides across logged-over or blow-down areas**.

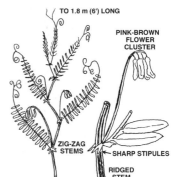

TO 1.8 m (6') LONG

PINK-BROWN FLOWER CLUSTER

ZIG-ZAG STEMS

SHARP STIPULES

RIDGED STEM

RANGE BC: Coastal areas northwards to Alaska.

RANGE WA: Coastal areas southwards to California.

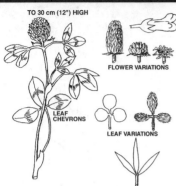

TO 30 cm (12") HIGH

FLOWER VARIATIONS

LEAF CHEVRONS

LEAF VARIATIONS

Pea Family, *Fabaceae (Leguminosae)*

RED CLOVER
Trifolium pratense

Note: Yellow clovers are on p. 232.

There are several dozen clovers in our area; half were introduced from Eurasia. A few native ones are quite rare. How do you distinguish between wildflowers, cultivated flowers and aliens? It becomes a matter of personal judgement. Though they might be expected to be of most interest, **native clovers tend to have small flowers and be very modest in size**, and are thus largely omitted. Rather than pick out a dozen or so species that a keen observer might find, I decided to provide illustrations that should enable anyone to recognize a clover.

Red clover, really deep pink, is the best known. Characteristic **3 leaflets per leaf, each with a whitish chevron on it**. Has **leaves right below flower**, instead of more-usual large, single bract (shape helps identify some species). European.

RANGE BC: Widespread, particularly in farming areas; roadsides, pastures, ditches. Mostly east of Cascades but also abundant in Lower Mainland and on Vancouver Island.

RANGE WA: Habitats as for B.C.

SPRINGBANK CLOVER [301], *T. wormskjoldii*: **Most common native clover**; small, slender, sprawling. Flowerhead is a tuft of **tiny flowers, magenta to pale purple, but white-tipped. Flower sits on spike-pointed circular bract**. Finely toothed leaflets. Long, fleshy roots were highly prized by coastal aboriginals. Stream banks and damp meadows of coastal B.C. and Washington.

WHITE-TIPPED CLOVER, *T. variegatum*: **Flowers red-purple with white or pink tips. Leaflets coarsely toothed. Leaf stipules with ragged leaves**. Mostly west of Cascades, from B.C. southwards to California.

WHITE CLOVER, *T. repens*: Introduced perennial with **white-to-pinkish flowerhead to 4 cm (1 ½")across. Finely toothed leaflets**. Wide range in fields and along roadsides.

CLOVER SPECIES, *Trifolium* spp.: **Flowerhead shape varies**; consists of **small, often 2-toned flowers. Leaves vary greatly**: may be smooth or hairy; some are toothed.

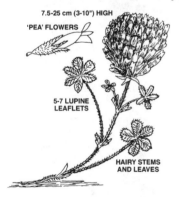

7.5-25 cm (3-10") HIGH

'PEA' FLOWERS

5-7 LUPINE LEAFLETS

HAIRY STEMS AND LEAVES

Pea Family, *Fabaceae (Leguminosae)*

BIG-HEAD CLOVER [300]
Trifolium macrocephalum
Giant's Head Clover

Another of those special plants that raises a thrill when seen for the first time. Perhaps it is the combination of lovely colours, large size and sudden appearance among low growth. **Only 7.5–25 cm (3–10") high**. Usually there are several blooms on short, branching stems. The clover flowerhead is distinctive, being similar to those of the common red and white clovers. The **pinkish ball, to 5 cm (2") across, is made up of tiny individual pea flowers coloured in shades of white, cream and pink**. Fine hairs clothe stems and leaves. The usual blooming time is mid-April to mid-May.

Clovers usually have 3 leaflets per leaf, as suggested by *Trifolium*, but big-head has **5–7 leaflets**; they closely resemble the whorl seen on lupines (see pp. 324–27). Spreads, as do most clovers, by extending its root system, often forming a loose mat perhaps 1.8 m (6') across.

RANGE BC: Not in B.C.

RANGE WA: Thin soils under Garry oak and ponderosa pine in central and eastern parts of Columbia Gorge. Open terrain in ponderosa pine and sagebrush areas of eastern Washington. Sporadic at 1200 m (4000') in Blue Mountains. Especially abundant about 13 km (8 mi.) southwest of Bickelton on road to Goldendale—main spring vegetation, with Howell's triteleia, along roadsides and on rocky plains. Dry roadsides near Ft. Simcoe, Yakima County.

Heather Family, *Ericaceae*
GNOME-PLANT [305]
Hemitomes congestum
Cone-plant

2.5-12.5 cm (1-5") HIGH

PINKISH, SCALY LEAVES

Use imagination to make sense of 'cone-plant,' but 'gnome-plant' may mislead you. **Pinkish overall; dries yellow or brownish.** Saprophytic, nevertheless its **thick, fleshy stem is covered with pale, scaly leaves.** Shape varies from 1 stem with 1 flowerhead to a mass of flowers seemingly bursting from the ground, somewhat fungus-like. Bulky cluster of well-defined flowers: 4 narrow sepals, petals flare into 4-part pattern. Blooms July–August.

RANGE BC: Dry, mossy coniferous forests; middle elevations. Mostly in southwest.

RANGE WA: Olympics and southwards in coastal forests. Mid-Columbia Gorge.

Buckwheat Family, *Polygonaceae*
WATER SMARTWEED [306]
Polygonum amphibium
Water Knotweed

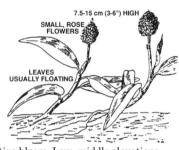

7.5-15 cm (3-6") HIGH

SMALL, ROSE FLOWERS

LEAVES USUALLY FLOATING

May grow so thickly that **short, rose-pink flower spikes blush water surface pinkish.** Roots in mud; lake edges, ponds or backwaters. Stems to 6 m (20') long. Leathery, oblong leaves often not noticed—usually flat on the water. Blooms July to August—can be seen from afar. Know water smartweed for its unusual aquatic nature and attractive bloom. Low–middle elevations.

RANGE BC: Across B.C.

RANGE WA: Across Washington and Oregon.

Figwort Family, *Scrophulariaceae*
THIN-LEAVED OWL-CLOVER
Orthocarpus tenuifolius
Thin-leaved Orthocarpus

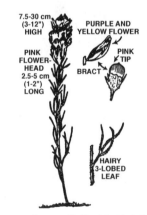

7.5-30 cm (3-12") HIGH

PURPLE AND YELLOW FLOWER

PINK FLOWER-HEAD 2.5-5 cm (1-2") LONG

PINK TIP

BRACT

HAIRY 3-LOBED LEAF

Note: Yellow owl-clovers are on p. 235.

Heads of coloured bracts; **similar to paintbrushes** (pp.235 & 291–92) but generally less colourful. **Usually under 30 cm (12") high. Tiny flowers almost hidden from sight behind bracts. Typically, an erect or curving upper lip and a lower one with 3 divisions.** Of the 10 or so species in our area, a number have become rarities.

This owl-clover, at least in early spring, has **distinctive yellowish green foliage** that stands out against **rabbit-brush-and-sagebrush habitat.** Narrow lower leaves quite hairy. Stem leaves with 3–5 narrow-fingers. Delicately tinted pink **bracts hide tiny yellow-and-purplish flowers 1.3 cm (½") long; very tiny hooked nose or beak.** Blooms May–June. Dry, exposed soils. Bunchgrass and sagebrush zones.

RANGE BC: Keremeos, Cranbrook.

RANGE WA: Okanogan, Douglas and Spokane counties, Blue Mountains.

DWARF OWL-CLOVER, *O. pusillus*: Easily overlooked. Only **5–15 cm (2–6") high; inconspicuous pink-purple flowers behind bracts (may be purplish) often resembling tiny leaves. Finely divided leaves cover stem.** Locally sporadic; damp ground of southern Vancouver Island, Gulf and San Juan islands; southwards through coastal Washington.

MOUNTAIN OWL-CLOVER [309], *O. imbricatus*: 10–20 cm (4–8") high. **Colourful pink-purplish bracts almost hide tiny pink-and-white flowers.** Usually branched near top. Slender, alternate leaves. Late summer bloomer. Gulf Islands region. Common in Cascades.

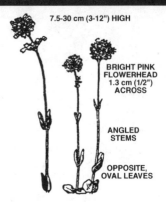

7.5-30 cm (3-12") HIGH

BRIGHT PINK
FLOWERHEAD
1.3 cm (1/2")
ACROSS

ANGLED
STEMS

OPPOSITE,
OVAL LEAVES

Valerian Family, *Valerianaceae*

SEA BLUSH [307]

Plectritis congesta

Rosy Plectritis

Note: A close relative, but with white flowers, is long-spurred plectritis (see p. 196).

Sea blush is an **early spring flower** that blooms during April and May, along with blue-eyed Mary, satin-flower, stonecrops, camases, shooting-stars and white fawn lily. It favours rocky knolls capped by moss.

In better soils, the **opposite, oval leaves** on taller plants may be 5 cm (2") long, but in rocky places they may be only 6 mm (¼") long. **Tiny, pink flowers are clustered together to form a compact, rounded head.**

RANGE BC: Abundant on Vancouver Island. Gulf Islands.

RANGE WA: Southwards through the San Juan Islands and Puget Sound to Columbia River and the Oregon coast.

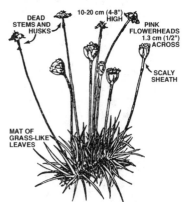

DEAD
STEMS AND
HUSKS

10-20 cm (4-8")
HIGH

PINK
FLOWERHEADS
1.3 cm (1/2")
ACROSS

SCALY
SHEATH

MAT OF
GRASS-LIKE
LEAVES

Sea-pink Family, *Plumbaginaceae*

THRIFT

Armeria maritima

Sea-Pink, Sea Lavender

Anyone who grows the garden variety of thrift will be struck by how similar the wild species is. There is the same **compact clump of needle-like leaves**, about 5 cm (2") long, and **stiff, leafless flower stems**, each with its **pinkish dome of tiny flowers**. Flowers are very small but you might see the 5 petals with their delicate papery lobes. After the flower has faded, its parchment-textured petals stay intact for months, leading to some use for thrift in 'everlasting' decorations. Successive new shoots keep thrift in bloom for several months, beginning in late spring. Flowerheads are 1.3 cm (½") across.

RANGE BC: Open, grassy bluffs along the coast and grassy meadows inland; northwards to Alaska.

RANGE WA: Habitats as for B.C.; southwards to California.

5-15 cm (2-6") HIGH

PINK HEADS
1.3 cm (1/2")
ACROSS

UNBRANCHED,
WOOLLY
STEMS

LEAVES
DENSELY
WOOLLY

Sunflower Family, *Asteraceae (Compositae)*

ROSY PUSSYTOES [310]

Antennaria microphylla

Most of the pussytoes are white (see p. 194). All have the **whitish, matted stems and leaves** characteristic of plants that live in bright sunshine and withstand drought. The fanciful name comes from the resemblance of the soft, rounded flower-heads to the hairy toes of a cat. Rosy pussytoes, which usually forms a mat, is sometimes dwarfed to a few centimeters (inches) in height but may otherwise reach 15 cm (6"). The alternate leaves are very narrow and are almost white with a soft wool. The **flowers bunch in a rounded head of rosy-and-white balls** that get their colour from the papery outer bracts. If the flowers are picked and dried, they will retain their colours for years.

RANGE BC: Widespread throughout, from valley bottom to timberline.

RANGE WA: Habitats as for B.C.

Buckwheat Family, *Polygonaceae*
MOUNTAIN SORREL [311]
Oxyria digyna

Note: Sheep sorrel, with showy spikes of reddish flowers, is on p. 333.

Like the leaves of the closely related sorrels of the *Rumex* spp., those of mountain sorrel have a **sour, acid taste** that is not too pleasant. But they contain oxalic acid, which can interfere with calcium absorption if eaten in excess. Generally from 10 to 35 cm (4 to 14") high, with about 6 thin, rounded, **bright green basal leaves** from 1.3 to 4 cm (½ to 1 ½") wide. In very poor situations it **may reach only 5 cm (2") in height** and have leaves hardly larger than 1.3 cm (½") across.

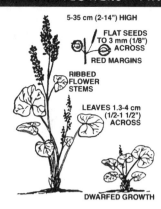

5-35 cm (2-14") HIGH

FLAT SEEDS TO 3 mm (1/8") ACROSS
RED MARGINS

RIBBED FLOWER STEMS

LEAVES 1.3-4 cm (1/2-1 1/2") ACROSS

DWARFED GROWTH

The **small, pink-to-green flowers** are inconspicuous and unattractive; they give rise to distinctive seed masses. Each seed is flat, circular and about 3 mm (⅛") across. The **margin or wing is a pale red and imparts a blush to the entire seed mass.** Blooming July to September.

RANGE BC: Subalpine and alpine terrain across our area.
RANGE WA: Subalpine and alpine terrain across our area.

Figwort Family, *Scrophulariaceae*
ELEPHANT'S HEAD [312]
Pedicularis groenlandica

Note: Other pink *Pedicularis*es are on p. 288, while the yellow ones are on p. 236.

This plant will easily be recognized as one of the louseworts by its **strangely shaped flowers of pink to reddish purple clustered on a spike** and by its **regular sawtooth leaves**. Once seen, the **little elephant's-head flowers with their up-thrown trunks** will not be forgotten, nor mistaken for anything else. The fringed leaves are most abundant low on the stem. **Note that all leaf segments are toothed.**

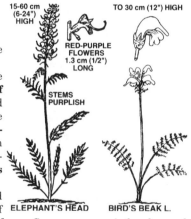

15-60 cm (6-24") HIGH

RED-PURPLE FLOWERS 1.3 cm (1/2") LONG

STEMS PURPLISH

TO 30 cm (12") HIGH

ELEPHANT'S HEAD   BIRD'S BEAK L.

With a height up to 60 cm (2'), elephant's head holds its own as an integral member in the galaxy of summer flowers found in **moist subalpine meadows**. Grows among paintbrushes and lupines, which it blooms a little ahead of, and arnicas.

RANGE BC: On mountains throughout; moist places near timberline.
RANGE WA: Habitats as for B.C.

BIRD'S BEAK LOUSEWORT, *P. ornithorhyncha*: A relative to elephant's head (above), but **only to 30 cm (12") high. Flowers of pink to reddish purple are clustered at the top of the stem. The hood is the shape of a bird's head** and the **beak or trunk is straight; it looks like a bird's bill.** Leaves are fern-like. Subalpine and alpine meadows in the Cascades of B.C. Vancouver Island and southwards through Washington.

SUDETEN LOUSEWORT, *P. sudetica*: A sturdy plant to 45 cm (18") high. **Reddish to purple flowers form a dense, hairy, spiralling spike. Upper petal twisted**; beakless but with **2 small spurs near top of the flower**. Subalpine and alpine terrain north of Quesnel, B.C. Not in Washington.

LANGSDORF'S LOUSEWORT, *P. langsdorfii*: A single stout stem to 25 cm (10") high. Narrow leaves with sawtooth edges. Thick, **short spike of pink–purplish flowers**. A **curving hood petal with 2 tiny teeth near the tip**. Alpine mountains north of Quesnel, B.C. Not in Washington.

PINK 'BELL' FLOWERS
6 mm (1/4") LONG

SEED
PODS
7.5-12.5 cm
(3-5")
LONG

THICK LEAVES
2.5-4 cm
(1-1 1/2")
LONG

Dogbane Family, *Apocynaceae*
SPREADING DOGBANE [317]
Apocynum androsaemifolium

Dogbanes die down and send up new stems each year and so could be considered flowers, but with their fibrous, branching stems (to 75 cm (2 ½') high) they look much like shrubs. The name reflects an unfounded superstition that they repel dogs.

If the smooth, reddish stem is broken, a sticky, milky juice appears. Dry, exposed habitat along road edges and prolonged blooming of end clusters of small, pinkish 'bell' blossoms help distinguish this plant. By early August it turns yellow, the first plant to do so, flaunting its distribution with brilliant colour.

Sharp-pointed, opposite, rather thick, egg-shaped leaves hang down in summer's heat as if tired and lifeless. Light pink flowers have noticeable darker veins. Fragrant; yields a very fine grade of honey. Blooms May to July. Amazingly big smooth, curved, 'green bean' pods 7.5–12.5 cm (3–5") long; they become brownish husks that split to release silky seeds.

RANGE BC: Patchy distribution on dry, exposed soils throughout B.C. Sea level to 900 m (3000') at the coast and to 1650 m (5500') in the interior. Common in Peace River parklands.

RANGE WA: Patchy distribution on dry, exposed soils throughout ponderosa pine ecosystem. From sea level to 900 m (3000') in the Cascades.

HEMP or COMMON DOGBANE [313], *A. cannabinum*: My mystery plant—I have found it only 3 times. Unlike spreading dogbane, except flowers and seed pods. Instead, a spreading patch on a very dry and sun-exposed steep bank, an unremarkable, light green, erect shrub, to 1.2 m (4') tall. Reddish stems; opposite, narrow leaves point upright.

Recorded as having been collected by the aboriginals, shredded and woven into lines and baskets. Very early botanists list it at Spence's Bridge, Sicamous, Keremeos and in the Kootenays. My first find was in Farwell Canyon, a remote part of the arid B.C. interior, on the lower reaches of the Chilcotin River. Then I found it again near Merritt, B.C., and once more, at Colville, Washington. It is widespread across the United States. Limited occurrences in sagebrush, bunchgrass and ponderosa pine ecosystems throughout eastern Washington. Colville.

CLASPING-LEAVED DOGBANE, *A. sibiricum*: Listed as rare, but scattered throughout southern half of B.C. and throughout interior of Washington. Easily mistaken for hemp dogbane, but look for clasping leaves; those of hemp dogbane have short stems.

UMBEL FLOWERS

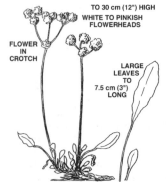

TO 30 cm (12") HIGH
WHITE TO PINKISH
FLOWERHEADS

FLOWER
IN
CROTCH

LARGE
LEAVES
TO
7.5 cm (3")
LONG

Buckwheat Family, *Polygonaceae*
STRICT BUCKWHEAT
Eriogonum strictum

Note: Also written up with white buckwheats (p. 202). Yellow ones are on p. 240.

Obvious umbel design; tiny 6-petalled flowers often pink—might be white, cream or yellow in some locations, or a beautifully mottled white and pink. Often a flower cluster where stems branch. Variable umbel branching in flowerhead; long stems or short. 15–30 cm (6–12") tall. Leaf pattern helps identify. Greyish leaf blades oval to rounded, 0.6–2.5 cm (¼–1") long, on thin stems to 4 times as long. Can be very abundant on thin, rocky soil. June is middle of blooming.

RANGE BC: Rare; south-central. Similkameen and Ashnola river valleys.

RANGE WA: Southwards through sagebrush and ponderosa pine areas to northern California. Toppenish to Goldendale and eastern section of Columbia Gorge. Blue Mountains.

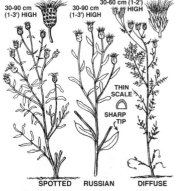

Sunflower Family, *Asteraceae (Compositae)*
SPOTTED KNAPWEED [308]
Centaurea maculosa

All the many **knapweeds recorded in our area are aliens—weedy plants from Europe.** Several are very rare, while **some threaten vast areas of rangeland.** Some are considered as pleasant garden escapes: cornflower or bachelor's button, for example. Knapweeds with rose or purplish flowers may be mistaken for thistles. However, knapweed **leaves are not spiny.** Another key to identification is the **swelling beneath the flowerhead—formed by triangular scales,** sometimes smooth, sometimes bristly. On **spotted knapweed, the tips of these scales are black and tufted with hair,** a good identification feature. Flower of spotted knapweed is about 2 cm (¾") across. Note very narrow leaves. 30–90 cm (1–3') tall.

Chemical sprays are not feasible for controlling knapweeds, except over limited areas. Studies are concentrating instead on biological controls such as insects and diseases that prey on this plant from eastern Europe. Methods include a seed-displacing fly, a seed-eating moth and a leaf fungal disease. Researchers hope for a good measure of control when 6 such planned bio-control agents are widely introduced.

RANGE BC: A serious weed pest in dry, open areas of southern B.C., primarily east of Cascades. Abundant roadside plant in many areas.

RANGE WA: Habitats as for B.C.

RUSSIAN KNAPWEED, *C. repens*: Generally **large numbers of small pink-purple flowers** on well-branched stems. **Scales broad, thin, papery**—quite different from black-tipped ones of spotted knapweed. Sharp-tipped, narrow, alternate leaves. Open wastelands and roadsides of southern B.C., east of Cascades; through Washington in drier areas.

DIFFUSE KNAPWEED, *C. diffusa*: **Commonly has a showy head of smallish, pink-purplish flowers but is often white-flowered!** The flower is supported by a series of **sharp-pointed scales fringed with bristles.** A twiggy plant with leaves to 15 cm (6") long, bearing **leaflets with many narrow lobes.** Usually 30–60 cm (1–2') high. Dry valleys and low mountain slopes of southern B.C., east of the Cascades. Locally common and spreading in various regions of Washington.

'DAISY' FLOWERS

Sunflower Family, *Asteraceae (Compositae)*
SUBALPINE DAISY [314]
Erigeron peregrinus

Note: Purple *Erigeron*s are on p. 310; white on pp. 208 & 210; yellow on p. 250, blue ones on p. 330.

A beautiful flower, a **key component of great subalpine flower meadow displays**; people are certain to want to know it. Common name is very descriptive. Fits well the popular conception of a daisy: an erect **stem usually has 1 handsome bloom** with 30–50 ray flowers, **much wider than on most fleabanes. Pink or red-purple flower, sometimes almost white,** stands out among lupines, paintbrushes, arnicas and mountain valerians.

Basal leaves may form thick cluster that supports several stems. Variably sized, broad 'paddle' leaves, quickly becoming smaller up climb the stem. High meadows and slopes with good soil. Blooming peaks in first week of August in subalpine, but **also at lower elevations** among coniferous forest glades.

RANGE BC: Most mountains. All high mountain parks.

RANGE WA: Habitats as for B.C.

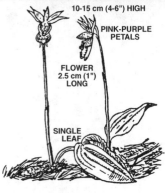

10-15 cm (4-6") HIGH

PINK-PURPLE
PETALS

FLOWER
2.5 cm (1")
LONG

SINGLE
LEAF

Orchid Family, *Orchidaceae*
FAIRY-SLIPPER [315]
Calypso bulbosa
False Ladyslipper, Calypso

Calypso, the sea nymph, long ago waylaid Odysseus. Today Calypso, the delicately tinted jewel, can still waylay you. Once found, in **quiet seclusion of a mossy forest**, its beauty can hardly be forgotten. **Style, stigma and stamens form 'column.'** Little, white bulb shallowly cradled in a protective layer of moss and needles. **Only 1 shiny green leaf** grows at base of reddish stem; oddly it is produced in fall, persists over winter, acts as background for colourful orchid flower April to June, then withers.

RANGE BC: Throughout; cool, shady coniferous forests. Sea level to 1350 m (4500').
RANGE WA: Habitats as for B.C.

UNUSUAL FLOWERS

20-50 cm
(8-20") HIGH

Figwort Family, *Scrophulariaceae*
SICKLETOP LOUSEWORT
Pedicularis racemosa
Parrot's Beak

Note: More pink louseworts, p. 285; yellow, p. 236.

'Lousewort' reflects an old-world belief that animals grazing on these plants would become infested with lice. Of the dozen or more louseworts, a number are only in the far north. Striking **flowers typically grow singly from stem, one above another. 2 lips: upper lip is hooded and encloses 4 stamens; broad lower lip, generally 3-lobed**, is insects' landing pad as they contact stamens above.

SICKLETOP   CAPITATE

Sickletop's **flower is generally in shades of pink and white**, but variable; **twisted, curved hood**. Resembles hand sickle or parrot's beak. Finely toothed, lance-like leaves.

RANGE BC: Middle-mountain forests to alpine heights in south. Manning Park.
RANGE WA: East and west, in same high habitat as for B.C.

CAPITATE LOUSEWORT, *P. capitata*: 'Capitate' means 'head'—a **spray of 2–4 large, showy cream-to-pink flowers** just above **tuft of 'fern' basal leaves**. Most abundant high mountains–subalpine. Common Prince George, B.C., to Alaska. Rockies. Not in Washington.

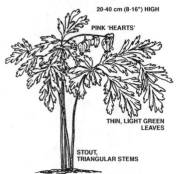

20-40 cm (8-16") HIGH

PINK 'HEARTS'

THIN, LIGHT GREEN
LEAVES

STOUT,
TRIANGULAR STEMS

Bleeding Heart Family, *Fumariaceae*
PACIFIC BLEEDING HEART [319]
Dicentra formosa

Note: Closely related *Corydalis* spp. on p. 233.

Delicately fringed leaves sweep upwards, almost hiding cluster of **drooping, pinkish 'hearts'** (occasionally white). Flower is a symmetrical jewel—**petals held together near tips**. Matching flowers' timid outlook on world is their **faint but fragrant perfume**. Rich, shady woods; sea level–middle-mountain heights. May bloom May or June.

RANGE BC: West of Cascades.
RANGE WA: Habitats as for B.C.

STEER'S HEAD [318], *D. uniflora*: 1 basal leaf; 3 lobed lobes. 1 odd **flower; 2 twisted sepals like steer's horns. Rare**; well-drained soils; middle-mountain elevations to subalpine. Southwestern to south-central B.C.; mostly east of Cascades. Throughout Washington.

Purslane Family, *Portulacaceae*
RED MAIDS [320]
Calandrinia ciliata
Desert Rock Purslane

15 cm (6") HIGH

HAIRY
SEPALS

THICK, FLESHY
LEAVES

One might omit this plant because of its **limited range** and the fact that **a bloom lasts only a day.** Then again, there could be several blooms, of which one or more could be out. With such a charming red flower, red maids is a **late spring bloomer** that deserves to be recognized.

A number of stems branch from base but tend to sprawl, so flowers are seldom more than 15 cm (6") from the ground. **Stems and leaves are fleshy and have 'ciliata' (tiny hairs) along their edges.** Leaf stems get progressively shorter up the stem. The **small, red–to–crimson-purple flowers are less than 1.3 cm (½") across.** Each has **only 2 broad sepals.**

RANGE BC: Moist, gravelly soil found in lowland valleys and on slopes of southeastern Vancouver Island and of the Gulf Islands.

RANGE WA: Habitats as for B.C. Common along the Washington coast.

Heather Family, *Ericaceae*
CANDYSTICK [316]
Allotropa virgata
Sugar Stick, Barber's Pole

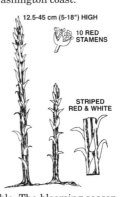

12.5-45 cm (5-18") HIGH

10 RED
STAMENS

STRIPED
RED & WHITE

The vividness of this plant's common names might alert you that it is something very unusual. What else would have a **stout stem 12.5–45 cm (5–18") high striped in white and red?**

A **saprophyte**, it can do without green leaves. As is usual with saprophytes, however, there are series of narrow, **pale scales along the stem**, the **upper ones each protecting a small, attractive, urn-shaped flower.** Flower has 5 small, white sepals, but the **cluster of 10 red stamens** is most noticeable. The blooming season is July and August. Found in rich humus of shady coniferous forests, from low to medium elevations.

RANGE BC: Southern Vancouver Island and adjacent mainland.

RANGE WA: Generally west of the Cascades, from B.C. to the Columbia Gorge, but also on the eastern slope of the Cascades.

Heather Family, *Ericaceae*
PINEDROPS [323]
Pterospora andromedea

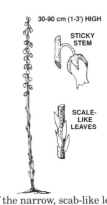

30-90 cm (1-3') HIGH

STICKY
STEM

SCALE-
LIKE
LEAVES

NOTE: Pinesap (p. 214), also saprophytic and usually yellow, is often grouped with candystick and pinedrops; Indian-pipe is white (p. 173).

Like candystick, pinedrops is also a floral curiosity. A **saprophyte** as well, it too does not need green leaves. It gets enough nourishment from dead organic matter to raise a **handsome orange-red stem** 30–90 cm (1–3') high and adorn it with a loose scattering of **pretty 'bell' flowers. Usually reddish, sometimes white to yellow. The globular flower has 5 recurved lobes and a 5-part calyx.** Most of the narrow, scab-like leaves are on the lower part of the stem, which may persist for several months after flowering. The whole plant has a glandular stickiness to it. Generally at low to medium elevations in rich humous soils of ponderosa pine and Douglas-fir forests.

RANGE BC: Across southern B.C.

RANGE WA: Through western and eastern Washington.

20-45 cm (8-18") HIGH

DOTTED
SCARLET
TRUMPETS
2.5 cm (1") LONG

THIN, DRAB
LEAVES

Phlox Family, *Polemoniaceae*
SCARLET GILIA [321]
Ipomopsis aggregata
Gilia aggregata

Note: Globe gilia (heads of blue flowers) p. 327.

Near-scarlet 'trumpets' vivid beyond belief. A great rarity? No, actually quite common in certain localities. Thin stems, usually straight; a number of slender branches near top. Drab, narrow leaves are hardly noticed. Thin **flowers 2.5 cm (1") long flare into 5 petals with small, white marks.**

Favours **dry, exposed places, usually with sandy soil.** Often brightens south-facing roadsides. Under favourable conditions, to 1800 m (6000'). Favourite plant of hummingbirds, also eaten by grazing animals. Blooms mid-May to mid-July.

RANGE BC: East of Cascades; commonly a part of ponderosa pine forest ecosystem, but ranging high on various mountains.

RANGE WA: Habitats as for B.C.

25-80
(10-32")
HIGH

'BURGUNDY'
FLOWERS
6 mm (1/4")
ACROSS

Forget-me-not Family, *Boraginaceae*
COMMON HOUND'S-TONGUE [322]
Cynoglossum officinale
Burgundy Hound's-tongue

European—does not look at home here. **Dotted here and there along roadsides** and dry ground as clumps of **large, soft, green basal leaves**—perhaps to 30 cm (12") long—with **long, tapering, clasping stems.** Why do they go uneaten? The answer is their disagreeable odour and taste.

5-petalled flowers, a very unusual red-purple or burgundy, about 6 mm ($\frac{1}{4}$") across, have darker centre—a circle of short teeth. Grouped in tight whorl clusters along the top 15 cm (6")
of stem; only a few bloom at a time. Blooms May–June. Short, multi-hooked bristles on seeds.

RANGE BC: At lower elevations in dry interior areas. Osoyoos, Cranbrook, Fernie.

RANGE WA: Mostly east of Cascades, in roadsides and fields.

GREAT HOUND'S-TONGUE, *C. grande*: Leafy. **To 90 cm (3') tall.** Basal leaves, oval blades to 15 cm (6") long, equally long stems. **Blue or violet flowers** in spring. Not recorded in southern B.C., but fairly common in Washington at lower elevations in mixed forests west of Cascades; extends southwards to include Columbia Gorge.

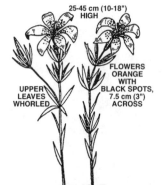

25-45 cm (10-18")
HIGH

FLOWERS
ORANGE
WITH
BLACK SPOTS,
7.5 cm (3")
ACROSS

UPPER
LEAVES
WHORLED

Lily Family, *Liliaceae*
WOOD LILY [324]
Lilium philadelphicum
Mountain Lily

Despite modest height of 25–45 cm (10–18"), has very large, colourful bloom compared to most native wildflowers. **Orange-red flaming 'trumpets' with dark markings**, often 7.5 cm (3") across. Several may be on one branching stem. Narrow, pointed leaves. Prefers moist soil. Blooms June–July. Their special habitat is also very suitable for mosquitoes.

RANGE BC: Eastern B.C., along western slope of Rockies as far north as Peace River.

RANGE WA: Apparently not in Washington.

Buttercup Family, *Ranunculaceae*
BROWN'S PEONY [325]
Paeonia brownii

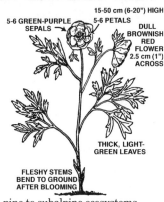

15-50 cm (6-20") HIGH

5-6 GREEN-PURPLE SEPALS

5-6 PETALS

DULL BROWNISH RED FLOWER 2.5 cm (1") ACROSS

THICK, LIGHT-GREEN LEAVES

FLESHY STEMS BEND TO GROUND AFTER BLOOMING

This plant and 1 in California are the only species of peony native to North America. Dull flowers and habit of growing as scattered individuals keep it relatively obscure. Heavy flower, about 2.5 cm (1") across, **appears to have double row of petals.** Actually, **outer row consists of greenish-to-purple sepals**, while **inner sheaf is formed by 5 or 6 slightly larger, dull brownish red petals**. In flower centre is an irregular, fleshy bulb on which stamens are set. Leaves are fleshy and have a waxy feel. Soon after blooming in April or May, the whole plant droops to the ground. Grassy slopes of ponderosa pine to subalpine ecosystems.

RANGE BC: Not in B.C.

RANGE WA: East of Cascades. Spokane, Leavenworth.

CLUSTERED FLOWERS

Figwort Family, *Scrophulariaceae*
PAINTBRUSHES
Castilleja spp.

TO 60 cm (2') HIGH   TO 30 cm (12") HIGH

SCARLET   ELMER'S

Paintbrushes are confusing if you are looking for a recognizable flower, for it is hidden by large, showy bracts. It is these bracts that give credence to the name 'paintbrush.' The actual **flowers are thin tubes, usually greenish red**. The **upper half of each tube has on its outer side a slit** that is quite evident to a hummingbird or long-tongued insect. The **tiny stamens and style show at the lip of the flower**. The precise size and arrangement of these flowering parts are used by the professional botanist to recognize some species and varieties. There are over 3 dozen paintbrushes in our area, with **colours ranging through white, cream, yellow, pink and various shades of red** (also see p. 235). Paintbrushes are **partially parasitic** on the roots of other plants and so do not lend themselves to transplanting.

SCARLET PAINTBRUSH [326], *C. miniata*: Fairly distinctive because of its **very bright red or scarlet colour** and **lower mountain habitat**. The tiny, green flowers project from the showy, hairy bracts. Note that the **leaves are alternate, narrow and have 3 veins**. This floral gem is from 30 to 60 cm (1 to 2') tall. Ranges from lowlands to medium elevations. From Alaska southwards, on Coast Mountains and Cascades. Blue Mountains, Rockies.

HARSH PAINTBRUSH, *C. hispida*: The colour is a **distinctive orange-red**! The 'harsh' part of the name came about because **stem and leaves are covered with short, stiff hairs**. Most stem leaves divide into 5 fingers. A plant 20–40 cm (8–16") high. **Usually the first paintbrush seen as you climb into the mountains**. It shows at about 900 m (3000') elevation. Common at middle-mountain elevations throughout southern B.C. and Washington.

ALPINE PAINTBRUSH, *C. rhexifolia*: A close relative of scarlet paintbrush (above); flowers are the same colour, but **leaves are broader**. This is the **common subalpine paintbrush of eastern Washington**—as far eastwards as the Rocky Mountains, in fact—but it is infrequent in southern B.C.

ELMER'S PAINTBRUSH, *C. elmeri*: Another crimson paintbrush of the **high mountains** that is closely related to those above. It differs, however, in having its **upper leaves and bracts very hairy and sticky**. Generally only 15–30 cm (6–12") tall. Found in the subalpine ecosystem. Local occurrences in the Cascades. Also from the Wenatchee Mountains northwards to south-central B.C.

25-35 cm (10-14")
HIGH

FLOWER
2-LIPPED

CENTRAL
LEAF

Figwort Family, *Scrophulariaceae*
SMALL-FLOWERED PAINTBRUSH
Castilleja parviflora

Common at subalpine elevations. Several named varieties, as described below. Showy **bracts commonly vary from red (not crimson) to soft pink but also magenta and even white**; they hide greenish flowers. In general, under 30 cm (12") tall. Note that **stem leaves are alternate, 3–5 lobed**.

OLYMPIC MAGENTA PAINTBRUSH, *C. parviflora* var. *olympica*: Common on drier slopes of the Olympics. Blooms during August.

MAGENTA PAINTBRUSH [327], *C. parviflora* var. *oreopola*: A close resemblance to Olympic magenta paintbrush (above), but plants shading more to purple and red. A common paintbrush of subalpine meadows and along higher roadsides on Mt. Rainier and southwards.

WHITE SMALL-FLOWERED PAINTBRUSH, *C. parviflora* var. *albeda*: Only to 25 cm (10") high; white to pinkish. Southern B.C. Manning Park; southwards to Mt. Rainier.

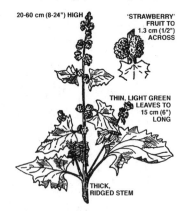

20-60 cm (8-24") HIGH

'STRAWBERRY'
FRUIT TO
1.3 cm (1/2")
ACROSS

THIN, LIGHT GREEN
LEAVES TO
15 cm (6")
LONG

THICK,
RIDGED STEM

Goosefoot Family, *Chenopodiaceae*
STRAWBERRY-BLITE [332]
Chenopodium capitatum
Indian Paint

Note: A related plant with greenish flowers is lamb's-quarters (see p. 338).

A goosefoot—so-named for shape of large, triangular leaves. May grow upright or be sprawling.

Bright red 'strawberry' fruit, to 1.3 cm (½") across—most noticeable feature—forms July to September. Berries (edible raw or cooked) were made into red dye by aboriginals. Inconspicuous small, greenish flowers. Grows best on disturbed ground.

RANGE BC: Across B.C. east of Cascades. Abundant north of Prince George to Yukon.

RANGE WA: Habitats as for B.C. Extends southwards to California.

UNUSUAL FLOWERS

60-90 cm (2-3')
HIGH

ORANGE
SPURS

YELLOW
PETALS

Buttercup Family, *Ranunculaceae*
RED COLUMBINE [333]
Aquilegia formosa
Sitka Columbine, Western Columbine

Surely leads the way among Nature's amazing floral works. 'Columbine,' from Latin *columba,* 'dove.' **5 scarlet petals arch backward**, do look like 5 perched doves. Each 'head' is a honey gland reachable only by humming birds or long-tongued butterflies. Leaves mostly basal—3 stems, each with 3 much-divided leaflets. Prefers moist, partly shaded roadsides and glades. Subalpine meadows can be rich with this prolific bloomer. Flowers May–August.

RANGE BC: Throughout; moist, shady places.
RANGE WA: Throughout; moist, shady places.

BLUE COLUMBINE, *A. brevistylis*: Has **blue-and-cream flowers** and is quite commonly found from valley bottom to timberline in central B.C. and northwards.

YELLOW COLUMBINE, *A. flavescens*: Rarer than blue one. **Flowers mostly yellow**. High elevations east of Cascades. Southern B.C., northern Washington. Sherman Pass.

Lily Family, *Liliaceae*
SAGEBRUSH MARIPOSA LILY [330]
Calochortus macrocarpus
Green-banded Mariposa Lily

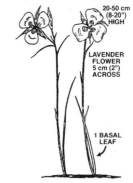

20-50 cm (8-20") HIGH

LAVENDER FLOWER 5 cm (2") ACROSS

1 BASAL LEAF

Note: White-flowered mariposa lilies on pp. 158–59.

Simplicity imparts an air of rarity, further enhanced by **apparently random appearances on drab rangeland**. No other plant like it. Stout stem with **1–3 pale purple or lavender blooms, often 5 cm (2") across**. **3 large petals, each marked on inside with dark blotches near base and down outside with centred green band**. 1 or 2 thin leaves grow from near stem base; several more small stem leaves. Often escapes notice because of delicate flower shading. Blooms during late May—into late July in northern part of range. Commonly with sagebrush and rabbit-brush.

RANGE BC: Seldom in quantity. Lower slopes of ponderosa pine and sagebrush areas. Okanagan Valley, Osoyoos to Kamloops. Lillooet, Clinton, Windermere Valley.

RANGE WA: Very abundant in certain lowlands east of Cascades. Benches of Okanogan and Columbia rivers. Nighthawk to Loomis, northern Okanogan County.

Lily Family, *Liliaceae*
ROUND-LEAVED TRILLIUM [328]
Trillium petiolatum
Long-stemmed Trillium, Wake Robin

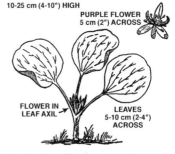

10-25 cm (4-10") HIGH

PURPLE FLOWER 5 cm (2") ACROSS

FLOWER IN LEAF AXIL

LEAVES 5-10 cm (2-4") ACROSS

Note: Common white trillium is on p. 160.

3 leaves branch from short stem attached to creeping rootstock—possible momentary confusion with white trillium (otherwise little resemblance). **Leaf stems almost as long as round blades.**

Cradled in cup formed where leaves join is a **delicate purple or maroon flower with 6 dark purple anthers—other flower parts in 3s**. Blooms April—May. Shady, moist places.

RANGE BC: Not in B.C.

RANGE WA: Bunchgrass ecosystem in eastern Washington. Spokane, Pullman.

MOTTLED or **GIANT TRILLIUM**, *T. chloropetalum*: To 60 cm (2') tall; characteristic **3 large leaves—stemless; dark green mottled with purple**. Unmistakable erect, stemless flower; **3 narrow white-to-pink-to-purple petals**. Lowlands west of Cascades, southwards through Pierce County and Oregon's Willamette Valley to California. Vicinity Mt. Rainier.

4 PETALS

Mustard Family, *Brassicaceae (Cruciferae)*
BLUE MUSTARD [329]
Chorispora tenella
Chorispora, Purple Cross Flower

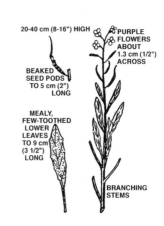

20-40 cm (8-16") HIGH

PURPLE FLOWERS ABOUT 1.3 cm (1/2") ACROSS

BEAKED SEED PODS TO 5 cm (2") LONG

MEALY, FEW-TOOTHED LOWER LEAVES TO 9 cm (3 1/2") LONG

BRANCHING STEMS

Fairly recent Asian arrival, spreading rapidly in drier regions; a potential nuisance. **Usually masses—spacious carpets of soft colour.** Much-branched. Loose clusters of **4-petalled purple flowers**. **Petals have narrow base. Curved seed pods stand abruptly from stem**; long, thin beak-like tips (beakless pods on purple rockcresses, p. 294). Lower leaves have stems; scattered teeth make margins wavy. Upper leaves stemless, toothless.

RANGE BC: South-central B.C. Keremeos.

RANGE WA: Blooms in early spring in bunchgrass ecosystem, with intrusions into sagebrush and ponderosa pine ecosystems. Spokane, Pullman.

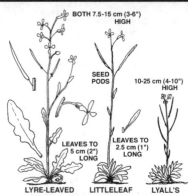

BOTH 7.5-15 cm (3-6") HIGH

SEED PODS

10-25 cm (4-10") HIGH

LEAVES TO 5 cm (2") LONG

LEAVES TO 2.5 cm (1") LONG

LYRE-LEAVED　　LITTLELEAF　　LYALL'S

Mustard Family, *Brassicaceae (Cruciferae)*
LYRE-LEAVED ROCKCRESS
Arabis lyrata

Note: Pink rockcress on p. 263, white on p. 162. Other purple ones can be over 30 cm (12") high, with abundant leaves; may cover meadows and roadsides.

Low roadside vegetation with purple sheen in early spring. Actually a number of erect little plants, 7.5–15 cm (3–6") high—twice that with age. Open cluster of purple flowers. **4-petal cross; long, thin seed pods. Tiny flowers 1 cm (⅜") across, in showy cluster.** Half-dozen leaves on lower stem; several shorter-stemmed ones climb part way up. **Very thin seed pods—some curve slightly.**

RANGE BC: Prefers moist ground, gravel banks and road ditches. Low–high elevations.
RANGE WA: Habitats as above. At subalpine elevations on Mt. Baker.

LITTLELEAF ROCKCRESS, *A. microphylla*: **Flowers purple–pink.** Like *A. lyrata*; more tufted base. **Narrow leaves 2.5 cm (1") long.** Pods 4–6 cm (1 ½–2 ½") long. Infrequent in eastern B.C. Dry, rocky places, valley bottoms–subalpine, in eastern Washington.

LYALL'S ROCKCRESS, *A. lyallii*: Rocky areas, **dry slopes; subalpine–alpine.** Low, shrubby. **Basal tufts of fleshy leaves** 2.5 cm (1") long; a few smaller, alternate stem leaves. **Small, purple flowers in loose clusters.** Erect pods to 7.5 cm (3") long. East of Cascades.

10-15 (4-6") HIGH

VIOLET-BLUE FLOWER, 1-1.3 cm (3/8-1/2") ACROSS, WITH CREAM CENTRE

Figwort Family, *Scrophulariaceae*
ALPINE SPEEDWELL [334]
Veronica wormskjoldii
Veronica

Note: Other *Veronica*s on p. 312.

Small in stature—draws attention with orderliness and brightness. **Several opposite pairs of small, ovalish leaves, neatly arranged** along thin, hairy stem; flowers and buds cluster at very top. **4 violet-to-blue petals surround round cream centre. 2 stamens and long style protruding beyond petals** make flower distinctive. Often found near Jacob's ladder; also blooms mid–late summer. Moist meadows and on stream banks.

RANGE BC: Throughout, middle-mountain to alpine heights.
RANGE WA: Habitats as for B.C.

5-10 cm (2-4") HIGH

DARK PURPLE VIOLET

FLOWER 1.3 cm (1/2") ACROSS

HOOKS

LEAVES 3-5 PARTED

Violet Family, *Violaceae*
SAGEBRUSH VIOLET [331]
Viola trinervata
Three-Nerved Violet

Note: White violets on p. 175, yellow p. 219, blue p. 315.

Grows where **sagebrush is typical ground cover**, instead of in damp, shady areas preferred by most *Violas*. **Deeply cut bluish leaves** also quite different. Some **larger leaves have 3 very distinct veins**; outer ones close to edges. Tiny flower distinctly 2-toned: **2 deep purple upper petals with the rest mauve–white, vein-streaked.** Yellow spot often in throat. Contrasting colours, **like small pansy**. Blooms March–May. Dry, gravelly soils of bunchgrass ecosystem.

RANGE BC: Not in B.C.
RANGE WA: Widespread southwards from Grand Coulee. Yakima Canyon, Fort Simcoe.

Saxifrage Family, *Saxifragaceae*
PURPLE MOUNTAIN SAXIFRAGE [338]
Saxifraga oppositifolia

2.5-7.5 cm (1-3") HIGH

CRIMPED
PETALS

OVER-
LAPPING
LEAVES

Note: White-flowering saxifrages are on pp. 165–67 and p. 195, and there is a yellow one on p. 218. This is the **only saxifrage with purple flowers**—but see p. 165 to make sure you have a saxifrage. Usually this plant is a low, **thick clump of tiny leaves** with short, erect **stems that each carry a single flower**. It wedges into the rocks and rises only a few centimetres (inches). The **tiny leaves grow as opposite pairs arranged spirally to overlap and cover the stem in scale-like fashion**. The **flowers range from purple to pink**. Petals appear crimped. One of the earliest-blooming flowers of the high country.

RANGE BC: A widely distributed alpine beauty that favours moist, rocky places. An identical plant grows in the Alps, posing the question of where this species originated. From Alaska southwards.

RANGE WA: Habitats as for B.C. Olympics and Cascades. Wallowa Mountains, Oregon.

Broomrape Family, *Orobanchaceae*
NAKED BROOMRAPE [335]
Orobanche uniflora
One-flowered Cancer Root

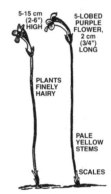

5-15 cm
(2-6")
HIGH

5-LOBED
PURPLE
FLOWER,
2 cm
(3/4")
LONG

PLANTS
FINELY
HAIRY

PALE
YELLOW
STEMS

SCALES

This **parasitic** plant without green leaves obtains nourishment by stealing from the roots of other plants, such as saxifrages and stonecrops. The thin stems, to 15 cm (6") tall and naked except for several scales, support a **single purplish flower with a white throat**—it resembles that of a penstemon. The flower is about 2 cm (¾") long and has a faint fragrance. On occasion the bloom may be yellowish. Watch for it in April and May, at low elevations, in open forests and on hillsides.

RANGE BC: Moist, grassy places on southeastern Vancouver Island. Spotty eastwards to the Rockies, in moist meadows and shady forests.

RANGE WA: Rather uncommon; in lowlands and foothills across state. Columbia Gorge.

CLUSTERED BROOMRAPE [336], *O.fasciculata*: Often **parasitic on sagebrush and buckwheats**, and other dryland species. The short visible section of the stem is very scaly, as well as coarsely hairy and glandular. The typical flower, from 1.3 to 2 cm (½ to ¾") long, has **2 broad upper lobes and 3 lobes below**. Although **usually purple-tinged, it may be yellow**. May and June are the months in which it blooms. In B.C. it is on dry, sandy soils, mostly east of the Cascades, in the dry interior; Chase, Chilcotin, Pouce Coupe. Also on Savary Island. Across Washington, especially in sagebrush areas.

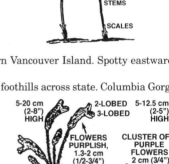

5-20 cm
(2-8")
HIGH

2-LOBED
3-LOBED

5-12.5 cm
(2-5")
HIGH

FLOWERS
PURPLISH,
1.3-2 cm
(1/2-3/4")
LONG

CLUSTER OF
PURPLE
FLOWERS
2 cm (3/4")
LONG

FLOWERS
AND STEMS
STICKY
AND
HAIRY

SCALY
STEMS

CLUSTERED   BUNCHED

CALIFORNIA or **BUNCHED BROOMRAPE**, *O. californica*: **Small, to 12.5 cm (5") high**, with a cluster of **purple flowers** each about 2 cm (¾") long and with long, thin sepals. Blooming time is from June to September. Exposed bluffs on southeastern Vancouver Island. Gulf Islands. In eastern Washington, where it is parasitic on sagebrush.

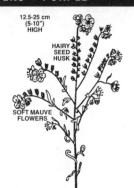

12.5-25 cm
(5-10")
HIGH

HAIRY
SEED
HUSK

SOFT MAUVE
FLOWERS

Waterleaf Family, *Hydrophyllaceae*
THREAD-LEAVED PHACELIA
Phacelia linearis

Distinctive. **Common in arid areas**; great mullein and sagebrush could be companions. **Usually 1 main stem**, but becomes much-branched to display showy **soft mauve or lavender flowers to 2 cm (¾") across**. Tightly coiled flowerhead becomes a row of seed capsules as buds in turn unfold, bloom and wither. Sparse and narrow, **lower leaves reflect 'thread-leaved,'** but upper ones are larger and 5-pronged, with centre prong by far the longest. Blooms May 15 to June 15. (White phacelias on p. 182)

RANGE BC: Occasional in Gulf Islands ecosystem, but common east of Cascades. Fraser Canyon, Cariboo, Okanagan and Kootenays.

RANGE WA: Sporadic west of Cascades, but more common eastwards in open lowlands and foothills. Valley floors and lower slopes in Okanogan County.

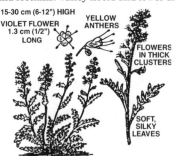

15-30 cm (6-12") HIGH

VIOLET FLOWER
1.3 cm (1/2")
LONG

YELLOW
ANTHERS

FLOWERS
IN THICK
CLUSTERS

SOFT,
SILKY
LEAVES

Waterleaf Family, *Hydrophyllaceae*
SILKY PHACELIA ³⁴¹
Phacelia sericea
Mountain Phacelia

Beautiful spikes of fuzzy, violet flowers, in sharp contrast to pure, bright colours of most alpine blooms. Leaves are **silky with fine hairs**, can appear silvery. Generally fuzzy appearance comes from flowers' **long protruding anthers, each with a yellow stamen at tip. Long anthers are characteristic of phacelias**. Most phacelias have a coiled flowerhead, but this one has flowers in clusters, on a profusion of stems. Blooms late summer.

RANGE BC: Dry, rocky places near or above timberline.

RANGE WA: Dry, rocky places near or above timberline.

30-60 cm (1-2') HIGH

PALE
PURPLE
FLOWERS

OPPOSITE
LEAVES TO
7.5 cm (3")
LONG

SQUARE STEMS

Mint Family, *Lamiaceae*
FIELD MINT ³⁴²
Mentha arvensis
Canada Mint

There is **only one native mint in our area**, but several related escapes have established themselves around cities. Mint will be found around edges of damp places and might often pass unnoticed except for the clean, spicy smell when it is stepped on. Unobtrusive, **tiny 5-petalled flowers of pale purple to pink form in tight clusters in axils between opposite, saw-toothed leaves and stem**. Leafy; erect, **stout, squarish stems**—may be quite hairy on upper sections. Blooms July–August. You can make a fair cup of herb tea by the usual method of using some fresh or dried leaves.

RANGE BC: Throughout, in suitable habitat.

RANGE WA: Throughout, in suitable habitat.

SPEARMINT, *M. spicata*: To 90 cm (3') tall. **Pale lavender flowers in whorls** above opposite, **crinkly, stemless leaves**. Like other introduced mints, has a **tall flower spike above leaves,** rather than with each pair of upper leaves. Wet places, lower elevations—very sporadic. Southern Vancouver Island and Fraser Valley in B.C. Scattered in Washington.

PEPPERMINT, *M. piperita*: Much like field mint, but **roundish leaves have definite stems. Crushed leaves have a sharp peppermint smell**. Garden escape in southwestern B.C. Cowichan Lake. Coastal areas of Washington.

Nightshade Family, *Solanaceae*
EUROPEAN BITTERSWEET [339]

Solanum dulcamara

Bittersweet Nightshade

Note: White-flowered nightshades are on p. 183.

Vigorous twining climber to 2.1 m (7') high.
By late summer, flaunting not only unusual **purple flowers**, but also successively **green, yellow and bright, shiny red berries.** Flowers, like those of tomatoes and potatoes, somewhat **resemble small shootingstars**: anthers form a yellow pointed column, whitish ring at base. Clusters of about a dozen, about 1.3 cm (½") across. **5 reflexed purple petals.**

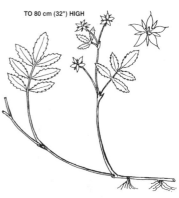

2 kinds of leaves: those lower on stem are alternate, entire and have a twist to stem that holds them upright; otherwise similar higher ones each have 1–2 pairs of ear-like lobes at base. European. Berries considered mildly poisonous.

RANGE BC: Very abundant in local occurrences in southern B.C.

RANGE WA: Very abundant in local occurrences throughout.

Rose Family, *Rosaceae*
MARSH CINQUEFOIL [340]

Potentilla palustris

Most cinquefoils have yellow 'buttercup' flowers and thrive in alpine areas (see p. 221). This one is quite different. **Horizontal stems; may extend into shallow water.** From each such creeping and rooting runner, vertical leaf stems and flower stems may branch upwards as high as 80 cm (32"). Lower leaves have 5–7 toothed leaflets. Higher stem leaves are smaller, may have only 3 leaflets. Beautiful and symmetrical **bowl-shaped purplish red flowers about 2 cm (¾") across**. Strong fetid odour attracts pollinating insects. **Large, sharp-pointed sepals are showiest feature**. Smaller projections between them are petals. Summer bloomer.

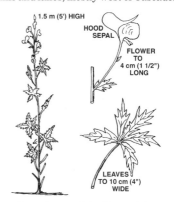

RANGE BC: Common throughout; swamps, backwaters, marshes; seashore–subalpine.

RANGE WA: Bogs, meadows and margins of streams and lakes, mostly west of Cascades.

Buttercup Family, *Ranunculaceae*
COLUMBIA MONKSHOOD [343]

Aconitum columbianum

Tall. Usually in moist areas, along creeksides or in high meadows. A **main stem to 1.5 m (5') tall** carries a **showy head of flowers,** each on its own upward-arching stem. **Large flower, 4 cm (1 ½") long, has large, forward-pointing spur.** Appears to have 5 petals—look carefully to find 3 sepals and 2 petals (same colour). **Topmost sepal forms both a hood and a spur. 2 large side-sepals arch outwards and upwards, join to form characteristic domed 'monk's hood,' and enclose cluster of short stamens. 2 small petals form bottom lip.**

Colour varies: blue–violet–purple. Blooming time depends on altitude: mid-June to August. Alternate leaves deeply divided, 3–5 sharp-pointed sections, smaller up stem.

RANGE BC: Moist, shady areas east of Cascades, valley bottoms–subalpine.

RANGE WA: Wideranging east of Cascades, but as spotty occurrences, suitable habitats.

MOUNTAIN MONKSHOOD, *A. delphiniifolium*: Differs from Columbia: **leaves cut to leaf stem**; flowers differently proportioned. North of Quesnel, B.C., moist forests to alpine.

10-35 cm (4-14") HIGH

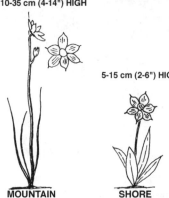

MOUNTAIN   SHORE

Iris Family, *Iridaceae*
MOUNTAIN BLUE-EYED GRASS [350]
Sisyrinchium montanum

Note: Although a number of these related plants have been generalized as blue-eyed grasses, most are **5-15 cm (2-6") HIGH** purplish and have been grouped here. There is also a golden-eyed grass (see p. 226). Another close relative, satin-flower, is pink-purple (p. 275).

The experts have stressed for decades that the different species of blue-eyed grasses are most confusing to separate and label. The fact that the ranges overlap, especially in the dry interior, does not help. The complicated technical jargon about flower parts that is the basis for identification does not alleviate the confusion.

It is of some help to know that this species is the one **most frequent east of the Cascades. Flowers of pale to deep purple are usually held erect or horizontally.** The filaments (supports for the anthers) are held together until near their tips. The plant's stems are usually broader than its very thin leaves.

RANGE BC: Common in moist areas east of the Cascades; favours dry interior region.

RANGE WA: Habitats as for B.C.

SHORE BLUE-EYED GRASS, *S. littorale*: First note that, as the name implies, this is a **coastal** plant. Unlike so many of its relatives, this one has distinctive flowers; they are large, with **dark purplish blue petals and a bright yellow centre.** Leaves are wide and form a basal tuft. It is rather rare; found along coastal bluffs and down to the edges of beaches and brackish marshes. From southeastern Alaska southwards to the B.C.–Washington border.

10-40 cm (4-16") HIGH

BLUE FLOWER 1.3-2 cm (1/2-3/4") ACROSS YELLOW CENTRE

IDAHO BLUE-EYED GRASS

IDAHO BLUE-EYED GRASS, *S. idahoense*: This plant is a **common spring flower** of moist meadows in the **sagebrush and bunchgrass ecosystems.** Leaves are narrow and shorter than the flower stem. Possibly varying in colour, the **3–9 short-lived flowers are sometimes purple but often shading to blue.** Moist, grassy meadows from lowlands to mountain slopes in southern B.C. Same habitat for Washington, east of the Cascades. Ellensburg, Pullman.

Iris Family, *Iridaceae*
NORTHERN BLUE-EYED GRASS
Sisyrinchium septentrionale
Grass Widow

10-25 cm (4-10") HIGH

This blue-eyed grass has **flowers that appear to protrude from a seam in the stem.** They are 2.5 cm (1") wide. Although a flower lasts only 1 day, a new one will pop out to replace it. The **flower colour varies through shades of blue and purple. Leaves are stiff and grass-like, about the same height as the flowers**, 10–25 cm (4–10") in this species. An early spring bloomer.

RANGE BC: Spotty occurrences at lower elevations in wet meadows of southern B.C.

RANGE WA: Not recorded.

Lily Family, *Liliaceae*
HARVEST BRODIAEA [346]
Brodiaea coronaria

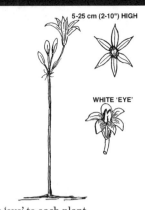

5-25 cm (2-10") HIGH

WHITE 'EYE'

Note that there are also 1 white relative (see p. 186) and 2 blue ones (p. 318). *Triteleia* is an alternative name for several of the *Brodiaeas*.

A **low, purple 'lily' flower with a white 'eye,'** often found in great numbers in parts of Washington. This little plant, only 5–25 cm (2–10") high, usually has a very showy cluster of **purple-pink flowers about 4 cm (1 ½") across.** Each petal is striped with a dark purple line. Notice how in the flower's centre the **3 white, petal-like anthers stand vertically,** forming a protective sheath for the **3-part yellowish stigma.** This design gives a projecting white 'eye' to each plant.

Blooms during August at subalpine elevations, but by late May or early June closer to sea level. Much of surrounding grass will have dried, allowing these flowers, often in a patch, to stand out clearly. No leaves will be seen, for the **1–3 leaves wither before flower appears.**

The plant looks as if it tried to form a true umbel design but fell short, for while flowers and buds originate from the same place, the stems vary greatly in length.

RANGE BC: Rather scarce. On coastal grasslands and rocky hillsides. Mt. Douglas, Cattle Point, Macauley Point and Malahat, all on southern Vancouver Island, and northwards as far as Campbell River; Gulf Islands and adjacent mainland.

RANGE WA: Common, sometimes by the thousands, in open, grassy places on light, rocky soils of western Washington. Penetrating eastwards into the Columbia Gorge and northwards into Yakima County. Southwards through Oregon west of the Cascades.

BALL-HEAD CLUSTER LILY or **OOKOW,** *B. congesta*: This *Brodiaea* has an **exceptionally long stem, sometimes to 1.2 m (4')** in length. The **dark purple flowers appear in a tight cluster** (a result of the very short stems) after the leaves have withered. *Ookow* is an aboriginal name for this plant, whose bulb was once used for food. It blooms during June and July, in dry places in Washington's coastal forest ecosystem. Columbia Gorge as far east as the Dalles, and also common in Oregon's Willamette Valley. Not in B.C.

Primrose Family, *Primulaceae*
PURPLE LOOSESTRIFE [344]
Lysimachia salicaria

0.9-1.8 m (3-6') HIGH

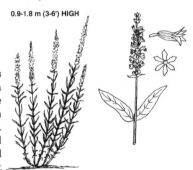

Note: 2 yellow loosestrifes are on p. 225.

You might call this plant a beautiful villain! Its **massed spikes of pink-purple flowers** can flood a **marsh or slough** with bright colour—so attractive is it that an unknowing person might pull up a stem or take some seed to sow in a wet area closer to home. But just one seed could start a severe environmental upheaval. This European introduction has spread from coast to coast, invading ponds and ditches. It takes over shallow waters to such an extent as to eliminate fish, muskrats, nesting birds and indigenous plants. Chemical controls are both too expensive and likely to harm bordering ecosystems. A start at practical control measures has begun in western Canada, with the introduction of the brown purple-loosestrife-eating beetle, a natural enemy. Several other insects may also prove valuable.

In the meantime, all one can do is to try and keep it from spreading—pull up newly established plants and burn all parts of them. This plant varies **from 0.9 to 1.8 m (3 to 6') in height** and grows with an **abundance of stems** in a clump. Flowers are long and pointed, with 5 or 6 petals that range from pink to violet. They bloom all summer.

Leaves are opposite, stemless, without teeth and branch from a stiff, 4-sided stem.

RANGE BC: Lowland sloughs, marshes and ditches. Okanagan River. Vancouver Island.

RANGE WA: Lowland sloughs, marshes and ditches. Sporadic.

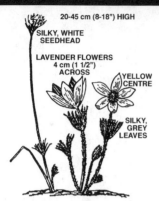

20-45 cm (8-18") HIGH

SILKY, WHITE SEEDHEAD

LAVENDER FLOWERS 4 cm (1 1/2") ACROSS

YELLOW CENTRE

SILKY, GREY LEAVES

Buttercup Family, *Ranunculaceae*
PASQUEFLOWER [345]
Anemone patens
Prairie Crocus

Note: White anemones are on pp. 176 & 185. Yellow anemone is on p. 217, blue on p. 313.

Bright and showy. A stranger to people at the coast but a **popular harbinger of spring** in eastern parts of our area. In April and May, **grows abundantly**. Brings **blushes of violet or purple-blue** to nooks and hollows of landscape. Cup-shaped blossom formed by **6 or 7 large sepals; golden centre**. Stems covered with long, silky hairs. After blooming, flower stems extend to 45 cm (1 ½'); **long-plumed seeds** are blown away by wind. Prefers well-drained soils, lower elevations.

RANGE BC: From B.C.–Yukon border southwards through Peace River region and flanking the Rockies. Windermere Valley, Fernie to Crow's Nest Pass.

RANGE WA: Wenatchee Mountains eastwards to Rockies. Northeastern Washington.

MANY PETALS

YELLOW OR PURPLE FLOWERS

5-12.5 cm (2-5") HIGH

CLUSTERS OF SPINES 0.6-2 cm (1/4-3/4") LONG

Cactus Family, *Cactaceae*
SIMPSON'S CACTUS [348]
Pediocactus simpsonii
Hedgehog Thistle, Pediocactus

Note: Yellow-flowering cacti are on p. 228.

There are 3 cacti in Washington—tell them apart by their shapes and well-separated ranges. Simpson's cactus is a **single blunt sphere**, but sometimes several plants form a bulbous unit. Usually 5–12.5 cm (2–5") in height, several centimetres (inches) across. Covered with a pattern of prominent bumps, each with a cluster of needle-like spines. There are 1 or several 2.5–5 cm (1–2") wide blooms atop each bulbous pincushion; each has **several rows of delicate, tissuey petals**. **Wide colour variation: purple–rose, even weak yellow**. Blooms from April or early May. **Exposed, rocky soils**, with yellow line-leaved fleabane (p.250), before bitterroot (p.277).

RANGE BC: Not in B.C.

RANGE WA: Stony ridges and knolls of sagebrush ecosystem; Kittitas and Douglas counties. Height of land between Vantage and Ellensburg on side-road, 'The Vantage Highway.'

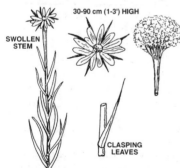

30-90 cm (1-3') HIGH

SWOLLEN STEM

CLASPING LEAVES

Sunflower Family, *Asteraceae (Compositae)*
COMMON SALSIFY [351]
Tragopogon porrifolius
Oyster Plant

Note: The 2 most common salsifies, or oyster plants, have large, yellow flowers (see p. 243).

Large, fleshy tap roots have been used as food—with a fancied oyster taste. All salsifies are from Europe. Notice a salsify by its very large 'dandelion' seedhead, to 7.5 cm (3") across. Each seed is a miniature umbrella. This **dryland plant** shuns a full day's sunlight by **closing at midday**—long green bracts close in to shield the flower. **Flower, composed of many purple ray flowers, is about 6 cm (2 ½") across**. Slender, pointed leaves grasp stem. Blooms during early summer.

RANGE BC: Southern B.C. A common plant of roadsides and hillsides.

RANGE WA: Throughout. A common plant of roadsides and hillsides.

Bladderwort Family, *Lentibulariaceae*
COMMON BUTTERWORT [349]
Pinguicula vulgaris

5-12.5 cm (2-5") HIGH

SLIMY, YELLOW-GREEN LEAVES

The simplicity of this little plant belies its uniqueness. Looks something like a showy violet, but notice those **peculiar basal leaves—yellowish green** might suggest they do not perform the usual functions. In fact, they are **covered with sticky slime that traps insects**, which are then digested. Very often, a number of these semi-carnivorous plants will be found growing together.

Single stems 5–12.5 cm (2–5") high each carry a **lovely violet-purple flower**. Looks as if it might have 5 petals, but these are actually lobes at flaring mouth of flower tube. Flower extends into a long spur, where nectar is stored. Notice how upper stem clamps onto flower spur.

RANGE BC: Wideranging, from Alaska southwards. Most abundant on edges of bogs and in seepage areas and wet meadows, from lower mountain slopes to subalpine.

RANGE WA: Habitats as for B.C. Southwards to California.

Mint Family, *Lamiaceae*
MOUNTAIN MONARDELLA
Monardella odoratissima

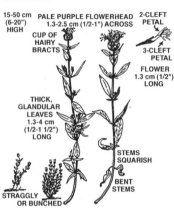

15-50 cm (6-20") HIGH
PALE PURPLE FLOWERHEAD 1.3-2.5 cm (1/2-1") ACROSS
2-CLEFT PETAL
CUP OF HAIRY BRACTS
3-CLEFT PETAL
FLOWER 1.3 cm (1/2") LONG
THICK, GLANDULAR LEAVES 1.3-4 cm (1/2-1 1/2") LONG
STEMS SQUARISH
BENT STEMS
STRAGGLY OR BUNCHED

Sometimes straggly, with several stems, or a compact cluster. **Leaves generally thickish, densely dotted with glands on underside; great shape variation.** Smaller leaves, branchlets often in leaf axils. **Strong minty smell when crushed**.

Pale purplish flower in greenish or purplish globe of fuzzy, overlapping scales. 2 lips: upper is 2-cleft, lower 3-cleft. Ornamental **purple stamens** on purple stalks. Blooms June–July.

RANGE BC: Very rare occurrences in southwestern B.C. Also near Nelson.

RANGE WA: Sagebrush, bunchgrass and ponderosa pine ecosystems. Republic, Cle Elum, Yakima, Blue Mountains.

Mint Family, *Lamiaceae*
COOLEY'S HEDGE-NETTLE [352]
Stachys cooleyae

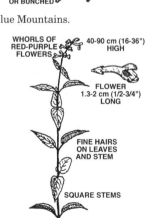

WHORLS OF RED-PURPLE FLOWERS
40-90 cm (16-36") HIGH
FLOWER 1.3-2 cm (1/2-3/4") LONG
FINE HAIRS ON LEAVES AND STEM
SQUARE STEMS

Coarse, weedy plant to 90 cm (3') high. Forms extensive masses of green until **red-purple flowers** dress it in gay colour. **Trumpet-like flowers, with protruding lower lip, in whorls of about 6** at tops of stems. Blooms June–July. Leaves suggest wild mint—or even stinging nettle—but they have a rank smell and are finely hairy. **Main stems are square** (characteristic of mint family), which may come as a shock to those people who believe all flower stems are round. First described in 1891, when it was collected at Nanaimo, B.C., by Grace Cooley, a professor from eastern U.S.A.

RANGE BC: At low elevations, in damp ground west of Cascades.

RANGE WA: Habitats as for B.C.

SWAMP HEDGE-NETTLE, *S. palustris*: May be under 60 cm (2') high. **Light-purple flowers, sometimes with lighter spots**. Blooms during summer. Wet ground east of Cascades, southwards from Fort St. John, B.C. Throughout Washington in similar habitats.

Figwort Family, *Scrophulariaceae*
PENSTEMONS
Penstemon spp.
Beard Tongue

Note: Some are white or yellow (see p. 230), or blue (p. 322); some are with shrubs (p. 133).

All penstemons have a **funnel-like flower with a 5-lobed calyx**. Flowers are cleverly designed for pollination by bees, bumblebees especially. Petals fuse into a tube with 5 lobes: 2 tilt upwards while lower 3 slope outwards and downwards to create a landing strip. Often, coloured lines or bright hairs give directions to the nectar. The bee lands on the pad and crawls into the tube, where its back brushes against the overhanging stigma, which receives fertilizing pollen deposited by another penstemon. More pollen falls on the bee, to be carried to the next flower. The reward for this performance is the nectar at the tube's rear. There are 5 stamens but only 4 anthers are fertile. In some species, throat or tongue is very hairy.

These beautiful plants are found on **rocky bluffs** and **slide areas** across our area, but become fairly common on gravel or sandy soils in higher mountains. Of the dozen or more species that might be found, fewer than half are commonly recognizable as penstemons by the average amateur naturalist. *P. fruticosa* and *P. davidsonii* are treated as shrubs in this book, because of their woody stems and evergreen leaves, but they look like flowers to most people.

SMALL-FLOWERED PENSTEMON [353], *P. procerus*: **Upright, to 60 cm (2') tall— does not have mat-like habit of Davidson's penstemon (p. 133) nor tufted growth. Flowers dark blue-purple, often tinged with pink**, and about 1.3 cm (½") long. In **distinctive massed whorls in higher leaf axils**. Blooms June–August, depending on elevation.

Favours dry ground, from low, open forests to rocky ridges near the subalpine ecosystem. Ranges east of the Cascades and southwards from north-central B.C. through Washington.

CARDWELL'S PENSTEMON, *P. cardwellii*: Another penstemon that deserves mention, for it often forms **sprawling, colourful mats of bright purple to blue**. Shrubby, to 30 cm (12") tall, with curved stems; **leaves with almost no stem. Flowers are hairy at mouth and have long, woolly anthers**. Favours open forest, rocky slopes and road edges. West of the Cascades at middle elevations in Washington. Skamania County; abundant on new volcanic rubble on Mt. St. Helens. Columbia Gorge. Not in B.C.

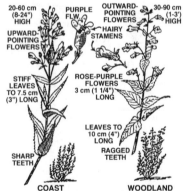

20-60 cm (8-24") HIGH · PURPLE FLW. · OUTWARD-POINTING FLOWERS · 30-90 cm (1-3') HIGH · UPWARD-POINTING FLOWERS · HAIRY STAMENS · STIFF LEAVES TO 7.5 cm (3") LONG · ROSE-PURPLE FLOWERS 3 cm (1 1/4") LONG · LEAVES TO 10 cm (4") LONG · SHARP TEETH · RAGGED TEETH · COAST · WOODLAND

Figwort Family, *Scrophulariaceae*
COAST PENSTEMON [354]
Penstemon serrulatus
Spreading Penstemon, Cascade Penstemon

Showy, with a **bright head of purplish blue 'trumpets.'** Often forms a cluster of erect stems, but sometimes branching and almost mat-like. **Stiff, sharp-toothed leaves**, 3–7.5 cm (1 ¼–3") long.

The **typical penstemon flower, with a lovely purplish blue**, has a 2-lobed upper lip and 3-lobed lower one. To 2.5 cm (1") long, it is **particularly full-throated, giving a wide face**. Unlike many other penstemons, it **has neither hairy throat nor woolly anthers**. This abundant flower blooms all summer and favours moist areas such as stream banks, at low to medium elevations.

RANGE BC: Generally west of the Cascades.

RANGE WA: Habitats as for B.C. Columbia Gorge.

WOODLAND PENSTEMON, *Nothochelone nemorosa*: Once considered a *Penstemon*, now stands as one of a kind because of a minor variation in flower structure. **1 or several low, often-arching stems**. Large, opposite leaves, to 10 cm (4") long; coarsely toothed. In clusters of 2–5 at stem end, the **rose-purple flowers are ridged and have hairy stamens**. The **upper lip is considerably shorter than the lower one**. Blooms from June to September.

Unlike most penstemons, which favour open, rocky places, this is a **woodland plant**, usually found in shade, from lowlands to subalpine terrain. However, it may appear on an exposed bank. Rare on moist and rocky slopes of mountains on Vancouver Island and adjacent mainland, but quite common on western slope of Cascades in Washington. Olympics.

Pea Family, *Fabaceae (Leguminosae)*
PEAVINES
Lathyrus spp.

The peavines and vetches (*Vicia* spp., see pp. 281 & 304) are very similar in appearance and include many introduced species. Both **peavines and vetches have weak stems and most depend largely on their twining tendrils for support.** But whereas the **purple peas have 3–5 pairs of opposite leaflets on each leaf, most of the vetches have 4–13 pairs.** Both bear typical 'sweet pea' flowers, but those of the vetches are smaller. The clincher, however, if you have a good eyes or a hand lens, is to take a flower apart and look at the style, the single central tube. **Peavine flowers have fine hairs arranged on 1 side of the tip of the style, like a brush; vetches have hairs that surround the tip.**

Pea Family, *Fabaceae (Leguminosae)*
PURPLE PEAVINE
Lathyrus nevadensis

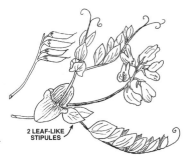

0.3–1.2 m (1–4') LONG

CLIMBING TENDRIL

2–5 PAIRS OF LEAFLETS

FLOWERS 1.3 cm (1/2") LONG PURPLE WINGS CREAMY BLUE KEEL

LEAFLETS 2.5–5 cm (1–2") LONG

Note: Creamy peavine, p. 191, 2 with pink flowers, p. 281.

While purple peavine bears a close resemblance to American vetch (p. 304), it has **straight, unbranched tendrils**, whereas the vetch has branching and curling ones. However, it is a stout vine and can reach a length of 1.2 m (4').

Purple peavine has **4–10 oval leaflets per leaf.**

A **string of 2–7 blossoms per cluster, ranging from red-purple to blue**, almost hidden in salal or other supporting shrubs. Common, particularly in moist, coastal forests. Look in open forests, lowlands to middle-mountain slopes. Blooms May to June.

RANGE BC: Common south of Peace River. Absent on the Queen Charlotte Islands, northern Vancouver Island and adjacent coast.

RANGE WA: Shady woods of coastal regions; Olympics. Also Cascades and ponderosa pine ecosystem bordering the Cascades. Columbia Gorge.

OREGON PEAVINE, *L. polyphyllus*: Much like purple peavine, but note the unexpected **10–16 alternating leaflets and the large, oval stipules at base of leaf stem.** Not in B.C. Coastal Washington northwards to and around Puget Sound. Also coastal mountains.

Pea Family, *Fabaceae (Leguminosae)*
BEACH PEA [356]
Lathyrus japonicus

Beach pea blends very well with other vegetation on **coastal shores** as it climbs over and through shrubbery, so look carefully for the light green leaves and **dark purple flowers**. The **flowers are in clusters of 2–8**. If there is any doubt about identification, look at the base of a leaf stem. There should be **2 leaf-like stipules, about the same size as a leaflet and of unusual triangular shape.** Each

2 LEAF-LIKE STIPULES

leaf has **3–6 pairs of leaflets and is tipped with curling tendrils.** Beach pea's hairy pods are shaped like garden pea pods and are likewise edible. Blooming time is late spring through summer.

RANGE BC: A common plant on coastal beaches and banks all the way to Alaska.

RANGE WA: Habitats as for B.C., through Washington and Oregon.

GREY BEACH PEAVINE, *L. littoralis*: This **small peavine, to 60 cm (24") long**, enjoys the same **coastal habitat** and range as the one above. It wards off chill winds and hot sun with a **silvery covering of fine hairs**, giving it a grey appearance. It **lacks tendrils.** Small leaflets number 1–4 pairs. The 3–6 very attractive **flowers are purple, with white wings and keel.** A coastal plant, it is rare on the Queen Charlottes and Vancouver Island, but it is more common along Washington's coast.

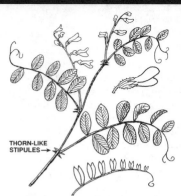

THORN-LIKE
STIPULES →

Pea Family, *Fabaceae (Leguminosae)*
AMERICAN VETCH [347]
Vicia americana

Note: See 'Peavines' (*Lathyrus* spp., p. 303), for a comparison of vetches and peavines.

Perennial, like most peavines. Climbs all over other roadside vegetation. **Flowers of purple-pink to lavender** can be a real eye-catcher when massed. **Flowers 2 cm (¾") long, in heads of 3–9.** Squarish stem. **Leaf has 8–18 opposite leaflets, usually terminates in branching tendrils.** Small, jagged leaf stipules. Can spread branching stems to a distance of 1.2 m (4'), or may be more subdued. Often unsupported and scattered, growing here and there as individual plants to 40 cm (16") high in dry ponderosa pine forests. Blooms April–May.

RANGE BC: Meadows and open forest in lowlands. Hedley.

RANGE WA: Widespread, mainly in ponderosa pine zones. Chelan, Klickitat counties.

TUFTED VETCH, *V. cracca*: Introduced perennial. **Dark bluish purple flowers with touch of white grow in row** from 1 side of stem. **12–18 leaflets, twice length of flowers.** Seed pods 2 cm (¾") long. Blooms June to August. Masses in fields and lower mountain slopes in extreme southern B.C., but infrequent northwards. Southern Vancouver Island and Gulf Islands. Common Yahk to Fernie. In Washington, roadsides in sagebrush areas, Ephrata.

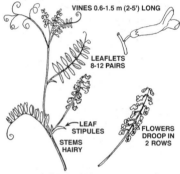

VINES 0.6–1.5 m (2-5') LONG

LEAFLETS
8-12 PAIRS

LEAF
STIPULES

STEMS
HAIRY

FLOWERS
DROOP IN
2 ROWS

WOOLLY VETCH [355], *V. villosa*: Easily identified, especially with sharp eyes or magnifying glass. The only vetch you will see with **fine white hair on stem and leaves and running part way up keels of light purple flowers, which are often fringed with white. Leaflets slightly longer than flowers**, which are about 1.3 cm (½") long. A European introduction. Has become very common on roadsides and fields. Branching form creates **thicket of vine** as it climbs over and through adjoining vegetation. Makes colourful display July–August when beautifully mixed with common St. John's-wort and pink knapweed. Ranges from Vancouver Island and adjacent mainland, and extreme southern B.C., southwards through Washington. Very abundant in Okanogan and Ferry counties.

TINY VETCH, *V. hirsuta*: **3–8 white to pale-blue flowers** per stem. **Small, hairy pods** with 2–3 seeds. Leaflets (10–18) and flowers, about 5 mm (³⁄₁₆") long, both tiny when compared with other vetches. Trails or climbs; to 75 cm (30") long, but can form tangled mass. Alien. Coastal; on southern Vancouver Island, Gulf and San Juan islands, and southwards.

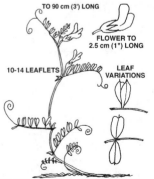

TO 90 cm (3') LONG

FLOWER TO
2.5 cm (1") LONG

10-14 LEAFLETS

LEAF
VARIATIONS

Pea Family, *Fabaceae (Leguminosae)*
COMMON VETCH
Vicia sativa

Common vetch keeps popping up here and there—a little **dab of bright red-purple colour amid green moss and grasses of spring**, where 1–3 'pea' flowers branch from each leaf axil. Usually that is about all you might see, for the rest, a slender stem with a number of alternate leaves, hides in the grass. Each **leaf has about 10 small leaflets; thread-thin stem ends in twisting tendrils; tiny bristle at end of each leaflet**. Widely distributed. European. Much variation in leaves and flowers.

RANGE BC: Southeastern Vancouver Island, Gulf Islands and mainland coastal areas.

RANGE WA: San Juans and southwards. Columbia Gorge. Occasionally in east.

Pea Family, *Fabaceae (Leguminosae)*
SHOWY LOCOWEED [357]
Oxytropis splendens

Note: General locoweed information on p. 190.

An attractive tufted, silvery plant. Flower stems, a little higher than leaves, bear a **dense cluster of purple-to-blue flowers**—hence 'showy.' Know also by **hairy, silvery leaves, in whorls of 3–6.**

RANGE BC: Dry, gravelly areas at lower elevations in Rockies; northeastern B.C. Peace River.

RANGE WA: Not in Washington.

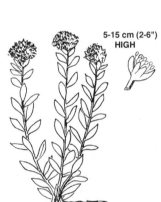

BLACKISH LOCOWEED, *O. nigrescens*: Shrubby, tufted plant carries several **large, purple flowers on short stems. Short leaves, silvery with white hair.** Amazingly **huge seed pods somewhat resemble small sausages**; pointed at both ends, red when really ripe. Hardy alpine plant; north of Ft. St. John, B.C., westwards to Alaska. Not in Washington.

STALKED-POD LOCOWEED, *O. podocarpa*: Closely resembles blackish locoweed in form and range. Flowers 1–2 per stem. **Big, oval pod has a stalk**. Dry, rocky, alpine terrain. Rare; in southeastern B.C. Rocky Mountains. Not recorded in Washington.

CLUSTERED FLOWERS

Sedum Family, *Crassulaceae*
ROSEROOT [360]
Sedum integrifolium
King's Crown

Note: Most *Sedum*s are yellow (see p. 218)

Recognize general *Sedum* form, but **lacks a cluster of basal leaves. Instead, all leaves (stemless) are thick on stem**; may be pink-tinged. Thick, fleshy leaves, a way of storing moisture, indicate exposed rocky habitat. Crowded into tight cluster, small, **4-petalled flowers are typically deep wine-purple**, but can vary to lighter shades. No hesitation in recognizing this unique wildflower. Blooms during summer. That bruised roots supposedly smell like roses was discovered by native peoples in Alaska, who ate the young stalks.

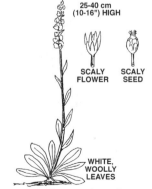

RANGE BC: Rocky places with some moisture; subalpine to alpine areas. Widespread.

RANGE WA: Throughout, in habitats as for B.C.

Sunflower Family, *Asteraceae (Compositae)*
PURPLE CUDWEED
Gnaphalium purpureum
Everlasting, Fragrant Everlasting

Alternative names lend charm. Has attractive features, but is part of a very large group of mostly weeds. Might first see **woolly leaves, in small, silvery clumps**, among dried grass on coastal bluffs. How could anything grow on such exposed, dry soils?

Most **small, spatula-shaped leaves** on first 2.5 cm (1") of stem. **Leaves, stem finely woolly**—whiteness makes small plant quite eye-catching. Short flower stem, narrow column of dry seeds poke through cottony tufts. Looks as if they could stand unchanged until next year—'everlasting' is significant. Appears similar even in spring, except for **head of small purplish scaly flowers** where tufts later form. On exposed ground.

RANGE BC: Southwestern B.C. Gulf Islands.

RANGE WA: San Juans, southwards; west of Cascades. Western part of Columbia Gorge.

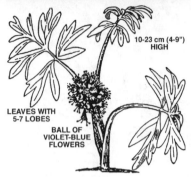

LEAVES WITH 5-7 LOBES

BALL OF VIOLET-BLUE FLOWERS

10-23 cm (4-9") HIGH

Waterleaf Family, *Hydrophyllaceae*
BALLHEAD WATERLEAF [361]
Hydrophyllum capitatum

Like wildflowers such as shootingstars, yellow bells and buttercups, **blooms in early spring** while ground is still moist. **Leaves divided into 5–7 lobes; each may be further lobed**. When young, these succulent 'waterleaves' make excellent greens. Appealing **flower mass: a ball of violet blue made misty by a large number of protruding stamens**. While usually a shade of blue, **flower can vary from nearly white to light pink** in some localities. Very abundant in right conditions. Most common in ponderosa pine zone, but ranges high into more open mountain forests.

RANGE BC: Southern B.C.

RANGE WA: Habitats as for B.C. Cle Elum, Winthrop, Columbia Gorge.

BLUE-PURPLE FLOWERS 1.3 cm (1/2") LONG

7.5-40 cm (3-16") HIGH

Mint Family, *Lamiaceae*
SELF-HEAL [362]
Prunella vulgaris
Heal-All

An introduced plant with square stems that will turn up in a myriad of places. Once you recognize its cluster of flowers, you will welcome its many unexpected appearances. Learn its intriguing Latin name, for there will be many chances to drop it on helpless friends. Long heads of **flowers vary from purple to pink**. Each flower is about 1.3 cm (½") long and quite attractive if examined closely—rather **resembles a small orchid bloom**. Thin, alternate leaves 2.5–6 cm (1–2 ½") long. Self-heal will bloom throughout entire spring and summer. Widespread use for medicinal purposes started a long time ago in Europe. Coastal natives have used it for various purposes in less distant times.

RANGE BC: Open places, fields and roadsides; throughout most of our area.

RANGE WA: Habitats as for B.C.

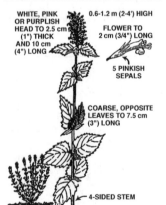

WHITE, PINK OR PURPLISH HEAD TO 2.5 cm (1") THICK AND 10 cm (4") LONG

0.6-1.2 m (2-4') HIGH

FLOWER TO 2 cm (3/4") LONG

5 PINKISH SEPALS

COARSE, OPPOSITE LEAVES TO 7.5 cm (3") LONG

4-SIDED STEM

Mint Family, *Lamiaceae*
NETTLE-LEAVED GIANT-HYSSOP
Agastache urticifolia
Nettleleaf Horsemint

As common names suggest, **looks much like a nettle or mint**, but **grows in dry, open places**. From 0.6 to 1.2 m (2 to 4') high, it has coarsely toothed, alternating, opposite pairs of leaves to 7.5 cm (3") long. Good deal of variation in size of leaves and flowers.

Dense flowerheads may overall be white, pink or purplish; to 2.5 cm (1") thick and 10 cm (4") long. Each **long, thin 'trumpet' flower**, which is about 2 cm (¾") long, has a **notched upper lip and a 3-lobed lower one**. 5 narrow sepals, often tinged with pink. Stamens protrude. Blooms late April onwards; dry, open slopes.

RANGE BC: Rare; in southern B.C.

RANGE WA: Eastern part. Spokane, Whitman, Columbia, and Walla Walla counties.

WESTERN GIANT-HYSSOP, *A. occidentalis*: Like nettle-leaved, but with smaller leaves and flowers. Distinguished by **leaves, which are hairy beneath**, and range—foothills and eastern slope of Washington's Cascades. Not in B.C.

Sunflower Family, *Asteraceae (Compositae)*
BURDOCK [363]
Arctium minus
Dock

BRISTLY, PURPLISH FLW.

0.9-1.5 m (3-5') HIGH

BROWN BUR

Exceptionally large leaves and **spiny burs** arouse curiosity. Much-branched, thick stem. **Leaves somewhat resemble those of rhubarb but are white-woolly on underside. Tiny flowers, tinged purplish to pink, surrounded by rough, hooked bristles.** By fall, top of this sturdy plant is well covered with hard, round burs that cling tenaciously to clothing or animal hair and occasionally even trap small birds. These **bur hooks are the origin of Velcro!** Many fine fibres make branches unbelievably tough. Usually in damp ground or fair soils.

RANGE BC: Lower elevations in coastal and southern B.C.

RANGE WA: Lower elevations throughout.

UMBEL FLOWERS

Carrot Family, *Apiaceae (Umbelliferae)*
FERN-LEAVED DESERT-PARSLEY [368]
Lomatium dissectum
Carrot Leaf

TO 1.2 m (4') HIGH

Note: More information on this desert-parsley is with the yellow-blooming ones, on p. 239.

Very vigorous. Blooms April–May. Rounded outline, fern-like leaves. Usually under 60 cm (2') high. Very decorative on rocky ridges and slopes. By early summer, mass of **foliage is noticeable olive-green.** Thick, aromatic roots.

RANGE BC: Common in south-central and southeastern B.C., on dry rocky places.

RANGE WA: Frequent on roadsides east of Cascades; rocky areas and exposed soils. Vantage, Yakima Canyon, Columbia Gorge.

'DANDELION' FLOWERS

Sunflower Family, *Asteraceae (Compositae)*
COMMON THISTLE [364]
Cirsium vulgare
Pasture Thistle, Bull Thistle

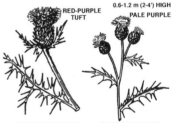

RED-PURPLE TUFT

0.6-1.2 m (2-4') HIGH

PALE PURPLE

COMMON THISTLE   CANADA THISTLE

Note: Hooker's thistle (p. 207) has white flowers. Contrast with sow-thistles (yellow flowers, p. 246).

There are about a dozen thistle species in our area; a quarter are from Europe. **Most thistles have reddish or purplish flowers.**

Single purple blossoms of common thistle, each on its bristly green bur, may be 5 cm (2") across. Notice how **leaf stem clasps main stem and continues down stalk.** Also, **stem has spiny wings.** Blooms July–August. Seeds favoured by American goldfinch.

RANGE BC: Across southern B.C. at low to middle elevations.

RANGE WA: Across Washington at low to middle elevations.

CANADA THISTLE, *C. arvense*: **Lacks stiff, spiny leaves and spiny stem wings of common thistle;** flower less bright. **Leaves white-hairy beneath. Clusters of pale purple-pink (occasionally white) flowers**, only 1.3 cm (½") across. Female and male (showiest) flowers on different plants; considerable variation in appearance. Introduced. Underground stems partly explain extensive dense patches. Summertime blooms may form long stretches of road border in eastern B.C. and Washington. Peace River, B.C., southwards.

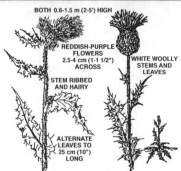

BOTH 0.6-1.5 m (2-5') HIGH

REDDISH-PURPLE
FLOWERS
2.5-4 cm (1-1 1/2")
ACROSS

WHITE WOOLLY
STEMS AND
LEAVES

STEM RIBBED
AND HAIRY

ALTERNATE
LEAVES TO
25 cm (10")
LONG

EDIBLE THISTLE   WAVY-LEAVED THISTLE

Sunflower Family, *Asteraceae (Compositae)*
EDIBLE THISTLE [365]
Cirsium edule
Indian Thistle

Note: More thistles on previous page.

Each flower's 'ball' is a white, woolly mass with many spines. These balls form attractive pedestals for the **several rose-purple blooms.** The thick stem, 0.6–1.5 m (2–5') high, carries leaves to 25 cm (10") long. **Leaves are deeply toothed**, more than halfway to the central rib. They are very ragged; those near the top are often twice-lobed. Thin spines provide armament. It is the plump roots, which are about 30 cm (12") long and can be boiled and eaten, that give this thistle its common name. In fact, the peeled stems and roots of most thistles can be used as emergency food. The blooming period is from late spring to late summer. An attractive plant often in mountain- and subalpine meadows that can also be found at lower elevations.

RANGE BC: Widely distributed throughout our area.

RANGE WA: Widely distributed throughout our area.

WAVY-LEAVED or **WOOLLY THISTLE** [366], *C. undulatum*: Has an overall **silvery white colour to the leaves and stems** and a beautiful display of **flowerheads in rose to lavender to white.** Older plants branch freely and carry **a single flower at the tip of each branch.** Leaves are to 20 cm (8") long, deeply cut, and protected by **long, yellowish spines.** Poor, dry soils east of the Cascades, from southern B.C. southwards through Washington to northern Oregon.

SILVER THISTLE, *NODDING* or *MUSK, Carduus nutans*: Failing to find a common name for this giant silver plant, the author has used 'silver thistle.' A **height to 1.8 m (6')**, **overall silvery colour and very spiny appearance** should provide quick identification, even from a car. Closer by, one sees a stout, winged and **ferociously spiny stem rising from a basal rosette of equally spiny leaves.** They are to 30 cm (12") long and half as wide, making them far broader than those of most thistles. The large, purple flowerhead may be to 6 cm (2 ½") across. In bloom during late spring and early summer.

This plant, a native of Europe and western Asia, was recorded as being 'sparingly abundant' here several decades ago, but is now a wideranging pest, especially along the Snake River and its tributaries. A rare subspecies has been found at Alexis Creek, B.C.

'DAISY' FLOWERS

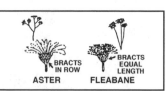

BRACTS
IN ROW

BRACTS
EQUAL
LENGTH

ASTER   FLEABANE

Sunflower Family, *Asteraceae (Compositae)*
ASTERS and FLEABANES

Few people will profess to tell many asters apart. With about 50 different species each of asters and fleabanes in our area, with similar attractive flowers, there is bound to be confusion in identification. However, despite some overlapping features, there are valuable clues to separate them. For the amateur, the first one could be that **asters flower in late summer while fleabanes do so from spring to the middle of summer. In general, the bracts of aster flowers are in several rows, like shingles, while those of fleabanes are of in a single series, of equal length** and less numerous. In addition, **fleabanes usually have ray flowers (petals) that are more numerous and thinner.** Finally, **asters tend to be taller and more leafy and have more flowers per stem than do fleabanes.**

ALPINE ASTER, *A. alpigenus*: Note that this is a dwarf aster, only **10–25 cm (4–10")** **tall**, that is found in the **subalpine ecosystem.** A stout tap root anchors it in rocky soil. Narrow leaves are mostly basal and from 2 to 15 cm (¾ to 6") long. 1 per stem, the showy **flowers are a pale purple to lavender, with bright yellow 'button' centres.**

Ray flowers are rather tousled and some overlap. Blooms July–September. High mountains and rocky slopes. Alaska southwards, through Coast and Cascade mountains. Olympics.

Sunflower Family, *Asteraceae (Compositae)*

ARCTIC ASTER

Aster sibiricus

Large, showy flowers typical of purple asters. Erect stems sometimes branch. Stem leaves decrease in size up stem. **Stemless leaves, fine hairs on undersurface.** Purple ray flowers, yellow centre, 1 or a considerable cluster per stem. Best identification feature: flower's **outer bracts are green and overlap an inner row with purplish tips.**
RANGE BC: Northern B.C. and from Alaska southwards in high mountains.
RANGE WA: High mountains, Mt. Baker northwards.

15-40 cm (6-16") HIGH

INNER BRACTS PURPLE-TIPPED

Sunflower Family, *Asteraceae (Compositae)*

SHOWY ASTER [367]

Aster conspicuus

Large Purple Aster

Very widespread; valley bottom to middle elevations. To 90 cm (3') high on gravelly soils in fairly open forests. Shape often quite ragged. Flowers, in widespreading heads, exceptionally colourful: golden centres, purple rays. Captures its share of attention during late summer. Distinguish by **large, thick, rough, sharp-toothed leaves**.
RANGE BC: East of Cascades to Alberta; drier areas. Through northeastern B.C. to Ft. St. John.
RANGE WA: Open forests east of Cascades.

40-90 cm (16-36") HIGH

DAISY-LIKE FLOWER 2.5 cm (1") ACROSS, PURPLE OUTSIDE AND YELLOW CENTRE

ROUGH, COARSE LEAVES TO 15 cm (6") LONG

Sunflower Family, *Asteraceae (Compositae)*

DOUGLAS'S ASTER [372]

Aster subspicatus

Representative of several widespread species with **showy heads of many bright purple flowers** on coast and in interior. Confirm as Douglas's by **overlapping bracts; outer areas have thin, transparent margin**. Usually, stout stems branch a number of times, to form rounded outline. **Coarse lower leaves have stems; upper ones do not**. May be toothed. This aster (and similar ones) grow on well-drained soil—under open fir forests, along roadsides and beside streams and salt water. Blooms August–September.

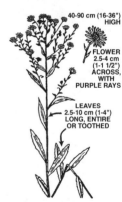

40-90 cm (16-36") HIGH

FLOWER 2.5-4 cm (1-1 1/2") ACROSS, WITH PURPLE RAYS

LEAVES 2.5-10 cm (1-4") LONG, ENTIRE OR TOOTHED

RANGE BC: Central, sporadically to east; most common west of Cascades. Victoria.
RANGE WA: Open lowland forests. Most common west of Cascades. Puget Sound area.

LEAFY ASTER, *A. foliaceus*: Also called 'leafy-headed aster.' **Many large, green, leaf-like bracts support flowerhead.** Unusually **numerous large, finely reticulate leaves** (network of veins) clasp stem; mostly entire. Contrast leaves and flower bracts with Douglas's aster. **Flower varies: blue to rose purple or violet**. Commonly a dozen showy blooms per plant. Moderate height, 30–75 cm (1–2 ½'). Throughout, south of Dawson Creek, B.C. Moister forests; valley bottoms–subalpine (blooms late summer). Hart's Pass, Washington.

COMMON CALIFORNIA ASTER, *A. chilensis*: Originally believed to originate in Chile. 30–90 cm (1–3') tall. **Rather small flowers cluster near top; 15–40 quite broad ray flowers**. Purple or pink; yellow centre. **Showy green, shingled cup; blunt-tipped outer bracts**. Basal leaves soon wither. Middle and upper leaves stemless, entire; 2–10 cm (2–4") long. Low mountain slopes on Vancouver Island. In Washington, range expands into dry, open areas across state. *A. chilensis* var. *hallii* has white rays; common in Washington.

PURPLE-VIOLET FLW. 2.5 cm (1") ACROSS
5-15 cm (2-6") HIGH
25-30 RAY FLOWERS
HAIRY LEAVES AND STEMS
LEAVES 1.3-4 cm (1/2-1 1/2") LONG

Sunflower Family, *Asteraceae (Compositae)*
CUSHION FLEABANE [369]
Erigeron poliospermus
Cushion Erigeron

Note: For differences between fleabanes and asters, see p. 308. Cut-leaved daisy (p. 208) may be purple and look similar. *Erigerons* can be white (pp. 208 & 210), yellow (p. 250), pink (p. 287) or blue (p. 330).

'Fleabane' reflects a belief that these plants repel fleas. Cushion fleabane is a good example of a group of low fleabanes. **Dull green, finely hairy leaves and stems**, like many dryland plants; use hand lens to see hairs clearly. **About 10 cm (4") high. Narrow 'paddle' leaves.** Very attractive, singly borne flowers, typically **soft purple–violet**; may be near-white. **25–30 ray flowers, yellow centre**. Blooms April–May. Often with rigid sagebrush in Washington.

RANGE BC: Southern Okanagan Valley.

RANGE WA: Widely distributed in sagebrush areas. Spokane, Wenatchee, Ellensburg, Vantage, Ft. Simcoe, Sunnyside, Walla Walla.

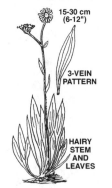

15-30 cm (6-12")
3-VEIN PATTERN
HAIRY STEM AND LEAVES

Sunflower Family, *Asteraceae (Compositae)*
LONG-LEAVED FLEABANE
Erigeron corymbosus

Rather small; most easily recognized by **long leaves—half as long as 15–30 cm (6–12") height**. Coarsely hairy leaves and stem. Stems branch from woody base—a compact clump effect. **Usually 1 light purple flower per stem; rays are wider than on other fleabanes**. Blooms in early summer.

RANGE BC: Favours both dry ground and gravel slopes in sagebrush areas; extends up lower mountain slopes. South-central and southeastern.

RANGE WA: Habitats as for B.C. Across eastern Washington to eastern Oregon.

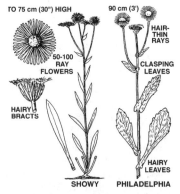

TO 75 cm (30") HIGH
90 cm (3')
HAIR-THIN RAYS
50-100 RAY FLOWERS
CLASPING LEAVES
HAIRY BRACTS
HAIRY LEAVES
SHOWY   PHILADELPHIA

Sunflower Family, *Asteraceae (Compositae)*
SHOWY FLEABANE [358]
Erigeron speciosus
Large Purple Fleabane

Rather bushy. Long, thin, entire leaves climb almost to top. Lower leaves have stems, upper ones do not. (Contrast leaves with Philadelphia fleabane, below.) Showy flowers mass in loose clusters; **yellow centre with nearly 100 thin, purple rays**. Under favourable conditions, flower may be 5 cm (2") across, but more often only half that. Rather prolonged blooming, late May–September.

RANGE BC: Open mountain forests. Rare on western slope of Cascades; more common eastwards.

RANGE WA: Habitats as for B.C.

PHILADELPHIA FLEABANE [359], *E. philadelphicus*: You might imagine that there must be a great number of people in Philadelphia and relate it to Philadelphia fleabane's **great number of ray flowers: 150–400,** say experts. **Colour varies to pink or even white**. Tall: often 90 cm (3') or more. **Very showy massed on roadsides**. May have 2 dozen blooms per stem. **Narrow-oblong bracts; wide, clear margins**. Bottom leaves have long, tapering stems and scattered, rough teeth. Hairy leaves and stem. Clasping bases on higher leaves. Favours moist ground; valley bottom to middle-mountain heights. Relatively common, except for the most arid regions. Blooms June–July.

Buttercup Family, *Ranunculaceae*

WESTERN MEADOWRUE [371]

Thalictrum occidentale

The very delicate, **blue-green foliage** of meadowrue will catch your attention. Ranges from 30 to 90 cm (12 to 36") in height. The graceful, airy design of small, **3-lobed leaves, each with a thread-like stem**, is quite distinctive. You may find them reminiscent of columbine.

Look closely and you may see **tiny, upright, purple flowers on one plant and dangling, green-and-purple filmy clusters on another**, as each plant (respectively) bears either female flowers or male ones. The male flowers are particularly delicate and attractive, with their tassels of silky purple stamens and anthers. This species is wind-pollinated.

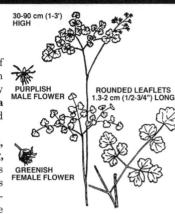

30-90 cm (1-3') HIGH

PURPLISH MALE FLOWER

GREENISH FEMALE FLOWER

ROUNDED LEAFLETS 1.3-2 cm (1/2-3/4") LONG

Almost everywhere one finds Indian paintbrush growing, there will also be meadowrue. Shady, fairly moist places at middle elevations are preferred. In bloom late spring to August.

RANGE BC: Throughout.

RANGE WA: Throughout.

FEW-FLOWERED MEADOWRUE, *T. sparsiflorum*: Like western meadowrue, except with **male and female flowers on the same plant**. Moist mountain areas from north of Dawson Creek, B.C., to Alaska. Rockies and southwards to Oregon and California.

VEINY MEADOWRUE, *T. venulosum*: Like western meadowrue, except with showy, **raised veins on undersides of leaves**. East of the Cascades, in moist meadows and on bordering mountains.

ALPINE MEADOWRUE, *T. alpinum*: A **dwarf alpine plant with glossy leaves**. Otherwise like western meadowrue. Mostly north of Ft. St. John, B.C., westwards to Alaska. Rocky Mountains. Wallowa Mountains of Oregon.

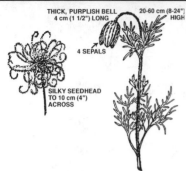

THICK, PURPLISH BELL
4 cm (1 1/2") LONG

20-60 cm (8-24")
HIGH

4 SEPALS

SILKY SEEDHEAD
TO 10 cm (4")
ACROSS

Buttercup Family, *Ranunculaceae*
SUGARBOWLS
Clematis hirsutissima

Note: Also noted with other *Clematis*es (p. 127).

No petals, just **4 ragged-edged sepals**. Unusual design—**heavy, nodding, dark blue-purple 'bell'**; ornamented by **rounded ridges padded with white, woolly hair**; appears leathery. **Tightly clustered stamens make throat yellow.** Like other *Clematis*es, beautiful **feathery white seedhead. Leaves deeply cut into shred-like pattern.** Generally hairy, 5–12.5 cm (2–5") long, with strychnine taste. One aboriginal group said to have spurred on tired horses with smell.

Showy flowers appear suspended in space because **slender vine** is so delicate. Often branched; **usually finds support on other vegetation**; will twine for 75 cm (30") or more. Technically a shrub. Look in shady forest borders.

RANGE BC: Your chance to have a 'first,' close to border in south-central B.C.

RANGE WA: Eastern Washington; sagebrush flats to ponderosa pine forests.

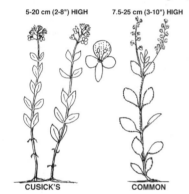

5-20 cm (2-8") HIGH   7.5-25 cm (3-10") HIGH

CUSICK'S   COMMON

Figwort Family, *Scrophulariaceae*
CUSICK'S SPEEDWELL [370]
Veronica cusickii
Mountain Veronica

Note: Alpine speedwell (purple flowers) on p. 294.

A *Veronica* has **pairs of opposite leaves** and **small, 4-petalled flowers**, up to 6 mm (¼") across; **2 stamens**. Of 12+*Veronica*s in our area, many have limited range, several are relatively inconspicuous.

This *Veronica* has **single stems**, topped by **cluster of large, blue-violet flowers**. Flowers 1.3 cm (½") long. Rather broad **upper petal beautifully marked with lines of purple dots**. Stemless, bright green leaves; oval, fleshy, toothless. Often clustered. Adaptable: moist meadows–rocky areas. Mountain forests to alpine heights.

RANGE BC: Southern B.C.

RANGE WA: Abundant in Cascades. Olympics.

COMMON SPEEDWELL, *V. officinalis*: An alien. **Often flat on forest floor**, leaves obscured by conifer needles. **Flower stems upright** as they branch off main stem. Tiny flowers are miniature gems. **Light blue petals often tinged with lavender stripes**. **Finely toothed, opposite leaves have stems**. Dry places, roadsides, fields. Most common in southern B.C.; low elevations–higher mountainsides. In Washington, west of Cascades.

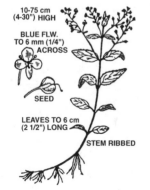

10-75 cm
(4-30") HIGH

BLUE FLW.
TO 6 mm (1/4")
ACROSS

SEED

LEAVES TO 6 cm
(2 1/2") LONG

STEM RIBBED

Figwort Family, *Scrophulariaceae*
AMERICAN BROOKLIME [373]
Veronica americana
American Speedwell

Looks **much like dainty blue garden forget-me-not**. Fleshy, creeping stem, usually **3–5 pairs of oval leaves** regardless of height. Bottom ones are oval-shaped and scarcely toothed; upper ones are more-sharply pointed and have distinct teeth.

Once considered of high medicinal value in Europe. Edible—can be used in salads. Wide range across North America; **wet places, especially ditches. Flowers in airy clusters**, May–July.

RANGE BC: Low–medium elevations in south.

RANGE WA: Wet places, especially ditches. Low to medium elevations throughout.

Gentian Family, *Gentianaceae*
WHITESTEM FRASERA
Frasera albicaulis
Shining Frasera

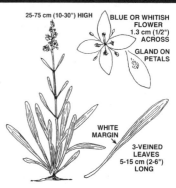

25-75 cm (10-30") HIGH

BLUE OR WHITISH FLOWER
1.3 cm (1/2") ACROSS

GLAND ON PETALS

WHITE MARGIN

3-VEINED LEAVES
5-15 cm (2-6") LONG

This frasera, along with clustered frasera (below) and the jumbo of the family, giant frasera (p. 335), catch one's attention possibly as much by **large, shining green leaves** as by flowers. One would never guess that they are members of the gentian family, except perhaps by noticing that **leaves are smooth and opposite.** They differ from their smaller cousins both by much larger size and by having a **gland at the base of each petal. Flowers are pale blue, or sometimes white,** and despite being in a large cluster, do not draw attention from afar. The flower has both pistils and stamens, the latter alternating with the petals.

Whitestem has a stout stem, to 75 cm (2 ½') high, and a large number of unusual **basal leaves with white margins and 3 veins** that provide positive identification. Stem leaves are smaller and fewer, but continue paired pattern. Watch for this plant late spring to August.

RANGE BC: Reported in B.C., but no specimens are on record, so keep your eyes open.

RANGE WA: Comparatively common east of the Cascades; on lower slopes and foothills.

Gentian Family, *Gentianaceae*
CLUSTERED FRASERA [374]
Frasera fastigiata

TO 1.5 m (5') HIGH

4 BRACTS

4-PETAL 'CUP' FLOWER

LEAVES TO 25 cm (10") LONG

Tall and quite distinctive, with **narrow, erect plumes of pale-blue to mauve flowers**. Usually a number grow in a patch. Dense flower clusters appear almost colourless from a distance but at close range, flowers are quite beautiful. 4-petalled flower is about 1.3 cm (½") across. Wide open, forms a cup, with 1 short, spiny bract between each pair of adjacent petals. 4 erect stamens decorate petal base.

The stout, smooth, green stems have a series of **large whorls of 3 or 4 leaves**; each leaf base partly encircles stem. Now look at another plant. Likely the leaves are in whorls of 3. Roughly oval leaves to 25 cm (10") long. Reminiscent of related giant frasera (p. 335) and of the *Veratrum*s (pp. 188 & 336).

RANGE BC: Not in B.C.

RANGE WA: High, open meadows and forest openings in Blue Mountains and adjacent areas. Big Butte Lookout near Anatone.

5 PETALS

Buttercup Family, *Ranunculaceae*
OREGON ANEMONE
Anemone oregana
Blue Anemone

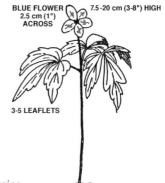

BLUE FLOWER
2.5 cm (1")
ACROSS

7.5-20 cm (3-8") HIGH

3-5 LEAFLETS

This fragile bloom of April and May has **sepals of soft sky-blue, unlike the white tinged with pale blue of most anemones** (white, pp. 176 & 185; yellow, p. 217; purple, p. 300). Unfortunately, petal-like sepals are short-lived, thus soon robbing this slender plant of its main attraction. The **3 leaves form a whorl toward the top of the stem**. Each of the 3–5 leaflets is coarsely toothed. Moist woods.

RANGE BC: Not known in B.C.

RANGE WA: Mostly east of Cascades. Blue Mountains.

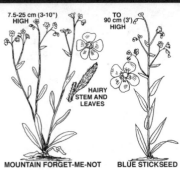

MOUNTAIN FORGET-ME-NOT   BLUE STICKSEED

Forget-me-not Family, *Boraginaceae*
MOUNTAIN FORGET-ME-NOT
Myosotis alpestris
Mouse Ear

Wild forget-me-nots resemble garden species. About a dozen *Myosotis*es in our area; 3 are widespread European introductions.

Forget-me-nots have **weak, slender stems covered with soft hairs. Alternate leaves never toothed. Small, blue, 5-petalled flowers; usually a pale yellow centre**; at stem tops.

Mountain forget-me-not is only high-elevation native one (**but check blue stickseed, a showy imitator**). Slender, hairy stems, **7.5–25 cm (3–10″) long**, often clumped. Each stem has a **spray of flowers**. 'Mouse ear' refers to hairiness of stem and leaves. Moist meadows; subalpine–alpine tundra.

RANGE BC: Common in northern B.C., but becoming scarcer southward.

RANGE WA: Not in Washington.

BLUE STICKSEED or **FALSE FORGET-ME-NOT** [376], *Hackelia micrantha*: **Like a tall forget-me-not, to 90 cm (3′) high**. 'Forget-me-not' flowers; yellow or whitish eye. '**Paddle' basal leaves**; long, narrow, alternate stem leaves; smaller up stem. Flower clusters stem from upper leaf axils. Summer bloomer. 'Stickseed' reflects **barbed seeds**. Wideranging; east of Cascades; **lowlands–subalpine**. Central B.C. southwards. Hart's Pass.

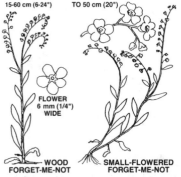

WOOD FORGET-ME-NOT   SMALL-FLOWERED FORGET-ME-NOT

Forget-me-not Family, *Boraginaceae*
WOOD FORGET-ME-NOT
Myosotis sylvatica

Widespread European; presence in many valleys and low mountains gives sense of authenticity. **Small, pale flowers; usually blue, may be white or pink; typically pale yellow centres. 6 mm (¼″) across. Petals lie flat, plate-like.** Early summer bloomer. Small, oval, finely bristly seeds.

RANGE BC: Most common in southern B.C., at low elevations and in damp places.

RANGE WA: Habitats as for B.C. Cascades and western Washington.

SMALL-FLOWERED FORGET-ME-NOT, *M. laxa*: Prefers **ditches and watery places. Usual features of forget-me-nots**, but has **trailing stems and smaller flowers**, to 5 mm (³⁄₁₆″) across. Wet places at lower elevations throughout area.

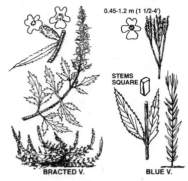

BRACTED V.   BLUE V.

Vervain family *Verbenaceae*
BRACTED VERVAIN
Verbena bracteata
Bracted Verbena

Most eye-catching: **spreading stems and spikes of blue flowers** (may be pinkish or whiteish). **Thin flowering spike with protruding flowers** is characteristic of *Verbena*s. **Long, leaf-like bract at flower base**, hence 'bracted.'

Sprawling; radiating stems to 60 cm (24″) long. Finely hairy, lobed leaves; coarsely toothed. Prefers **dry, sandy soils**; roadsides, banks; low–medium elevations. Blooms late spring through summer.

RANGE BC: South-central B.C. Okanagan Valley

RANGE WA: Eastern and western Washington.

BLUE VERVAIN, *V. hastata*: **Many thin flowering spikes branch from erect, single stem**. Small flowers may be tinged purplish. Prefers damp soils. Range as for bracted.

Violet Family, *Violaceae*
EARLY BLUE VIOLET [377]
Viola adunca
Blue Violet

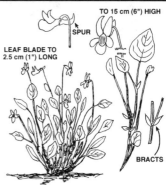

TO 15 cm (6") HIGH

SPUR

LEAF BLADE TO
2.5 cm (1") LONG

BRACTS

Note: White ones, p. 175; yellow, p. 219; purple, p. 294.
Violets are usually quickly recognized. Wild *Violas* **lack fragrance**. This one **blooms early**: April–May. Highly variable. **Pale–dark blue flower, less than 1.3 cm (½") across. Lower petals, with deep purple lines, form pouch; white inside. Upper ones form slender, backward-projecting spur half as long as lower petal.** Attractive leaf clump. **Round–'heart' leaves, long stems.** 2 sharp bracts where leaves join main stem. Forget lunch? Violet leaves and flowers are edible.

RANGE BC: Wideranging, from low to high elevations.
RANGE WA: Wideranging, from low to high elevations.
MARSH or **BLUE SWAMP VIOLET**, *V. palustris*: **Little or no main stem.** Leaves and flowers often on runner from nearby plant. 1 of 6 native species with **bluish mauve–white flowers. Cool, wet places—stream banks**; lowlands to mountain elevations; throughout.

Bluebell Family, *Campanulaceae*
COMMON HAREBELL [375]
Campanula rotundifolia
Bluebell

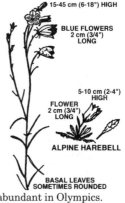

15-45 cm (6-18") HIGH

BLUE FLOWERS
2 cm (3/4")
LONG

5-10 cm (2-4")
HIGH

FLOWER
2 cm (3/4")
LONG

ALPINE HAREBELL

BASAL LEAVES
SOMETIMES ROUNDED

Real 'bluebell of Scotland.' Widely distributed—circles northern part of world. Slender, unbranched stems, several beautiful blue 'bells' near tops. **5 distinct petals. Thin, narrow stem leaves**; younger plants have rounded leaves at base—usually disappear by time of flowering. Height varies greatly with elevation. Blooms during August at higher elevations; by early summer on valley slopes.

RANGE BC: Spotty occurrences, drier ground; roadsides, coastal cliffs; valley bottoms–subalpine.
RANGE WA: Habitats as for B.C. Large-flowered and abundant in Olympics.
MOUNTAIN or **ALPINE HAREBELL**, *C. lasiocarpa*: **Very short. Small, toothed leaves under 2.5 cm (1") long; single large flower.** Rocky places; high elevations throughout central–northern B.C.; Rockies. Washington's northern Cascades. Also see arctic harebell.

Bluebell Family, *Campanulaceae*
SCOULER'S HAREBELL
Campanula scouleri
Scouler's Bluebell

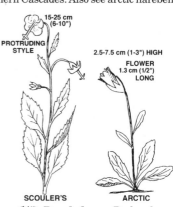

15-25 cm
(6-10")

PROTRUDING
STYLE

2.5-7.5 cm (1-3") HIGH

FLOWER
1.3 cm (1/2")
LONG

SCOULER'S

ARCTIC

Widely spaced flowers. **Thick, protruding yellow style, 3-cleft tip**, extends well out from the **5 recurved, pale blue petals (may be white).**
Sharply toothed, sharp-pointed, alternate leaves; lower ones egg-shaped and short-stemmed. Sometimes branches. Dry forests–rocky exposures.

RANGE BC: Lower elevations west of Cascades. Vancouver Island, Gulf Islands.
RANGE WA: Habitats as for B.C. Near-coastal.
ARCTIC HAREBELL, *C. uniflora*: Tiny. Like mountain harebell but single **flowers half size—1.3 cm (½"). Basal clump.** Rocky places; high elevations; spotty; east of Cascades. Trophy Mtns., Rockies in B.C. Not in Washington.
OLYMPIC HAREBELL [378], *C. piperi*: **Only in Olympics**; high-elevation cracks, crevices. Often abundant on particular rock face. August **flowers 2.5 cm (1") across; 'bowl,' not 'bell.'**

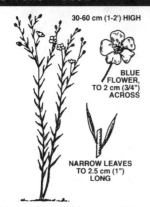

30-60 cm (1-2') HIGH

BLUE FLOWER, TO 2 cm (3/4") ACROSS

NARROW LEAVES TO 2.5 cm (1") LONG

Flax Family, *Linaceae*

WESTERN BLUE FLAX

Linum perenne

A person may encounter this plant growing over a rather large area, and then not see it again for a long time.

Its **saucer-like flowers of brightest blue** are bound to arouse a desire to know more about it. These blooms are carried at the top of a single slender plant to 60 cm (2') high. Each **sky-blue bloom is about 2 cm (¾") across** and has its **flower parts in 5s**. Seedheads form brownish balls almost 6 mm (¼") across. June to August is blooming time.

The leaves are neatly staggered in alternate fashion and are seldom over 2.5 cm (1") long. The flax family's **fibrous stems have been made into linen thread** since time immemorial.

RANGE BC: Mostly east of the Cascades. Spotty occurrences in association with sagebrush and ponderosa pine. Most common in Cariboo. Chilcotin. Windermere Valley.

RANGE WA: Across the state in open, dry places, from lowlands to mountainsides.

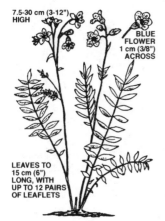

7.5-30 cm (3-12") HIGH

BLUE FLOWER 1 cm (3/8") ACROSS

LEAVES TO 15 cm (6") LONG, WITH UP TO 12 PAIRS OF LEAFLETS

Phlox Family, *Polemoniaceae*

SHOWY JACOB'S-LADDER [380]

Polemonium pulcherrimum

Polemonium, Blue Jacob's-ladder

Note the **ladder-like arrangement of the 11–25 leaflets, each about 1.3 cm (½") long**. With its **tufted, fern-like foliage and bright blue flowers**, this plant looks like an escape from a rock garden. At subalpine elevations it may be a bushy mat with leaves to 5 cm (2") long, but at valley elevations it can be a loose plant to 30 cm (12") high. The flowers and buds cluster at the end of the several stems and are in bloom from May to July. They are about 1 cm (⅜") across, **light blue with a yellow throat—a background for the 5 white stamens**.

RANGE BC: Common on eastern slope of Cascades, from valley bottoms to alpine ecosystem. Manning and Cathedral Lakes parks, Hedley.

RANGE WA: Habitats as for B.C. Hart's Pass, Olympics.

TALL JACOB'S-LADDER, *P. caeruleum*: Very similar to the Jacob's-ladder above, but this one has **unbranched stems** and tends to be a taller plant, to 60 cm (2') high. Also prefers damper sites: **swamps and moist meadows**, from middle-mountain to subalpine. Northern B.C. Not in Washington.

SKUNK or **STICKY JACOB'S-LADDER**, *P. viscosum*: A **tiny plant only a few centimetres (inches) high**. **Tufts of small 'ladder' leaves** surround a few short stems that carry a tight cluster of showy blue flowers. There is a **musky fragrance to the leaves**. Quite rare; in dry areas in the alpine ecosystem of the Cascades. Sporadic in southeastern B.C. and Rockies. Manning Park, B.C. Mountains in Okanogan County, Washington, and Wallowa Mountains of Oregon.

Buttercup Family, *Ranunculaceae*
LARKSPURS
Delphinium spp.

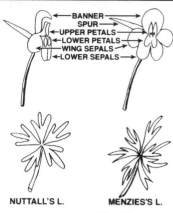

Most people recognize larkspurs because of the unusual flower and its eye-catching intense blue or purple colour. Sepals on most other plants are on outside floral ring, giving support to petals. But in larkspurs, the 5 **large 'petals' you see are actually sepals**. The **uppermost sepal is prolonged backwards into a spur; it encloses 2 small nectar-bearing petals—usually appear to be 1.** Lower 2 petals, both quite small, can be seen clearly; behind them is a **cluster of stamens.** A skilled botanist may make identification based on whether these small petals are lobed or hairy. Most authorities call **flowers bluish purple**, so they are coded here as blue; they can vary considerably.

NUTTALL'S L.   MENZIES'S L.

Larkspurs range over much of North America. B.C. is blessed (possibly the reverse, if you consider the poisonous nature of larkspurs) with 6 species and Washington with double that. They can be very confusing, for there is not only a similarity in leaf shapes but there is considerable variation within the same species. Eventually you come to realize that you are seeing the **same species over and over again, not different ones,** across the same wide range.

The **following key uses range** as the main factor in easy identification of the most common species. Some other larkspurs have particular features that help in identification.

MENZIES'S LARKSPUR [374], *D. menziesii*: **West of Cascades!** From Vancouver Island southwards through Washington to California. The **common** larkspur on bluffs and in rocky openings and dry meadows. Usually less than 40 cm (16") high with stems very finely hairy. **Straight spur.** Blooms mid-May to mid-July.

NUTTALL'S LARKSPUR, *D. nuttallianum*: **East of Cascades!** Widespread from southern B.C. through eastern Washington. Favours dry, rocky soil associated with **sagebrush, grasslands and ponderosa pine;** to lower mountain areas. This is the **common larkspur of early spring to early summer.** (But contrast with Burke's, below.) Usually less than 40 cm (16") high. Stems finely hairy. An impression arises of a **'dirty face' because of the central whitish smudge of small petals amid the bluish sepals.** If these petals are a clear white, see rockslide larkspur (below). On some plants the spur is upward-tilting.

BURKE'S LARKSPUR, *D. burkei*: **Range, size and spur very similar to those of Nuttall's.** Both have deeply divided leaves. **Some leaves are 3-lobed, others 5-lobed.** Leaf stems are no longer than width of a bottom leaf and **flower stem is not hairy** as is that of Nuttall's. The **flower is blue;** showy sepals may be streaked with lighter areas. **Sepals cup-shaped rather than flaring. Spur is straight.** General range as for Nuttall's, but **prefers moist ground** instead of dry ground. Blooms in spring and early summer. Not common in southeastern B.C. but more abundant along easterly foothills of Cascades and eastwards through Washington.

2 TYPES OF LEAVES   OVERLAPPING LEAF LOBES   BASAL LEAVES TO 15 cm (6") WIDE

BURKE'S L.   BICOLOURED L.   TALL L.

BICOLOURED or **LOW LARKSPUR**, *D. bicolor*: **Flowers a deep bluish purple overall, but with a light blue face.** Plant to 30 cm (12") high. **Blooms from late spring to August. Rare;** in southeastern B.C. and extending into northeastern Washington. Wide range from valley to subalpine.

TALL or **PALE LARKSPUR**, *D. glaucum*: **Recognizable by its height, to 2.1 m (7').** **Flowers 2-toned with sepals dark purple and petals pale blue.** In moist forest to subalpine heights, ranging from northern B.C., through the province east of the Cascades, into Washington. Central and southern Cascades. Olympics

ROCKSLIDE LARKSPUR, *D. glareosum*: A **small** larkspur, usually less than 40 cm (16") high. Leaves bushy; 3-lobed, with further divisions. **Clear white petals; otherwise much like Nuttall's.** Distinctive because of range: **on gravel or in rocky places, at subalpine elevations.** Olympics and Cascades of Washington. Hart's Pass. Not known in B.C.

TO 50 cm (20") HIGH

UMBEL
HEAD

BLUE
LINES
ON
PETALS

Lily Family, *Liliaceae*
HOWELL'S TRITELEIA
Triteleia howellii

Note: This *Triteleia* can also be white (p. 186). Related purple *Brodiaeas* are on p. 299.

Lovely cluster lily, **blooms often drooping; dark blue tubes and sometimes pale blue lobes. Dark blue line down each tube and lobe.** Flowers to 2.5 cm (1") long. Up to a dozen flowers and buds per head. **Usually 2 leaves, about half plant height**, during blooming, May–June.

RANGE BC: Southeastern Vancouver Island and nearby Gulf Islands.

RANGE WA: San Juans, Puget Sound; southwards to Oregon. Also east of Cascades; Columbia Gorge, eastern flank of Cascades; Chelan area southwards. Brewster, Bridgeport.

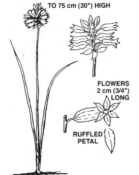

TO 75 cm (30") HIGH

FLOWERS
2 cm (3/4")
LONG

RUFFLED
PETAL

Lily Family, *Liliaceae*
LARGE-FLOWERED TRITELEIA
Triteleia grandiflora
Brodiaea douglasii
Douglas's Brodiaea

To 75 cm (30") high. Slender stem. **2 thin leaves, green at flowering, reach almost to flowerhead.** 3 chaffy bracts just below flowers. **Blue 'trumpets' point outwards horizontally when mature; erect or slightly nodding before.** **Darker lines on tubes and petals.** Short stems. **Ruffled margins on inner petals.**

RANGE BC: East of Cascades: Okanagan, Columbia, Kootenay valleys. Also on Gulf Islands, but rare on southeastern Vancouver Island.

RANGE WA: Common on roadsides April-May; valley bottom to middle-mountain. Common on San Juans. Sagebrush–ponderosa pine habitat. Chelan County, Steptoe Butte. Columbia River benchlands south of Pasco. Columbia Gorge. Idaho's U.S. 95.

PURPLISH BLUE
FLOWER
2.5-4 cm
(1-1 1/2")
ACROSS

30-60 cm
(1-2')
HIGH

GRASS-LIKE
LEAVES

Lily Family, *Liliaceae*
COMMON CAMAS [382]
Camassia quamash

Note: Be sure to distinguish from death-camases— white flowers (p. 196) or green ones (p. 336).

Large, deep-seated bulb was important food of natives. Growing areas were carefully tended; could be inherited. Bulbs were roasted or boiled, ground and then processed and stored as thin cakes.

Long, narrow leaves; like coarse grass. Flower-bearing stem a little longer than leaves. 3 sepals and 3 petals similar; not usually differentiated. **Blue-purple flower to 4 cm (1 ½") across;** occasionally white. 10–30 flowers or buds per stem. **Near coast, range nearly matches that of arbutus and Garry oak.** Was so abundant that explorers Lewis and Clark wrote, 'from the colour of its bloom...it resembles lakes of fine, clear water.' **May cover meadows in blue bloom**, April–June.

RANGE BC: Most abundant near Victoria. Gulf Islands–Campbell River, westwards to Port Alberni. Columbia River valley and Trail–Nakusp.

RANGE WA: Sporadic; widespread; moist, open, grassy areas; west and east.

GREAT CAMAS, *C. leichtlinii*: **Much like common camas.** Slightly larger; blooms later, **only 1–3 flowers open at once. Later, tepals (sepals and petals) twist together.** Southern Vancouver Island, Gulf and San Juan islands; southwards on western lowlands.

Iris Family, *Iridaceae*
WESTERN BLUE IRIS
Iris missouriensis
Blue-Flag, Rocky Mountain Iris

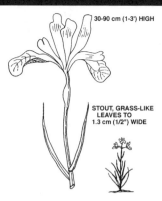

30-90 cm (1-3') HIGH

STOUT, GRASS-LIKE LEAVES TO 1.3 cm (1/2") WIDE

Note: A yellow iris is on p. 226.

A plant beyond confusion when in bloom, for it is **shaped much like a typical garden iris**—perhaps it even started as an escape. There are **1–4 lovely violet-blue flowers, to 7.5 cm (3") across,** on a stout stem that can be 90 cm (3') high. While the **flower appears to have 6 petals, 3 are actually sepals.** The broad sepals usually stand upright. There is a **yellowish intrusion in the throat of the flower, with spreading violet lines extending to the edges of the sepals.** Wide possible range in **wetlands**, along **lake edges** and in meadows, from valley bottoms to middle-mountain heights.

RANGE BC: Infrequent occurrences in north-central B.C. (Prince George) and along coast southwards.

RANGE WA: Infrequent in northern Puget Sound. Common, from sagebrush to ponderosa pine habitats, but usually on the edges of marshes, seepage areas and wet meadows. Ellensburg, Yakima Canyon, Steptoe Butte, Pullman.

OREGON or **PURPLE IRIS**, *I. tenax*: A plant very similar to the iris above, but smaller, with **flowers of blue to purple**. The **3 showy, large petals are marked with yellow, but the lines do not come to the edge of the petal**. The word *tenax* means 'tough' or 'strong' and refers to the leaf fibres, which natives used for cording in centuries past. The range—pastures, meadows and open forests in western Washington—should identify this flower of spring to early summer. Abundant along roadsides in Oregon's Willamette Valley. Not in B.C.

Sunflower Family, *Asteraceae (Compositae)*
CHICORY [383]
Cichorium intybus
Blue Sailors

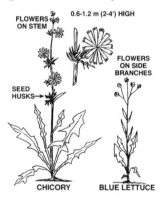

0.6-1.2 m (2-4') HIGH

FLOWERS ON STEM

FLOWERS ON SIDE BRANCHES

SEED HUSKS→

CHICORY   BLUE LETTUCE

A wildflower...or an introduced weed escape—as you see it. Chicory's **bright blue flowers** here and there along roadsides is one thing, but running wild through farm and meadow, as in some areas, is something else. The name 'chicory' is familiar to many through the use of its deep taproot as a coffee substitute. 'Blue sailors' comes from an old legend concerning a sailor's sweetheart who was deserted, but nevertheless kept faithful watch for him. The gods took pity on her and turned her into a blue-eyed plant to haunt roadsides from July to September.

Clumps of 2–3 tissuey, light blue flowers in clumps of 2–3 in leaf axils (or often 1, with 2 seed husks). They are 2.5–4 cm (1–1 1/2") across and have **toothed petals**.

RANGE BC: Sporadic occurrences along roadsides of southern B.C. and coastal areas. Salmon Arm, Sicamous. A serious pest in some areas of B.C.

RANGE WA: Roadsides and waste places across Washington.

BLUE LETTUCE, *Lactuca tatarica*: Note: Wall lettuce, yellow flowers, is on p. 224. **Same flower colour and general size as chicory; also exudes a milky juice from broken stem.** However, **flowers are at ends of short side branches, in leaf axils.** Supposedly fairly common at lower elevations in all but northwestern B.C.—I consider it rather rare. More common east of Cascades. Manning Park. Not recorded for Washington.

TALL BLUE LETTUCE, *Lactuca biennis*: Shape of leaves and flowers very **similar to blue lettuce, but more robust; 1.8 m (6') high. Flowers generally light blue–dull white.** Leaves 10–30 cm (4–12") long. Considered a wide-ranging weed on moist meadows in eastern and western Washington. Across B.C. in similar habitat.

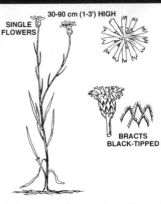

SINGLE
FLOWERS

30-90 cm (1-3') HIGH

BRACTS
BLACK-TIPPED

Sunflower Family, *Asteraceae (Compositae)*
BACHELOR'S BUTTON [379]
Centaurea cyanus

This Mediterranean plant is plentiful in some areas as an escaped garden flower. Its usual bright blue colour is eye-catching, but it **can also range to white, pink and purple**. It is closely related to the troublesome knapweeds with their pink-to-purple heads. And notice the **similarities to chicory** and to blue lettuce.

This plant ranges from 30 to 90 cm (1 to 3') in height and carries a single flower at the tip of each branching stem. The **head is a spray of ray flowers, with no central button. Outer rays are much larger and give the effect of fringed petals**. This motif is perpetuated by a **fine, black fringe on the bracts**. Leaves are very narrow and alternate; when young, fine hairs give them a grey sheen.

RANGE BC: Turns up here and there in waste areas and on roadsides across southern B.C.

RANGE WA: Locally throughout, in waste areas and on roadsides. Columbia Gorge.

TUBULAR FLOWERS

TO 30 cm (12")

TO 35 cm (14")

RIGHT
ANGLE

WIDE
ANGLE

LARGE-FLOWERED   SMALL-FLOWERED

Figwort Family, *Scrophulariaceae*
LARGE-FLOWERED BLUE-EYED MARY [384]
Collinsia grandiflora
Innocence, Blue Lips

Open, rocky places, green knolls, spring sunshine...and blue-eyed Mary, **often the earliest flower to bloom**. The many branched and **spreading stems often produce a mat effect**. They lengthen considerably as the flowering season progresses; the flower stems can reach 30 cm (12") in height. The **tiny flowers, about 1.3 cm (½") long, grow in a series of whorls** up the stem. Notice that the **upper petals are white—the others are blue**—and almost square. A decisive identification feature is the way the **flowers attach to the calyx at a right angle**, as shown in the drawing. Leaves are alternate and vary considerably in shape.

RANGE BC: Most common west of the Cascades, in open places from low to middle-mountain elevations.

RANGE WA: Habitats as for B.C.

SMALL-FLOWERED BLUE-EYED MARY, *C. parviflora*: The **blue flowers, their upper lips tipped with white and purple, are quite tiny, about 6 mm (¼") long**. They **tend to hide among the thin leaves at the base of the cluster**. The **angle of calyx to flower is only about half of that for the blue-eyed Mary above**. However, the shape of the 2 plants and the difference in size of flowers should allow for quick identification. The range is open places throughout our area from valley to middle-mountain heights. Primarily east of the Cascades.

Gentian Family, *Gentianaceae*
GENTIANS
Gentiana and *Gentianella* spp.

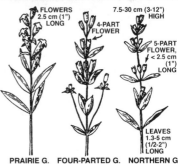

PRAIRIE G.   FOUR-PARTED G.   NORTHERN G

Gentians prefer shady glades or moist alp-lands to make their shy appearances. All have **opposite or whorled leaves** and **most have erect, blue-purple flowers balanced in an artistic, symmetrical fashion**.

There are about a dozen gentians in our area; several are quite rare and are therefore not included. **Gentians with folds between the flower petals are *Gentianas*; those without are *Gentianellas*.**

Most gentians are famous for their beautiful shades of blue, which have been described as 'Heaven's own blue.' 'Gentian' comes from Gentius (180–167 B.C.), King of Illyria. He must have been very young, and yet he receives credit for popularizing gentians as having medicinal values

PRAIRIE or **LARGE GENTIAN**, *Gentiana affinis*: This erect, branched plant is from 12.5 to 30 cm (5 to 12") high and has **tubular flowers 2.5 cm (1") long**, either borne singly or in a cluster of several at stem ends. Notice that there is a **single, short tooth between each pair of short petal lobes**, and no fringe of hairs across the inside of the petal base, as there is in northern gentian (below). Favours **moist meadows** and **grasslands**, from lowlands to lower mountains. From Cranbrook to the Rockies in B.C. East of the Cascades in Washington; southwards to northern California.

FOUR-PARTED GENTIAN, *Gentianella propinqua*: A **small, much-branced plant to 15 cm (6") tall**, with a **flower stem from almost every leaf axil. Flowers single and varying greatly in size, with the larger ones at the top of the plant. Petals have 4 sharp-pointed lobes**. No fringe of hairs across the inside of the petal base, as there is in northern gentian (below). Moist areas, lowlands to subalpine heights. Most common in northern B.C. but extending southeastwards along the Rockies. Not in Washington.

NORTHERN GENTIAN, *Gentianella amarella*: To 30 cm (12") tall. **Often with 3 or more flowers in each cluster.** Narrow, tube-like **flowers are less than 2.5 cm (1") long and are parted into 5 short lobes. Fringe of hairs on inside of each petal base.** Most common east of the Cascades at lower elevations, in moist meadows and woods. From the far north southwards through Washington.

Gentian Family, *Gentianaceae*
KING GENTIAN [385]
Gentiana sceptrum
Swamp Gentian

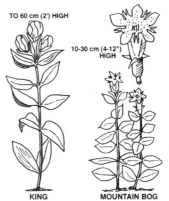

KING   MOUNTAIN BOG

King gentian has **numerous dark blue flowers**, on a **robust plant to 60 cm (2') tall**. Several per cluster, **showy flowers, 2.5–4 cm (1–1 ½") long**, are an appropriate size. Notice the **dark green dots on the insides of the petals**. Leaves are stemless and climb in opposite pairs to the tops of the stems. **Wet ground** habitat and **late summer blooming** are other distinguishing features.

RANGE BC: Boggy places, wet meadows, lake edges at lower elevations. Southern B.C., west of Cascades

RANGE WA: Habitats as for B.C. West of Cascades, southwards to southern California.

MOUNTAIN BOG GENTIAN, *Gentiana calycosa*: **Wet-ground**, like king gentian, but prefers **subalpine–alpine**. Look for **several erect stems**, each supporting a **single deep blue flower**, though occasionally 2. **Stemless, erect, 'bluebell' flowers**. Unusual floral design: **forked filament between each pair of petal lobes**. Opposite, **egg-shaped leaves**. Similar to species found in Alps and Himalayas. Blooms all summer; into September in wet places at high elevations. Common in our area. Most high mountain parks. Olympics.

TO 40 cm (16") HIGH

'FUZZY' TONGUE

FLOWERS TO 2.5 cm (1") LONG

LOWER LEAVES TO 7.5 cm (3") LONG

Figwort Family, *Scrophulariaceae*
FUZZY-TONGUED PENSTEMON
Penstemon eriantherus

Note: Purple penstemons and more information on p. 302. Yellow–white ones on p. 230. Shrubs on p. 133.

Penstemons generally have very attractive, generously sized, abundant flowers. But this one is special: **extra-large, extra-vibrant bulbous flowers with varicoloured faces.** My memory takes me back to open ponderosa pine country, dry, exposed banks and this lovely plant—**to 40 cm (16") high.** **'Fuzzy-tongue' is very descriptive.** Unbelievably luxuriant and colourful. Many stems, topped by **thick clusters of flowers; light blue to pinkish or purple, plus other-coloured streaks. 4 dark stamens above tongue.** Narrow, irregularly toothed, long-stemmed lower leaves. Blooms late spring onwards; low–medium elevations.

RANGE BC: Infrequent in south-central and southeastern B.C.

RANGE WA: Common in Okanogan, Ferry counties. Southwards east of Cascades.

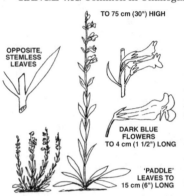

TO 75 cm (30") HIGH

OPPOSITE, STEMLESS LEAVES

DARK BLUE FLOWERS TO 4 cm (1 1/2") LONG

'PADDLE' LEAVES TO 15 cm (6") LONG

Figwort Family, *Scrophulariaceae*
ELEGANT PENSTEMON
Penstemon venustus

Very eye-catching, because of height, **to 75 cm (30"),** and **very dark blue, large, clustered flowers.** Usually branches from base; stems arch outwards then upwards, terminating in **30 cm (12") flower spike.** Twin leaf axils each give rise to a short stem with a **cluster of 3–5 flowers, each to 4 cm (1 ½") long.** All flowers on stem face same way. **2 small, whitish pads in throat** contrast with blue petals. 4 anthers held close to arch of top petal. Sterile **fifth anther has finely haired, yellowish tip.**

Cluster of long-stemmed leaves may sprout at base, soon giving way to opposite, clasping, serrated leaves. Neither stem nor leaves are hairy.

Usually in **sagebrush and bunchgrass ecosystems.** Blooms in June, with white and purple fleabanes and Hooker's onion as balsamroot fades.

RANGE BC: Not recorded.

RANGE WA: Mostly in south (Walla Walla County), but also in Douglas County.

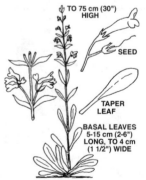

TO 75 cm (30") HIGH

SEED

TAPER LEAF

BASAL LEAVES 5-15 cm (2-6") LONG, TO 4 cm (1 1/2") WIDE

Figwort Family, *Scrophulariaceae*
TAPER-LEAVED PENSTEMON
Penstemon attenuatus

To distinguish this penstemon from several dozen in our area can take a long, involved botanical key. As well, nearby plants of same species may have various shades of flowers and different-sized leaves. This general description covers several varieties.

To 75 cm (30") high. Can be locally abundant. Attractive. **Whorls of blue, purple or pinkish flowers.** Bright green leaves, smooth edges. **Base tends to appear tufted because of numerous 'paddle' leaves.** Pairs of opposite, stemless leaves climb stem. **Flower clusters on twin stems from leaf axils.** Stems shorten with each clearly spaced whorl, disappearing in top few clusters. **Flowers 1–2 cm (³⁄₈–³⁄₄") long;** leaves **long, thin tendril** on going to seed. Blooms mid-May to mid-June. Lowland forests.

RANGE BC: Not recorded.

RANGE WA: Common; eastern slope of Cascades. U.S. 12 west of Yakima.

Forget-me-not Family, *Boraginaceae*
LONG-FLOWERED BLUEBELL [386]
Mertensia longiflora

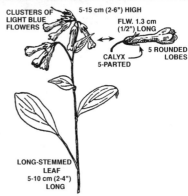

It is fairly easy to **recognize the *Mertensia*s by their clusters of 'bluebells.'** Most of the taller ones grow in humid situations or at mountain elevations. All are characterized by **broad, alternate leaves** and by **seeds with a long, protruding appendage.**

Especially treasured **among the abundance of early spring blooms**—buttercups, yellow bells, shootingstars, prairie stars, western springbeauty and, in certain areas, avalanche lilies—for its infusion of deep blue colour into the floral picture. A **small plant, from 5 to 15 cm (2 to 6") high,** it bears on its erect stem a **top-heavy clump of long-tubed blue flowers. Usually buds and young flowers have a pinkish tinge.** The **long flower tube flares out for the last third of its length.** The few leaves are long-stemmed at the base of the plant but stemless above.

RANGE BC: East of Cascades; in meadows, but more often in the shade of ponderosa pine. Chase. South Okanagan.

RANGE WA: East of Cascades, at lower elevations. Most common under ponderosa pine but extending into sagebrush and bunchgrass ecosystems. Lower slopes of Blue Mountains.

Forget-me-not Family, *Boraginaceae*
TALL BLUEBELL [390]
Mertensia paniculata

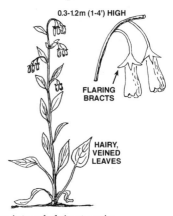

Tall bluebell almost calls its name, for it may indeed be very tall, **up to 1.2 m (4'),** and it has **clusters of bell-shaped blue-to-pink flowers that droop gracefully from short stems.** The main stem, sometimes branched, **droops at top so flowers hang down. Flowers are about 2 cm (¾") long, pinkish when young.** After blooming, a **long style remains protruding.** Basal leaves are long-stemmed, but leaf stems rapidly shorten further up stem. **Leaves have prominent veins.** Often grows in vigorous abundance, especially on wetter ground. Its striking blue is a beautiful addition to the profusion of spring wildflowers. Ranges from moist forests up into subalpine terrain.

RANGE BC: Cascades. Most abundant north of Quesnel.

RANGE WA: Cascades and Olympics. Eastern slope of Steven's Pass. Ellensburg.

Forget-me-not Family, *Boraginaceae*
WESTERN MERTENSIA
Mertensia platyphylla

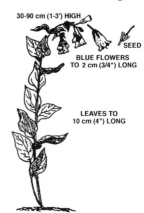

One of the taller *Mertensia*s, this plant ranges from 30 to 90 cm (1–3') in height. It **tends to have a single stem,** in contrast to the bluebells described above.

Has the usual **long-stemmed basal leaves** and almost-clasping upper leaves. The **tightly packed clusters of buds** gradually evolve into beautiful blue 'bells' during the early summer months.

RANGE BC: Not recorded.

RANGE WA: Moist places and rich soils west of Cascades. Grays Harbor County and Puget Sound region.

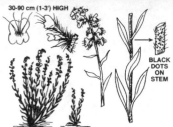

30-90 cm (1-3') HIGH

BLACK DOTS ON STEM

Forget-me-not Family, *Boraginaceae*
VIPER'S BUGLOSS [387]
Echium vulgare
Blueweed

Echis is Greek for 'viper.' Forms a showy roadside community. Eye-catching **whorls of dark blue flowers in coiled cluster**. A dozen or more stout stems per **bush-like** plant. Commonly **to 90 cm (3') high**; may be only half that. Distinguish from phacelias (p. 296) by **4 protruding stamens plus 1 shorter one** (some phacelias have many) and by **2 petals larger than other 3** (phacelias have 5 uniform ones). Distinctive **stems marked with black dots and short, bristly hairs**. Basal leaves have stems; higher ones do not. European. Weed pest in eastern U.S.; not yet very common in B.C. or Washington.

RANGE BC: Sporadic; coastal, south-central, southeast. Merritt, Crow's Nest Pass.

RANGE WA: Southwards from B.C. border into northeast. Spokane region.

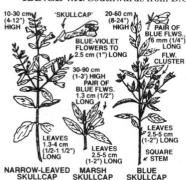

10-30 cm (4-12") HIGH • 'SKULLCAP' • 20-60 cm (8-24") HIGH

PAIR OF BLUE FLWS., 6 mm (1/4") LONG

BLUE-VIOLET FLOWERS TO 2.5 cm (1") LONG

FLW. CLUSTER

30-90 cm (1-3') HIGH PAIR OF BLUE FLWS. 1.3 cm (1/2") LONG

LEAVES 2.5-5 cm (1-2") LONG

LEAVES 1.3-4 cm (1/2-1 1/2") LONG

LEAVES 2.5-5 cm (1-2") LONG

SQUARE STEM

NARROW-LEAVED SKULLCAP — MARSH SKULLCAP — BLUE SKULLCAP

Mint Family, *Lamiaceae*
MARSH SKULLCAP [389]
Scutellaria galericulata
Bugleweed

Skullcaps named for curious little **projection or crest on calyx of long, funnel-like flowers. Flowers arise as 'twins' from leaf axils. Flowers are pale blue—almost mauve—and usually white-streaked**. About 2.5 cm (1") long. Slender, square stem. Narrow, **opposite leaves**. Wet habitats: lake and stream edges, damp meadows and damp forests. Blooms July–August.

RANGE BC: Lowlands across our area.

RANGE WA: Habitats as for B.C.

BLUE SKULLCAP, *S. lateriflora*: **Paired, long-stemmed, elongate clusters of small, pale blue flowers from leaf axils; easily overlooked**. Square stem. Blooms in late summer. Moist ground; lowlands to middle-mountain heights. Most common in central and southern B.C. and southwards through Washington, both sides of Cascades.

NARROW-LEAVED SKULLCAP, *S. angustifolia*: Like marsh skullcap but flowers blue-violet. Dry places from southern B.C. southwards.

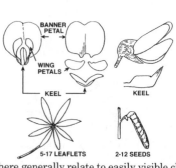

BANNER PETAL

WING PETALS

KEEL — KEEL

5-17 LEAFLETS — 2-12 SEEDS

Pea Family, *Fabaceae (Leguminosae)*
LUPINES
Lupinus spp.

Note: Variable; all **lupines that may be blue, violet, magenta or purple are here**. A large shrub (yellow) is on p. 149; white, p. 192; yellow, p. 234.

Lupines—locally at least 24 species and many subspecies—are common and widespread, but very confusing. Form hybrids easily. Species with limited range (about half) are omitted. Distinctions given here generally relate to easily visible clues or range rather than floral detail.

5 petals take unusual shapes: **Banner petal (large, at top) generally grooved in centre, sides turned upwards**. However, **shape and hairiness of keel (bottom) are important in identification. 10 stamens unite at base. Typically blue–purple flowers**. Pod dries out and twists open to scatter 2–12 seeds—contain toxic alkaloids.

5–17 leaflets radiate from central point (palmately compound); **very hairy on some species, smooth on others; never toothed**. They 'sleep' at night—they fold umbrella-like. As legumes, lupines enrich the soil with nitrogen from the air.

Pea Family, *Fabaceae (Leguminosae)*
SMALL-FLOWERED LUPINE
Lupinus polycarpus
Lupinus micranthus

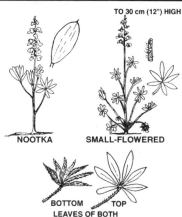

TO 30 cm (12") HIGH

NOOTKA   SMALL-FLOWERED

BOTTOM   TOP
LEAVES OF BOTH

Look for a **small annual to 30 cm (12") high**, though on dry hillsides they may be half that size, much branched and abundant with leaves. Usually there are **7 leaflets, smooth above but finely hairy beneath**; the stems are also finely hairy. Nootka lupine (below) has similar stems and leaves. Bicoloured lupine (p. 327) may be in the area, but its leaves are hairy above. Leaves, on long stems, average about 4 cm (1 ½") across.

The flower stem protrudes a short way above the leaves and carries a cluster of **small, dark blue flowers, each about 6 mm (¼") long**, with twin white blazes on the face. It may be blooming with spring gold and shootingstars.

RANGE BC: Locally common in dry areas on southern Vancouver Island, Gulf Islands and adjacent mainland.

RANGE WA: San Juan Islands and southwards through the state to California, but generally west of the Cascades. Central section of Columbia Gorge.

NOOTKA LUPINE, *L. nootkatensis*: Very similar to small-flowered lupine (above) in general appearance. **Leaf stems branch from part way up the main stem. Tips of leaflets are rounded, with a slight point**. Coastal Vancouver Island and mainland. Also in the Cascades of B.C. Not recorded in Washington.

SEASHORE LUPINE, *L. littoralis*: The **only lupine found on seashores**. Usually with **prostrate, hairy stems** and often forming mats. Leaf has 6–8 leaflets. Small, clustered flowers, pink to mauve to purple. Scattered locations in southern coastal B.C. and southwards through Washington to California.

Pea Family, *Fabaceae (Leguminosae)*
SPURRED LUPINE [391]
Lupinus laxiflorus

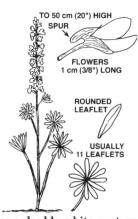

TO 50 cm (20") HIGH
SPUR

FLOWERS
1 cm (3/8") LONG

ROUNDED
LEAFLET

USUALLY
11 LEAFLETS

Look for the spur! Another lupine with great possibilities for initial confusion, as the **flower can be whitish, pinkish, blue, purple or violet.** A redeeming feature is that it is a pleasant companion along many of the forested roadsides of northwestern or south-central Washington—U.S. 12 west of Yakima, for example. Another is that despite its variable colour and habitat, there is one feature of the small flower that allows positive identification: the **spur on the top side of the calyx** (see drawing).

This is a slim, erect lupine, to 50 cm (20") high, with its 12.5 cm (5") plume of **small, purple-violet flowers**. Each flower averages 1 cm (⅜") long and carries a **double white spot on the banner petal**—the large, central, upper one. Note that arctic lupine (p. 326) also has these white markings, but lacks the tell-tale spur. **Pale purple keel** (bottom part of the flower). Leaves are dark green, usually with 11 pointed leaflets. Leaves climb high on the stem. Watch for taper-leaved penstemon blooming nearby. Spurred lupine is 1 of 2 lupines that Fendler's blue butterfly feeds on. This rare butterfly appeared to have vanished in 1931, but has been found recently (1993) in a state forest in Oregon.

RANGE BC: Rare; in extreme southern B.C., east of the Cascades. Rockies.

RANGE WA: Northeastern part of the state and along the eastern base of the Cascades. Edges of Garry oak and ponderosa pine forest. Klickitat County; Goldendale. Yakima County.

30-60 cm (1-2') HIGH

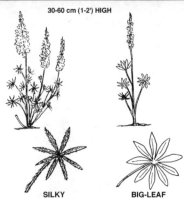

SILKY   BIG-LEAF

Pea Family, *Fabaceae (Leguminosae)*
SILKY LUPINE [392]
Lupinus sericeus

2 features allow for easy identification: First, silky lupine is by far the **most abundant lupine in the sagebrush and ponderosa pine regions**. Second, an **overall grey or silvery sheen** caused by silky white hairs on stem and leaves. **Soft blue flowers with lavender tinge, thickly clustered on erect spikes that reach well above leaves**. **Flowers have 2 white markings** on them, as do those of arctic lupine (below), but are otherwise quite different. They seem a perfect complement to sagebrush and rabbit-brush. Abundant leaves, **5–9 very narrow leaflets. Blooms in early summer**, before landscape becomes parched.

RANGE BC: Upper borders of the sagebrush ecosystem and throughout the ponderosa pine ecosystem.

RANGE WA: Habitats as for B.C. Note that a white-flowered variety is quite common south of the Snake River and in open uplands of Asotin County.

BIG-LEAF LUPINE, *L. polyphyllus*: **Largest of lupines, except for tree lupine** (p. 149), big-leaf is to 1.2 m (4') high. **Commonly seen along roadsides**. From this plant came many of the very showy cultivated varieties. **Big leaves, to 15 cm (6") across, form almost a perfect circle**. Typically **10–17 sharp-pointed leaflets**, more numerous than on other lupines. **Flower spike, to 50 cm (20") long, stands erect above leaves**. Often a pinkish hue to banner petal. **Keel petal has sharp curve in long spine**.

David Douglas discovered big-leaf in Washington or Oregon in 1825 and wrote, 'Nor can I pass the beauty, not to say grandeur of *Lupinus polyphyllus*...on the lowland and borders of streams....' Found in damp meadows, on streamsides and in seepage areas; at lower elevations throughout our forested areas.

30-60 cm (1-2') HIGH

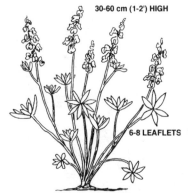

6-8 LEAFLETS

Pea Family, *Fabaceae (Leguminosae)*
ARCTIC LUPINE
Lupinus arcticus
Lupinus latifolius
Broad-leaved Lupine

For most people doing a fair amount of travel, will be **seen over and over** for vast stretches. A **near-constant roadside companion in mountain forest ecosystem**, from south Cascades of Washington to far north of B.C. When you reach the glorious wildflower meadows **at timberline, there it is again! Provides the dominant blue colour**—sometimes a plant here and there mingled with other flowers, at other times a solid **carpet of blue with little white dots** to set it off. These dots are **double rectangular white blazes on each flower's otherwise blue face**. By this marking, and by **white keel** (take a close look), you can quickly recognize it, though some plants in Olympics do not show blazes so distinctly. Blooms in Columbia Gorge in mid-April, but into late summer in high mountains.

This lupine is from 30 to 60 cm (1 to 2') high. The basal leaves are abundant and have very long stems. Leaves have **6–9 light green leaflets with sharp-pointed tips**.

RANGE BC: 3 varieties, based on slightly differing floral features; combined range includes much of our area, up to northern B.C. Often with trees; middle to subalpine elevations.

RANGE WA: Habitats as for B.C. Columbia River northwards. Mt. Rainier, Olympics.

DWARF MOUNTAIN LUPINE, *L. lyallii*: You might put a hand or a hat over this **tiny** plant that flourishes on **high, dry ridges**. Beautiful dwarf with tiny, **silky-hairy leaves** and **bright blue flowers**. Size, less than 15 cm (6") tall, will identify it. Spotty occurrences in subalpine and alpine ecosystems of Cascades in southern B.C. and northern Washington.

Pea Family, *Fabaceae (Leguminosae)*
BICOLOURED LUPINE [393]
Lupinus bicolor

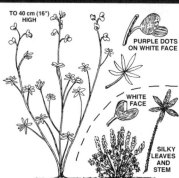

TO 40 cm (16") HIGH

PURPLE DOTS ON WHITE FACE

WHITE FACE

SILKY LEAVES AND STEM

BICOLOURED LUPINE PRAIRIE LUPINE

Recognize with confidence: **contrasting dark blue and white should be sharply defined on each flower**. To make sure, check for **dainty purple dots on white banner petal**.

May be 40 cm (16") tall; often less than half that. Leaves about 2.5 cm (1") across; usually **7 leaflets, finely hairy tops**. Small-flowered, p. 325, overlaps range, but leaves smooth above. Blooms April–May.

RANGE BC: Scarce; southeastern Vancouver Island, Gulf Islands and adjacent mainland. John Macoun, a pioneer botanist, found it in Victoria's Beacon Hill Park in 1908.

RANGE WA: Dry lowlands of west; Columbia Gorge to The Dalles. Klickitat County. David Douglas found it 'in the interior of the country about the Columbia River,' *c.* 1825.

PRAIRIE LUPINE, *L. lepidus*: **Many blue-and-white flowers densely clustered on short, projecting spikes tight on top of mat** about 30 cm (12") across. **Long-stemmed, hairy leaves only 15–30 cm (6–12") tall**. Rich colour after spring rains. Blooms early June–summer. Rare; several varieties. Southeastern Vancouver Island, across B.C.; low–medium elevations. Dry places, both sides of Washington's Cascades. Okanogan County, Olympics.

Pea Family, *Fabaceae (Leguminosae)*
ALFALFA
Medicago sativa

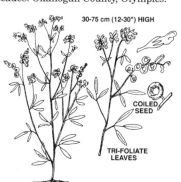

30-75 cm (12-30") HIGH

COILED SEED

TRI-FOLIATE LEAVES

Common farm plant but may not be recognized by everyone. **Small, pea-like flowers usually blue, may be yellow, white, mauve or even dark purple**—immediate recognition as relative to clovers, lupines, etc. 3 leaflets reinforce connection. Eliminate any confusion by **tightly coiled seed pods**. High protein content; very nutritious for livestock. Like clovers, roots fix atmospheric nitrogen into organic compounds that enrich soil. Common escape from farm areas; long stretches of roadside on good soil with some moisture. *Medicago*'s name relates to ancient Persia, where it was grown.

RANGE BC: Main irrigated valleys and roadsides.

RANGE WA: Main irrigated valleys and roadsides.

CLUSTERED FLOWERS

Phlox Family, *Polemoniaceae*
GLOBE GILIA
Gilia capitata
Blue Gilia, Globeflower

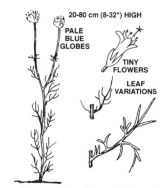

20-80 cm (8-32") HIGH

PALE BLUE GLOBES

TINY FLOWERS

LEAF VARIATIONS

Note: Scarlet gilia (red 'trumpets') is on p. 290.

One of a kind. **Tightly packed globe, 2.5 cm (1") across**, on slender stem. **Small, light-to-dark-blue flowers, less than 6 mm (¼") across. 5 petals**. Anthers barely protrude. Alternate **leaves 'thread' thin; irregular branching arms**; size and shape can vary greatly. Blooms by mid-June in Columbia Gorge, later elsewhere. Very adaptable; prefers dry, open areas at lower elevations.

RANGE BC: Rare on Vancouver Island and in Peace River region. Sporadic in south.

RANGE WA: Sporadic occurrences across state. Most common west of Cascades. Columbia Gorge. David Douglas collected seeds from near Ft. Vancouver *c.* 1825.

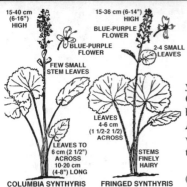

15-40 cm (6-16") HIGH

15-36 cm (6-14") HIGH
BLUE-PURPLE FLOWER

BLUE-PURPLE FLOWER

FEW SMALL STEM LEAVES

2-4 SMALL LEAVES

LEAVES 4-6 cm (1 1/2-2 1/2) ACROSS

LEAVES TO 6 cm (2 1/2") ACROSS 10-20 cm (4-8") LONG

STEMS FINELY HAIRY

COLUMBIA SYNTHYRIS   FRINGED SYNTHYRIS

Figwort Family, *Scrophulariaceae*
COLUMBIA KITTENTAILS [394]
Synthyris stellata
Columbia Synthyris

This is a **flower you can see from your car** if you drive slowly in the Multnomah Falls area on the Oregon side of the Columbia Gorge. It is an early bloomer, so the **best time is mid-March to mid-April.** While shady, north-facing banks are rich in vegetation, the **columnar clusters of small, blue-to-purple flowers** are abundant and eye-catching.

Although this plant is usually only 15–30 cm (6–12") high, its erect stem of light purple flowers is easily distinguished. The flowers have **4 rather square-cornered petals.** Closer inspection reveals a fuzzy outline to the flowers, because each has **2 long, protruding anthers and a style.**

Distinctive leaves arise from the base of the plant. They are **long-stemmed and kidney-shaped, and have large lobes each with 3 teeth.** Fifty years ago, Columbia kittentails was not known to occur in Washington. Fortunately it is an example of a plant expanding its range rather than disappearing.

RANGE BC: Not recorded.

RANGE WA: Most abundant along the scenic highway on the Oregon side of the Columbia Gorge. Extends to eastern Washington in moist, shady areas. Blue Mountains.

Figwort Family, *Scrophulariaceae*
FRINGED SYNTHYRIS
Synthyris schizantha

The 'fringed' part of the name relates to the **small flower, which has 4 fringed petals.** The flowerhead is a **fuzzy cluster of blue-purplish blooms** on a stem to 15 cm (6"). Leaves closely resemble those of Columbia kittentails (above) but have **hairy stems and leaf ribs.** However, this one will likely be blooming 1 or 2 months later, from April into May. It prefers moist banks and bluffs at middle-mountain elevations.

RANGE BC: Not recorded.

RANGE WA: Cascades, Mt. Rainier and lower slopes of the Olympics.

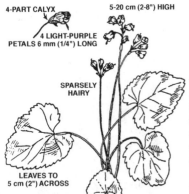

4-PART CALYX

4 LIGHT-PURPLE PETALS 6 mm (1/4") LONG

5-20 cm (2-8") HIGH

SPARSELY HAIRY

LEAVES TO 5 cm (2") ACROSS

Figwort Family, *Scrophulariaceae*
ROUND-LEAVED SYNTHYRIS
Synthyris reniformis
Snow Queen, Spring Queen

'Snow queen' is a good descriptive name, for these shy little flowers make a **welcome appearance as soon as the snow is off the ground.** Often the plants are widespread on the forest floor—the only blooms to be seen. Although most plants are **only 5–15 cm (2–6") high,** the **erect, leafless stem** with its **cluster head of light purple flowers** is easily distinguished.

The **large, dull green leaves stand as high as the flowers** and almost overshadow them. Leaves are heart- to kidney-shaped and palmately veined. The **tiny, purple, bell-shaped flowers have 4 sepals, 4 petals, 2 purple protruding stamens and a long style.** The whole flower is only 6 mm (¼") long. Blooms from March to May. At higher elevations, this plant may be found as the first trilliums appear.

RANGE BC: Not recorded.

RANGE WA: Open, well-drained coniferous woods west of the Cascades, at lower elevations. Columbia Gorge.

Teasel Family, *Dipsacaceae*
TEASEL [395]
Dipsacus sylvestris

This **large, coarse weed** is especially noticeable in April and May, when its **tall, prickly dead
stem and great spiny burs** tower above the ver-
dant spring greenery. The dead stalks with their
burs are a far more familiar feature than the plant in
summer dress.

Its large, oposite leaves often reach 30 cm (12")
in length. During the summer months, it displays
dense, spike-like heads of 4-petalled pale purplish
flowers, each suported by an involucre with long
bracts. Then, **brown burs of imposing proportions** form. When dry, they are a perfect symmetry of

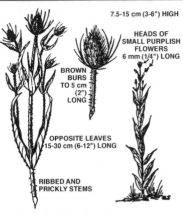

7.5-15 cm (3-6") HIGH

HEADS OF
SMALL PURPLISH
FLOWERS
6 mm (1/4") LONG

BROWN
BURS
TO 5 cm
(2")
LONG

OPPOSITE LEAVES
15-30 cm (6-12") LONG

RIBBED AND
PRICKLY STEMS

long, curving spines. So stiff and evenly spaced are the spines that the large burs were once
used to card (or tease, and hence the name 'teasel') wool—thus its introduction from Europe
by early pioneers. The burs make an attractive winter decoration.

RANGE BC: Rare; in southwestern B.C. and on southeastern Vancouver Island.

RANGE WA: Mostly in damp places in the bunchgrass ecosystem. Whitman County,
Steptoe Butte (lots!). Also Garfield, Pullman, Dayton, Prosser.

'DAISY' FLOWERS

Sunflower Family, *Asteraceae (Compositae)*
LINDLEY'S ASTER
Aster ciliolatus
Fringed Aster

A rather modest aster with small flowers, but
with **erect stems from 0.3 to 1.2 m (1 to 4') high**.
Very often it masses, thereby producing a concentrated blaze of colour; a creeping root system is responsible. The **dark blue flowers, with bright
yellow centres**, are very **welcome in late summer**, when many of the other mountain flowers are
finished. Although the **flowers are only about
1.3 cm (½") across**, their colour contrast and numbers make them eye-catching. The toothed leaves are
unusual, for they change shape as they progress up

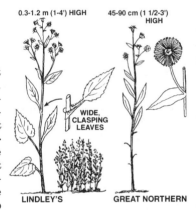

0.3-1.2 m (1-4') HIGH   45-90 cm (1 1/2-3')
HIGH

WIDE,
CLASPING
LEAVES

LINDLEY'S   GREAT NORTHERN

the stem. Notice the **long, wide, flat stems on the lowest heart-shaped leaves** and how
the stems on **higher leaves become progressively shorter**, until they disappear alto-
gether. Blooms July to October.

RANGE BC: A mountain species common in southern B.C. It extends northwards
through open forests to Dawson Creek.

RANGE WA: Rare.

GREAT NORTHERN ASTER, *A. modestus*: The names 'great' and *modestus* appear to
conflict. This aster is not any higher than many other asters, at 45–90 cm (1 ½–3'). The flowers are mildly spectacular and have v**ery narrow rays more characteristic of fleabanes**.
Together, the numerous rays and the bright yellow centre should suffice for identification—if
not, note the **sharp, narrow bracts, which are spotted with small glands**. The **blooming season, late summer and into fall, is significant**.

Grows on moister ground and stream banks, from lowlands to forested mountainsides,
from Prince George, B.C., southwards. Extends through Washington, mostly in Cascades.

TO 25 cm (10") HIGH

Sunflower Family, *Asteraceae (Compositae)*
THREAD-LEAVED FLEABANE
Erigeron filifolius

Note: *Erigerons* also come in white (pp. 208 & 210), yellow (p. 250), pink (p. 287), and purple (p. 310).

This fleabane is confusing as to colour! While the **flower is usually considered blue, in some areas pink or white** would be correct. The distinguishing feature of this showy plant is, as the name implies, **extremely thin, string-like leaves**. They tend to cluster at the base, but a few also climb higher. There may be several flowers on a stem. They are about **2.5 cm (1") across, with a yellow centre**. Blooms April-May. **Associated with sagebrush** and usually from 10 to 25 cm (4 to 10") tall, though often dwarfed on hard, stony ground; common in south-central and southeastern B.C. and also ranges southwards into Washington. Vantage.

RANGE BC: Common in south-central and southeastern B.C.

RANGE WA: Sagebrush habitats.

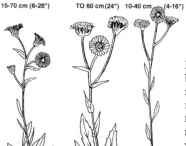

15-70 cm (6-28")   TO 60 cm(24")   10-40 cm (4-16")

BITTER   SMOOTH DAISY   DIFFUSE

Sunflower Family, *Asteraceae (Compositae)*
BITTER FLEABANE
Erigeron acris

While it appears that for the 3 fleabanes listed here the best description of the **flower colour is blue, there can be considerable variation, mostly to pink or white**. Bitter fleabane—with a name mystery that goes back to 1753 when it was first described—has a fine colour combination; its **many thin, blue ray flowers surround a showy yellow centre**. Remember that **fleabanes have much narrower ray flowers than asters, which bloom in late summer, after them**. The **paddle-shaped leaves** also help in identification. It ranges in height from 15 to 70 cm (6 to 28") and **prefers damp areas**, such as lake and stream borders, but does range to high elevations.

RANGE BC: Combined, the 3 recognized varieties of bitter fleabane have a wide range throughout B.C., especially along streams and in open forest, from low to middle elevations. Collected by Macoun, a pioneer botanist, at Kicking Horse Lake in 1890. Rockies.

RANGE WA: Olympics. Also in Oregon's Siskiyou Mountains and in the Cascades south of Mt. Hood.

SMOOTH DAISY, *E. glabellus*: Handsome both in colour contrast and size, but very variable in colour, the **flower ranges from blue to pink, purple or white; it is sometimes to 5 cm (2") across**. The stems, to 60 cm (2') tall, usually branch near the top to carry a number of flowers. **Leaves are long and broadly lance-like**. Common on lower elevations in southeastern B.C. and the Rockies. Not in Washington.

DIFFUSE FLEABANE, *E. divergens*: **Some stems lie along the ground** before turning upwards, a habit that gives rise to the alternative name of 'spreading fleabane,'. The **flower colour can be blue, pink, or white**. Blooms in June and July. Usually from 10 to 40 cm (4 to 16") in height. **Dry soils in sagebrush areas** of southern B.C. and Rockies. A variety, *E. divergens* var. *divergens*, continues into eastern Washington.

Birthwort Family, *Aristolochiaceae*
WILD GINGER [396]
Asarum caudatum

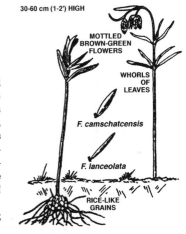

5-20 cm (2-8") HIGH

HEART-SHAPED LEAVES TO 10 cm (4") ACROSS

PURPLISH BROWN FLOWER 5 cm (2") ACROSS

RUNNERS ROOTING IN DEAD LEAVES AND MOSS

Large, heart-shaped leaves much like others on forest floor, with inconspicuous flowers near ground. **Creeping-and-rooting habit** means **leaves generally occur in small patches. 2 leaves at every node.** Fine hairs on stem, lower leaf margin. Single curious **purplish brown 'bell' flowers. 3 widespreading lobes.** Not closely related to true ginger (*Zingiber officinale*), but whole plant has **faint ginger smell**, most pronounced in crushed root. Roots and leaves saw wide medicinal use by natives. Blooms spring to August. Coastally, rich bottomlands to 1050 m (3500') elevation. Inland, damp, shady places and beside creeks.

RANGE BC: Across southern half of B.C. in suitable habitats.

RANGE WA: Throughout. Douglas collected it in 1831, near Vancouver, Washington.

6 PETALS

Lily Family, *Liliaceae*
CHOCOLATE LILY [398]
Fritillaria lanceolata
Fritillaria affinis
Rice Root

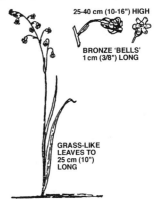

30-60 cm (1-2') HIGH

MOTTLED BROWN-GREEN FLOWERS

WHORLS OF LEAVES

F. camschatcensis

F. lanceolata

RICE-LIKE GRAINS

Peculiar **dark brown flower, with green and yellow mottlings**, hardly seems real and might be missed among grass or low brush, on grassy bluffs and open sidehills. **Cup-like 6-petalled flower, about 2.5 cm (1") across, hangs down. 1–2 whorls of narrow-pointed leaves.** 'Rice root' named because large, white bulb appears covered with rice-like nodules; edible, said to taste somewhat like rice when cooked. But please do not disturb chocolate lily! **A most-prized spring wildflower.**

RANGE BC: Across south; lower elevations; Coastal Mtns., Cascades. Inland, to 900 m (3000').

RANGE WA: Both sides of Cascades; to 1500 m (5000') level. West of Oregon Cascades.

RICEROOT FRITILLARY, *F. camschatcensis*: Sturdier than chocolate lily; **leaves in 3 main whorls of 5–10; more open purple-brown flowers—not mottled.** Both *Fritillaria*s have disagreeable smell that attracts flies and beetles. Moist areas, sea level–alpine. Largely west of Cascades; Alaska to Snohomish County, Washington. Sporadic east of Cascades.

Lily Family, *Liliaceae*
WESTERN MOUNTAINBELLS [399]
Stenanthium occidentale
Bronze Bells, Western Stenanthium

25-40 cm (10-16") HIGH

BRONZE 'BELLS' 1 cm (3/8") LONG

GRASS-LIKE LEAVES TO 25 cm (10") LONG

Shy, but **distinguished by flowers**. Lily traits include bulb, **grass-like leaves to 25 cm (10") long**.

Aptly called 'bronze bells,' **drooping flowers are greenish, streaked with purple. Flower parts in 6s, with petals recurved at tips. Stamens cluster in throat as tiny, golden dots.** Elusive, tangy perfume. Blooms April–August; depends on exposure and elevation. Amidst lush growth of shady mountain creeks. Most common middle-mountain–subalpine; moist, open forests.

RANGE BC: Vancouver Island; southern B.C and through Rockies to Mt. Robson.

RANGE WA: Mostly in moist forests west of Cascades. Olympics and Columbia Gorge.

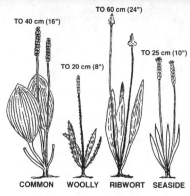

TO 60 cm (24")

TO 40 cm (16")

TO 20 cm (8")

TO 25 cm (10")

COMMON   WOOLLY   RIBWORT   SEASIDE

Plantain Family, *Plantaginaceae*
PLANTAIN
Plantago spp.

Plantains are weedy-looking. Most are aliens. **Cylindrical flowerhead and seedhead attract attention.** Tiny, white flowers; brown seeds.

COMMON PLANTAIN [397], *P. major*: **Distinctive, large, smooth, oval leaves.** A common weed of roadsides, fields and gardens, with wide distribution throughout our area.

WOOLLY PLANTAIN, *P. patagonica*: Recognize by **silky hairs on slender stems and leaves.** Latin name is significant, for it also grows in Argentina and Chile. Most common east of Cascades.

RIBWORT PLANTAIN, *P. lanceolata*: **Long, lance-like leaves rise from base.** Very dense flowerheads. Common over much of southwestern B.C. and western Washington.

SEASIDE PLANTAIN, *P. maritima*: To 25 cm (10") high. **Leaves narrow, to 1.3 cm (½") wide. Favours beaches and salt-marshes.** Fairly common near coast.

30-90 cm (1-3') HIGH

CLUSTERS OF TINY
WHITE-BROWN
FLOWERS

2 ROWS
WHITE-HAIRY
BRACTS

LOWER LEAVES
VARIABLE

DENSELY
WHITE-HAIRY
AND TO
7.5 cm (3") LONG

STEM
WHITE-HAIRY

Sunflower Family, *Asteraceae (Compositae)*
WESTERN MUGWORT [400]
Artemisia ludoviciana
Western Wormwood, Cudweed Sagewort

Note: A green-flowering *Artemisia* is on p. 337. Others are shrubs (pp.129 & 131).

A range plant, included because of 2 characteristics: a **slight resemblance to sagebrush—leaf shape and aromatic 'sage' smell**—and **overall silvery colour**, which makes it stand out. Unlike sagebrush (twisted stems), **grows erect, 30–90 cm (1–3') high**. Often consists of 1 stem, but on occasion branches freely. **Leaves to 7.5 cm (3") long; silvery sheen created by dense mat of soft, white hairs**, which are denser on upper side. Most leaves are entire, but lower ones may be lobed. **Dense clusters of tiny flowers** form in August. On dry soils, usually fully exposed to sun.

RANGE BC: Dry regions east of Cascades. Nicola and Okanagan Valleys. Cranbrook.
RANGE WA: Open, arid lands east of Cascades; favours foothills and mountain slopes.

0.6-1.2 m (2-4') HIGH

FLOWERS
OLIVE GREEN
TO BROWN,
3 mm (1/8") LONG

DULL GREEN,
ALTERNATE
LEAVES,
3-7.5 cm (1 1/4-3")
LONG

Sunflower Family, *Asteraceae (Compositae)*
TARRAGON [401]
Artemisia dracunculus
Dragon Sagewort

Coarse and weedy; on **dry, gravelly banks** or road slopes. A **dozen or so thick, woody stems form a 90 cm (3') high spray of narrow, dull-green leaves**; alternate, to 7.5 cm (3") long. Lower leaves may be deeply 3-lobed. **Tiny flowers, less than 3 mm (⅛") long, are olive-green or brownish**; on thin stems, near top, during late summer.

Tarragon from eastern Europe has been known for its **fragrant smell and aromatic taste** since the 16th century, but there has been little such recognition in the Pacific Northwest. Culinary varieties are now in some seed catalogs, but flowers are barren, so propagation is by roots or cuttings. Leaves and twig tips are used in salads and pickles. Also, try leaves to improve the appetite.

RANGE BC: Common in the dry interior. Occasional on Vancouver Island. Chilcotin.
RANGE WA: In dry, open places of eastern Washington, from lowlands to foothills.

Sunflower Family, *Asteraceae (Compositae)*
WORMWOOD
Artemisia campestris
Silky Field Wormwood

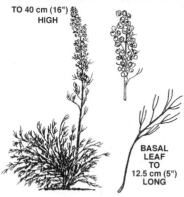

TO 40 cm (16")
HIGH

BASAL
LEAF
TO
12.5 cm (5")
LONG

The confusing wormwoods remind me of a friend who, forced to look at several very young babies, confided, 'They all look like little weasels to me.' Substitute 'wormwoods' for 'weasels' and you have the picture. **Wormwoods are *Artemisias*, so note typical sage-green colour. Most are aromatic**; some have medicinal properties. **Small flowers among highest stems attract little attention.**

Only 2 subspecies are described here.

NORTHERN WORMWOOD, *A. campestris* ssp. *borealis*: Some hard, dry valley bottoms east of Cascades are decorated with **light green, fern-like tufts of silky leaves.** I once noticed some single flower stems that looked as if they hardly belonged there, well off-centre compared to leaves. And how could a plant in such an arid place appear so much as if from the moist tropics? One might guess 'sagebrush' from leaf colour and **spike of tiny, dull green flowers,** but there is no 'sage' smell. I next saw this wormwood, which **averages about 30 cm (12") high,** 725 km (450 mi.) northwards, on the Peace River's north bank in northern B.C. A strange place to find this plant—there is a suitable microclimate at the top edge of this great canyon. Cactus and prairie sagewort, plants more associated with southern Okanagan, also grow here. Hardy; on suitable exposures and soils. Dry, exposed lands east of Cascades.

PACIFIC WORMWOOD, *A. campestris* ssp. *pacifica*: Similar to northern wormwood, but **30–90 cm (1–3') tall.** Leaves mostly basal; deeply divided, lightly hairy. Southern Vancouver Island, adjacent mainland. Probably on Gulf Islands and southwards in coastal areas.

Buckwheat Family, *Polygonaceae*
CURLY DOCK [402]
Rumex crispus

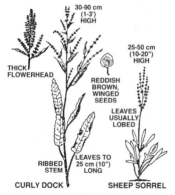

30-90 cm
(1-3')
HIGH

25-50 cm
(10-20")
HIGH

THICK
FLOWERHEAD

REDDISH
BROWN,
WINGED
SEEDS

LEAVES
USUALLY
LOBED

RIBBED
STEM

LEAVES TO
25 cm (10")
LONG

CURLY DOCK

SHEEP SORREL

Most docks (and sorrels) are introduced species with a weedy appearance. Recognize by their **narrow lower leaves with curly edges,** and by **clusters of small, russet flowers or winged seeds.** In some species seeds measure 6 mm (¼") across. **Foliage and seeds have sharply acidic taste.**

'Curly' because **leaves have a tendency to curl along edges.** Long-stemmed lower leaves; no stems or short ones on upper ones. Flowerhead atop unbranched stem may be open or in a dense cluster. **To 90 cm (3') tall.** Blooms June–August. Thereafter, **coarse plant turns russet-brown, stands out clearly.** A dock leaf is supposed to have remedial properties when rubbed on skin suffering from stinging nettle's irritation.

RANGE BC: Low to middle elevations, mostly west of Cascades. Sporadic inland.

RANGE WA: Habitats as for B.C.

CLUSTERED DOCK, *R. conglomeratus*: Much like, and often with, curly dock, but **flowers clustered, in spaced whorls on stem.** Same range. Ladysmith, B.C. Strait of Georgia–Puget Sound area and southwards.

SHEEP SORREL or **SOUR WEED** [403], *R. acetosella*: A weed throughout most of world. **Usually under 30 cm (12") tall. Very small, reddish flowers or seeds.** Slender plants, tend to bunch; become noticeable with overall rusty-red colour. As kids, we used to chew on these plants to get a **sour taste.** While leaves may be pleasant and refreshing to snack on, they contain oxalic acid, which can create problems if eaten in excess. Sorrels (*R. acetosa* and *R. scutatus*) are cultivated, in Europe especially, for use in salads and as pot herbs. **Long-stemmed leaves to 7.5 cm (3") long may be entire or lobed—fanciful 'arrowheads.'** Poor, exposed soils throughout most of area. Steptoe Butte, Washington, at 900 m (3000').

30-60 cm (1-2') HIGH — HAIR-LIKE PETALS

GREENISH FLOWER, 6 mm (1/4") LONG, WITH PURPLE LINES

NEW LEAVES

HAIRY LEAVES AND STEMS

Saxifrage Family, *Saxifragaceae*
YOUTH-ON-AGE [404]
Tolmiea menziesii
Piggy-Back Plant

Often passed up because of **inconspicuous flowers**. Examine leaves in late summer to see reason for odd name—**small leaves grow from bases of old leaf blades**. Withering old leaves slowly drop to ground, allowing new leaves to root. **Heart-shaped leaves are broadly lobed and round-toothed, with a fine covering of white hairs**.

April to June, stem sports small, odd-looking flowers: **greenish, purple-streaked. Petals are 4 thread-like projections**. If flowers are absent, check fringecup (p. 178) and mitrewort (below). An attractive house-plant. Moist, shady, low-elevation coniferous forests, streamsides.

RANGE BC: Common plant west of Cascades.

RANGE WA: Habitats as for B.C. throughout Washington and Oregon.

5 PETALS

10-25 cm (4-10") HIGH — TINY FLOWERS

FLOWER

SEED

YELLOW-GREEN STEMS AND LEAVES

Pink Family, *Caryophyllaceae*
SEABEACH SANDWORT [405]
Honkenya peploides

Tough and fleshy; thrives on coastal beaches. Sandwort's whole appearance seems to say, 'It is tough here, but I can do it.' If well rooted, forms a **mound of thick, trailing stems**. Otherwise, may be a single erect stem. **Stout shoots, 10–25 cm (4–10") tall, poke up**. A few small, **greenish white flowers, each from a leaf base; 5 sepals, 5 widely spaced petals**; sometimes as twins. Sepals are left holding **small, oval, yellowish seed**. Distinctive **yellowish green stems and leaves**—short, stiff, opposite; about 2.5 cm (1") long. By mid-August, plant has become completely yellow and many leaves have blown off, leaving stems with scattering of seeds.

At times, sandwort and searocket (p. 262) dominant long stretches of sandy coastal beach.

RANGE BC: Scattered; up to Alaska.

RANGE WA: Scattered; southwards to Oregon. Abundant near Westport, Long Beach.

15-40 cm (6-16")

Saxifrage Family, *Saxifragaceae*
BREWER'S MITREWORT [406]
Mitella breweri

There are 6 mitreworts in B.C.; 2 more in Washington. 5 have greenish or greenish yellow flowers. **Slender flower stems each with a number of unusually delicate, tiny, symmetrical flowers**. Unique floral design helps distinguish from related *Heucheras* (p. 181). **Threadlike petals protrude further than other flower parts**. Mitreworts have unusual but **effective seed dispersal**. Tiny seeds remain in bottom half when **top breaks off rounded seed caspsule; a falling raindrop splashes seeds out**. Usually, **Brewer's has over 20 flowers per stem. 5 stamens rise between petals**, as on oval-leaved mitrewort (facing page). But Brewer's has **rounded, kidney-shaped leaves, with rounded teeth**. Look for this plant from late spring into summer.

RANGE BC: Southern; wetter valley areas—subalpine. Manning Park, Mt. Revelstoke.

RANGE WA: A mountain plant of moist woods and meadows across the state.

Saxifrage Family, *Saxifragaceae*
FIVE-STAMENED MITREWORT
Mitella pentandra

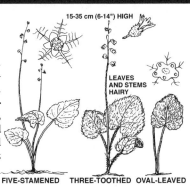

15-35 cm (6-14") HIGH

LEAVES AND STEMS HAIRY

FIVE-STAMENED   THREE-TOOTHED   OVAL-LEAVED

Small, greenish flowers, like most mitreworts. 'Five-stamened' should be significant and it is—except that 5 species of mitreworts all have 5 stamens. However, its **stamens are opposite petals, instead of alternating** (other mitreworts). **Fringed, hair-like petals; sepals quite broad and recurved.** Several **round-toothed, hairy, oval basal leaves. Summer bloomer.** Moist woods, wet meadows; middle-mountain forest–subalpine.

RANGE BC: Common in our area.

RANGE WA: Habitats as for B.C. Western part of Columbia Gorge.

THREE-TOOTHED MITREWORT [407], *M. trifida*: Name helps identify—**3 teeth on each of 5 petals** that protrude from **closed, narrow, bowl-shaped flower. Many blooms point upwards** (downwards for most other mitreworts). **White petals, sometimes pinkish, instead of green. Stems and leaves hairy.** Wetter places; mid-elevation forests; most abundant on southern Vancouver Island and in southern B.C. but fairly common on San Juans, in foothills of Puget Sound and southwards to Columbia Gorge.

OVAL-LEAVED MITREWORT, *M. ovalis*: Notice that the **5 stamens alternate between the petals. Flower much different** than on five-stamened. **Coarse, hairy oblong–oval leaves.** Locally common; southern Vancouver Island and southwards; mainland coastal forests; Gulf and San Juan islands.

COMMON MITREWORT, *M. nuda*: **Slender stem to 20 cm (8″) tall, tiny flowers** and **kidney-shaped basal leaves**—all features typical of mitreworts. But there is an important distinction: **10 stamens.** Summer bloomer. With that flower design, enough said! Common across B.C. east of Cascades; wet areas; valleys–lower mountain slopes. Extends into Cascades of northern Washington.

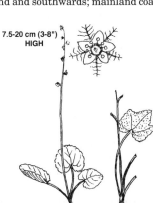

7.5-20 cm (3-8″) HIGH

COMMON   LEAFY

LEAFY MITREWORT, *M. caulescens*: **'Maple' leaves; coarse hairs on upper surface.** Usually **3 basal leaves; 1–3 much smaller stem leaves**—a good distinguishing feature. Strangely, **flowers bloom from top downwards,** a reversal of usual method; spring–early summer. Shady, dampish areas, sea level to middle-mountain heights. Spotty; southern Vancouver Island, Fraser Valley. More abundant across Washington on lower slopes.

Gentian Family, *Gentianaceae*
GIANT FRASERA
Frasera speciosa

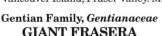
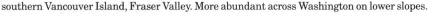

0.9-2.1 m (3-7') HIGH

FLOWERS 2.5 cm (1") WIDE

TWIN GLANDS ON PETALS

WHORLS OF 3-5 LEAVES

Flower is **'greenish yellow with purple spots.'** This 'monument plant' can be **to 2.1 m (7') tall. Could rate as our area's tallest wildflower,** but white veratrum (p. 188) beats it for sheer bulk and floral display. Also see whitestem frasera (p. 313).

Generally a stout single stalk. **Narrow leaves to 23 cm (9″) long; whorls of 3–5.** They decrease rapidly in size up stem. **Lovely flowers, about 2.5 cm (1″) across, in open clusters,** arise from upper leaf axils. **Purple-spotted petals, each with matched pair of hairy glands toward its base.** Open slopes from valley to mountainside.

RANGE BC: Not in B.C.

RANGE WA: Wideranging in east; favours ponderosa pine zone. High in Blue Mtns.

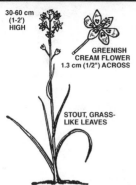

30-60 cm
(1-2')
HIGH

GREENISH
CREAM FLOWER
1.3 cm (1/2") ACROSS

STOUT, GRASS-
LIKE LEAVES

Lily Family, *Liliaceae*
MOUNTAIN DEATH-CAMAS
Zygadenus elegans
Elegant Death-camas, Green Lily

Note: Meadow death-camas (white) on p. 196.

A charming sight sure to be noticed. Symmetrical arrangement of fascinating flowers on a spike. Despite beauty, they **smell bad. 6 waxy, greenish white petals with green–yellow, arch-like spot near base.** Large bulb sends up cluster of **heavy, grass-like leaves,** mostly from base; also a few small stem leaves. Suspected, like meadow death-camas, to be poisonous. Blooms late June–August.

RANGE BC: Fairly abundant in low to high mountains east of Cascades. At higher elevations in Rockies and northwards to Alaska. Golden.

RANGE WA: Mostly on western slope of Cascades, at middle elevations. Olympics.

GREENISH YELLOW
FLOWERS
1.3 cm (1/2") ACROSS

RIBBED LEAVES
TO 30 cm (12")
LONG

Lily Family, *Liliaceae*
INDIAN HELLEBORE [408]
Veratrum viride
Green False Hellebore

Note: White false hellebore is on p. 188.

Tall, rank, single stem; largest lily family member in our area. Large, heavily ribbed leaves. Exotic appearance seem to fit tropics. Large size of **0.9–1.5 m (3–5')** attracts attention. Nutrients in thick, scaly roots enable vigorous spring growth. Best known as an **eye-catching part of lush subalpine vegetation** of temperate mountain slopes and meadows, but also in **shady, moist forests near sea level.** Varied blooming period, May–August. **Small, yellowish green flowers,** unexpected for so dramatic a plant, in **thin spikes or drooping tassels;** little contrast with surroundings. Leaves turn yellow, dotting mountain slopes with attractive fall colour. **Strong poison in roots and leaves;** used in medicines and insecticides. **No part of this plant should be eaten!**

RANGE BC: Most common in subalpine ecosystem, but drops to lower coastal elevations. Moist mountains throughout. Rocky Mountains.

RANGE WA: Habitats as for B.C. Olympics. Cascade Mountain to northern Oregon.

CLUSTERED FLOWERS

5-15 cm (2-6") HIGH

GREENISH
YELLOW
CONE

Sunflower Family, *Asteraceae (Compositae)*
PINEAPPLE WEED [409]
Matricaria discoidea
Matricaria matricarioides

Exotic name may relate to **1 cm (⅜") high, compact, greenish yellow 'pineapple' flowerhead,** or to purported pineapple smell. Unusual; attracts attention, but seems to fall into weed category. **Lacy, dark green leaves** give velvety, succulent look. Sharp but **pleasant odour**—not surprising for a relative of the chamomiles. To 40 cm (16") tall, but usually **5–15 cm (2–6").** On **hard, dry ground** (e.g., parking lot edges) tolerated by few other plants.

RANGE BC: To middle elevations.

RANGE WA: To middle elevations, ranging southwards to Mexico.

Sunflower Family, *Asteraceae (Compositae)*
SILVER BUR-WEED
Ambrosia chamissonis
Franseria chamissonis
Sand Bur

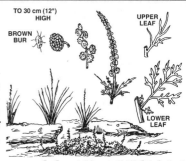

TO 30 cm (12")
HIGH

BROWN
BUR

UPPER
LEAF

LOWER
LEAF

It is difficult to find some attractive phrase to describe any part of this **sprawling, drab olive-green plant**. At the very least, it is conspicuous by its range, which is limited to the **upper edge of coastal beaches**, among the sand and logs. Coarse, blue-green dunegrass likely will be close by.

A **twisted mat** of bur-weed **may be 3 m (10') across and about 30 cm (12") high**, but lift a leaf or flower stem and you will find a recumbent stem 60 cm (2') long or more. If you explore further, you will find that the lower end is **buried in the sand**. The alternate leaves, toothed and divided, are **given a drab silvery sheen by fine hairs**—a magnifying glass is necessary to see all the details.

The **flowering stem has a spike to 15 cm (6") high that is covered with round knobs** that show touches of **yellow-green because of the tiny flowers**. The flowers will form into **burs with horrendous small spines**.

The *Ambrosia* part of the name refers to the food of the Greek gods. While bur-weed is related to plants with a sweet smell, its own smell and eating qualities are doubtful. The *chamissonis* part of the name comes from Chamisso, California, where this plant was collected in 1891.

RANGE BC: Common along coastal beaches. Queen Charlotte Islands southwards.

RANGE WA: Common along coastal beaches.

Sunflower Family, *Asteraceae (Compositae)*
MOUNTAIN SAGEWORT [410]
Artemisia norvegica
Mountain Wormwood, Longstem Greencaps

30-60 cm (1-2')
HIGH

SPONGY,
LIGHT GREEN
FLOWER,
0.6-1.3 cm
(1/4-1/2")
ACROSS

HAIRY
STEMS

Note: Some *Artemisia*s have brown flowers (p. 332), while others are shrubs (pp. 129 & 131).

One of the joys of plants is in the associations they can bring back. While sitting on a high ridge in the Trophy Mountains in B.C.'s Wells Gray Park in 1992, I put my lunch bag down on several of these plants. Forty years before, some similar incident had brought these plants to my attention. Since I could find no book that offered a common name, I gave them one: longstem greencaps. And while this name does give a fair description, the plant is now relegated to its 'sagebrush' classification, with a less appealing name.

There is no sagebrush smell to the foliage and the flowers are relatively quite large in comparison to most *Artemisia*s. However, in its range it is a common plant and not unattractive in late summer, with its **fringed leaves** and **hanging caps consisting of minute, spongy, green flowers**. Each round cap, 0.6–1.3 cm (¼–½") wide, is supported by a long stem. There is no other plant to confuse with this one.

RANGE BC: Subalpine slopes from the coast to the Rockies. In northerly parts of the range, it is found at lower elevations. Abundant in central and northern Rockies.

RANGE WA: Rocky places and dry meadows, from subalpine into alpine regions. Sporadic in mountain systems, southwards to California.

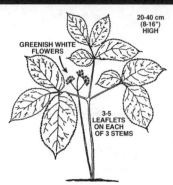

20-40 cm
(8-16")
HIGH

GREENISH WHITE
FLOWERS

3-5
LEAFLETS
ON EACH
OF 3 STEMS

Ginseng Family, *Araliaceae*
WILD SARSAPARILLA [414]
Aralia nudicaulis

Sarsaparilla shows little variety in form, being a **short stem that branches into 3 arms, each carrying from 3–5 oval leaflets 5–10 cm (2–4") long**. The flower stalk is shorter than the leaf stem, so the umbrella cluster of tiny, **5-petalled greenish white flowers** is generally hidden from sight. Plump but **inedible green or purple-brown, ribbed berries** form later. Although a common forest ground cover plant, it is easily passed by until its different form makes an impression on you. Thereafter, its frequent occurrences will make it a bit special.

The root has medicinal properties similar to those of true sarsaparilla, which is a tropical plant. The peculiar word 'sarsaparilla' is a close approximation to that of its tropical relative, *zarzaparrilla*, 'a prickly vine' in Spanish.

RANGE BC: Semi-open and open forests at medium elevations. Edges of the dry interior. Fraser Canyon, Nelson, Quesnel, north-central B.C.

RANGE WA: Moist, shady forests of northeastern area. Low to middle elevations.

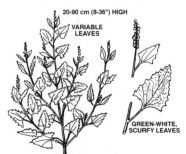

20-90 cm (8-36") HIGH

VARIABLE
LEAVES

GREEN-WHITE,
SCURFY LEAVES

Goosefoot Family, Chenopodiaceae
LAMB'S-QUARTERS
Chenopodium album
Pigweed, Goosefoot

Note: A related plant with red fruits is strawberry-blite, p. 292.

The fact that lamb's quarters looks about as weedy as a plant could be is balanced by an intriguing name. As a child in the Okanagan, I can still remember someone telling me that it was good to eat. The mystery of whether it could taste like a leg of lamb remained with me for years. Reference as to its edibility is usually omitted in botanical texts. Dr. Nancy Turner, however, in *Food Plants of B.C. Indians*, notes that it was eaten as greens by various tribes of south-central B.C. But, because it contains oxalate salts, it should not be eaten in large quantities or over long periods of time.

This **common garden and backyard plant** grows from **20 to 90 cm (8 to 36") high**. The most noticeable feature is its **overall grey-green colour** with **leaves covered by a light, granular white powder**—really only small scales. Leaves vary greatly in shape, but the lower ones are triangular and coarsely toothed. The plant **often streaks with red as it ages**. The tiny, green flowers are closely clustered in terminal spikes.

RANGE BC: A widespread naturalized weed, common at lower elevations throughout, except for the Queen Charlotte Islands.

RANGE WA: Habitats as for B.C.

RED GOOSEFOOT, *C. rubrum*: Sometimes covers a **moist, saline bottomland** with a uniform **low, yellowish green blanket**. Here, red goosefoot may be only 30 cm (12") tall, with leaves to 2.5 cm (1") long. However, the same plant, or a variety, growing on better soil along the edge of bottomlands, may be a spindly growth to 75 cm (2 ½') tall, with leaves to 10 cm (4") long.

Leaves have the characteristic 'goosefoot' shape and there is a reddish tinge to lower stems. Flowers are inconspicuous clusters of green nodules. Saline bottomlands of southern B.C. (Nicola Valley) southwards to California.

Nettle Family, *Urticaceae*
STINGING NETTLE [411]
Urtica dioica

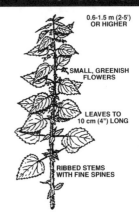

0.6-1.5 m (2-5')
OR HIGHER

SMALL, GREENISH
FLOWERS

LEAVES TO
10 cm (4") LONG

RIBBED STEMS
WITH FINE SPINES

The chances are that a person will first discover stinging nettle the hard way, for there is little to tell of its presence in the thickets. It frequently **grows luxuriantly around abandoned habitations**, where it usually marks old garden and barn areas. Often, great **masses of nettle, 1.5 m (5') or taller**, grow together, making it almost impossible to pass through without full body cover. The **coarsely toothed leaves are covered with fine stinging hairs** that cause severe irritation by giving tiny jolts of formic acid. After a few encounters, one watches for the rather ragged, opposite leaves and the **inconspicuous drooping clusters of greenish or whitish flowers**.

Many years ago, after each day spent timber cruising around Harrison Lake, B.C., we would lie on our narrow canvas cots with hands hanging in pails of cold water to ease the pain from this nettle. I have heard that for every plant that causes a problem, Nature provides a remedy, and not far away. An old-time remedy for stinging nettles is to rub the area with a dock leaf (see pp. 307 & 333), but try to find one when you need it!

Young nettle makes excellent steamed greens. Aboriginals used fibres from mature stems for weaving cord. Some European species produce excellent fibre for the best linen.

RANGE BC: In shady places across the province and into north-central B.C. Particularly abundant on abandoned house sites and stable areas.

RANGE WA: Common in shady areas with fair soils.

'ORCHID' FLOWERS

Orchid Family, *Orchidaceae*
ALASKA REIN-ORCHID [412]
Platanthera unalascensis
Piperia unalascensis, Habaneria unalascensis

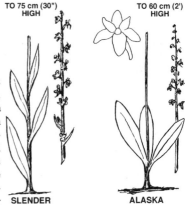

TO 75 cm (30")
HIGH

TO 60 cm (2')
HIGH

SLENDER

ALASKA

Note: For whiter rein-orchids, see p.212.

Even botanists find these names confusing! But all of them are in use someplace or other. **Tiny, green flowers, each with a twist, tend to form a loose spiral up a slender, leafless stem to 60 cm (2') high**. Possible faint fragrance. **Flowers sometimes almost white or tinged purplish. Lip and spur are same length. 2–4 narrow leaves**; bases wrap around stem. Usually leaves are drying before plant blooms, in summer. Rein-orchid's name is puzzling—in part refers to Latin *habane*, 'a bridle or rein,' alluding to narrow lip of some species. Dry woods, open mountainsides and meadows.

RANGE BC: Most common east of Cascades. Also abundant along coast.

RANGE WA: Habitats as for B.C. Common along the coast southwards to Baja.

SLENDER REIN-ORCHID, *Platanthera stricta* or *Habaneria saccata*: Like Alaska rein-orchid (above), with **small, green flowers scattered along the stem, but not in a spiral arrangement and without fragrance**. This rein-orchid **may reach 75 cm (30'") in height**. To definitely distinguish it from other rein-orchids probably requires a study of the flower. In this case, the **spur petal is a broad tongue (sac) and shorter than the lip petal**. The several leaves high on the stem are narrow while lower leaves are ovalish.

Swamps, bogs, wet meadows in middle-mountain forests. Widespread throughout B.C. and Washington. East and west of Cascades. Mt. Revelstoke, B.C. Rockies. Columbia Gorge.

7.5-20 cm (3-8") HIGH

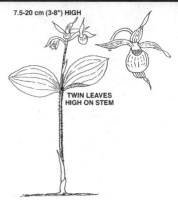

TWIN LEAVES
HIGH ON STEM

Orchid Family, *Orchidaceae*
CLUSTERED LADYSLIPPER
Cypripedium fasciculatum
Note: White ladyslippers, p. 211; yellow, p. 257.

Attractive, with unique **flower colour: a very unusual green-brown streaked with purple veining. Narrow, pointed sepals and petals.** Several flowers in cluster; may tend to droop. **Unusual pair of large, oval, veined leaves well above stem midpoint. To over 30 cm (12") high.** **RANGE BC**: Not recorded.

RANGE WA: Upper elevations of ponderosa pine forests, along eastern slopes of Cascades, from Wenatchee to near Columbia River.

7.5-50 cm (3-20") HIGH

Orchid Family, *Orchidaceae*
ELEGANT REIN-ORCHID
Platanthera elegans
Piperia elegans, Habenaria elegans

The 'why' of a name may arouse one's first interest in a plant. Why 'elegant?' Considering this rein-orchid was first found, or named, around 1835, and that this occurred in the eastern U.S., it could have looked very elegant. Nevertheless, compared with many other rein-orchids in our area, it does not display anything very special.

Height, to 50 cm (20"), does not give great stature, nor do usual **2 leaves, which quickly wither,** bring fame. Spike of quite-small flowers is outclassed by other rein-orchids; however, worth checking. Notice shape and placement of leaves, even if withered. **Each small, greenish white flower has a very long spur, about twice length of lip petal.**

RANGE BC: Low- to middle-elevation forests of southwestern B.C.
RANGE WA: Throughout, in habitats as for B.C. Klickitat County, Columbia Gorge.

20-60 cm (8-24")
HIGH

10-20 cm (4-8")
HIGH

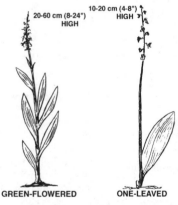

Orchid Family, *Orchidaceae*
GREEN-FLOWERED BOG-ORCHID [413]
Platanthera hyperborea
Habenaria hyperborea
Northern Green Bog-Orchid
What is distinctive, without analyzing flower parts? Perhaps the **dense cluster of small flowers at stem top, not scattered along it. Basically green, stemless flowers, often tinged or veined with purple. Spur cylindrical or tube-like. Upper sepal and 2 lower petals arc together to form a head above the lip petal. Most leaves are on lower half of stem**, which is 20–60 cm (8–24") high. Tends to lose itself among surrounding rich vegetation. Boggy places at lower elevations.
RANGE BC: Mostly east of Cascades.

GREEN-FLOWERED   ONE-LEAVED

RANGE WA: Habitats as for B.C. Southwards to California.
ONE-LEAVED REIN-ORCHID, *P. obtusata*: Common name is twice blessed—it is both simple and accurate. The **1 basal leaf is narrowly egg-shaped, widest above middle**. **3–15 loosely spaced flowers are pale greenish, but with tinges of white and yellow**. A smaller plant, **usually 10–20 cm (4–8") tall**. East of Cascades; **wet places** from middle-mountain to alpine meadows. Extends into northern B.C. but not known in Washington.

Orchid Family, *Orchidaceae*

HEART-LEAVED TWAYBLADE [417]

Listera cordata

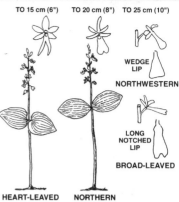

TO 15 cm (6") TO 20 cm (8") TO 25 cm (10")

WEDGE LIP
NORTHWESTERN

LONG NOTCHED LIP

BROAD-LEAVED

HEART-LEAVED NORTHERN

There are 4 twayblades in B.C., of which 3 extend into Washington. They all look very much like the rein-orchids and coral roots in that they have **single stems** and **small orchid flowers**. They should, for they all belong to the orchid family. Twayblades have an unpleasant odour that attracts certain types of insects. They are distinguished by having **paired green leaves (twayblades) that clasp at their bases** and a **lip petal (longest and lowest) without the spur common to the rein-orchids.**

Heart-leaved twayblade has a **narrow lip petal with a very noticeable forked tip**. It has 2 heart-shaped leaves near the centre of the stem. Flowers vary from light green to purplish or brown. Moist forests, bogs and streamsides, from low to subalpine.

RANGE BC: Common throughout B.C., except for arid regions.

RANGE WA: From coastal forests to middle-mountain elevations. Mt. Baker, Olympics.

NORTHWESTERN TWAYBLADE, *L. caurina*: Very much like heart-leaved (above). **Leaves oval rather than heart-shaped**. Orchid-like **flowers greenish yellow. Lip wedge-shaped, broad and not lobed or notched**. Range as for heart-leaved.

BROAD-LEAVED TWAYBLADE, *L. convallarioides*: As for heart-leaved (above), but **lip petal projects horizontally, is broadly wedge-shaped and notched**. Same range and occurrences as for heart-leaved.

NORTHERN TWAYBLADE, *L. borealis*: Very similar to broad-leaved (above). The **pair of leaves are not heart-shaped, but oval; located a little above centre. Lip petal with 2 shallow lobes**. Not common, but in wet forest areas to alpine heights in southern B.C. Lower elevations in northern B.C. Not in Washington.

Lily Family, *Liliaceae*

GIANT HELLEBORINE

Epipactis gigantea

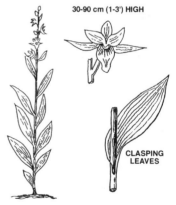

30-90 cm (1-3') HIGH

CLASPING LEAVES

A **showy** plant by reason of its **large size, attractive parallel-veined leaves**, and **long spire of flowers** almost impossible to describe because the colours are so varied. A habitat clue is that giant helleborine **loves water**: streamsides, pools and **particularly seepage areas, especially those near hotsprings**. Its height varies from 30 to 90 cm (1 to 3'), depending on the quality of the site. Leaves range from 7.5 to 12.5 cm (3 to 5") long and have bases that sheath the stem.

You could use half the colours you know in trying to describe the **large, orchid-like flower, which is to 5 cm (2") across**: 'flowers are green and purple, touched here and there with yellow or gold,' 'sepals coppery-green, lightly brownish-veined...purplish lines...greenish yellow,' and so on as people wrestle with a description. For now it will be placed in the 'Green' section, with apologies to those who see it quite differently.

Look for giant helleborine in bloom from spring to late summer.

RANGE BC: Rare; in southern B.C.

RANGE WA: Sporadic here and there, extending to California.

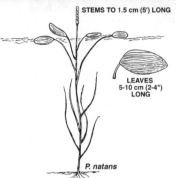

STEMS TO 1.5 cm (5') LONG

LEAVES
5-10 cm (2-4")
LONG

P. natans

Pondweed Family, *Potamogetonaceae*
PONDWEEDS [418]
Potamogeton spp.

Of 100 or so pondweeds in North America, we are blessed with around 25 in our area; 17 are quite common. **Generally in marshes, ponds and lakes, at lower elevations**; several range into subalpine.

Some have floating leaves; some are submerged. Leaves vary: string-like to tropical-looking growths. Stems to 3 m (10') long. Most plants raise a flowering stem above the water. Note: **if leaf stem is attached to leaf centre, it is likely watershield** (p. 343), which can look much like a pondweed.

FLOATING-LEAVED PONDWEED, *P. natans*: Roots in quiet, shallow water to 1.5 m (5') deep. Easy-to-see, leathery, long-stemmed floating leaves; attractive shape, green; blades 5–10 cm (2–4") long. Grasss-like submerged leaves. Flower stem soon withers. Widespread, wide range; ponds and lakes; valley bottom–subalpine; northwards to Alaska.

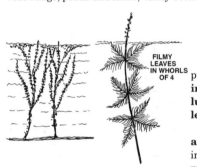

FILMY
LEAVES
IN WHORLS
OF 4

Water-milfoil Family, *Haloragaceae*
EURASIAN WATER-MILFOIL [415]
Myriophyllum spicatum

Worst villain among many milfoils. Unlike purple loosestrife, which flaunts showy flowers, is **invisible to most people: anglers may lose lures, boaters may find weed-choked propellers, swimmers may dismay at soggy foliage.**

Peak growth in **warmer, recreational lakes and rivers of southern B.C. and Washington.** An import, it reached Okanagan Lake in 1970. Spreads rapidly from any bits and pieces; control is extremely difficult. May root in lake bottoms to 7.5 m (25') deep, but prefers 1.8–3 m (6–10'). **Leaves often form filmy floating mat**; on stems in whorls of 4; fine and abundant (botanical name means 'myriad of leaves'). **Tiny, pinkish flowers on spike to 15 cm (6") above water.**

Boaters: Take time to make sure you do not spread this intrusive weed!

RANGE BC: Okanagan Lake and lower Fraser River systems; Shuswap, Cultus lakes.

RANGE WA: Scattered occurrences; warm-water lakes and brackish ponds throughout.

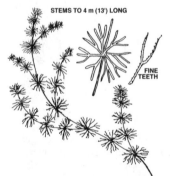

STEMS TO 4 m (13') LONG

FINE
TEETH

Hornwort Family, *Ceratophyllaceae*
COMMON HORNWORT
Ceratophyllum demersum
Coontail

Far from being a 'flower to know' or even from resembling the usual concept of a plant. An interesting, **mostly submerged, fragile tangle of green leaves; very fine, in whorls on a long stem**, like a skinny but attractive raccoon tail. Each whorl has 5–13 forked spokes; it takes a sharp eye to see the distinct fine teeth. Rootless stems, to 4 m (13') long.

Tiny, green flowers, borne singly in leaf axils. Named for 2 thin horns protruding from tiny seed. Can be a pest in its abundance—**can clog ditches and drains, and mat in ponds**.

RANGE BC: Most common in southern B.C.

RANGE WA: Throughout, lower elevations. Ponds, sluggish streams, lake margins.

SPRING HORNWORT, *C. echinatum*: Closely resembles common hornwort in general form, except **leaves are thread-like, not noticeably toothed**. Ponds and marshy places from lowland to middle-mountain elevations. Locally frequent on southern Vancouver Island, Gulf and San Juan islands. Spotty occurrences northwards to Alaska.

Waterlily Family, *Nymphaeaceae*
WATERSHIELD [416]
Brasenia schreberi

This unusual plant belongs to the waterlily family, as one would logically guess from the type of leaves and its habitat.

Watershield is **one of a kind, and has no close look-alikes**, so it can be positively identified. Its **leaves are 5–10 cm (2–4") wide**; by contrast, our common yellow waterlily (p. 227) has very large leaves. Also, **watershield's pads are not split**, whereas those typical of true waterlilies are.

**UNDIVIDED LEAVES
5-10 cm (2-4") WIDE**

The name 'watershield' could derive from the (shield) shape of some of the **thick, variably shaped, dark green leaves**. They are **attached at the centre of the underside by a long, gelatin-coated stem that reaches to the bottom** of **shallow ponds, lake margins** and **sluggish streams**. While pondweeds (p. 342) have quite similar leaves, they have their stem attached to the edge of the leaf.

Branching from near the top of the leaf stem, the **flower stems poke up through the mass of leaves, each carrying a single purplish flower**. Watershield is a late summer bloomer, not too showy, but made more interesting by the **great number of purplish stamens**. In October, the carpet of leaves turns into a fine display of many fall colours. They are a special delight to slowly canoe through.

RANGE BC: Mostly on Vancouver Island and adjacent mainland. Thetis Lake, Vancouver Island. Okanagan and Shuswap Lakes area.

RANGE WA: Widespread in ponds and sluggish streams in west and east.

Duckweed Family, *Lemnaceae*
DUCKWEED [421]
Lemna minor
Common Duckweed

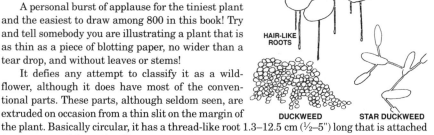

TINY, FLOATING PADS

HAIR-LIKE ROOTS

DUCKWEED   **STAR DUCKWEED**

A personal burst of applause for the tiniest plant and the easiest to draw among 800 in this book! Try and tell somebody you are illustrating a plant that is as thin as a piece of blotting paper, no wider than a tear drop, and without leaves or stems!

It defies any attempt to classify it as a wildflower, although it does have most of the conventional parts. These parts, although seldom seen, are extruded on occasion from a thin slit on the margin of the plant. Basically circular, it has a thread-like root 1.3–12.5 cm (½–5") long that is attached to the centre of the underside.

Picture a mallard foraging in the shallow arm of a lake or pond. How beautiful it looks as it parts a bright green floating carpet of duckweed. Somehow we appreciate this green display as a natural part of the outdoor scene and know it as duckweed. Hopefully it now also has an added dimension of interest.

There are 2 duckweeds, this one and star duckweed (below); both are common on backwaters, edges of ponds and marshes.

RANGE BC: Wide occurrence throughout our area but not common above 1000 m (3000') elevation, where low temperatures limit its growth.

RANGE WA: Habitats as for B.C.

STAR or **IVY-LEAVED DUCKWEED**, *L. trisulca*: One might suppose that the 'star' name comes from the little pads and connecting arms that form as they drift about. A casual look might not take this in, since the plants usually form partially submerged mats. Range is as for the duckweed above.

STRANDS TO 3 m (10') LONG

Eel-grass Family, *Zosteraceae*
COMMON EEL-GRASS
Zostera marina

It takes a real quirk of the imagination to think that eel-grass should go into a wildflower book. However, the limits have been stretched before and personally I find eel-grass a very attractive plant. Look down through a metre (a few feet) of sea water and observe how crabs dart into it. Watch eel-grass swirl its long, green stems around starfish and anemones as the tide and current keep it in constant motion. And at low tide, brant and other geese find it a favourite food in muddy bays and on sandy shores.

A **perennial that roots in sand and gravel, so beds of eel-grass are fairly stationary year to year**. Although on casual inspection eel-grass looks like 1 long ribbon, it is a **land plant, not a seaweed—it has defined stems and leaves**. A sheath opens on a grasslike leaf and exposes tiny flowers, which fertilize by water currents and produce **very tiny, gourd-like seeds**. Leaf ribbons branch off branching stems; may grow to a length of several metres (10'). Think you **see eel-grass in fresh water? Try bur-reed** (p. 196).

RANGE BC: Salt water; sheltered coastal waters to mudflats and beaches. Wide range.

RANGE WA: Habitats as for B.C.

SCOULER'S SURF-GRASS, *Phyllospadix scouleri*: In shape, closely resembles common eel-grass but is a **bright green, in contrast to the dull green strands of eel-grass**. Moreover, it is **generally found as a thick mat on boulders and rocky coastal shores** pounded by strong currents and waves; throughout our area.

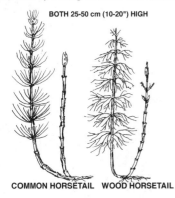

BOTH 25-50 cm (10-20") HIGH

COMMON HORSETAIL   WOOD HORSETAIL

Horsetail Family, *Equisetaceae*
COMMON HORSETAIL
Equisetum arvense

There are at least a dozen *Equisetum*s in our area, in 2 groups: **horsetails, easily recognized by whorls of thin branches, and scouring-rushes, with unbranched, evergreen stems**.

Horsetails are named for a fanciful resemblance between a horse's tail and the plant's sterile green stems with whorls of wire-like branches. These **sterile stems and branches function like leaves**. **Horsetails have hollow, jointed stems and fringed scales around each joint**, a pattern duplicated, though reduced, by the branches. **The sterile stems are most commonly seen**, for they last all summer, whereas the pale, unbranched fertile stalks occur early in the season only. Each **pale brown fertile stem bears a small cone at its tip**, which holds the spores. Small particles of silica give *Equisetum*s their scouring quality, which sometimes sees them used in cleaning camp cookware.

Horsetail usually grows in a group, rather than as stray individuals; in moist, sandy soils. Often found in wet places along road edges.

Common horsetail has hollow stems with 10–12 ridges, with whorls of branches. **Fertile stems are unbranched, thick and succulent; they soon wither**. The **central cavity takes up about half of the stem**.

RANGE BC: Damp, sandy places throughout our area at low to middle elevations.

RANGE WA: Habitats as for B.C.

WOOD HORSETAIL, *E. sylvaticum*: To 50 cm (20") high, with whorls of fine fronds. The **only horsetail in which the branches divide several times**. Moist places from low to middle elevations. From Alaska southwards to southern B.C. Not in Washington.

GIANT HORSETAIL [419], *E. telmateia*: A **giant copy of common horsetail**, with **branching sterile stems possibly to 1.8 m (6') tall**. Fertile stems unbranched, to 60 cm (2') high. Wet areas southwards from Alaska, but west of the Coast and Cascade mountains only.

Horsetail Family, *Equisetaceae*
SCOURING-RUSH [420]
Equisetum hyemale

Scouring-rush is easily recognized by its **stout, green stem, which is to 1.2 m (4') tall, and is marked into definite sections by narrow, ash-coloured and black bands.** Stems are all alike. **Stems are hollow and will pull apart into short sections.** The finely ridged stems are gritty to the touch, thus leading to the name. Many campers know that a handful of these stems can be a great help in scouring out the porridge pot and fry pan. Long ago, aboriginals already prized these plants for their polishing qualities in smoothing their wooden objects.

0.3-1.2 m (1-4') HIGH

BLACK→
ASHY→
BLACK→

HOLLOW, DARK GREEN, RIBBED STEMS 0.6-1.3 cm (1/4-1/2") THICK

RANGE BC: Common along the edge of interior lakes and ponds. Also sporadic occurrences at the coast. Vancouver Island, Gulf Islands.

RANGE WA: Habitats as for B.C. San Juan Islands.

NORTHERN SCOURING-RUSH, *E. variegatum*: **Much thinner stems than the scouring-rush above and marked into segments by a sheath of black teeth**; to 50 cm (20") tall. Lower elevations throughout our area.

DWARF SCOURING-RUSH, *E. scirpoides*: The smallest and most delicate of our scouring-rushes. **Rarely above 20 cm (8") in height**. Easily identified by its **thin, kinked stems**, which give a group of these plants a very untidy appearance. It ranges east of the Coast Mountains.

Goosefoot Family, *Chenopodiaceae*
EUROPEAN GLASSWORT
Salicornia europaea

TO 25 cm (10") HIGH

The glassworts are an oddity in the plant world. You can quickly see that they have **succulent, jointed stems**, but where are the leaves and flowers? They are there, but so minute you really have to look closely. **Flowers are sunk so deeply in the joints as to be unrecognizable, but they do have conventional flower parts**. The flowers come in 3s, the central one being slightly larger and in this species, slightly above the other. A slit-like opening in each one allows stamens and style to barely protrude.

3 MINUTE FLOWERS

EUROPEAN   AMERICAN

The unusual leaves and the opposite, scale-like bracts wrap around the flower bases.

The real curiosity European glasswort arouses is because it **usually grows in ground almost pure white from accumulations of coastal marine deposits, or in the alkaline margins of ponds and lakes in the driest zones of our area**. One subspecies, *Salicornia europaea* ssp. *rubra*, an annual, forms **mats of bright red colour to 25 cm (10") tall**—it is a startling contrast against white salts and blue skies.

RANGE BC: At the coast it grows along saline marshes, ponds and estuaries. Also east of the Cascades, around saline and alkaline marshes and lakes.

RANGE WA: Habitats as for B.C.

AMERICAN GLASSWORT [424], *S. virginica*: A perennial that forms **thick mats of erect, succulent, green stems** that grow **to 25 cm (10") high**. The central stem flower is slightly below the others. Young growth can be cooked as a vegetable. **Often the land plant closest to the ocean**, where it usually gets flooded by high tides. Saline edges of ponds and bays along the coast. B.C.'s Pacific Rim National Park, Gulf Islands and San Juan Islands. Continues southwards to Mexico.

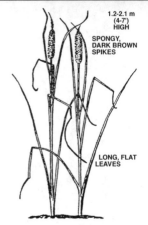

1.2–2.1 m
(4–7')
HIGH

SPONGY,
DARK BROWN
SPIKES

LONG, FLAT
LEAVES

Cattail Family, *Typhaceae*
COMMON CATTAIL [425]
Typha latifolia
Bulrush, Tule

Cattail, because of its wide distribution throughout North America and characteristic form, is known to almost everyone. There is **only 1 species of cattail. The spongy, dark brown spike, 10–20 cm (4–8") long and 2.5 cm (1") in diameter, is the female part of the flower.** The **lighter-coloured male flowers are in the withered tangle above.** The **cattail turns into a fluffy white seedhead** that releases thousands of tiny parachutes to float long distances. Evidence of this propagation is the fact that cattails grow in ditches and small pockets of wet soil along roadsides and in very remote areas.

Leaves are long, flat and about 2.5 cm (1") wide. Aboriginal peoples in the interior used the thin leaves to weave mats for use inside their summer shelters and also to make the walls of the shelters. The seed fluff was used in various ways where soft padding was needed.

Cattail jungles are beloved by red-wing blackbirds and muskrats. A cattail stalk in seed provides a good source of nesting material for hummingbirds and chickadees.

RANGE BC: Throughout our area, but most plentiful around ponds and swamps east of the Cascades. Usually at low and middle elevations.

RANGE WA: Habitats as for B.C.

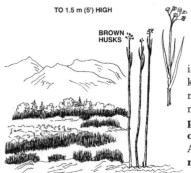

TO 1.5 m (5') HIGH

BROWN
HUSKS

Sedge Family, *Cyperaceae*
SMALL-FLOWERED BULRUSH
Scirpus microcarpus

Let me clarify the possible confusion surrounding such names as 'cattail,' 'tule' and 'bulrush.' As kids, we thought they were all interchangeable names for the vigorous growth around the edges of marshes. **'Cattail' (above) refers to the only plant of genus *Typha* in our area, and 'tule' is another name that should be reserved for cattail.** As should be apparent, **cattail is quickly recognized by its thick, dark brown spikes** (the cattails) on the stem tops.

On the other hand, **'bulrush' refers to the genus *Scirpus*, which has a dozen local representatives**, a few of which are called 'clubrush.' *Scirpus*es have rather sparse spikes of small, brownish flowers.

So, **'bulrush' is the logical name for the thin, round-stemmed rushes seen in nearly every swamp, marsh and shallow lake, especially those east of the Cascades.** They vary in height from only a few centimeters (inches) to the large small-flowered bulrush, described here, which may be 1.5 m (5') high. As the name suggests, this **rush has very small, brown flowers**. It **always stands in water**, where its dark green growth reaches a density made possible by new shoots from spreading roots. It **covers hectares (or acres) along the edges of shallow waterways** and in this way it and its relatives can be dominant parts of the landscape. **Several grass-like leaves near the top are hardly noticed.**

RANGE BC: In ponds, shallow lakes and rivers throughout our area, except northwestern B.C., but most common east of the Cascades. Macoun collected it in the Chilliwack River valley in 1901.

RANGE WA: Doubtful—there is a lack of specific information, but it should occur since it is found both north and south of the state.

APPENDICES

COMMON FERNS

CLIFF FERNS, *Woodsia* spp.: There are 5 cliff ferns in the Pacific Northwest, of which 3 are rare and not noted. The most common is mountain cliff fern, *W. scopulina*, which has hairy fronds. Look for it in dry, rocky places from lowlands to subalpine heights and across B.C. In Washington, Coulee City and Pullman.

Western cliff fern, *W. oregana*, is the common fern on rockslides east of the Cascades. Its bunchy tuft of fronds is from 10 to 30 cm (4 to 12") high.

PARSLEY FERN, *Cryptogramma acrostichoides*: A densely tufted little fern 15–30 cm (6–12") tall. There are 2 distinct types of frond, the fertile one being the taller and having its margins rolled over to enclose the spores. Dry, rocky areas from lowlands to alpine heights across southern B.C. In Washington, mostly in the subalpine of the Olympics and Cascades, Mt. Rainier, Steven's Pass, Chelan County.

LICORICE FERN [422], *Polypodium glycyrrhiza*: The thickish roots have a licorice flavour, which accounts for the common name. Often found on mossy cliffs, logs or tree trunks. A sparse fern with fronds to 30 cm (12") long and 2.5–7.5 cm (1–3") wide. Spores are large and round. Frond leaves are lobed to the midrib. It ranges from lowlands to middle-mountain heights along coastal B.C. Subalpine heights of the Olympics and Cascades in Washington.

WESTERN POLYPODY, *Polypodium hesperium*: Rather similar to licorice fern (above) but the pinnules (smallest frond leaves) are rounded. Note that they continue about halfway down the stem. Compare to deer fern (below), with pinnules almost to the base. Dry, rocky places across B.C. and possibly in northeastern Washington.

LEATHERY POLYPODY [426], *Polypodium scouleri*: Leathery leaves, pinnae rounded. Notable because of its tolerance to the salt-spray zone. West coast of Vancouver Island and southwards through Washington. Infrequent northwards to the Queen Charlotte Islands.

LADY FERN, *Athyrium filix-femina*: A large, graceful fern common in damp, shady woods to 1200 m (4000'). Fronds to 1.2 m (4') long and 25 cm (10") wide arch outwards; widest below the centre, tapering in both directions. Common throughout B.C. in dampish places from lowlands to alpine heights. Ranges in rocky places at subalpine and alpine elevations throughout Washington.

ALPINE LADY FERN, *Athyrium distentifolium*: A smaller, bushier edition of lady fern. Rock slides and scree slopes in subalpine and alpine terrain.

MAIDENHAIR FERN [423], *Adiantum aleuticum* or *A. pedatum*: The most delicate of our ferns, with each tiny leaf frond fringed along the upper edge. Strong, shiny black stems with fronds usually not more than 60 cm (2') long. Often clings to wet rock faces or masses in shady, damp places. Most common in the coast forest and interior cedar–hemlock ecosystems. Northwards to Dawson Creek. Coast forest ecosystem in Washington to elevations of 1500 m (5000'). Also occurs on eastern slopes of southern Cascades.

DEER FERN [427], *Blechnum spicant*: A common fern easily recognized by its 2 distinct types of fronds. The fertile or spore-bearing fronds that shoot from the plant's centre are often 90 cm (3') high. Non-fertile or vegetative fronds are evergreen and form a low rosette. They taper toward both ends and have pinnules (the smallest leaf divisions) that are rounded and range almost to the base of the stem. Most common in damp, shady coastal forests at low to medium elevations. Sparse occurrences in south-central B.C. Coast forest ecosystem of Washington.

SWORD FERN, *Polystichum munitum*: Look for symmetrical dark green sprays of fronds to 90 cm (3') long. Pinnules (the smallest leaf divisions) are sharp-pointed and sharp-toothed. The underside is almost orange because of twin rows of spore cases. These attractive fronds are shipped east for use as florists' decorations. Most common in southwestern B.C., but sparse occurrences inland as far north as Prince George. In Washington, a major part of the forest flora of the coast forest ecosystem.

BRACKEN, *Pteridium aquilinum*: Several subspecies are recognized but are not segregated here. Bracken is the most widespread and luxuriant fern in the region. The stout stems, often to 2.1 m (7') high, grow singly instead of clustering from a compact base, as with most ferns. A line of spore cases follows the margin of each leaf division. Found in damp ground from lowlands to subalpine slopes. Common in B.C., except for the dry interior areas. Found at lower elevations in Washington, in coniferous forests west of the Cascades, and also in shady places from the bunchgrass ecosystem to mountain elevations. Blue Mountains.

COMMON FERNS

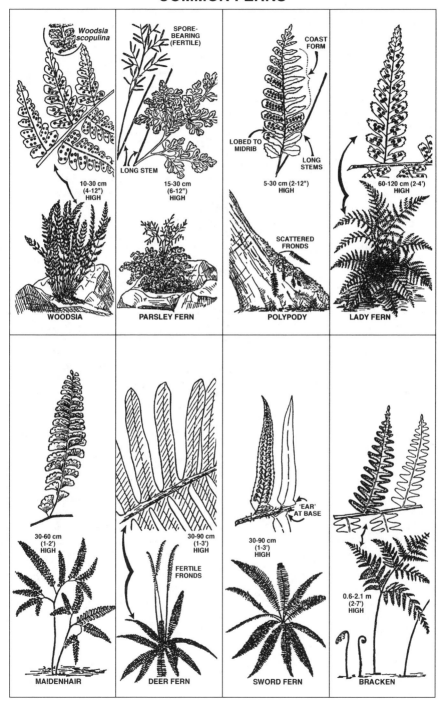

Woodsia scopulina

WOODSIA
10-30 cm (4-12") HIGH

PARSLEY FERN
SPORE-BEARING (FERTILE)
LONG STEM
15-30 cm (6-12") HIGH

POLYPODY
COAST FORM
LOBED TO MIDRIB
LONG STEMS
5-30 cm (2-12") HIGH
SCATTERED FRONDS

LADY FERN
60-120 cm (2-4') HIGH

MAIDENHAIR
30-60 cm (1-2') HIGH

DEER FERN
30-90 cm (1-3') HIGH
FERTILE FRONDS

SWORD FERN
'EAR' AT BASE
30-90 cm (1-3') HIGH

BRACKEN
0.6-2.1 m (2-7') HIGH

Nature's Calendar for Some Trees and Shrubs

Note: Average times shown. Plants in southern Washington may be several weeks ahead of those in B.C. Conversely, each increase of 300 m (1000') causes a week's delay.

▓▓▓ in flower ━━━ in fruit

p.#	SPECIES	in flower	in fruit	REMARKS
	TREES			
98	arbutus	MAY	JUL–OCT	orange-red berries
91	cherry, bitter	APR–MAY	JUL–AUG	bitter, red cherries
92	cherry, choke	APR–MAY	JUL–SEP	dense blossom cluster
89	crab apple, Pacific	MAY	JUL–OCT	white 'apple' blossoms
97	dogwood, w. flowering	APR–MAY	AUG–SEP	flower with 4-6 bracts
90	hawthorn, black	APR–MAY	AUG	smelly, white blossoms
93	maple, bigleaf	APR	AUG–SEP	pale yellow flowers
	SHRUBS			
118	blackberry, trailing	MAY–JUN	JUL	glossy black fruit
149	broom, Scotch	MAY–JUN	JUN–JUL	yellow 'pea' flowers
127	clematis, white	JUL–AUG	SEP–OCT	fluffy seedhead
114	currant, red-flowering	APR–MAY	JUN–JUL	showy red flowers
140	devil's club	JUN	AUG	pyramid of red berries
143	dogwood, red-osier	MAY–JUN	AUG	lead-white berries
137	elderberry, blue	JUN–JUL	AUG–SEP	blue berries
137	elderberry, red	MAY	JUN–JUL	red berries
143	goatsbeard	JUN	JUL	pencils of white flowers
126	hardhack	JUL–AUG		fluffy, pink plumes
141	hazelnut, beaked	MAR	JUL–OCT	edible, husked nut
128	honeysuckle, w. trumpet	MAY–JUN	AUG	orange-red flowers
145	honeysuckle, Utah	MAY–JUN	JUL	'red jelly' berries
151	Indian-plum	MAR–APR	JUN–JUL	early white flowers
119	kinnikinnick	APR–MAY	JUL–OCT	bright red berries
124	Labrador tea	MAY–JUN		head of white flowers
142	mock-orange	JUN		white 'orange' blossoms
146	ninebark	JUN		white flower clusters
142	oceanspray	JUN–JUL		creamy flower plumes
139	Oregon-grape, tall	APR–MAY	JUL	bright yellow flowers
138	rose, wild	JUN	SEP	edible, blackish berries
150	salal	MAY–JUN	AUG	edible, blackish berries
151	salmonberry	APR–MAY	JUN	large, pinkish flower
141	Saskatoon	APR–MAY	JUL	seedy, blackish berries
147	snowbrush	JUN		plumes of white flowers
135	soopolallie	APR	JUN	bitter, red berries
140	thimbleberry	MAY–JUN	JUL	domed, red berries
145	twinberry, black	MAY–JUN	JUL–AUG	black 'twin' berries
268	twinflower	JUN–JUL		tiny 'twin' flowers
144	waxberry; snowberry	JUN–JUL	AUG–OCT	waxy white berries
136	wolf-willow	JUN	JUL	silvery berry

Nature's Calendar for Some Coastal Flowers

Note: Average times shown. Plants in southern Washington may be several weeks ahead of those in B.C. Conversely, each increase of 300 m (1000') causes a week's delay.
Most of these flowers are also found east of the Cascades.

▬▬ in flower

p.#	SPECIES	MAR	APR	MAY	JUN	JUL	AUG	SEP	OCT	REMARKS
181	alumroots			▬	▬	▬				several species
309	aster, Douglas's					▬	▬	▬		forest edges
288	bleeding heart		▬	▬						pink 'heart' flowers
315	bluebells; harebells				▬	▬				low to high elev.
320	blue-eyed Mary	▬	▬	▬						tiny, blue flowers
299	brodiaea, harvest					▬	▬			purple flower, white eye
162	bunchberry			▬	▬					white flower, red berries
216	buttercups		▬	▬	▬	▬				many species
246	butterweeds			▬	▬	▬				many species
318	camas		▬	▬						2 species
205	carrot, wild					▬	▬			flat umbel head
241	cat's-ear, hairy				▬	▬	▬			widesprd, dandelion-like
209	chamomile, corn				▬	▬	▬	▬		white 'daisy' flowerhead
169	chickweed, field		▬	▬						5-cleft petals
319	chicory					▬	▬			flowers on long stems
198	coltsfoot	▬	▬							large flowerhead
292	columbines			▬	▬					red, blue, yellow spp.
260	coralroots			▬	▬					small, 'orchid' flowers
195	cotton-grass			▬	▬	▬				low to high elevations
206	cow-parsnip			▬	▬					very large flowerhead
209	daisy, oxeye				▬	▬				common along roadsides
196	death-camas		▬	▬						poisonous plant
186	fairybells, Hooker's			▬	▬					pairs of drooping flowers
288	fairy-slipper		▬	▬						small forest 'orchid'
201	false bugbane				▬	▬				bristly, white flowers
187	false Solomon's-seal			▬	▬					plume of creamy flowers
264	fireweed					▬	▬			often in masses
177	foamflowers			▬	▬					small, white 'star' flowers
237	goldenrod					▬	▬			low elev. to alpine
246	groundsel, common			▬	▬					blacktipped bracts
252	gumweeds					▬	▬			sticky flowerheads
336	hellebore, Indian			▬	▬					low elev. to subalpine
173	Indian-pipe				▬	▬				waxy white plant
317	larkspur, Menzies's			▬	▬					the coastal larkspur
331	lily, chocolate		▬	▬						unusual flower colour
261	lily, tiger				▬	▬				a nodding, orange lily
185	lily, white fawn	▬	▬							the 'Easter lily'
324	lupines			▬	▬					many species
171	miner's lettuce		▬	▬						used in salads
296	mint, field					▬	▬			noticeable mint smell
279	monkey-flower, pink			▬	▬	▬				'snapdragon' flowers
229	monkey-flower, yellow					▬	▬			crimson dots in throat
270	montia, small-leaved		▬	▬						5 pale pink petals
339	nettle, stinging			▬	▬	▬				clusters of greenish flws.
276	onion, Hooker's			▬	▬					erect flower cluster
276	onion, nodding				▬	▬				nodding flowerhead
198	pathfinder				▬	▬				tiny, white flowers
303	pea, purple			▬	▬	▬				strings of purple flowers
199	pearly everlasting					▬	▬			heads of white 'ball' flws.
302	penstemon, coast				▬	▬				favours moist areas
211	pyrolas; wintergreens				▬	▬				flowers single, waxy

Nature's Calendar for Some Coastal Flowers (cont'd.)

Note: Average times shown. Plants in southern Washington may be several weeks ahead of those in B.C. Conversely, each increase of 300 m (1000') causes a week's delay. Most of these flowers are also found east of the Cascades.

▬ in flower

p.#	SPECIES	In flower	REMARKS
184	queen's cup	MAY–JUN	6 white petals
211	rattlesnake-plantain	JUN–JUL	small, green-white flws.
224	St. John's-wort	JUL–AUG	many protruding stamens
275	satin-flower	APR	6 'satiny' petals
165	saxifrages	APR–JUN	many species
284	sea blush	MAY–JUN	tiny, clustered flowers
306	self-heal	JUN–JUL	blue-purple flowers
266	shootingstars	APR–MAY	distinctive; many species
221	silverweed	MAY–JUN	single 'buttercup' flowers
236	skunk cabbage	APR	large, yellow hood
238	spring gold	APR–MAY	among earliest blooms
277	starflowers	JUN–JUL	6- or 7-petalled flowers
218	stonecrops	JUN–JUL	many species
238	tansy	AUG	clusters of compact flws.
307	thistle, common	JUL–SEP	flower on bristly bur
284	thrift	MAY–JUN	dome of tiny flowers
160	trillium, white	APR–MAY	3 large petals
187	twistedstalk	MAY	twist in flower stem
197	vanilla-leaf	MAY	spike of small, white flws.
304	vetch, American	JUN–JUL	clusters of 3-9 flowers
315	violet, early blue	MAY–JUN	upper petals form spur
219	violet, yellow	MAY–JUL	several species
204	water-parsnip	JUL–AUG	umbel of flower clusters
174	woodland stars	MAY–JUN	small, fringed petals
198	yarrow	JUN–SEP	flat cluster head
334	youth-on-age	JUN–JUL	tiny, greenish flowers

Month columns (header): MAR APR MAY JUN JUL AUG SEP OCT

Nature's Calendar for Some Interior Flowers

Note: Average times shown. Plants in southern Washington may be several weeks ahead of those in B.C. Conversely, each increase of 300 m (1000') causes a week's delay. Many of these flowers also occur west of the Cascades.

▓▓▓ in flower

p.#	SPECIES	MAR	APR	MAY	JUN	JUL	AUG	SEP	OCT	REMARKS
181	alumroot, round-lved.				▓	▓				spike of tiny, cream flws.
185	anemone, western			▓	▓					early subalpine plant
251	arnica, heart-leaved			▓	▓					often 3 flws. on stem
309	aster, showy					▓	▓			like Michaelmas daisy
254	balsamroots		▓	▓						several species
277	bitterroot			▓						ground-hugging flower
249	brown-eyed Susan				▓	▓				brown-centred 'sunflower'
202	buckwheats			▓	▓					umbels; many species
228	cactus				▓					yellow 'tissue' flowers
221	cinquefoils				▓					5-petalled flowers
310	fleabane, showy				▓	▓				like Michaelmas daisy
269	geranium, sticky			▓	▓					petals finely veined
290	gilia, scarlet			▓	▓	▓				dotted, scarlet 'trumpets'
316	Jacob's ladder, skunk			▓	▓					clusters of blue flowers
212	ladies' tresses				▓	▓				spike of waxy flowers
257	ladyslipper, yellow				▓					streaked yellow 'slipper'
317	larkspur, Nuttall's			▓	▓					common in interior
305	locoweed, showy			▓	▓					dense clusters of flowers
185	lily, white glacier		▓	▓						high elevations
293	lily, sagebrush			▓	▓					the 'mariposa' lily
290	lily, wood				▓					large, reddish lily flower
326	lupine, silky				▓	▓				long flower spires
311	meadowrue				▓	▓				male and female flowers
273	milkweed					▓				knobby flowerhead
297	monkshood						▓	▓		1 petal forms a hood
253	mule's-ears				▓	▓				a Washington 'sunflower'
274	old man's whiskers			▓	▓					flower has several colours
291	paintbrush, harsh				▓	▓				distinctive orange-red
133	penstemon, Scouler's				▓	▓				massed blooms on cliffs
296	phacelia, silky					▓				dwarf; high elevations
271	phlox, long-leaved		▓	▓	▓					5-petalled pink flower
271	phlox, showy		▓	▓	▓					notched pink petals
266	shootingstars		▓	▓	▓					many species
283	smartweed, water					▓				aquatic; rosy plumes
294	veronicas;speedwell				▓	▓				4 blue petals
306	waterleaf, ballhead			▓	▓					fuzzy flower ball
227	waterlily, yellow				▓					common waterlily
217	yellow bell		▓	▓						nodding yellow flower

Some Edible Plants
(other than well-known berries)

Note that while those plants used as greens are usually abundant, many of those with edible bulbs are best protected and left to produce flowers for next year. As access to wilderness areas becomes easier, edible species can be placed at risk by over-eager gathering, so consider this subject as one more for the mind than for the stomach. With dried foods so readily available and flavourful, it makes sense to leave the bulbs and roots of our native plants in the ground, except in case of emergency. Be certain of identification before eating any plant! Mention in this list means only that these plants have been eaten in the past; it does not guarantee safe eating of any particular specimen by any particular person. Also, look-alikes may be definitely poisonous or harmful. The authors and the publisher do not recommend experimentation by readers.

P. #	NAME	EDIBLE PARTS	PREPARATION	SEASON
277	bitterroot	thick roots	peel and boil, but note that plant needs to be protected!	April–May
348	bracken fern	(roots (rhizomes))	***roots and other parts no longer considered edible***	***Avoid!***
162	bunchberry	red berries	raw	August–September
318	camas	bulb	boil or roast	April–May
92	choke cherry	berries	raw; very sour unless fully black	July–August
206	cow-parsnip	young stalks and stems	peel; raw or boiled	May–June
115	currant, squaw	berries	raw; dry and tasteless	July
203	desert-parsley	roots	raw, but better if boiled or roasted	May–June
264	fireweed	stem centres	split young stalks, eat raw	June–August
90	hawthorn, black	berries	raw; dry and seedy	July–August
119	kinnikinnick	berries	boil; poor taste	August–October
331	lily, chocolate	bulb	boil or steam	May–August
261	lily, tiger	bulb	boil in 2 waters; bitter	May–August
290	lily, wood	bulb	boil in 2 waters; bitter	May–August
171	miner's lettuce	leaves	raw, as salad greens	March–June
339	nettle, stinging	leaves	cook as pot greens	May–August
275–77	onions, wild	bulbs	raw or steamed	May–July
139	Oregon-grape	berries	raw (when ripe)	mid-August
138	rose, wild	outer part of fruit (hip)	raw; rich in vitamin C	August–December
170	springbeauty	small bulbs (corms)	boil like potatoes; sweet	May–June
140	thimbleberry	berry/young shoots	raw; poor flavour/raw	May–July
308	thistle	roots	boiled	July–August
217	yellow bell (scarce)	bulb	raw or boiled	April–May

GLOSSARY

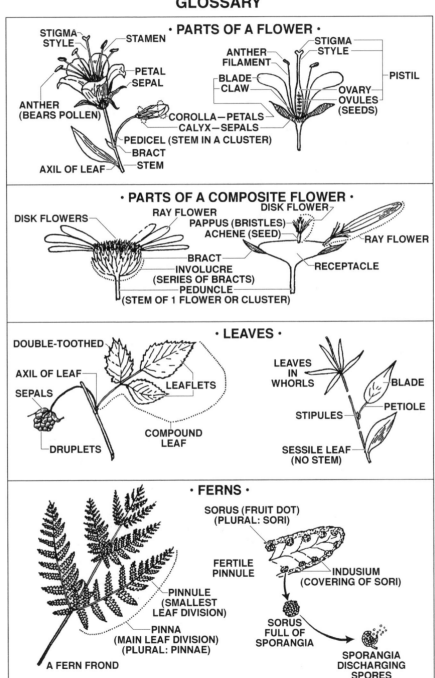

• PARTS OF A FLOWER •

STIGMA
STYLE
STAMEN
PETAL
SEPAL
ANTHER (BEARS POLLEN)
AXIL OF LEAF
STEM
BRACT
PEDICEL (STEM IN A CLUSTER)
CALYX—SEPALS
COROLLA—PETALS

ANTHER
FILAMENT
BLADE
CLAW
STIGMA
STYLE
PISTIL
OVARY
OVULES (SEEDS)

• PARTS OF A COMPOSITE FLOWER •

DISK FLOWERS
RAY FLOWER
DISK FLOWER
PAPPUS (BRISTLES)
ACHENE (SEED)
RAY FLOWER
BRACT
INVOLUCRE (SERIES OF BRACTS)
PEDUNCLE (STEM OF 1 FLOWER OR CLUSTER)
RECEPTACLE

• LEAVES •

DOUBLE-TOOTHED
AXIL OF LEAF
SEPALS
DRUPLETS
LEAFLETS
COMPOUND LEAF

LEAVES IN WHORLS
BLADE
PETIOLE
STIPULES
SESSILE LEAF (NO STEM)

• FERNS •

SORUS (FRUIT DOT) (PLURAL: SORI)
FERTILE PINNULE
INDUSIUM (COVERING OF SORI)
PINNULE (SMALLEST LEAF DIVISION)
PINNA (MAIN LEAF DIVISION) (PLURAL: PINNAE)
A FERN FROND
SORUS FULL OF SPORANGIA
SPORANGIA DISCHARGING SPORES

REFERENCES CITED

Brayshaw, T.C. 1976. *Catkin Bearing Plants of British Columbia*. Occasional Paper No. 18. Victoria: Royal British Columbia Museum.

Clark, L.J. 1973. *Wildflowers of British Columbia*, Sidney: Gray's Publishing Ltd.

Craighead, John J., Frank C. Craighead, Jr. and Ray J. Davies. 1963. *A Field Guide to Rocky Mountain Wildflowers*. Boston: Houghton Mifflin Company.

Douglas, G.W., G.B. Straley and D. Meidinger. 1990–1994. *The Vascular Plants of British Columbia*. Special Report Series 1–4. Victoria: British Columbia Ministry of Forests, Research Branch.

Haskin, Leslie L. 1934. *Wild Flowers of the Pacific Coast*. Portland: Binfords and Mort.

Henry, J.K. 1915. *Flora of Southern British Columbia and Vancouver Island*. Toronto: W.J. Gage and Co.

Hitchcock, K.C.L., A. Cronquist, M. Ownbey and J.W. Thompson. 1955, 1959, 1961, 1964, 1969. *Vascular Plants of the Pacific Northwest*. Parts 1–5. Seattle: University of Washington Press.

Horner, Chester E. and Ernest S. Booth. 1953. *Spring Wild Flowers of Southeastern Washington and Northeastern Oregon*. College Place: Walla Walla College, Department of Biological Sciences.

Hosey, R.C. 1979. *Native Trees of Canada*. Don Mills: Fitzhenry and Whiteside.

Jolley, Russ. 1988. *Wildflowers of the Columbia Gorge*. Portland: Oregon Historical Society.

Kirk, Ruth and Camelia Alexander. 1990. *Exploring Washington's Past: A Road Guide to History*. Seattle: University of Washington.

MacKinnon, Andy, Jim Pojar and Ray Coupe. 1992. *Plants of Northern British Columbia*. Edmonton: Lone Pine Publishing.

Niehaus, Theodore F. and Charles L. Ripper. 1976. *Pacific States Wildflowers*. Boston: Houghton Mifflin Company.

Pojar, Jim and Andy MacKinnon. 1994. *Plants of the Pacific Northwest Coast: Washington, Oregon, British Columbia & Alaska*, also published as *Plants of Coastal British Columbia, Including Washington, Oregon & Alaska*. Edmonton: Lone Pine Publishing.

Taylor, Ronald J. 1992. *Sagebrush Country, a Wildflower Sanctuary*. Missoula: Mountain Pass Publishing Company.

Turner, N.J. 1975. *Food Plants of British Columbia Indians*. Victoria: Royal British Columbia Museum.

Underhill, J.E. and C.C. Chuang. 1976. *Wildflowers of Manning Park*. Victoria: British Columbia Provincial Musuem and British Columbia Provincial Parks Branch.

Index

Note: Main entries are in **bold**; secondary species and aliases are in normal type. Botanical names are in *italics*

Lone Pine Nature

More Lone Pine Books on the Natural Beauty of the Northwest

PLANTS OF THE PACIFIC NORTHWEST COAST—REVISED SECOND EDITION
Washington, Oregon, British Columbia and Alaska

by Jim Pojar and Andy MacKinnon

This newly revised second edition of our best-selling field guide features 794 species of plants commonly found along the Pacific coast from Oregon to Alaska, including trees, shrubs, wildflowers, aquatic plants, grasses, ferns, mosses and lichens, and covers the entire length of the British Columbia coast, from shoreline to alpine. The book includes 1100 color photographs, more than 1000 line drawings and silhouettes, clear species descriptions and keys to groups, descriptions of each plant's habitat and range, and 794 color range maps. Rich and engaging notes on each species describe aboriginal and other local uses of plants for food, medicine and implements, along with unique characteristics of the plants and the origins of their names.

Softcover • 5.5" x 8.5" • 528 pages • 1100 color photographs, 1000 line drawings, 794 maps
ISBN10: 1-55105-530-9 • ISBN13: 978-155105-530-5 • $28.95 CDN • $22.95 USD

PLANTS OF SOUTHERN INTERIOR BRITISH COLUMBIA
AND THE INLAND NORTHWEST

by Roberta Parrish, Ray Coupé and Dennis Lloyd

Over 675 species of trees, shrubs, wildflowers, grasses, ferns, mosses and lichens commonly found in the region from the crest of the Rockies to the Coast Mountains, including the interior of Washington and Idaho. Detailed species descriptions are combined with concise drawings and color photographs to make plant identification easy.

Softcover • 5.5" x 8.5" • 464 pages • 1000 color photos, 700 illustrations
ISBN10: 1-55105-219-9 • ISBN13: 978-1-55105-219-9 • $26.95 CDN • $19.95 USD

HIKING THE ANCIENT FORESTS OF BRITISH COLUMBIA AND WASHINGTON

by Randy Stoltmann

Detailed trail information is accompanied by natural history and ecology along the way.

Softcover • 5.5" x 8.5" • 192 pages • 92 B/W photos, 37 maps
ISBN10: 1-55105-045-5 • ISBN13: 978-1-55105-045-4 • $19.95 CDN • $15.95 USD

BEST HIKES AND WALKS OF SOUTHWESTERN BRITISH COLUMBIA—REVISED

by Dawn Hanna

The author updates her popular 1997 guide to the hiking trails and pathways of the greater metropolitan Vancouver area and the Lower Mainland.

Softcover • 4.25" x 8.25" • 360 pages • color photos
ISBN10: 1-55105-455-8 • ISBN13: 978-1-55105-455-1 • $19.95 CDN • $15.95 USD

REPTILES OF THE NORTHWEST: BRITISH COLUMBIA TO CALIFORNIA

by Alan St. John

Interpretive naturalist, writer, educator and reptile specialist Alan St. John has crafted a richly photographed guide to the reptiles, snakes and turtles found in Alaska, British Columbia, Washington, Oregon and parts of Montana, Idaho, Nevada, Utah, Wyoming, Colorado and California. Each of the book's 42 accounts features photographs of the animal, as well as range maps and notes on identification, variation, distribution, habitat, behaviour and similar species. The guide also includes information about three introduced species. The book has a strongly personal and accessible quality from St. John's field notes of his own encounters with each of the featured species.

Softcover • 5.5" x 8.5" • 272 pages • 373 color photographs
ISBN10: 1-55105-349-7 • ISBN13: 978-1-55105-349-3 • $24.95 CDN • $18.95 USD